TABLE 3-1   FOURIER SERIES SYMMETRY CONDITIONS

Type of Symmetry	DC-term, a_0	Cosine terms, a_n	Sine terms, b_n
Even	a_0	a_n	$b_n = 0$
Odd	$a_0 = 0$	$a_n = 0$	b_n
Half-wave odd	$a_0 = 0$	$a_{2n} = 0$	$b_{2n} = 0$
		a_{2n+1}	b_{2n+1}

TABLE 3-2   FOURIER SERIES OF PRIMITIVE SIGNALS OF PERIOD T

Periodic Signal	Fourier Series Coefficients	Remark
$x(t) = \text{rect}\left(\dfrac{t}{\tau}\right), t \in \left[-\dfrac{T}{2}, \dfrac{T}{2}\right]$	$c_k = \dfrac{\tau}{T} \text{sinc}\left(\dfrac{k\pi\tau}{T}\right)$ where $\text{sinc}(x) = \dfrac{\sin(x)}{x}$	pulse train
$x(t) = 1 - \left(\dfrac{2t}{T}\right), t \in \left[-\dfrac{T}{2}, \dfrac{T}{2}\right]$	$c_0 = \dfrac{1}{2}, c_k = \dfrac{2}{(k\pi)^2}$	triangle train
$x(t) = \dfrac{t}{T}, t \in [0, T]$	$c_0 = \dfrac{1}{2}, c_k = j\dfrac{1}{2k\pi}$	ramp
$\cos\left(2\pi t \dfrac{K}{T}\right)$	$c_{\pm k} = \dfrac{1}{2}\delta(k \pm K)$	cosine
$\sin\left(2\pi t \dfrac{K}{T}\right)$	$c_k = \dfrac{j}{2}\delta(k - K)$ if $k > 0$	sine
	$c_k = \dfrac{-j}{2}\delta(k + K)$ if $k < 0$	

PRINCIPLES OF SIGNALS AND SYSTEMS

McGraw-Hill Series in Electrical and Computer Engineering

Senior Consulting Editor

Stephen W. Director, Carnegie Mellon University

Circuits and Systems
Communications and Signal Processing
Computer Engineering
Control Theory
Electromagnetics
Electronics and VLSI Circuits
Introductory
Power and Energy
Radar and Antennas

Previous Consulting Editors

Ronald N. Bracewell, Colin Cherry, James F. Gibbons, Willis W. Harman, Hubert Heffner, Edward W. Herold, John G. Linvill, Simon Ramo, Ronald A. Rohrer, Anthony E. Siegman, Charles Susskind, Frederick E. Terman, John G. Truxal, Ernst Weber, and John R. Whinnery

Communications and Signal Processing

Senior Consulting Editor

Stephen W. Director, Carnegie Mellon University

Also Available from McGraw-Hill

Schaum's Outline Series in Electronics & Electrical Engineering

Most outlines include basic theory, definitions and hundreds of example problems solved in step-by-step detail, and supplementary problems with answers.

Related titles on the current list include:

Analog & Digital Communications
Basic Circuit Analysis
Basic Electrical Engineering
Basic Electricity
Basic Mathematics for Electricity & Electronics
Digital Principles
Electric Circuits
Electric Machines & Electromechanics
Electric Power Systems
Electromagnetics
Electronic Circuits
Electronic Communication
Electronic Devices & Circuits
Electronics Technology
Feedback & Control Systems
Introduction to Digital Systems
Microprocessor Fundamentals

Schaum's Solved Problems Books

Each title in this series is a complete and expert source of solved problems with solutions worked out in step-by-step detail.

Related titles on the current list include:

3000 Solved Problems in Calculus
2500 Solved Problems in Differential Equations
3000 Solved Problems in Electric Circuits
2000 Solved Problems in Electromagnetics
2000 Solved Problems in Electronics
3000 Solved Problems in Linear Algebra
2000 Solved Problems in Numerical Analysis
3000 Solved Problems in Physics

Available at most college bookstores, or for a complete list of titles and prices, write to:

Schaum Division
McGraw-Hill, Inc.
1221 Avenue of the Americas
New York, NY 10020

PRINCIPLES OF SIGNALS AND SYSTEMS

Fred J. Taylor

University of Florida

McGraw-Hill, Inc.

New York St. Louis San Francisco Auckland Bogotá Caracas
Lisbon London Madrid Mexico City Milan Montreal
New Delhi San Juan Singapore Sydney Tokyo Toronto

This book was set in Times Roman by Electronic Technical Publishing Services.
The editors were George T. Hoffman and John M. Morriss;
the production supervisor was Leroy A. Young.
The cover was designed by Joseph Gillians.
Project supervision was done by Electronic Technical Publishing Services.
R. R. Donnelley & Sons Company was printer and binder.

PRINCIPLES OF SIGNALS AND SYSTEMS

 This book is printed on recycled, acid-free paper containing a minimum of
50% total recycled fiber with 10% postconsumer de-inked fiber.

1 2 3 4 5 6 7 8 9 0 DOH DOH 9 0 9 8 7 6 5 4

P/N 063197-2
PART OF
ISBN 0-07-911171-8

Library of Congress Cataloging-in-Publication Data

Taylor, Fred J., (date).
    Principles of signals and systems / Fred J. Taylor.
       p.   cm. — (McGraw-Hill series in electrical and computer
engineering. Communications and signal processing)
    Includes bibliographical references.
    ISBN 0-07-911171-8 (set)
    1. Signal theory (Telecommunication)—Mathematics.   2. System
analysis.   I. Title.   II. Series.
TK5102.5.T36   1994
621.382′2—dc20   93-51502

ABOUT THE AUTHOR

Fred J. Taylor is a Professor of Electrical Engineering and Computer and Information Science at the University of Florida. Since graduating from the University of Colorado in 1968, he has specialized in digital signal processing and computer engineering. He is the author of *Advanced Digital Signal Processing*, with G. Zelniker; *Electronic Filter Design Handbook, 2d Edition*, with A. Williams; *Digital Filter Design for the IBM PC*, with T. Stouratis; *Residue Number System Arithmetic: Modern Applications in Digital Signal Processing*, with M. Soderstrand, K. Jenkins, and G. Jullien; *Digital Filter Handbook, Digital Signal Processing in FORTRAN*, with Smith; with contributions to *The Encyclopedia of Telecommunications*, and *The Encyclopedia of Physical Science and Technology*. He has authored over 100 papers, four patents, and serves as a consultant to a number of industrial, federal, and educational institutions.

Dedicated to my best friend
Lori

CONTENTS

Part V   Appendixes

PREFACE

"Where shall I begin your Majesty?" he asked. "At the beginning," the King said gravely, "and go on till you come to the end then stop."

—Lewis Carroll, *Alice in Wonderland*

The study of signals and systems is a fundamental element in virtually every electrical engineering program. It is often the first college course to integrate prerequisite mathematical and prior discipline-specific material into a meaningful and coherent body of knowledge. Furthermore, signals and systems are a gateway to many important continuation studies in the fields of communications, controls, and signal processing. Signals and systems are also discussed in power systems, electronics, photonics, electromagnetics, and other areas. In many instances, the practicing electrical engineer relies more on the material, concepts, and problem-solving skills developed in a signals and systems course than knowledge acquired in any other.

A classic signals and systems course generally is placed in electrical engineering curriculum, as shown in Figure P-1. The range of courses and topics presented in this curriculum diagram represents a major portion of a typical, modern electrical engineering undergraduate program. How a signals and systems course is interwoven into the curriculum will, of course, vary with programs and institutions. This text is designed to account for these differences by dividing the study into four basic sections. The first section introduces signals in the context of elementary continuous-time, discrete-time, and digital signal entities. The second section deals exclusively with continuous-time signals and systems. The third section re-

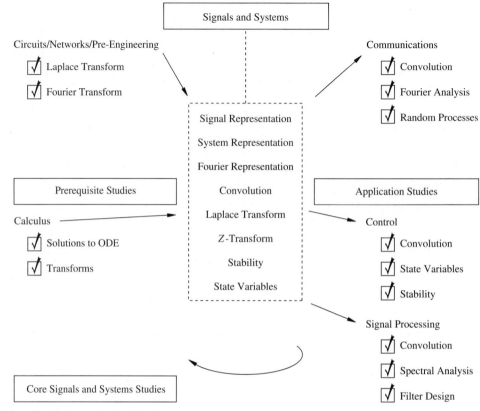

FIGURE P-1   Curriculum coverage diagram.

lates to discrete-time signals and systems. Finally, the fourth section develops specific application areas for focused studies. This is sometimes referred to as the sequential approach. The alternative, called the parallel approach, would introduce continuous-time and discrete-time topics concurrently. While both techniques have merit, we have found that the parallel approach often confuses students because they must simultaneously assimilate concepts being developed in two different domains. In addition, experience has shown that the sequential approach provides the instructor with more flexibility, since the pace and depth of each study can be more easily controlled.

We do, however, deviate from the traditional sequential paradigm of introducing signals and systems together. The rationale for the traditional approach is that the study of signals and systems uses the same analytical tools, namely transforms. We feel, however, that signals and systems are different subjects that simply share a common mathematical foundation. Of the two, signals are the easiest to model and analyze conceptually. For this reason, we introduce signals first and develop a mathematical infrastructure, namely transforms, that support their study. From this foundation, higher order, system-level concepts are built, including stability, convolution, and state-space. In this manner, we achieve a "load-balancing" of new concepts and procedures that we spread equally amongst the study signals and systems. During this process we have made a strong commitment to maintain an obvious mathematical and conceptual connection between the abstract and the concrete. In addition, the connection between analysis and synthesis is continually reinforced. Students must master this material if they are to comprehend the material studied in numerous follow-up courses required in a typical electrical engineering curriculum. The material provided here can be completed in a one-semester or two-quarter course. For a one-semester course, Part I should be presented in an accelerated manner. If Part IV is developed, it should be done so selectively. The architecture studies found in Part II and Part III may be omitted, if necessary.

The textbook is philosophically based on three main concepts. The first is that the text is viewed as a system in itself, having an input, body, and output. This is implemented by dividing the book into signals, followed by systems, and finally applications. The second philosophical cornerstone is the cultivation of active learning. While formal learning methods have been shown to be effective, they are often criticized as being too strongly based on a factual understanding of a discipline and, as a result, the development of higher-order reasoning abilities has suffered. In too many cases, students have apparently mastered a particular subdiscipline or topic, but fail to see how this can be extended, generalized, or integrated with other knowledge. In a pedagogical sense, this leads to what is called inert knowledge, which means knowledge difficult to recall on demand. We stress active learning by constantly building one idea on its predecessor. The third cornerstone of the text is its integration of computer-based instruction (CBI). The study of these signals and systems, whether emphasizing mathematical rigor, conceptual knowledge, applications, or all of these emphases, can be vastly improved with CBI. CBI is becoming increasingly popular because it makes the educational experience active rather than passive and allows faculty to become purveyors of knowledge instead of serving simply as lecturers. The use of CBI to enhance the text is optional however. The examples and exercises in the text can be interpreted and understood without CBI support. However, after 20 years of working with and without CBI, and after being part of a number of CBI development projects, I have reached the unquestioned conclusion that it is an invaluable asset, both for the student and teacher provided that:

1 The CBI experience is relevant and extensible to the total study of signals and systems (including its follow-up studies) and is not just an isolated computer-based tutorial of a single concept or problem.

2 It provides an avenue to integrate engineering concepts in the context of analysis, synthesis, and professional practice.

3 It provides an immediate, hands-on learning experience and does not trivialize the study with overly simplified examples. To repeat an ancient Chinese saying:

I hear, and I forget.
I see, and I remember.
I do, and I understand.

4 It saves labor, is easy to use (requiring no *a priori* skills), contains real-world mathematical and system objects, and is rich in display capabilities.

Without CBI, the textbook provides essentially the same baseline support to the study of signals and systems as found in most other texts on the subject, except the examples are more meaningful and rewarding than normally found in an introductory text. With CBI, higher-order problem-solving skills can be developed and a more comprehensive understanding of signals and systems gained.

The CBI software for the text is anchored with two, commercially available, low-cost packages, namely MONARCH and MATLAB Student Editions. The examples and tutorials presented in the text can be completed using MONARCH or MATLAB, with a preference given to MONARCH if design topics are to be emphasized. I also encourage the instructor to establish software copyright ethics now, in the classroom, rather than leave it to industry to enforce ethical behavior.

I fully assume responsibility for any of the shortfalls the reader may find with this approach to signals and systems. If there are kudos, they must certainly be shared with a small group of scholars on whom I have been intellectually dependent over the last half decade. To Dr. Glenn Zelniker, I give praise for being the keeper of the key to the gate of scientific correctness. To Ms. Monica Murphy, I give my thanks for adding a voice of literacy, elegance, and encouragement to this project. Without her, this project would never have been completed. To Erik Ström, I give my thanks for being a close collaborator from the beginning of this project. He realized early on that my typing was hopeless and, probably out of pity, has assisted in producing the final manuscript and developing MATLAB capabilities. Kudos to Ahmad Ansari, who took my preliminary MONARCH computer programs, tested them, and more often than not, improved them. Kudos to Scott Miller for his assistance on communication topics. The following reviewers supplied many helpful comments and suggestions: Nirmal K. Bose, Pennsylvania State University; John R. Deller, Michigan State University; Gary E. Ford, University of California, Davis; James A. Heinen, Marquette University; B. V. K. Vijaya Kumar, Carnegie Mellon University; V. John Mathews, University of Utah; Keshab K. Parhi, University of Minnesota; Steve F. Russell, Iowa State University; Andrew Sekey, University of California, Santa Barbara; Marvin Siegel, Michigan State University; Andreas S. Spanias, Arizona State University; Ahmed H. Tewfik, University of Minnesota; and Stephen Yurkovich, Ohio State University. Finally, I am grateful to Anne Brown of McGraw-Hill, the ideal editor, who provided the encouragement and opportunity to complete this project.

Fred J. Taylor

PRINCIPLES OF SIGNALS AND SYSTEMS

## PART ONE

INTRODUCTION

ONE

INTRODUCTION

SIGNALS

All this is amusing, though rather elementary Watson.

—Sir Arthur Conan Doyle
Sherlock Holmes

1-1 INTRODUCTION

We might ask ourselves why humans are placed so high in the food chain. It would seem, at times, that this lofty position cannot be justified, either on the basis of our ethics or our social behavior. Considering the perpetual state of global crisis, it appears that our highly touted ability to reason should be seriously questioned. In addition, we are anatomically inferior to many other species. What then is our advantage? One could argue that we enjoy a definite advantage over other animals by evolving a sophisticated signal processing capability. Our ancestors' experience with signals was limited by their natural sensory systems. Vision, for example, provided us with the signals needed to survive as hunter-gatherers. The signal source was the sun, which transmitted an energy signal to earth. The signal was then filtered (modified and reflected) by physical objects. The filtered signal was received by sensors (the eyes), and communicated (via the optical nerve) to a signal processing system (the brain). Today we engineer systems with a similar purpose. A radar system, for example, consists of a signal source (a pulse), a signal modifier (the target), a sensor (an antenna), a communication link (wires and cable), and a signal processor (a computer).

The signals processed by our early ancestors spanned a specific and limited range of frequencies. We processed high-frequency information located at optical frequencies, midrange frequencies occupying the audio spectrum, and low-frequency signals with our tactile systems. The brain processed these signals and from them developed an awareness of seasonal changes and the migration cycles of prey. Today our engineered systems process signals that span a wider range of frequencies, extending far above the frequencies of visible light. We sample and manipulate signals to develop models of the universe, examine the atom, probe the human body, provide entertainment, and so forth. Technologists have augmented our physical sensors with computers, electronics, and other apparatus that have added to our signal processing capabilities.

Our position at the top of the food chain was secured when we learned how to use signals and systems to communicate.[1] Initially this was nothing more than a set of grunts and gestures. Once signals were understood to be conveyers of information, we made steady progress to today at the dawn of the information age. Today technology provides the means to collect, assimilate, collate, and communicate vast amounts of information globally and (virtually) instantaneously. The signals and systems that are the result of our invention are now the foundations of commerce, medicine, transportation, defense, and entertainment. Today, we can view our favorite television show by displaying the received TV signal on a signal-processing system (TV receiver), albeit this is still usually nothing more than grunts and gestures!

1-2   SIGNAL TAXONOMY

The formal study of signals assumes that they can be represented as mathematical functions of one or more variables. A signal that is represented as a collection of M variables is referred to as an M-dimensional signal. A car-horn blast, for example, is defined by a switch state (i.e., a function of a single variable) and is, therefore, a function of a one-dimensional signal. A simple black-and-white video image can be expressed as a two-dimensional signal, where at each coordinate (x,y) the image intensity is $i(x, y)$. The Dow Jones industrial average, for example, is a scalar function of 30 economic variables, which is an example of a multidimensional or MD signal.

If a signal has an explicit mathematical representation, then it is said to be *deterministic*. For example, $x(t) = \sin(\omega_0 t)$ deterministically models a perfect sinusoid or, as it is sometimes referred to, a *harmonic oscillation*. A pendulum is often modeled as a harmonic oscillator. However, due to friction, drag, nonlinear effects, and so forth, the pendulum motion may deviate from the ideal. This deviation can be measured to be an amount $e(t) = x(t) - y(t)$, where $y(t)$ is the actual motion of the pendulum. Unless $e(t) \equiv 0$ for all time, $x(t)$ would be referred to as an *approximate* mathematical model.

If a signal is deterministic, theoretically it can be modeled exactly. There are times, however, when an exact, or even a meaningful approximate model, cannot be developed. A *random* signal, for example, defies accurate modeling by an

---

[1]"They received use of the five operations of the Lord, and in the sixth place He assigned them understanding, and in the seventh speech so they could express the meaning of their cognitations." — Ecclesiasticus 17:5

equation. Instead, we quantify our ignorance about the exact shape of a random signal with a set of parameters called *statistics*, which represent signal attributes.

Some signals are assumed to be produced by physical devices or systems and are called *causal*. If a physical signal generator is turned on at time $t = 0$, then the produced causal signal $y(t)$ satisfies

$$y(t) = \begin{cases} x(t) & \text{for } t \geq 0 \\ 0 & \text{for } t < 0 \end{cases} \tag{1-1}$$

That is, a causal signal can exist only at or after the time the signal generator is turned on. Signals that are not causal are called *noncausal*. Some signals are assumed to have existed forever. The sinewave $x(t) = \sin(\omega t)$, for example, has an infinite past and future. Such a signal would have had to preexist the history of the universe and, therefore, is physically unrealizable, noncausal. However, from a mathematical modeling viewpoint, the signal has meaning.

Signals can also be classified in terms of their mode, which can be *continuous-time*, *discrete-time*, or *digital*. In the next few sections these three important signal classifications will be developed.

1-3 CONTINUOUS-TIME SIGNALS

The study of *continuous-time signals* has always been an intrinsic part of science. The sound of a twig breaking in continuous-time may have warned our pre-historic ancestors that they were about to become a saber-toothed tiger's lunch. Continuous-time signals, also referred to as *analog* signals, are found in every facet of electronic communication, control, and instrumentation. The mathematical representation of a continuous-time signal is a continuum of points in both the independent and dependent variables. A signal is said to be *continuous* if its derivative is defined everywhere, and is said to be *discontinuous* if it is not. A signal is *piecewise-continuous* if it is discontinuous only at isolated points. Examples of continuous and piecewise continuous signals are shown in Figure 1-1.

Signals can be quantified in terms of partial specifications or statistics. The *average* or *mean* value of a continuous-time signal $x(t)$ is given by[2]

$$\langle x(t) \rangle = \lim_{T \to \infty} \frac{1}{2T} \int_{-T}^{T} x(t)\, dt \tag{1-2}$$

Another important signal statistic is called the signal's *mean squared* value, which is given by

$$\langle x^2(t) \rangle = \lim_{T \to \infty} \frac{1}{2T} \int_{-T}^{T} |x(t)|^2\, dt \tag{1-3}$$

The square root of the mean squared value is called the *root mean squared* value, or *rms* value, and is often used in instrumentation. The mean squared value also leads to another quadratic statistic, called *variance*. Variance has particular significance

---

[2]In more advanced courses on random processes a distinction will be made between *time* and *ensemble* statistics. For simplification, we shall assume that they are the same as those called the *ergodic* assumption.

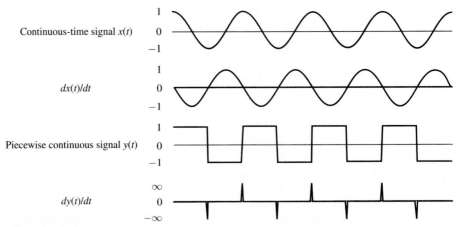

Continuous-time signal $x(t)$

$dx(t)/dt$

Piecewise continuous signal $y(t)$

$dy(t)/dt$

FIGURE 1-1   Graphs from top to bottom; continuous-time signal $x(t)$, $dx(t)/dt$, piecewise continuous signal $y(t)$, $dy(t)/dt$.

in signal analysis and is given by

$$\sigma^2_{x^2(t)} = \langle x^2(t) \rangle - \langle x(t) \rangle^2 \tag{1-4}$$

Whether a signal is deterministic or random, these statistical measures are valid parameters and can provide valuable information about the signal.

If a signal $x(t)$ is deterministic, then

$$E_x = \lim_{T \to \infty} \int_{-T}^{T} |x(t)|^2 \, dt \tag{1-5}$$

is the *total energy* in the continuous-time signal $x(t)$. From energy, the power in $x(t)$ can be defined to be

$$P_x = \lim_{T \to \infty} \frac{1}{2T} \int_{-T}^{T} |x(t)|^2 \, dt. \tag{1-6}$$

Observe that a signal having a finite energy E_x, must have a power metric $P_x = 0$ according to Equation 1-6. Also observe that the power in a deterministic signal is equal to the mean squared value given by Equation 1-3. The power contained in a signal can also be expressed over a finite interval of finite duration T_0, using the formula

$$P_{x[T_0:c]} = \frac{1}{T_0} \int_{c}^{c+T_0} |x(t)|^2 \, dt. \tag{1-7}$$

where c is an arbitrary constant. Generally, when using this formula, the value of $P_{x[T_0,c]}$ will vary as a function of the starting time $t = c$ and period T_0. The root mean squared, or rms power value, is given by

$$P_{x_{rms}} = \sqrt{P_x} \tag{1-8}$$

Estimates of all these statistics often are produced by electronic instruments.

Example 1-1   Signal power

The continuous-time signal $x(t) = A\sin(\omega_0 t + \phi)$ is differentiable everywhere and thus is continuous. The mean value of $x(t)$ is given by

$$\langle x(t) \rangle = \lim_{T \to \infty} \frac{1}{2T} \int_{-T}^{T} x(t)\, dt = \lim_{T \to \infty} \frac{A}{2T} \int_{-T}^{T} \sin(\omega_0 t + \phi)\, dt = 0$$

Since $\langle x(t) \rangle = 0$, the mean squared value and variance are equal, namely

$$\langle x^2(t) \rangle \; = \sigma_{x^2(t)}^2 = \lim_{T \to \infty} \frac{1}{2T} \int_{-T}^{T} A^2 \sin^2(\omega_0 t + \phi)\, dt$$

$$= \lim_{T \to \infty} \frac{A^2}{2T} \int_{-T}^{T} \left[\frac{1}{2} - \frac{1}{2} \cos(2\omega_0 t + 2\phi) \right] dt = \frac{A^2}{2}.$$

The power in the signal, computed over an interval $T_0 = 2\pi/\omega_0$, beginning at $t = 0$, is given by

$$P_{x[T_0:0]} \; = \frac{1}{T_0} \int_0^{T_0} A^2 \sin^2(\omega_0 t + \phi)\, dt = \frac{A^2}{T_0} \int_0^{T_0} \left[\frac{1}{2} - \frac{1}{2} \cos(2\omega_0 t + 2\phi) \right] dt$$

$$= \frac{A^2}{2}.$$

It can be seen that the power calculation is independent of phase and, in this case, equals the variance of $x(t)$.

1-4   DISCRETE-TIME SIGNALS

The set of time events $t = \{kT_s\}$, called *sample instants*, define the sampling instances at which the *sample values* of a continuous-time signal $x(t)$ are taken. The sample values at $t = \{kT_s\}$ are given by $x(t = kT_s) = x[k]$, where T is called the *sample period*. Furthermore, $f_s = 1/T_s$ is called the *sample rate* or *sample frequency* and is measured in samples per unit time. The collection of all sample values is called a *time-series* and is denoted $\{x[k]\}$. A discrete-time series is a continuum of points distributed along the dependent axis y and is discretely resolved along the independent axis. That is, the values of the time-series are only defined at the discrete sample instances $t = \{kT_s\}$, but the value of $y[k]$ can have any value. Discrete-time signals can be created by passing a continuous signal through an electronic *sampler* circuit. An ideal sampler, instantaneously captures and saves the value of a signal at each sample instant $t = kT_s$. Discrete-time signals can also be produced by computing algorithms, such as those studied in discrete mathematics. An early discrete data generator was provided in 1202 by the Italian mathematician Leonardo da Pisa (aka Fibonacci, or literally, Blockhead). He posed a formula that could count newborn rabbits, assuming that a mated adult pair would produce another pair. The Fibonacci sequence, given by

$$F_n = F_{n-1} + F_{n-2} \tag{1-9}$$

with the initial conditions

$$\begin{cases} F_0 = 1 \\ F_{-1} = 0 \end{cases} \tag{1-10}$$

produces a discrete sequence

$$\{1, 1, 2, 3, 5, 8, 13, 21, 34, 55, \ldots\} \tag{1-11}$$

which is a prediction of the rabbit population. Numerous types of discrete-time sequences arise in the fields of economics, biology, calculus, statistics, physics, and many others. The engineering study of physically meaningful discrete-time signals can be traced back to post-World War II telephone and sample data-control systems. The intent was to replace a slowly varying analog signal with a sparse set of samples. In the case of telephony, it was discovered that a number of distinct time-series could be interlaced (i.e., time-multiplexed) onto a common channel, thereby increasing the line's capacity in terms of subscribers per line (and increasing the billing potential).

Discrete-time signals can also be qualified in terms of their statistics. The sample average or mean value of a discrete-time signal $x[k]$ is given by

$$\langle x[k] \rangle = \lim_{N \to \infty} \frac{1}{2N+1} \sum_{-N}^{N} x[k] \tag{1-12}$$

The *mean squared* value is given by

$$\langle x^2[k] \rangle = \lim_{N \to \infty} \frac{1}{2N+1} \sum_{-N}^{N} |x[k]|^2 \tag{1-13}$$

The square root of the mean squared value is called the root mean squared value. The mean squared value is also used to compute the sample *variance*, which is given by

$$\sigma_{x^2[k]}^2 = \langle x^2[k] \rangle - \langle x[k] \rangle^2 \tag{1-14}$$

These statistics can be readily produced by a digital computer.

1-5   DIGITAL SIGNALS

Digital signals are discrete-time signals that are also quantized along the dependent axis, as shown in Figure 1-2. One way to produce digital signals is to pass a discrete-time sample signal $x[k]$ through an *analog-to-digital* converter (aka ADC or A/D). An n-bit ADC will quantize a sample value into one of 2^n finite values. If the dynamic range of a discrete-time signal $x[k]$ is $\Delta = \max(x[k]) - \min(x[k])$, then each bit must cover a range of $Q \geq \Delta/2^n$ volts/bit. Here Q denotes the *quantization step-size*. The quantized (digital) value of $x[k]$ is d if $dQ \leq x[k] < (d+1)Q$. A digital designer would interpret the quantized number as an n-bit fixed point signed-integer, binary, octal, hexadecimal, or other digitally formatted word. Others would interpret the same number as a rational or decimal number

Continuous-time (Analog) Signal $x(t)$

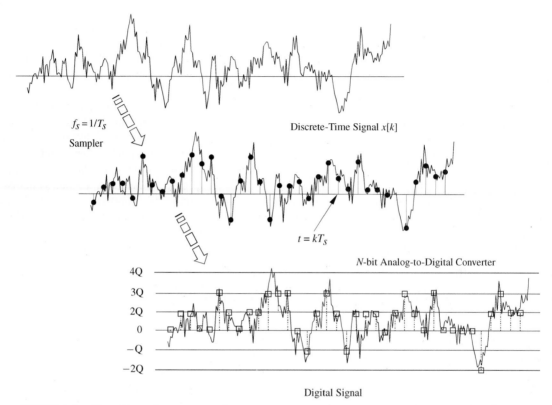

FIGURE 1-2   Signal hierarchy consisting of analog, discrete-time or sampled, and digital or quantized processes.

of finite precision. Examples of these three classes of signals can be found in Figure 1-2.

Example 1-2   Mixed signals

Many physical systems simultaneously contain analog, discrete-time, and digital signals. The inhabitants of the planet Fibonacci, for example, are searching the universe for evidence of intelligence. To the Fibonaccians, this means bunnies. Their galactic query begins with a digital signal representing the ASCII character sequence BUNNY, where

$$B = \{0, 1, 0, 0, 0, 0, 0, 0, 1\}$$
$$U = \{1, 0, 1, 0, 1, 0, 0, 0, 1\}$$
$$N = \{1, 1, 0, 0, 0, 1, 0, 0, 1\}$$
$$Y = \{0, 0, 1, 1, 0, 1, 0, 0, 1\}$$

The signal is transmitted throughout the galaxy as an analog transmission at radio frequencies. Between Earth and Fibonacci, analog noise is added. On Earth, a radio receiver, by chance, captures the analog signal, which is sampled then sent to an ADC, as shown in Figure 1-3. It can be seen that the original digital signal undergoes a number of changes as it interacts with the physical world. Similar effects can introduce uncertainty in terrestrial communication systems.

Typically, contemporary electronic systems contain a mix of analog, discrete-time, and digital signals and systems, but digital signals and systems are becoming increasingly predominant. Applications that were once exclusively analog, such as sound recording and reproduction, have become digital. Images and video are now routinely coded as digital signals. Discrete-time systems, as defined, are rarely used today. They, too, have been largely replaced by digital systems. At their zenith, discrete-time systems were commonly found in servomotor control loops and autopilots. Analog systems remain viable as optical, acoustic, and radio frequency communications systems, which operate at very high frequencies. Nevertheless, digital systems are found at the back end of these communication links where sophisticated signal and data processing is required. Here, analog electromagnetic, acoustic, and optical signals are converted to digital signals for further processing. From that point on, signals are processed digitally.

FIGURE 1-3   The digital Bunny signal. From top to bottom are the original Bunny message, message with noise added, received signal, discrete-time sampled signal, and quantized received message.

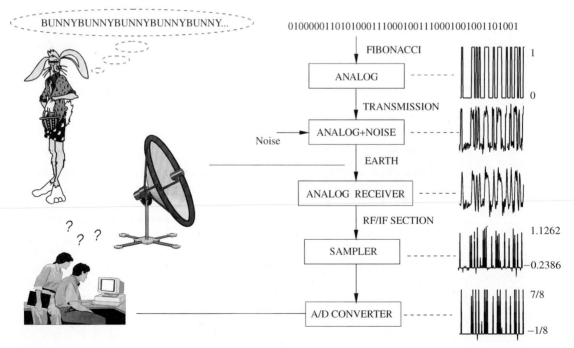

By comparing analog and digital-signal processing technologies, a number of broad conclusions can be drawn:

1 Both analog and digital systems generally can be fabricated as highly integrated semiconductor systems, but digital devices generally are more electronically dense, with an attendant economic advantage.

2 Digital systems do not have impedance-matching requirements; analog systems do.

3 Digital systems can operate at extremely low frequencies, requiring unrealistically large capacitor and resistor values if implemented as analog systems.

4 Analog systems can operate at extremely high frequencies (e.g., optical frequencies) that are in excess of the maximum clock rates of digital devices.

5 Digital systems can be programmed easily to change their functions. Reprogramming analog systems is difficult.

6 Digital signals can be delayed and/or compressed easily, which is difficult to achieve with analog signals.

7 Digital systems can work with dynamic ranges in excess of 60–70 dB (10–12 bits), which is the maximum limit of an analog system.

8 Analog systems need periodic adjustment (due to temperature-drift, aging, etc.) whereas digital systems do not require alignment.

9 Digital systems are less sensitive to additive noise, as a general rule.

Regardless of the mode of operation, many signals can be studied in their discrete-time form. If the sample rate f_s is sufficiently high, a discrete-time signal will assume the appearance, and often the properties, of its continuous-time parent. The mechanism by which a discretely sampled analog signal is converted back into a continuous signal is called *interpolation*. Digital signals are often modeled as discrete-time signals containing a measured amount of random uncertainty due to quantization. As a result, discrete-time signal analysis techniques will be used extensively to study all signals.

1-6 SAMPLING THEOREM

One of the most important scientific advancements of the twentieth century was due to Claude Shannon. Shannon's pioneering work in information theory laid the groundwork upon which many of the technological advancements in communication and signal processing have been built. One of his minor, but amusing, creations was a black box which, when activated with a switch, would extend a green hand outward and turn the switch off. Shannon is best known for his celebrated *sampling theorem*. Because Shannon worked for the telephone company he was interested in having as many billable subscribers as possible use a telephone line simultaneously. One approach was to sample the individual subscriber's conversations, interlace them together, put them on a common tele-

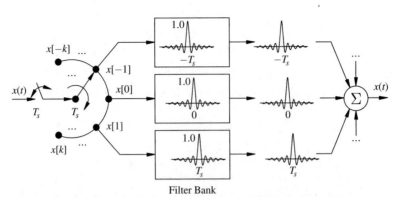

Filter Bank

FIGURE 1-4   Shannon interpolator showing sampler, sample values, interpolation filters, and reconstruction.

phone wire, and reconstruct the original message at the receiver. Shannon established the rules that govern the sampling, as well as the reconstruction procedure.

Sampling Theorem   Suppose the highest frequency contained in a continuous-time signal $x(t)$ is $f_{\max} = B$ Hz. Then, if $x(t)$ is sampled periodically at a rate $f_s = 1/T_s > 2B$, the signal can be exactly recovered from the sample values $\{x[k]\}$ using the interpolation rule[3]:

$$x(t) = \sum_{k=-\infty}^{\infty} x(kT_s)h(t - kT_s), \qquad (1\text{-}15)$$

where

$$h(t) = \frac{\sin(\pi t/T_s)}{\pi t/T_s} \qquad (1\text{-}16)$$

The lower bound on the sampling frequency, denoted $f_s = 2B$, is called the *Nyquist sample rate* or simply *Nyquist rate*. It is important to remember that sampling must take place above, and not (as is sometimes stated), at the Nyquist frequency. The frequency $f_N = f_s/2$ is generally referred to as the *Nyquist frequency*. Observe then $f_N > B$. This theory is elegant and critical to all discrete-time and digital-signal generation, analysis, and processing.

A Shannon interpolator is shown in Figure 1-4. The continuous-time signal $x(t)$ is sampled at a rate in excess of $2B$ samples per second by an ideal sampler. The sample values are presented to a bank of interpolating filters, after Equation 1-15, which convert the individual samples into a continuous-time signal. Upon summing together all the filtered outputs, the original continuous-time signal $x(t)$ is faithfully reconstructed. A detail of this reconstruction is found in Figure 1-5 where the signals being sent to the summer by the individual interpolating filters

---

[3]Equation 1-15 will be referred to as convolution, or linear filtering, in Chapter 4.

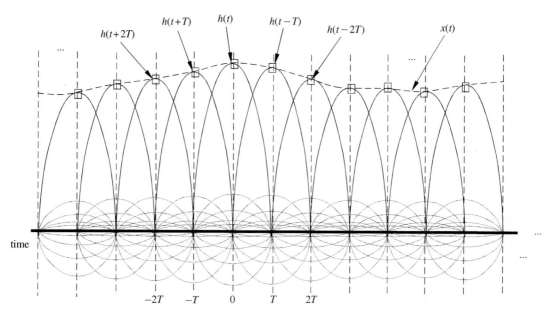

FIGURE 1-5 Shannon interpolation shown over a finite interval of time.

are shown over a finite interval of time. Observe that each interpolating filter provides a null (i.e., zero value) at sample instant $t = kT_s$ except for the kth interpolating filter, which has a value of unity. Therefore, at every sampling instant the value of the $x(t)$ is given by $x(kT_s) = x[k]$. Elsewhere, a complicated collection of weighted interpolation functions (i.e., $\sum x(kT_s)h(t - kT_s)$) are added together to reconstruct the original analog signal $x(t)$.

Example 1-3 Shannon interpolator

Suppose a continuous-time signal $x(t)$ is sampled at a high rate f_s, to form a dense time-series $\{x[k]\}$, so that $\{x[k]\}$ is a reasonably good facsimile of $x(t)$. Then resampling $x[k]$ to form $x'[k]$, where

$$x'[k] = \begin{cases} x[k] & \text{if } k \bmod p = 0 \\ 0 & \text{otherwise} \end{cases}$$

produces a sparse time-series that will be used to represent a discrete-time signal. That is, only every pth sample of $x[k]$ is kept by $x'[k]$. Shannon's work states that $x[k]$ can be reconstructed by interpolating the values of $x'[k]$ using the Shannon interpolator

$$h[k] = \frac{\sin(k\pi/p)}{k\pi/p}$$

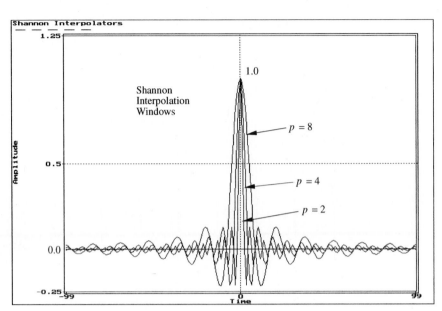

Shannon Interpolators

FIGURE 1-6   Shannon interpolation windows for different *p*. The width of the primary (center) lobe is directly proportional to *p*. The first zero crossings are found at sample $k = p$.

Computer Study[4]   S-file `ch1_3.s` and M-file `shannon.m` create a Shannon interpolation window (see Figure 1-6) using the format `shannon(p,N)`, where `p` is the decimation value and `N` is the number of samples used to form the window, rounded to the nearest odd integer. The overlay graph of several 199-sample windows can be displayed as follows.

SIGLAB	MATLAB
`>include "ch1_3.s"`	`>> [h2, k] = shannon(2,100);`
`>s2=shannon(2,200) #p=2`	`>> h4 = shannon(4,100);`
`>s4=shannon(4,200) #p=4`	`>> h8 = shannon(8,100);`
`>s8=shannon(8,200) #p=8`	`>> plot(k, [h2 h4 h8]); grid;`
`>3=<-99+rmp(199,1,199),zeros(199)>`	`>> title('Shannon windows`
`#zero axis`	`for different p');`
`># optional graph frame`	
`>box=[<-99,1.25>,<99,1.25>,`	
`<99,-.25>, <-99,-.25>]`	
`>ograph(s2,s4,s8,z,box)`	

[4]Throughout the textbook we shall be performing computer simulations and experiments to demonstrate key ideas. The computer-based instructional (CBI) programs are designed to run under MONARCH and MATLAB Student Versions described in Appendix B. It is highly recommended that the reader refer to Appendix B to better understand how CBI has been integrated with the text. *Be sure your directory paths are properly set!* Additional information can be found in README.TXT.

Observe that the Shannon window has a highly localized peak for $p = 2$ and is more spread for $p = 8$. Notice, also, that the window response falls off to nearly zero at the end points regardless of the value of p. Nevertheless, a longer window is needed to approximately reconstruct a signal resampled with a large p (low sample rate) than for a signal resampled with a small p (higher sample rate) if the same qualitative results are to be achieved.

S-file `ch1_3a.s` and M-file `ch1_3a.m` performs an approximate interpolation of a signal from its decimated time-series. The format for the S-file is `interp1(x,p)` where x is the original (undecimated) signal and p is the decimation value. Interpolate a cosine wave having a frequency $1/64$ of the sample rate as follows

SIGLAB	MATLAB
| ```
>include "ch1_3a.s"
># Nyquist frequency =>p<32
>x=mkcos(1,1/64,0,128)
>interp(x,6)
``` | ```
>> ch1_3a
``` |

Refer to Figure 1-7 where the middle area of the interpolated signal is seen to be a good facsimile of the original signal $x[k]$. The degradation at the outside edges is due to effects associated with discrete-time linear convolution (see Chapter 9).

FIGURE 1-7 Shannon interpolator experiment consisting of an analog signal $x(t)$, a $p = 6$ interpolator applied to the time-series $x[k] = x(kT_s)$, and the reconstructed signal.

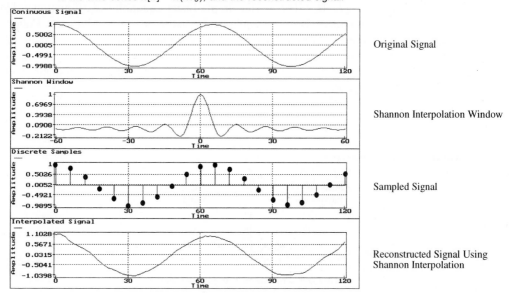

The Shannon interpolation window has a systemic weakness. It is impractical because, to be accurate, the window must extend in time out to $t = \pm\infty$. This range cannot be covered by a physically realizable interpolator. Even if this were not an issue, a different problem would be faced. In Chapter 9 it will be shown that convolution can be a time-consuming operation. As such, high data-rate real-time interpolation with a large window may be physically impossible.

A number of "practical" interpolation windows have been developed to approximate the Shannon window in shape, but are of finite duration. This simplest approximation is the *zero-order hold*, or *sample and hold*, shown in Figure 1-8. The zero-order hold interpolates the signal values residing between two adjacent samples $x[k]$ and $x[k+1]$ with a constant value. The interpolated signal, at time t, is defined to be the piecewise constant function

$$x(t) \approx x[k] = x(kT_s) \qquad (1\text{-}17)$$

for $t \in [kT_s, (k+1)T_s)$. An example of a zero-order hold interpolation of a sampled signal can be found in Figure 1-9.

Another important practical interpolation procedure is the *first-order hold*. The first-order hold interpolates the values between two adjacent samples $x[k]$ and $x[k+1]$, using a linear approximation given by

$$x(t) \approx x(kT_s) + \Delta_x(t - kT_s)(t - kT_s)$$

$$\Delta_x = (x((k+1)T_s) - x(kT_s))/T_s \qquad (1\text{-}18)$$

for $t \in [kT_s, (k+1)T_s)$. The simple linear equation (i.e., Equation 1-18), given in slope-intercept form, interpolates the value of $x(t)$ as shown in Figure 1-9.

A third interpolation method uses a lowpass filter (see Chapter 12), called a *smoothing filter*. The lowpass smoothing filter permits only small (i.e., slowly varying) incremental changes to appear at the filter's output process and, as a result, produces a smooth output.

FIGURE 1-8   Zero-order hold.

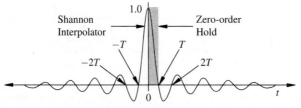

Time (zero-crossing on T second intervals)

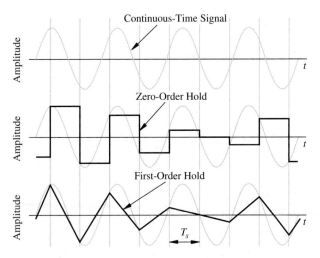

FIGURE 1-9   Continuous-time signal, zero- and first-order hold approximations.

Example 1-4   Interpolation

A discrete-time $x[k]$ signal is created by sampling an analog signal $x(t)$ at a rate above the Nyquist frequency. A zero-order hold, first-order hold, and a smoothing filter are used to interpolate the value of $x[k]$ from $x'[k] = x[k8]$ (i.e., every eighth sample of $x[k]$).

Computer Study   The interpolation algorithm is chosen from the following rules: zero-order hold, first-order hold, or smoothing filter. S-file `ch1_4.s` and M-file `ch1_4.m`, perform the interpolation on the signal $x(t)$. The signal $x(t)$ will be assumed to be obtained by decimating a sine wave by a factor $p = 8$, using all three interpolation options.

m	M-file Interpolation Method
0	Zero-order Hold
1	First-order Hold
2	Shannon Interpolator
3	Smoothing (Lowpass) Filter

SIGLAB	MATLAB
>include "ch1_4.s"	>> ch1_4

The results are displayed in Figure 1-10. It can be easily seen that there is a hierarchy of approximate interpolation schemes. The smoothing filter is best, followed by the first-order hold, and then by the zero-order hold. The smoothing filter is seen to produce a good, but slightly delayed reproduction of the original signal. Introducing temporal delays is a natural by-product of

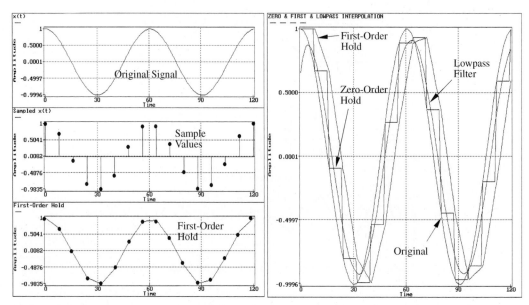

FIGURE 1-10   Interpolation demonstration: (left) sinewave, sampled sinewave and first-order hold reconstruction, (right) reconstruction of a sinewave using zero-order hold, first-order hold, and smoothing filter.

filtering and is generally not considered objectionable in signal reconstruction. Finally, the complexity, thus the speed at which these interpolation schemes can operate, has the reverse ordering.

If a signal is sampled at a rate at or below the Nyquist rate, *aliasing errors* can occur. Aliasing, as the name implies, means that a signal can be impersonated by another signal. For example, if the signals $x(t) = \cos(2\pi\omega_0 t)$ and $y(t) = 1$ are sampled at a rate $f_s = \omega_0/2\pi$, the time-series $x[k] = \{1, 1, 1, \ldots\}$ and $y[k] = \{1, 1, 1, \ldots\}$ will result. These two time-series are identical, therefore impersonate each other (i.e., they are aliases of each other). Note that, in this case, the cosine was sampled at half the Nyquist frequency rather than above it, as required by the theory.

Example 1-5   Aliasing

Consider the signals being sampled at a rate of 1 sample/second.

Signal	Minimum Sample Frequency	Aliased?
$\cos(0.2\pi t)$	> 0.2 samples/second	No
$\cos(0.4\pi t)$	> 0.4 samples/second	No
$\cos(\pi t)$	> 1.0 samples/second	Yes
$\cos(1.2\pi t)$	> 1.2 samples/second	Yes
$\cos(1.4\pi t)$	> 1.4 samples/second	Yes

Computer Study   S-file `ch1_5.s` and M-file `ch1_5.m` sample a cosine wave having a period of 32 seconds (i.e., $B = 1/32$ Hz). This establishes the minimum sample rate to be in excess of 1 sample per 16 seconds (i.e., $f_s > 2B = 1/16$ Hz).

SIGLAB	MATLAB
`>include "ch1_5.s"`	`>> ch1_5`

The results are displayed in Figure 1-11. Observe closely that when the signal is sampled above the Nyquist frequency, one can intuitively visualize the envelope of the original signal. However, once the sampling frequency is at or below the Nyquist frequency, the resulting time series takes on the appearance of a signal having a frequency below the Nyquist rate. In particular, when sampling at a rate of 1/4 samples per second, a good facsimile of $x(t)$ is generated. When sampled at 1/31 samples per second, an aliased signal of extremely low frequency appears. When sampling at 1/32 sample per second, a time-series equal to $\{1, 1, 1, \ldots\}$ is created, which is an aliased version of a 0 Hz signal of constant amplitude.

FIGURE 1-11   Original sinewave $x(t)$, $x(t)$ sampled at 1 sample per 4 seconds, aliased signal at a sample rate of 1 sample per 31 seconds, and $x(t)$ sampled at 1 sample per 32 seconds and the overlay of $x(t)$ and $x(t)$ sampled at 1 sample per 31 seconds.

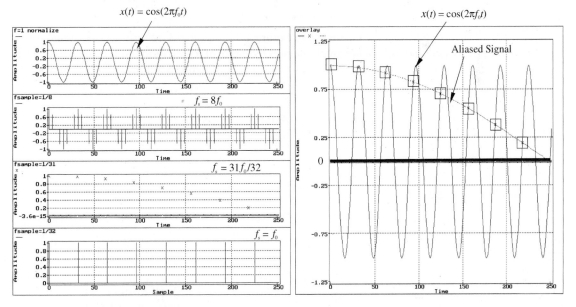

$f_s = 2f_0$: Nyquist Frequency

1-7   ELEMENTARY SIGNAL TRANSFORMATIONS

Deterministic signals can be simple or complicated. When analyzing complicated signals, it is often desirable to describe them in terms of simpler signals, called *elementary signals*, which have known properties. For example, a complicated signal may, in some cases, be expressed as a sum of sinusoids using the Fourier techniques developed in Chapter 3. Since each sinusoid has known properties, the original signal can be analyzed by studying individual sinusoids rather than the original signal itself.

Some important signal properties and transformations are listed in Table 1-1. For the sake of efficiency, continuous- and discrete-time signals are treated together. The first three transforms are introduced in Example 1-6.

Example 1-6   Signal properties—time reversal, time delay, and time scaling

Using a sampled EKG signal, examine its time reverse image, as well as a time delayed version, where the delay is set to 30 sample units of time. Each sample unit equals 1/135 seconds or $\Delta T = T_s = 7.407$ ms.
Computer Study   S-file ch1_6.s and M-file ch1_6.m perform time reversal and the linear shift of a signal $x(t)$. The S-file uses the format timerev(x).

SIGLAB	MATLAB
>include "ch1_6.s"	>> ch1_6
>x = rf("ekg1.imp")	
>timerev(x)	

The results are shown in Figure 1-12. In the discrete-time case, signal reversal and delays are implemented by elementary memory shifts. Next time scale $\cos(s\omega_0 t)$, for $f_0 = 0.05$ and $s = 1, 2, 3$, using S-file ch1_6a.s and M-file ch1_6a.m.

SIGLAB	MATLAB
>include "ch1_6a.s"	>> ch1_6a

TABLE 1-1   SIGNAL PROPERTIES

Transformation or Property	Continuous-time	Discrete-time
Reflection	$y(t) = x(-t)$	$y[k] = x[-k]$
Time Shift	$y(t) = x(t - t_0)$	$y[k] = x[k - k_0]$
Time Scaling	$y(t) = x(st)$	$y[k] = x[rk]$, r an integer
Even Symmetry	$x(-t) = x(t)$	$x[-k] = x[k]$
Odd Symmetry	$x(-t) = -x(t)$	$x[-k] = -x[k]$
Periodic	$x(t) = x(t + T)$, period T	$x[k] = x[k + K]$, period K

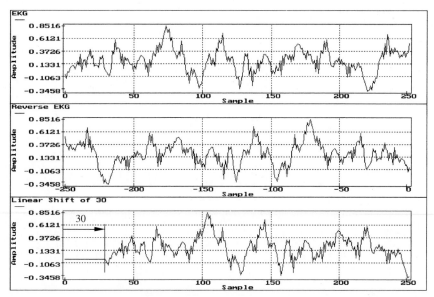

FIGURE 1-12   Time delay and time reversal of an EKG waveform.

Time scaling can be seen to have a temporal compression effect (see Figure 1-13).

The even and odd signal properties, reported in Table 1-1, can be very useful in generating and describing signals. The *even part* of an arbitrary continuous- or discrete-time signal is given by $x_e(t) = 1/2(x(t) + x(-t))$ or $x_e[k] = 1/2(x[k] + x[-k])$, respectively. The *odd part* is given by $x_o(t) = 1/2(x(t) - x(-t))$ or $x_o[k] = 1/2(x[k] - x[-k])$, respectively. In fact, any arbitrary continuous-time signal $x(t)$ can be represented as $x(t) = x_e(t) + x_o(t)$. In discrete-time, $x[k] = x_e[k] + x_o[k]$.

Example 1-7   Signal properties—even and odd symmetry

Compute and display the even and odd components of the EKG signal studied in Example 1-6.
Computer Study   S-file `ch1_7.s` and M-file `ch1_7.m` performs even and odd decompositions of a signal x using the format `oddev(x)`.

SIGLAB	MATLAB
>include "ch1_7.s"	>> ch1_7
>x = rf("ekg1.imp")	
>oddev(x)	

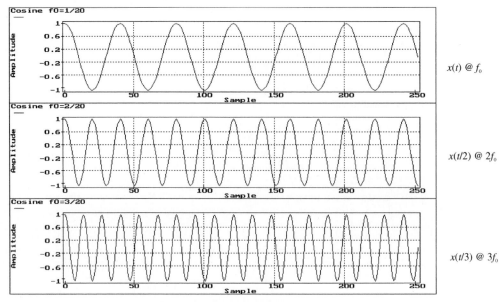

FIGURE 1-13   Time scaling where each sample equals 1 second.

The results are displayed in Figure 1-14. Note that the odd component exhibits asymmetry (antisymmetry) about the midpoint, while the even component shows symmetry.

Periodicity is a most important signal property. A signal $x(t)$ is *periodic* with period T, if T is the smallest nonzero value of time such that $x(t) = x(t + T)$. If a signal is not periodic, then it is said to be *aperiodic*. If $x(t)$ is a periodic signal having a period T, then $x(t) = x(t + kT)$ for all t, where k is an integer. Therefore, it is apparent that if $x(t)$ is periodic, then it must exist for all time (i.e., noncausal). If a periodic signal has the period T, then $f = 1/T$ is called its *natural* or *fundamental frequency*. Many signals found in nature periodically oscillate at specific natural frequencies for which values are established by the laws of physics. Since a periodic signal must exist for all time, technically, causal signals cannot be periodic.

In general, a signal of the form $x(t) = A\cos(\phi(t))$ has an *instantaneous frequency*, given by $\omega(t) = d\,\phi(t)/dt$. A classic sinusoidal signal given by

$$x(t) = A\sin(\omega_0 t + \theta) \qquad (1\text{-}19)$$

is defined in terms of the amplitude A, a fundamental frequency $d\,\phi(t)/dt = d\,[(\omega_0 t + \theta)]/dt = \omega_0$ in radians per second, and an initial phase offset θ in radians. The fundamental frequency is, in this case, $f_0 = \omega_0/2\pi$, for all time, and the fundamental period is $T = 1/f_0$.

In many cases, the phase angle $\phi(t)$ is nonlinearly related to time. PM (phase-modulated) signals (see Chapter 13), for example, have a phase angle that varies

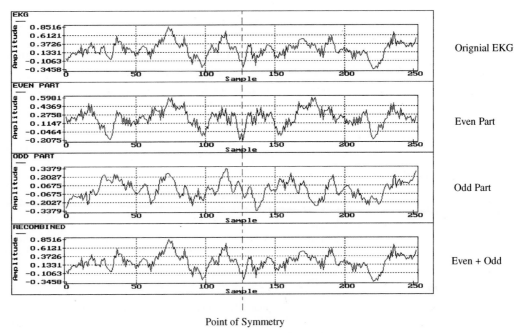

Point of Symmetry

FIGURE 1-14   From top to bottom: an EKG signal, even component, odd component, and reconstructed EKG signal from the sum of the even component and the odd component.

in accordance with some message process (e.g., speech, music). As such, the instantaneous frequency of a transmitted PM signal is constantly varying over a range of frequencies.

Example 1-8   Signal properties—periodicity

A signal described by

$$x(t) = \sin(m(t))$$

and defined by a time-varying phase angle, where $m(t)$ is a message process, will be called a PM signal in Chapter 13. The instantaneous frequency $\omega(t) = dm(t)/dt$ is, in general, time-varying. For example, if $m(t)$ is given by the time-integral of an EKG signal, then the instantaneous frequency would be the EKG signal itself.

Computer Study   S-file ch1_8.s M-file ch1_8.m performs a PM modulation of a message $m(t)$, which in this case is an EKG signal, and displays the results in Figure 1-15.

SIGLAB	MATLAB
>include "ch1_8.s"	>> ch1_8

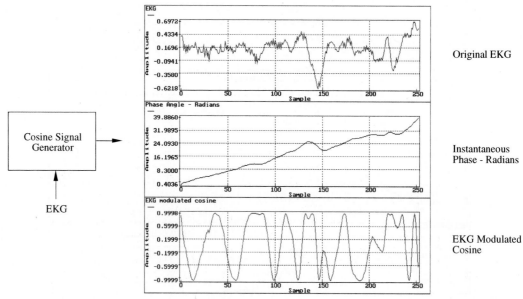

Cosine Signal
Generator

EKG

Original EKG

Instantaneous
Phase - Radians

EKG Modulated
Cosine

FIGURE 1-15   Phase modulation of EKG signal. EKG signal (top), instantaneous phase angle of PM signal (middle), and PM signal (bottom). Each sample interval corresponds to 1/135 seconds.

Observe that the PM signal appears to be highly dynamic without a well-defined period of oscillation and therefore is called an aperiodic signal.

A *pseudoperiodic* signal has verifiable periodic behavior over a finite interval of time. Technically, a periodic signal is of infinite duration (i.e., defined over the interval $t \in (-\infty, \infty)$). Signals of finite length T can self-replicate to create a pseudoperiodic signal of period T and of length kT. If $x(t)$ is known over $t \in [0, T]$, then its *periodic extension* is given by $x'(t) = x(t \bmod T)$. That is, $x'(t \pm kT) = x(t)$ for $t \in [0, T]$ for k, an integer. Many important signal analysis tools presume that a signal has a periodic extension in time. An example of this, as we shall discover in Chapter 8, is the discrete Fourier transform (DFT).

Example 1-9   Signal properties—periodic extension

A section of an EKG signal can be periodically extended using concatenation. *Computer Study*   S-file `ch1_9.s` and M-file `ch1_9.m` perform a periodic extension of a signal into the next four time intervals.

SIGLAB	MATLAB
>include "ch1_9.s"	>> ch1_9

The periodically extended signal, shown in Figure 1-16, is seen to be repeated with a period $T = 50/f_s$.

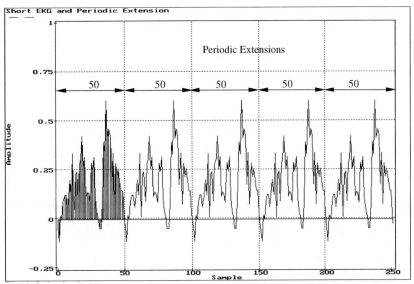

FIGURE 1-16   Periodic extensions of an EKG signal into periodic subintervals. Each sample interval corresponds to 1/135 seconds.

Signals often naturally or mathematically appear as a linear combination of sinusoids, which is called a *harmonic series*. For example

$$x(t) = \sum_{k=0}^{L} x_k(t) = \sum_{k=0}^{L} A_k \cos(\omega_k t + \phi_k) \tag{1-20}$$

for $\omega_k = k\omega_0$, represents a harmonic series where $x_k(t)$ is called the kth *harmonic*. The zeroth harmonic (i.e., $k = 0$) corresponds to a DC signal with amplitude $A_0 \cos(\phi_0)$. The first harmonic (i.e., $k = 1$) is called the fundamental harmonic and is a sinusoid of amplitude A_1, at frequency ω_1 and a phase offset ϕ_1. The second harmonic is at frequency $\omega_2 = 2\omega_1$, and so forth.

Example 1-10   Signal properties—linearity

The sum of sinusoids of the form $x(t) = \sum_{k=0}^{3} A_k \sin(k\omega_0 t + \theta_k)$ is displayed in Figure 1-17, for $\omega_0 = 2\pi/40$.

k	A_k	$k\omega_0$	θ_k
0	1	0	0
1	1.5	$2\pi/40$	$\pi/2$
2	0.5	$4\pi/40$	$-\pi/2$
3	0.25	$6\pi/40$	$\pi/4$

Computer Study   S-file `ch1_10.s` and M-file `ch1_10.m` produces $x(t)$ from its sinusoidal components.

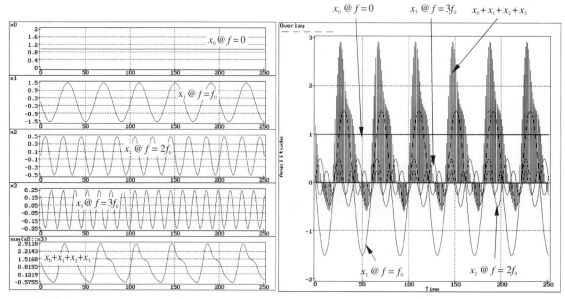

FIGURE 1-17   Harmonic time-series: (left) from top to bottom, x_0, x_1, x_2, x_3 and $x_0 + x_1 + x_2 + x_3$, (right) data from (left) in overlaid form.

SIGLAB	MATLAB
>include "ch1_10.s"	>> ch1_10

Referring to Figure 1-17, note that the original signal has been reconstructed from its individual harmonic components.

1-8   ELEMENTARY SIGNALS

The mathematical study of signals is often premised on the representation of complicated signals by a set of *elementary functions*, also called *signal primitives*, which have agreed-upon mathematical representations and well-defined properties and behavior. They can be combined linearly or nonlinearly, with imparted shifts or delays, reversals, and periodic extensions to create new, complicated signals. Their common feature is, however, that they are members of a primitive set of functions. While there is no universally accepted list, the most basic elementary signals are developed in this section. For the sake of efficiency, continuous- and discrete-time elementary signals will be developed concurrently.

Unit Step   A *unit step function* (see Figure 1-18) is defined to be

$$u(t) = \begin{cases} 0 & \text{if } t < 0 \\ 1 & \text{if } t \geq 0 \end{cases} \quad \text{continuous-time} \tag{1-21}$$

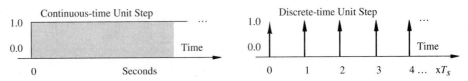

FIGURE 1-18   Continuous-time and discrete-time unit step functions.

$$u[k] = \begin{cases} 0 & \text{if } k < 0 \\ 1 & \text{if } k \geq 0 \end{cases} \quad \text{discrete-time} \tag{1-22}$$

and is often used as a test signal, with the resulting system response called the *step response*.

Unit Impulse   The *unit impulse*, also called a *Dirac impulse*, is an idealized mathematical model of an important, but physically unrealizable, continuous-time signal. The *Kronecker impulse* is its discrete-time counterpart and is physically realizable. Both are essential to the study of signals. The response of a system to an impulse is called the system's *impulse response*.

A mathematical definition of an impulse is often given by the derivative of the unit step function. That is

$$\delta(t) = \frac{du(t)}{dt} \tag{1-23}$$

Equivalently, the unit step function can be generated from an impulse using

$$u(t) = \int_{-\infty}^{t} \delta(\tau)\, d\tau \tag{1-24}$$

Graphically, a Dirac impulse can be modeled as shown in Figure 1-19. It can be seen that for either case, as $\varepsilon \to 0$, both models become infinitely tall and infinitely thin, but continue to maintain a unit area. Such signals do not exist in the physical world but do have significant importance in the mathematical

FIGURE 1-19   Rectangular and triangular models of the impulse function.

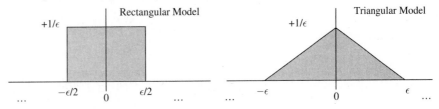

study of signals and systems. The properties common to all impulse models are summarized below

$$\begin{cases} \int_{-\infty}^{\infty} \delta(t)\,dt = 1 \\ \lim_{t \to 0} \delta(t) = \infty \\ \delta(t) = 0 & \text{for } t \neq 0 \\ \delta(t) = \delta(-t) & \text{i.e., } \delta(t) \text{ is an even function} \end{cases} \tag{1-25}$$

The impulse is unique among the elementary signals. Technically, an impulse is a *distribution* rather than a function of t. The distinction is that a function must be single-valued. That is, for any time t, a function has one and only one value. A distribution, however, can be multiple-valued at a point (i.e., $\delta(0) \to \infty = k\infty$).

The impulse distribution also exhibits the important *sifting* or *sampling property*, which is given by

$$\int_{-\infty}^{\infty} x(t)\delta(t - t_0)\,dt = \int_{-\infty}^{\infty} x(t_0)\delta(t - t_0)\,dt = x(t_0) \int_{-\infty}^{\infty} \delta(t - t_0)\,dt = x(t_0) \tag{1-26}$$

The sampling property provides a mathematical means of isolating the sample value of an analog signal at time $t = t_0$. Impulse distributions and the sampling property, as we shall see, can provide a mathematical framework that can be used to model a discrete-time signal derived from sampling a continuous-time signal.

In the discrete-time case, an impulse is given by the *Kronecker delta function*

$$\delta_K[k] = \begin{cases} 0 & \text{if } k \neq 0 \\ 1 & \text{if } k = 0 \end{cases} \tag{1-27}$$

The discrete-time impulse is seen to be bounded for all time, is physically realizable, and satisfies the conditions

$$\begin{cases} \sum_{k=-\infty}^{\infty} \delta_K[k] = 1 \\ \delta_K[k] = \delta_K[-k] & \text{i.e., } \delta_K[k] \text{ is even} \end{cases} \tag{1-28}$$

The Kronecker impulse function also exhibits a sampling property, in that

$$\sum_{k=-\infty}^{\infty} x[k]\delta_K[k - k_0] = \sum_{k=-\infty}^{\infty} x[k_0]\delta_K[k - k_0]$$

$$= x[k_0] \sum_{k=-\infty}^{\infty} \delta_K[k - k_0] = x[k_0] \tag{1-29}$$

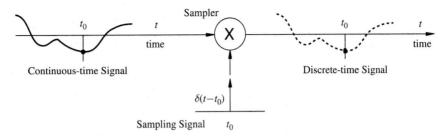

FIGURE 1-20   Sampling property of the impulse.

Equation 1-29 is potentially useful in modeling and studying discrete-time signals at selected sample instances. This process is interpreted in Figure 1-20.

Example 1-11   Sampling property

The previously studied EKG signal can be sampled at time instants $t = \{0, 1, 3, 6, 33, 47, 66, 88, 93, 96\}T_s$ using the sampling property of the discrete-time impulse function.

Computer Study   S-file `ch1_11.s` and M-file `ch1_11.\m` samples a signal $x(t)$ by shifting a Kronecker impulse to the desired sample instant and multiplying the shifted impulse with $x(t)$, as shown in Figure 1-21, to produce a time-series $x[k]$.

FIGURE 1-21   From top to bottom: EKG signal, Kronecker impulses shifted to the desired sample instances, and sampled EKG signal. Each sample interval corresponds to 1/135 seconds.

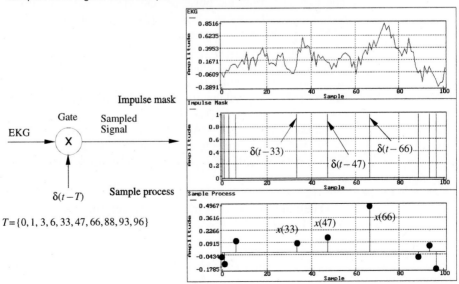

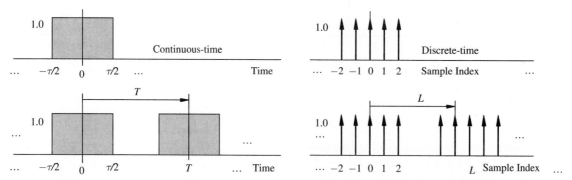

FIGURE 1-22 Rectangular pulses and pulse train of period T (Continuous-time) or L-samples (Discrete-time).

SIGLAB	MATLAB
>include "ch1_11.s"	>> ch1_11

Rectangular Pulse Continuous- and discrete-time rectangular pulses are displayed in Figure 1-22.

$$\text{rect}(t/\tau) = \begin{cases} 1 & \text{if } |t| < \tau/2 \\ 0 & \text{otherwise} \end{cases} \quad \text{continuous-time} \tag{1-30}$$

$$\text{rect}[k/K] = \begin{cases} 1 & \text{if } |k| < K/2 \\ 0 & \text{otherwise} \end{cases} \quad \text{discrete-time} \tag{1-31}$$

A rectangular pulse can be generated using two interlaced unit step functions. The first step turns the pulse on and the second turns it off. More specifically, $\text{rect}(t/\tau) = u(t + \tau/2) - u(t - \tau/2)$. If a continuous-time rectangular pulse of width τ begins at $t = 0$ (rather than at $t = -\tau/2$), then it can be represented as $\text{rect}((t - \tau/2)/\tau) = \text{rect}(t/\tau - 1/2)$.

A causal rectangular pulse-train consists of a string of nonoverlapping pulses distributed along regular time intervals as shown in Figure 1-23. The pulse-train can be parameterized by its period T and a parameter called the *duty cycle*, denoted d ($0 \le d \le 1$). The duty cycle is the ratio of the pulse's "on" time and its period (i.e., $d = \tau/T$). An infinitely long periodic pulse-train could be created by extending the production rule for the causal signal into negative time. The discrete-time version of a periodic pulse-train is simply the sampled version of $p_\tau(t)$.

Triangular Pulse A triangular waveform, of duration $2T$, is shown in Figure 1-24. It is given by

$$\text{tri}(t/T) = \begin{cases} 1 - \left|\dfrac{t}{T}\right| & \text{if } |t| \le T \\ 0 & \text{otherwise} \end{cases} \quad \text{continuous-time} \tag{1-32}$$

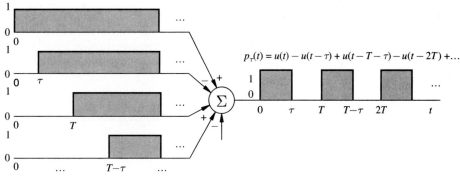

FIGURE 1-23   Pulse-train defined by the shift property.

$$\text{tri}[k/K] = \begin{cases} 1 - \left| \dfrac{k}{K} \right| & \text{if } |k| \le K \\ 0 & \text{otherwise} \end{cases} \quad \text{discrete-time} \qquad (1\text{-}33)$$

A triangular pulse-train can also be constructed in the same manner and be used to synthesize a rectangular pulse-train from a rectangular pulse.

Ramp Function   A ramp waveform of duration T is shown in Figure 1-25. It is given by

$$\text{ramp}(t/T) = \begin{cases} t/T & \text{if } 0 \le t < T \\ 0 & \text{otherwise} \end{cases} \quad \text{continuous-time} \qquad (1\text{-}34)$$

$$\text{ramp}[k/K] = \begin{cases} k/L & \text{if } 0 \le k < K \\ 0 & \text{otherwise} \end{cases} \quad \text{discrete-time} \qquad (1\text{-}35)$$

A ramp has many purposes, including generating a time-base that linearly sweeps out an interval of time, resets itself, and then repeats the process. The ramp-train $r(t) = \text{ramp}(t/T)$, will linearly sweep out the range $[0, 1]$ every T seconds.

FIGURE 1-24   Triangular pulse and triangular pulse-train.

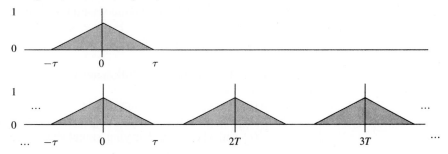

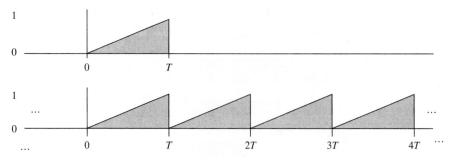

FIGURE 1-25   Ramp and ramp train.

Harmonic Oscillation   A fundamentally important signal is the familiar sinusoidal (harmonic) oscillation. A real harmonic oscillation at frequency ω_0 radians
per second is given by

$$x(t) = \sin(\omega_0 t) \qquad \text{continuous-time sine wave} \qquad (1\text{-}36)$$

$$x(t) = \cos(\omega_0 t) \qquad \text{continuous-time cosine wave} \qquad (1\text{-}37)$$

$$x[k] = \sin(\omega_0 k T_s) \qquad \text{discrete-time sine wave} \qquad (1\text{-}38)$$

$$x[k] = \cos(\omega_0 k T_s) \qquad \text{discrete-time cosine wave} \qquad (1\text{-}39)$$

Since these signals are defined for all $t \in (-\infty, \infty)$, they are seen to be noncausal.
In practice, we often use a continuous-time pseudoperiodic versions of these signals given by $y(t) = x(t)u(t)$, where the sinusoid is turned off in negative time
by the unit-step function and turned on in positive time by the same unit step. For
discrete-time signals, $y[k] = x[k]u[k]$ by analogy.

Exponential Functions   Real exponential waveforms can represent the energy decay of a natural system, or the energy expansion in the case of an instability. Exponential signals are fundamental building blocks for solutions to ordinary
differential or difference equations. A real causal exponential function is given by

$$x(t) = e^{\sigma t} u(t) \qquad \text{continuous-time exponential} \qquad (1\text{-}40)$$

$$x[k] = e^{\sigma k} u[k] \qquad \text{discrete-time exponential} \qquad (1\text{-}41)$$

The control parameter σ can have the following values

$$\begin{cases} \sigma = 0 & x(t) \text{ and } x[k] \text{ are constant at one} \\ \sigma > 0 & x(t) \text{ and } x[k] \text{ are an expanding signal (diverges to } \infty) \\ \sigma < 0 & x(t) \text{ and } x[k] \text{ are a decaying signal (converges to 0)} \end{cases} \qquad (1\text{-}42)$$

Complex Exponentials   Complex causal exponential signals have the form

$$x(t) = Ae^{\lambda t}u(t) \qquad \text{continuous-time} \tag{1-43}$$

$$x[k] = Ae^{\lambda k T_s}u[k] \quad \text{discrete-time} \tag{1-44}$$

where λ is the complex number

$$\lambda = \sigma + j\omega \tag{1-45}$$

where σ and ω are real numbers. If $\omega = 0$, a real exponential signal results. If $\sigma = 0$, then $x(t)$ can be described in polar form as a causal signal

$$x(t) = Ae^{j\omega t}u(t) \tag{1-46}$$

or, in the noncausal case

$$x(t) = Ae^{j\omega t} \tag{1-47}$$

where, in both cases, the harmonic oscillation has period $T = 2\pi/\omega$ and fundamental frequency $f = \omega/2\pi$. Using *Euler's equation*, the oscillator can also be expressed in cartesian coordinates as

$$x(t) = A\cos(\omega t) + jA\sin(\omega t) \tag{1-48}$$

Example 1-12   Complex exponentials

Causal complex exponentials can be used as test signals, as well as models of physically realizable signal processes. A 10-second record for assorted values of σ and ω, where $x(t) = e^{\sigma t}e^{j\omega t}u(t)$, can be approximated by a discrete-time time-series.

Computer Study   S-file ch1_12.s and M-file ch1_12.m create and display a complex exponential using the format cmpxeps(s, w, N), where s is σ, w is ω, and N is the length of the signal in samples. M-file ch1_12.m performs the same operation. For example, define $\sigma = -0.1$, $\omega = \pi/4$, then

SIGLAB	MATLAB
>include "ch1_12.s"	>> ch1_12
>cmpxeps(-.1,pi/4,128)	

produces the graphs shown in Figure 1-26.

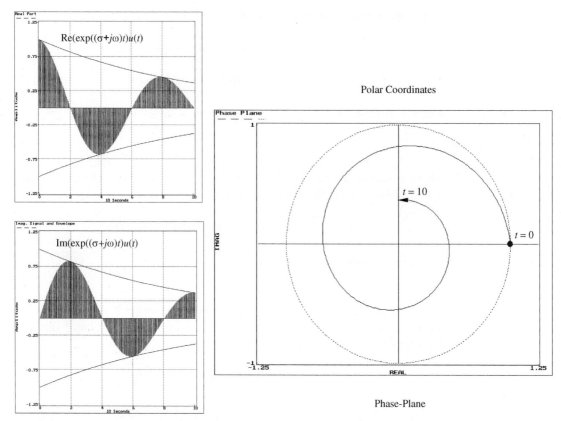

FIGURE 1-26   Complex exponential: (top left) real part, (bottom left) imaginary part, (right) polar plot.

1-9   SUMMARY

In Chapter 1, basic signals and their properties were developed. Basic signals, varying from the unit step to harmonic oscillation, were presented and defined mathematically. The signals form a primitive set of signals that can be used to define signals of greater complexity.

It was shown that signals can be classified on the basis of a number of attributes. The list presented is, admittedly, not exhaustive, but it is representative. The most important classification separates continuous-time or discrete-time signals. Discrete-time signals, when digitized, became digital signals. The fundamental concept underlying the conversion of continuous-time signals to discrete-time signals is the sampling theorem. Besides establishing conditions under which a continuous signal can be sampled and reconstructed, it also provides a gateway to the study of digital signal processing systems.

In Part II, a mathematical framework will be developed for the study of continuous-time signals. Initially the study will focus on developing methods to represent these signals mathematically.

1-10 COMPUTER FILES

These computer files were used in Chapter 1.

File Name	Type	Location
Subdirectory CHAP1		
ch1_3.s	C	Example 1-3
ch1_3a.x	C	Example 1-3
intpol.m	C	Example 1-3
ch1_4.x	C	Example 1-4
ch1_5.x	T	Example 1-5
ch1_6.x	C	Example 1-6
ch1_6a.x	T	Example 1-6
1shift.m	C	Example 1-3
ch1_7.x	C	Example 1-7
oddev.m	C	Example 1-3
ch1_8.x	T	Example 1-8
ch1_9.x	T	Example 1-9
ch1_10.x	T	Example 1-10
ch1_11.x	T	Example 1-11
ch1_12.x	T	Example 1-12
Subdirectory SFILES or MATLAB		
interp.s	C	Example 1-4
shannon.m	C	Example 1-3
interp1.m	C	Examples 1-3, 1-4
Subdirectory SIGNALS		
ekg1.imp	D	Examples 1-6, 1-7, 1-9, 1-11
ekg2.imp	D	Example 1-8
Subdirectory SYSTEMS		

Where "s" denotes S-file, "m" an M-file, "x" either/or, "T" a tutorial program, "D" a data file, and "C" a computer-based instruction (CBI) program. CBI programs can be altered and parameterized by the user.

1-11 PROBLEMS

1-1 Signal parameters   A Fibonacci number generator found in Equation 1-9 is designed to produce {1, 1, 2, 3, 5, 8, 13, 21, 34, 55} over and over again (i.e., periodic with period 10). What is the sequence's mean value, mean squared value, variance, and power? What would these values be if the signal was received in reverse order? Permuted order?

1-2 Signal parameters   In Example 1-6, an EKG signal was analyzed (Figure 1-12). Suppose, for convenience, that the EKG signal is scaled to form a new signal $x(t) = a \times ekg(t)$. Assume that the mean and variance of EKG(t) are M and V respectively, regardless of the length of the observation interval (i.e., $T > 0$). What is the power in the signal $x(t)$ for $a = 1$, $a = 1/2$, and $a = 2$?

1-3 Digital signals   The signal $x(t)$, studied in Problem 1-2, is sent to an 8-bit analog to digital converter having a dynamic range of ± 15 volts. What is the quantization step size, in volts per bit, if $a = 1$, $a = 1/2$, and $a = 2$? What is the maximum value of a that will result in nonsaturating analog to digital conversion?

1-4 Nyquist frequency   Compute the Nyquist frequency for

$$x(t) = \sum_{i=0}^{9} \cos(2\pi i t)$$

If $x(t)$ is sampled at a rate of 100 samples per second, what is the mathematical definition of the ideal Shannon interpolator? If $x(t)$ is sampled at a rate of 10 samples per second, what would be the mathematical representation of $x[k]$?

1-5 Signal taxonomy   In Example 1-6, an EKG signal was analyzed. The signal is actually a discrete-time signal sampled at a rate of 135 samples per second. Suppose that the signal is to be communicated to a remote facility for computer-based analysis. Unfortunately, the communication channel can only transmit 135/8 samples per second. Reconstruct a facsimile of the EKG at the remote location using Shannon, zero-order, and first-order interpolation methods. Determine the quality of reconstructed signal using subjective visual techniques plus computing the mean error and error variance. Explain any abnormalities.

1-6 Sum of sinusoids   Derive a formula to express $x(t) = A\cos(\omega_0 t + \phi)$ if $x(t) = a\cos(\omega_0 t) + b\sin(\omega_0 t)$.

1-7 Periodicity   Show that if $x(t)$ is a periodic signal of period T, as is also periodic with period kT, where k is an integer.

1-8 Periodicity   Determine the period (where applicable) of the following signals. Experimentally verify your results if $a = 0.1$ and $b = \pi$

$$x_1(t) = A\cos(at + b)$$

$$x_2(t) = A\cos\left((at + b)^2\right)$$

$$x_3(t) = Ae^{j\cos(at+b)}$$

1-9 Periodicity   A signal is given by $x(t) = A_0\cos(k_0\omega_0 t + \phi_0) + A_1\cos(k_1\omega_0 t + \phi_1)$. Show that if $k = k_0/k_1$ is a rational number, the signal is periodic. What is the period of $x(t)$?

1-10 Periodicity   Given $x(t) = e^{j\omega_0 t}$ sampled at a rate T_s to form $x(kT_s) = e^{j\omega_0 k T_s}$, show that if $x[k]$ is periodic with period T_0, then T_s/T_0 must be rational where $T_0 = 2\pi/\omega_0$. If $T_s/T_0 = 3/7$, what is the period of $x[k]$?

1-11 Even and odd signals   Given $x(t) = \exp(-at)u(t)$, where $a > 0$, express and sketch $x(t) = x_{\text{even}}(t) + x_{\text{odd}}(t)$, $x_{\text{even}}(t)$, and $x_{\text{odd}}(t)$.

1-12 Radar signals   A pulse radar signal is assumed to be an ideal train of impulses reflected by several targets to form a received signal

$$y(t) = p(t - \tau) + 0.1p(t - 2\tau) + 0.025p(t - 3\tau)$$

Let $p(t)$ be a pulse-train with a pulse repetition frequency of 1000 pulses per second, and a pulse width of 10 μs. Sketch $y(t)$ for $t \in [0, 10]$ms and $\tau = 10$ μs. Assuming that the radar beam propagates at a speed of 3×10^8m/s, how far away are the targets?

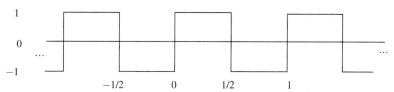

FIGURE 1-27 $x(t)$ for Problem 1-15.

1-13 Impulse distributions   Are the following relationships valid models for an impulse distribution?

$$\delta_1(t) = \lim_{\mu \to \infty} \frac{\mu}{\sqrt{\pi}} e^{-\mu^2 t^2}$$

$$\delta_2(t) = \lim_{\varepsilon \to \infty} \frac{1}{\pi} \frac{\sin(\varepsilon t)}{t}$$

1-14 Impulse distributions   Evaluate the following

$$x_1(t) = \int_{-\infty}^{\infty} [\tau^2 \cos(2\pi(\tau + 1) + \pi/8)]\delta(\tau + 1)d\tau$$

$$x_2(t) = \int_{-\infty}^{\infty} \delta(\tau - 1)[\tau^2 \cos(2\pi\tau + \pi/8)]d\tau$$

1-15 Harmonic series approximation   Given $x(t)$, shown in Figure 1-27, write a computer program to compute the error as a function of the number of terms N in a harmonic series approximation $x'(t)$ to $x(t)$ given by

$$x'(t) = \frac{4}{\pi} \sum_{\text{odd } n=1}^{N} \frac{1}{n} \sin(2\pi nt)$$

The approximation error is defined as $e(t) = x(t) - x'(t)$. Plot your results as an overlay display for $N = 1, 3, 5, 7, 9, 11$. At what instants of time do the maximum and minimum errors occur? What is the mean error? What is the error variance?

1-16 Fundamental frequency   If $x_1(t)$ has a period T_1 and $x_2(t)$ has a period T_2, is $x_3(t) = x_1(t) + x_2(t)$ periodic, and (if so) what is its fundamental frequency?

1-17 Complex signals   A complex number $z(t) = x(t) + jy(t) = r \exp(j\phi(t))$. Show $z(t)z(t)^* = r^2$, $z(t)/z(t)^* = \exp(j2\phi(t))$.

1-18 Complex exponential   A discrete-time signal is given by

$$x[n] = \sum_{i=0}^{N-1} x_i[n]$$

where $x_i[n] = \exp(j2\pi ni/N)$. What is the value of $x[0]$? What is the value of $x[n]$ for $n > 0$, and N a positive integer? Graph, in polar coordinates, $x_i[2]$ for all i and $N = 9$.

1-19 Euler's Equation   Use Euler's equation $\exp(j\phi) = \cos(\phi) + j\sin(\phi)$ to derive

$$\cos(\phi) = 0.5\left[\exp(j\phi) + \exp(-j\phi)\right]$$

$$\sin(\phi) = -0.5j\left[\exp(j\phi) - \exp(-j\phi)\right]$$

$$\sin(\phi)\sin(\theta) = 0.5\left[\cos(\phi - \theta) - \cos(\phi + \theta)\right]$$

and use computer simulations to verify these relationships.

1-20 2-D signals   A 2-D signal can be created from a symmetric 1-D signal by "spinning" it about a center-axis. For example, run the S-file `spin.s` and create a 2-D $n \times n$ signal, $n \le 16$, as follows:

```
SIGLAB>include "spin.s"; x=ham(16); y=spin(x) #16-sample
       Hamming window
```

Is the symmetry even, odd, or neither?

1-21 AM communication   In Chapter 13, a type of AM radio signal will be expressed as $x(t) = m(t)\cos(\omega_0 t + \phi)$. Sketch $x(t)$ if the message $m(t)$ is a pulse train of duty cycle $d = 1/2$, period $T = 10^{-3}$ seconds, and $\omega_0 = 2\pi 10^4$ radians per second. Over an interval of $T = 10^{-3}$ seconds, what is the power in $x(t)$, $m(t)$, and $\cos(\omega_0 t + \phi)$? How do they compare?

1-22 Data communications   A commercial voice-grade telephone channel will be assumed to pass frequencies up to 3.4 kHz. What is the minimum sampling rate that can be used to allow (under ideal conditions) the signal to be reconstructed from its sample values. If digitized voice requires at least 12-bits of precision to maintain speech quality, what would be the minimum channel-data rate requirements in bits per second?

1-23 Financial model   A discrete-time model for a savings account is given by

$$b[k] = d[k] + 1.01b[k - 1]$$

where $b[k]$ is the balance and $d[k]$ is the deposit for month k. The factor 1.01 represents the monthly interest. What is the monetary growth of the account started at $k = 0$ with a \$10,000 initial deposit and an additional \$100 deposited every month? What is the value of the account at 12 and 24 months?

1-24 Signals in noise   Assume that the ASCII "Bunny" message studied in Example 1-2 is denoted $x(t)$. Suppose that it is transmitted as $y(t) = xbi(t) + an(t)$ where $n(t)$ is uniformly distributed random noise (i.e., `rand(n)`) over the interval $-1/2 \le n(t) \le 1/2$ and $xbi(t)$ is a bipolar version of $x(t)$. At the receiver, the decoded message is given by $z(t) = \text{sign}(y(t)) = \{-1, 1\}$, where 1 corresponds to a logical "1" and -1 represents a logical "0". Conduct an experiment over 450 consecutive sample values of $y(t)$ to determine how large the value of "a" may be so that no decoding error takes place. In this case, what is the ratio of signal power to noise power?

1-25 Discrete-time modeling   Assume that the Dow-Jones average at the end of a 250 trading days, is modeled by

$$x[k] = x[k - 1] + 20n[k]; \; x[0] = 300$$

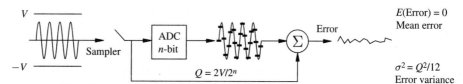

FIGURE 1-28   Quantization error production and values.

where $n[k]$ is a uniformly distributed random process distributed over $-1/2 \leq n[k] \leq 1/2$ (i.e., `rand(250)`). Suppose you check the market only on Mondays, which we shall assume are all days such that $k = 0 \bmod(5)$. Use Shannon and first-order interpolation to reconstruct a facsimile of $x[k]$. Compute the error statistics. How well did your interpolation scheme work? (Explain).

1-26 Quantization error   A digital signal is an approximation of a discrete-time signal. The quality of the approximation is determined by the number of bits used by the digital word $x_D[k]$ to represent a discrete-time sample $x[k]$. The difference between the digital and discrete-time value at sample instant k will be defined to be the quantization error $e[k] = x_D[k] - x[k]$. If the digital word contains a sufficient number of fractional bits of precision (≥ 6 bits typical), and if arithmetic rounding is used, then the error process is assumed to be zero mean with a variance of $\sigma_e^2 = Q^2/12$, where Q is called the quantization step-size and is measured in volts/bit (see Figure 1-28). For example, if a ± 5 volt range is digitized with an 8-bit analog-to-digital converter, then $Q = 10/2^8 = 0.39$ volts per bit. If a unit amplitude sinewave is digitized using the ± 5 ADC, what is the quantization error variance? Conduct a numerical experiment to test your hypothesis. Repeat using 12- and 16-bit ADCs.

1-12 COMPUTER PROJECTS

1-1 Digital Communication   The message from the planet Fibonacci was sent to earth in a 9-bit ASCII code format. The code for BUNNY consisted of 8 information bits, followed by an odd parity bit. The complement of the binary (modulo 2) sum of the 8 information bits defines the value of the odd parity bit. If there is an odd number of ones in the information field, then the parity bit is one. If there is an even number of ones in the information field, the parity bit is zero. If, upon reception, the transmitted binary bit disagrees with the locally generated parity bit, a transmission error is assumed to have occurred.

The Fibonaccians send 5 consecutive BUNNY requests that form a $5 \cdot 5 = 25$ character transmitted message. Electronically, each binary valued symbol (bit) in the Fibonacci message will be assumed to be converted into 10 discrete sample values. That is, a logical one is modeled as the time-series $[1,1,1,1,1,1,1,1,1,1]$. A logical zero would be modeled as a string of 10 zeros. Therefore, the BUNNY message is represented by $9 \cdot 10 \cdot 5 = 450$ sample time-series.

Devise a scheme by which the BUNNY message is sampled near the midpoint of each binary valued symbol frame (i.e., near the middle of a logical one or zero) and use the sampled value to represent the transmitted symbol (**hint**: use the decimation operator). The result will be a 45 ($9 \cdot 5$) bit decoded ASCII code representing BUNNYBUNNYBUNNYBUNNYBUNNY.

Devise a scheme to check the parity bit of the received message against the locally generated parity bit (**hint**: use a modulo 2 operator). Test your solution using an ideal (noise-free) communication channel.

Add noise $n(t)$ to message $m(t)$ to form $s(t) = m(t) + cn(t)$ where c is used to adjust the signal-to-noise power ratio to have values of 100, 10, 1. Perform the previous experiment and to determine if transmission errors have occurred. Evaluate the parity bit.

1-2 Filtering   In future chapters, filters will be introduced. A simple filter, which will be later classified to be an FIR lowpass system, is given by the Hamming window. Filter the 450-sample message used in Project 1-1 with a filter using a 30th-order Hamming window (see convolution). Discard the first and last 15 samples (for reasons that will become apparent in Chapter 9) and repeat the Project 1-1 exercise. Does this suggest that the performance of communication channels will be affected by (lowpass) filtering? Explain.

## PART TWO

CONTINUOUS-TIME SIGNALS AND SYSTEMS

MATHEMATICAL REPRESENTATION OF CONTINUOUS-TIME SIGNALS

If you have built castles in the air, your work need not be lost, that is where they should be—now put a foundation under them.

—H. D. Thoreau

2-1 INTRODUCTION

In the previous chapter, the basic properties of signals were presented. While some signals defy a mathematical description, many other signals are deterministic and, therefore, do have an exact mathematical representation. For example, many important continuous-time signals can be expressed as a solution to an ordinary differential equation, or ODE. In this chapter, mathematical transform methods are introduced. These methods can convert a signal defined in one domain (e.g., time) into another domain (e.g., frequency). The use of transforms can, in many cases, simplify the analysis of a signal, as well as quantify how signals and systems interact.

2-2 ORDINARY DIFFERENTIAL EQUATIONS

Linear constant coefficient ODEs are often used to model the signals produced by electronic, electromagnetic, acoustic, mechanical, biological, and optical systems. A *harmonic oscillation*, for example, is characterized by the solution to the second-order ODE $d^2x(t)/dt^2 + \omega_0^2 x(t) = 0$, with initial conditions given by $x(t)|_{t=0} = x_0$, $dx(t)/dt|_{t=0} = x_1$. If $x_0 = 1$ and $x_1 = 0$, then $x(t) = \cos(\omega_0 t)$, which can be verified by direct substitution. For initial conditions $x_0 = 0$ and $x_1 = \omega_0$, the solution becomes $x(t) = \sin(\omega_0 t)$. Mixing the initial conditions produces a harmonic oscillation of the form $x(t) = A \cos(\omega_0 t + \Phi)$. Thus, the defining ODE and initial conditions completely characterize the amplitude, period and phase of a simple harmonic oscillation. Extending this concept, higher order signals can be modeled as the *homogeneous (unforced) solution* to an *n*th-order ODE

$$c_n \frac{d^n x(t)}{dt^n} + c_{n-1} \frac{d^{n-1} x(t)}{dt^{n-1}} + \cdots + c_1 \frac{dx(t)}{dt} + c_0 x(t) = 0 \qquad (2\text{-}1)$$

with initial conditions

$$\left. \frac{d^i x(t)}{dt^i} \right|_{t=0} = x_i \qquad (2\text{-}2)$$

for $i = 0, 1, \ldots, n-1$. From calculus, it is known that the solution to Equation 2-1 can be expressed as a linear combination of exponentials called *eigenfunctions*, which are denoted $\phi_k(t)$. In particular, the homogeneous solution of Equation 2-1 satisfies the equation

$$x(t) = \sum_{k=1}^{n} a_k \phi_k(t) \qquad (2\text{-}3)$$

where the eigenfunction $\phi_k(t) = \exp(s_k t)$. The parameter s_k is, in general, a complex number having the form $s_k = \sigma_k + j\omega_k$, where σ_k and ω_k are real numbers. Equation 2-3 can be substituted into Equation 2-1 and solved for the a_k's using the method of undetermined coefficients. For example, the solution to the second-order ODE, $d^2 x(t)/dt^2 + \omega_0^2 x(t) = 0$, with initial conditions $x(t)|_{t=0} = 1$, $dx(t)/dt|_{t=0} = 0$, is defined by $\phi_1(t) = \exp(j\omega_0 t)$, $\phi_2(t) = \exp(-j\omega_0 t)$, with $a_0 = a_1 = 1/2$. That is, $x(t) = (\phi_1(t) + \phi_2(t))/2 = (\exp(j\omega_0 t) + \exp(-j\omega_0 t))/2 = \cos(\omega_0 t)$ for $-\infty \leq t \leq \infty$.

2-3 LAPLACE TRANSFORM

Given for one instant an intelligence that could comprehend all the forces by which nature is animated and ... sufficiently vast to submit these data to analysis—it would embrace in the same formula the movements of the greatest bodies of the universe and those of the lightest atom. For it, nothing would be uncertain and the future as the past, would be present to its eyes.

—Pierre Simon de Laplace

Equation 2-3 suggests the need for a tool to efficiently compute the coefficients $\{a_k\}$. Such a tool exists. It is called the *Laplace transform*, named after Pierre Simon de Laplace, an early influence in a field now called Probability Theory.[1]

---

[1]Laplace is also well-known for his partial differential equation $\partial^2 V/\partial x^2 + \partial^2 V/\partial y^2 + \partial^2 V/\partial z^2 = 0$, which has been applied to the study of the stability of the solar system and electric fields. In the electromagnetic case, $V(x, y, z)$ represents the electrical potential at the point (x, y, z).

The Laplace transform is a mathematical tool that maps a continuous-time signal $x(t)$, residing in the time-domain, into another function $X(s)$, residing in the so-called *s-domain*, where $s = \sigma + j\omega$ (complex). By itself, this may seem to be an unusual choice of variables until one recalls that the solution to an ODE, such as that found in Equation 2-1, can be expressed as

$$x(t) = \sum_{i=1}^{n} a_i e^{s_i t} \qquad s_i = \sigma_i + j\omega_i \tag{2-4}$$

after Equation 2-3, where s_i is called an *eigenvalue* of the ODE. The Laplace transform performs an efficient mapping of $x(t)$ into an eigenvalue solution of the ODE. In particular, the transform will directly produce the coefficients a_i found in Equation 2-4.

The Laplace transform of a signal $x(t)$ is formally defined to be $X(s)$, denoted $x(t) \overset{\mathcal{L}}{\leftrightarrow} X(s)$, where

$$X(s) = \int_{-\infty}^{\infty} x(t)e^{-st}\, dt \tag{2-5}$$

Equation 2-5 is referred to as the *two-sided Laplace transform* and is also sometimes referred to as the *spectrum of the Laplace operator*. Recall that causal signals have a finite time of origin, which shall be assumed to be $t = 0$ (e.g, $u(t)$). In such cases, the two-sided transform can be replaced by the *one-sided* Laplace transform, which is given by

$$X(s) = \int_{0}^{\infty} x(t)e^{-st}\, dt \tag{2-6}$$

The Laplace transform exists if Equation 2-5 (or 2-6), produces a bounded result. That is

$$\left| \int_{-\infty}^{\infty} x(t)e^{-st} dt \right| \le M < \infty \tag{2-7}$$

The integral is guaranteed to be bounded, for some s, if $x(t)$ can be dominated by an exponential of the form $\exp(-st)$. The exponential $\exp(-st)$ is said to dominate $x(t)$ if their product (i.e., $x(t)\exp(-st)$) converges to zero as $t \to \infty$. Exponential domination is graphically interpreted in Figure 2-1, where a continuous-time causal real signal $x(t)$, the real part of the exponential $\exp(-st)$, and their product are plotted against time. Exponential domination is shown by their product converging to zero as $t \to \infty$. Since $s = \sigma + j\omega$ (complex), it follows that $x(t)\exp(-st) = x(t)\exp(-\sigma t)\exp(-j\omega t)$ and note that $|\exp(-j\omega t)| = 1$. If $x(t)$ is a real signal, then $|x(t)\exp(-st)| = |x(t)\exp(-\sigma t)||\exp(-j\omega t)| = |x(t)\exp(-\sigma t)|$, which must converge to zero as $t \to \infty$ if the Laplace transform of $x(t)$ is to exist. Those values of s for which this is true define what is called the *region* (domain) *of convergence*, or ROC. Formally, the ROC of $X(s)$ is a subregion of the s-plane which is parallel to the $\pm j\omega$ axis, and is defined to be a real range of values of s for which the Laplace transform of $x(t)$ is guaranteed

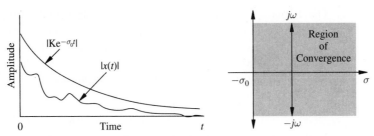

FIGURE 2-1   Graphical interpretation of exponential dominance.

to exist (i.e., ROC $= \{\sigma \,|\, R_{\text{lower}} < \text{Re}(s) = \sigma < R_{\text{upper}}, \forall j\omega\ \})$. The ROC shown in Figure 2-1 would be given by ROC $= \{\sigma \,|\, -\sigma_0 < \text{Re}(s) < \infty, \forall j\omega\}$.

Example 2-1   Exponential signal

For $x(t) = \exp(\sigma_0 t)u(t)$, where $u(t)$ is a unit step, the one-sided Laplace transform of $x(t)$ is given by

$$X(s) = \int_0^\infty e^{\sigma_0 t} e^{-st}\, dt = \int_0^\infty e^{(\sigma_0 - s)t}\, dt = \frac{1}{(s - \sigma_0)}$$

provided the integral is bounded. The exponential $\exp(\sigma_0 t)$ is dominated by $\exp(-st)$ if their product exponentially converges to zero as $t \to \infty$. In this case, $|\exp(-st)\exp(\sigma_0 t)| = |\exp((-\sigma + \sigma_0)t)||\exp(-j\omega t)| = |\exp((-\sigma + \sigma_0)t)|$, which will converge to zero as $t \to \infty$ if $\sigma > \sigma_0$ (i.e., $\text{Re}\,(s) = \sigma > \sigma_0$).

Some studies of Laplace transforms strongly emphasize the ROC; others barely address the question. Since most basic, or *elementary*, signals appear in standard tables of Laplace transforms along with their precomputed ROC, deriving an ROC will not be a major focus of our study. Nevertheless, knowledge of the ROC can be used to differentiate between two signals, say $x_1(t)$ and $x_2(t)$, which have identical Laplace transforms (i.e., $H_1(s) = H_2(s) = H(s)$), but differing ROCs. For purposes of simplification, assume that $x_1(t)$ is a causal signal beginning at $t = 0$ and continuing out to $t \to \infty$.[2] In the previous example, the causal $x(t) \doteq x_1(t) = \exp(\sigma_0 t)u(t)$ was shown to have a Laplace transform $X(s) = 1/(s - \sigma_0)$ if $\text{Re}(s) = \sigma > \sigma_0$. A variant of $x_1(t)$ is the signal $x_2(t) = -\exp(\sigma_0 t)u(-t)$, which is noncausal since its beginning is found at or before $t \to -\infty$.[3] Formally, the Laplace transform of $x_2(t)$ is given by

$$X_2(s) = \int_{-\infty}^\infty -e^{\sigma_0 t} e^{-st} u(-t)\, dt = \int_{-\infty}^0 -e^{(\sigma_0 - s)t}\, dt = \frac{1}{(s - \sigma_0)} \qquad (2\text{-}8)$$

[2] Such signals are also called right-sided signals since they extend to the right of the origin along the positive time axis.
[3] Such signals are also called left-sided signals since they extend to the left of the origin along the negative time axis.

provided the integral is bounded. Referring to the previous example, it follows that

$$|\exp{(-st)}\exp{(\sigma_0 t)}| = |\exp{((-\sigma+\sigma_0)t)}||\exp{(-j\omega t)}| = |\exp{((-\sigma+\sigma_0)t)}| \tag{2-9}$$

will converge to zero as $t \to -\infty$ (i.e., in negative time), if $\mathrm{Re}\,(s) = \sigma < \sigma_0$. Therefore, $X_1(s) = X_2(s) = 1/(s-\sigma_0)$, but the ROC for $x_1(t)$ is $\mathrm{Re}\,(s) > \sigma_0$ and for $x_2(t)$ is $\mathrm{Re}\,(s) < \sigma_0$.

The Laplace transforms of elementary signals and their ROCs are cataloged in Table 2-1. It can be noted that all the signals represented in the table are causal. If they were noncausal, the ROC must be suitably altered.

Example 2-2   Noncausal signal

The exponential $x(t) = \exp{(\sigma_0|t|)}$, for $\sigma_0 < 0$, is a noncausal signal and is graphically interpreted in Figure 2-2. It has been previously established that

$$x_1(t) = e^{\sigma_0 t}u(t) \overset{\mathcal{L}}{\leftrightarrow} X_1(s) = 1/(s-\sigma_0)$$

for $\mathrm{Re}\,(s) > \sigma_0$ and

$$x_2(t) = e^{-\sigma_0 t}u(-t) \overset{\mathcal{L}}{\leftrightarrow} X_2(s) = -1/(s+\sigma_0)$$

for $\mathrm{Re}\,(s) < -\sigma_0$. It then follows that for $-\sigma_0 < \mathrm{Re}\,(s) < \sigma_0$

$$x(t) = x_1(t) + x_2(t) = e^{\sigma_0|t|} \overset{\mathcal{L}}{\leftrightarrow} X(s) = 1/(s-\sigma_0) - 1/(s+\sigma_0)$$

Therefore, over the ROC strip given by $-\sigma_0 < \mathrm{Re}\,(s) < \sigma_0$ (see Figure 2-2), the Laplace transform converges and is equal to

$$X(s) = 1/(s-\sigma_0) - 1/(s+\sigma_0) = 2\sigma_0/(s^2 - \sigma_0^2)$$

FIGURE 2-2   Noncausal signal $x(t) = \exp{(\sigma_0|t|)}$ and its ROC.

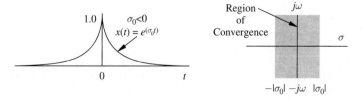

TABLE 2-1   LAPLACE TRANSFORMS OF ELEMENTARY SIGNALS

Time Function	Laplace Transform	Region of Convergence	Remark
$\delta(t)$	1	All s	Impulse
$u(t)$	$\dfrac{1}{s}$	Re $(s) > 0$	Unit step function
$tu(t)$	$\dfrac{1}{s^2}$	Re $(s) > 0$	Ramp function
$t^n u(t)$	$\dfrac{n!}{s^{n+1}}$	Re $(s) > 0$	
$e^{at} u(t)$	$\dfrac{1}{s-a}$	Re $(s) > a$	Exponential
$te^{at} u(t)$	$\dfrac{1}{(s-a)^2}$	Re $(s) > a$	
$t^n e^{at} u(t)$	$\dfrac{n!}{(s-a)^{n+1}}$	Re $(s) > a$	
$\sin(\omega_0 t)u(t)$	$\dfrac{\omega_0}{s^2 + \omega_0^2}$	Re $(s) > 0$	Causal sine wave
$\cos(\omega_0 t)u(t)$	$\dfrac{s}{s^2 + \omega_0^2}$	Re $(s) > 0$	Causal cosine wave
$e^{at} \sin(\omega_0 t)u(t)$	$\dfrac{\omega_0}{(s-a)^2 + \omega_0^2}$	Re $(s) > a$	Damped sine wave
$e^{at} \cos(\omega_0 t)u(t)$	$\dfrac{(s-a)}{(s-a)^2 + \omega_0^2}$	Re $(s) > a$	Damped cosine wave

The *inverse Laplace transform* of $X(s)$ is given by:

$$x(t) = \frac{1}{j2\pi} \int_{\sigma_0 - j\infty}^{\sigma_0 + j\infty} X(s)e^{st}\, ds \qquad (2\text{-}10)$$

where σ_0 is any real value, such that the contour $\sigma_0 - j\omega$, for $\omega \in [-\infty, \infty]$, is in the ROC of $X(s)$. However, the inverse Laplace transform is rarely computed in this manner. Rather, tables of standard transforms (such as Table 2-1) are normally used to look up $x(t)$ for a given $X(s)$, rather than to derive it using the integral definition found in Equation 2-10. The underlying assumption is that a complicated signal can be expressed as a combination of elementary signals with known and tabulated transforms. Building up complicated signals from elementary signals normally makes use of one or more of the properties of Laplace transforms listed in Table 2-2.

Example 2-3   Linearity and time delay

A unit duration unit amplitude pulse is defined by $x(t) = \text{rect}(t - 1/2) = u(t) - u(t-1)$, for $t \in [0, 1]$, and 0 elsewhere, is seen to be synthesized using two unit step functions. From the linearity and the time-shifting property found in Table 2-2, it follows that

$$u(t) \overset{\mathcal{L}}{\leftrightarrow} U(s) = \frac{1}{s}$$

$$w(t) = u(t-1) \overset{\mathcal{L}}{\leftrightarrow} W(s) = \exp(-s)U(s) = \frac{\exp(-s)}{s}$$

and,

$$x(t) = \text{rect}(t - 1/2) = u(t) - w(t) \overset{\mathcal{L}}{\leftrightarrow} X(s) = (1/s) - (\exp(-s)/s)$$
$$= (1 - \exp(-s))/s.$$

Example 2-4   Modulation and time scaling

Euler's identity states that a pure (noncausal) cosine wave can be expressed as $\cos(t) = (\exp(jt) + \exp(-jt))/2$ for all t. A causal cosine wave, therefore, satisfies the equation $c(t) = \cos(t)u(t) = \exp(jt)u(t)/2 + \exp(-jt)u(t)/2$ for $t \in [0, \infty]$. The Laplace transform of $c(t)$ can be derived using the modulation property as follows

$$x_1(t) = \frac{\exp(jt)u(t)}{2} \overset{\mathcal{L}}{\leftrightarrow} X_1(s) = \frac{U(s-j)}{2} = \frac{1}{2(s-j)}$$

$$x_2(t) = \frac{\exp(-jt)u(t)}{2} \overset{\mathcal{L}}{\leftrightarrow} X_2(s) = \frac{U(s+j)}{2} = \frac{1}{2(s+j)}$$

TABLE 2-2   LAPLACE TRANSFORM PROPERTIES

Time Function	Laplace Transform	Remark
$x(t)$	$X(s)$	
$y(t)$	$Y(s)$	
$ax(t) + by(t)$	$aX(s) + bY(s)$	Linearity
$x(t - t_0)$	$\exp(-st_0)X(s)$	Time shift for $t_0 \geq 0$
$x(at)$	$\frac{1}{a}X\left(\frac{s}{a}\right)$	Time scaling for $a > 0$
$\exp(\alpha t)x(t)$	$X(s - \alpha)$	Modulation
$t^n x(t)$	$(-1)^n \dfrac{d^n X(s)}{ds^n}$	For an integer, $n > 0$
$x(0) = \lim_{s \to \infty} sX(s)$		Initial value theorem[1]
$x(\infty) = \lim_{s \to 0} sX(s)$		Final value theorem[2]
$\dfrac{dx(t)}{dt}$	$sX(s) - x(0)$	Differentiation
$\dfrac{d^n x(t)}{dt^n} = x^{(n)}$	$s^n X(s) - s^{n-1}x(0) - s^{n-2}x^{(1)}(0)$ $- \cdots - x^{(n-1)}(0)$	Differentiation
$\int_{-\infty}^{t} x(\tau)\, d\tau$	$X(s)/s$	Integration

[1] If $x(t)$ is causal and has no impulses or higher order singularities at $t = 0$.
[2] If $X(s) = N(s)/D(s)$, all roots of $D(s)$ must have a negative real part, with the exception being a distinct root at $s = 0$.

Since $c(t) = x_1(t) + x_2(t)$, linearity provides

$$C(s) = X_1(s) + X_2(s) = \frac{1}{2(s-j)} + \frac{1}{2(s+j)}$$

$$= 0.5 \left(\frac{1}{(s-j)} + \frac{1}{(s+j)} \right) = 0.5 \left(\frac{2s}{(s-j)(s+j)} \right) = \frac{s}{s^2+1}$$

Continuing one step further, if $x(t) = \cos(\omega_0 t)u(t)$, then from the time-scaling property it follows that if $c(t) \overset{\mathcal{L}}{\leftrightarrow} C(s)$, and $x(t) = c(\omega_0 t) \overset{\mathcal{L}}{\leftrightarrow} X(s) = (1/\omega_0)C(s/\omega_0) = s/(s^2 + \omega_0^2)$ (for $\omega_0 > 0$).

Example 2-5   Multiplication by t

A causal ramp $r(t)$ is given by $r(t) = tu(t)$, where $u(t)$ is the unit step function. This immediately leads to:

$$R(s) = \mathcal{L}[tu(t)] = -\frac{dU(s)}{ds} = -\frac{d}{ds}\frac{1}{s} = \frac{1}{s^2}$$

Example 2-6   Integration theorem

The simple RC circuit shown in Figure 2-3 can be used as a signal generator. The control voltage $v(t)$ is obtained by integrating a rectangular pulse

$$v(t) = \int_0^t \text{rect}(t - 1/2)\, dt$$

Referring to Example 2-3, the Laplace transform of $p(t) = \text{rect}(t - 1/2)$ was given by $P(s) = (1 - \exp(-s))/s$. Therefore, from the integration property

$$v(t) \overset{\mathcal{L}}{\leftrightarrow} V(s) = \frac{P(s)}{s} = \frac{1 - \exp(-s)}{s^2}$$

The differential equation defining the loop current $i(t)$ is given by

$$v(t) = Ri(t) + \frac{1}{C}\int_0^t i(\tau)\, d\tau$$

FIGURE 2-3   Series RC circuit.

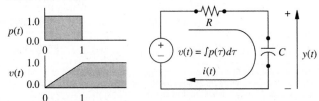

The loop equation, therefore, has a Laplace transform representation given by

$$V(s) = RI(s) + \frac{I(s)}{sC}$$

or

$$I(s) = \frac{sV(s)/R}{s + 1/RC}$$

Since $y(t) = (1/C) \int_0^t i(\tau)d\tau$, it follows that

$$Y(s) = \frac{I(s)}{sC} = \frac{V(s)/RC}{s + 1/RC}$$

where $V(s)$ has been previously defined.

It was previously stated that manually inverting a Laplace transform, using the integral definition (i.e., Equation 2-10), is awkward and rarely used. Instead, standard tables of Laplace transforms are employed wherever possible. This technique, however, requires that the elements of an arbitrary Laplace transform $X(s)$ be placed in direct correspondence with the elements from a standard table of Laplace transforms. Formal procedures exist to reduce a complicated transform $X(s)$ into a collection of identifiable elementary transforms. Among the foremost of these procedures is the *Heaviside Expansion Method*.

2-4 PARTIAL FRACTION (HEAVISIDE) EXPANSION

It shall be assumed that the Laplace transform of a signal $x(t)$ can be represented as a ratio of polynomials given by

$$X(s) = \frac{N(s)}{D(s)} = K \frac{\prod_{i=1}^{M}(s - z_i)}{\prod_{i=1}^{N}(s - p_i)} \tag{2-11}$$

where $N(s)$ is the Mth order numerator polynomial and $D(s)$ is the Nth order denominator polynomial. If $M < N$, then $X(s)$ is said to be *strictly proper* and $X(s)$ is said to be *proper*, if $M \leq N$. If $M \geq N$ then $X(s)$ is not strictly proper and $X(s)$ would normally be expressed in a *quotient form*, as shown in Equation 2-12. Using polynomial (long) division, $X(s)$ can be written in quotient form as

$$X(s) = Q(s) + \frac{R(s)}{D(s)} \tag{2-12}$$

where the degree of $Q(s)$, called the quotient polynomial, is $M - N$ and the degree of $R(s)$, called the remainder polynomial, is less than N. Since the order of $R(s)$ is now less than $D(s)$, $R(s)/D(s)$ is strictly proper. The parameters z_i (called the ith *zero* of $X(s)$) and p_i (called the ith *pole* of $X(s)$) are roots of $N(s) = 0$ and $D(s) = 0$ respectively. Referring to Table 2-1, it can be noted that the denominators of elementary signals are either first or second order (namely $D(s) = (s-a)$ or $(s^2+\omega_0^2)$ or $((s-a)^2+\omega_0^2)$. In fact, the second-order polynomial $D(s) = ((s-a)^2 + \omega^2)$ can always be expressed as the product of two first-order polynomials with complex coefficients, namely $D(s) = (s-a-j\omega)(s-a+j\omega)$. In general, it shall be assumed that $D(s)$ can be placed into the product of factors form

$$D(s) = \prod_{i=1}^{L}(s - p_i)^{n(i)} \tag{2-13}$$

where $N = \sum_{i=1}^{L} n(i) = \deg(D(s))$. If $n(i) > 1$, then p_i is said to be a *repeated root* of *multiplicity* $n(i)$; otherwise the root is said to be a *distinct root*. Based on Equation 2-13, the product-of-factors representation of $X(s)$ given in Equation 2-11 can be changed to a sum-of-factors model called a *partial fraction expansion*. The partial fraction expansion, or Heaviside expansion, of a strictly proper $X(s)$ is given by

$$X(s) = \sum_{i=1}^{L} \sum_{j=1}^{n(i)} \frac{a_{i,j}}{(s - p_i)^j} \tag{2-14}$$

The problem, of course, is one of computing the *Heaviside coefficients* $a_{i,j}$. The procedure presented is called the *Heaviside method*. To understand this technique, first consider creating the polynomial $X(s)(s - p_k)^{n(k)}$ as follows

$$X(s)(s - p_k)^{n(k)} = (s - p_k)^{n(k)} \sum_{j=1}^{n(k)} \frac{a_{k,j}}{(s - p_k)^j} + (s - p_k)^{n(k)} \sum_{\substack{i=1 \\ i \neq k}}^{L} \sum_{j=1}^{n(i)} \frac{a_{i,j}}{(s - p_i)^j}$$

$$= a_{k,n(k)} + (s - p_k)^{n(k)} \sum_{j=1}^{n(k)-1} \frac{a_{k,j}}{(s - p_k)^j}$$

$$+ (s - p_k)^{n(k)} \sum_{\substack{i=1 \\ i \neq k}}^{L} \sum_{j=1}^{n(i)} \frac{a_{i,j}}{(s - p_i)^j} = a_{k,n(k)} + \sum_{j=1}^{n(k)-1} a_{k,j}(s - p_k)^{n(k)-j}$$

$$+ (s - p_k)^{n(k)} \sum_{\substack{i=1 \\ i \neq k}}^{L} \sum_{j=1}^{n(i)} \frac{a_{i,j}}{(s - p_i)^j} \tag{2-15}$$

It can be seen that $a_{k,n(k)}$ is now an isolated term, devoid of any multiplier of the form $(s - p_k)$. If Equation 2-15 is evaluated at $s = p_k$, all the terms except $a_{k,n(k)}$ would contain a zero multiplier. Therefore

$$a_{k,n(k)} = \lim_{s \to p_k} X(s)(s - p_k)^{n(k)} \tag{2-16}$$

The coefficient $a_{k,n(k)-1}$ found in Equation 2-15 can be seen to be linear in $(s - p_k)$. Therefore, it can be isolated by using a simple derivative. In particular, differentiating Equation 2-15 with respect to s, one obtains

$$\frac{d}{ds} X(s)(s - p_k)^{n(k)}$$

$$= \frac{d}{ds} \left[a_{k,n(k)} + \sum_{j=1}^{n(k)-1} a_{k,j}(s - p_k)^{n(k)-j} + (s - p_k)^{n(k)} \sum_{\substack{i=1 \\ i \neq k}}^{L} \sum_{j=1}^{n(i)} \frac{a_{i,j}}{(s - p_i)^j} \right]$$

$$= a_{k,n(k)-1} + \sum_{j=1}^{n(k)-2} a_{k,j}(n(k) - j)(s - p_k)^{n(k)-j-1}$$

$$+ n(k)(s - p_k)^{n(k)-1} \sum_{\substack{i=1 \\ i \neq k}}^{L} \sum_{j=1}^{n(i)} \frac{a_{i,j}}{(s - p_i)^j} + (s - p_k)^{n(k)} \sum_{\substack{i=1 \\ i \neq k}}^{L} \sum_{j=1}^{n(i)} \frac{a_{i,j}(-j)}{(s - p_i)^{j+1}}$$

$$(2\text{-}17)$$

Therefore, the isolated Heaviside term $a_{k,n(k)-1}$ can be computed using the rule

$$a_{k,n(k)-1} = \lim_{s \to p_k} \frac{d}{ds}(s - p_k)^{n(k)} X(s) \qquad (2\text{-}18)$$

The production rule for $a_{k,n(k)-2}$ is a direct extension of this procedure. Notice that $a_{k,n(k)-2}$ appears in Equation 2-15 with a quadratic dependency on $(s - p_k)$. Extending the previously developed rule, the next Heaviside coefficient can be computed to be

$$a_{k,n(k)-2} = \lim_{s \to p_k} \frac{d^2}{ds^2} \frac{1}{2}(s - p_k)^{n(k)} X(s) \qquad (2\text{-}19)$$

Continuing this line of reasoning, the coefficient $a_{k,j}$ is defined by

$$a_{k,j} = \lim_{s \to p_k} \frac{d^{n(k)-j}}{ds^{n(k)-j}} \frac{1}{(n(k) - j)!}(s - p_k)^{n(k)} X(s) \qquad (2\text{-}20)$$

Therefore it can be seen that when repeated roots are encountered, successive derivatives must be used to compute the Heaviside coefficients. Once the Heaviside coefficients are known, the inverse Laplace of $X(s)$ can be determined by evaluating

$$x(t) = \mathcal{L}^{-1}[X(s)] = \mathcal{L}^{-1} \left[\sum_{i=1}^{L} \sum_{j=1}^{n(i)} \frac{a_{i,j}}{(s - p_i)^j} \right] = \sum_{i=1}^{L} \sum_{j=1}^{n(i)} \mathcal{L}^{-1} \left[\frac{a_{i,j}}{(s - p_i)^j} \right]$$

$$(2\text{-}21)$$

where the inverse Laplace transforms of the elemental terms are found in Tables 2-1 and 2-2.

Example 2-7   Distinct (nonrepeated) roots

If $X(s) = s^2/(s^2 + 1.5s + 0.5) = s^2/(s + 1)(s + 0.5)$, then all roots are nonrepeated (i.e., distinct). Note that $X(s)$ is not strictly proper, in that

$M = N = 2$. Upon reducing $X(s)$ using long division, the partial fraction expansion of $X(s)$ becomes

$$X(s) = 1 + \frac{-1.5s - 0.5}{s^2 + 1.5s + 0.5} = 1 + \frac{-1.5s - 0.5}{(s + 1)(s + 0.5)}$$

$$= a_0 + a_1 \frac{1}{(s + 1)} + a_2 \frac{1}{(s + 0.5)}$$

Upon applying Equation 2-20, the Heaviside coefficients are seen to be

$$a_0 = 1$$

$$a_1 = \lim_{s \to -1} (s + 1)X(s) = \lim_{s \to -1} \frac{-1.5s - 0.5}{s + 0.5} = \frac{1.5 - 0.5}{-1 + 0.5} = -2$$

$$a_2 = \lim_{s \to -0.5} (s + 0.5)X(s) = \lim_{s \to -0.5} \frac{-1.5s - 0.5}{s + 1} = \frac{0.75 - 0.5}{-0.5 + 1} = 0.5$$

Once the Heaviside coefficients are known, so is $X(s)$ and therefore $x(t)$. Using Table 2-1 (assuming $x(t)$ is causal), observe that

$$a_0 \overset{\mathcal{L}}{\leftrightarrow} a_0 \delta(t)$$

$$a_1 \frac{1}{(s + 1)} \overset{\mathcal{L}}{\leftrightarrow} a_1 \exp(-t)u(t)$$

$$a_2 \frac{1}{(s + 0.5)} \overset{\mathcal{L}}{\leftrightarrow} a_2 \exp(-t/2)u(t)$$

Thus the inverse Laplace transform can be written as $x(t) = a_0\delta(t) + a_1 e^{-t} u(t) + a_2 e^{-t/2} u(t)$. As a check, a set of simple tests can be performed on the candidate solution. Because $X(s)$ is not strictly proper, the delta distribution is present. If the degree of the quotient polynomial is $n = M - N > 1$, then $x(t)$ would contain derivatives of $\delta(t)$ up to order n (i.e., $d^n\delta(t)/dt^n$). Also the final value theorem states that $\lim_{t \to \infty} x(t) = \lim_{s \to 0} sX(s)$, or

$$\lim_{t \to \infty} x(t) = \lim_{s \to 0} \frac{s^3}{(s + 1)(s + 0.5)} = 0$$

which is consistent with the exponential decay of $x(t)$.

Computer Study   S-file "pf.s" performs a partial fraction expansion of $X(s) = N(s)/D(s)$ using the format `pfexp(n,d)` where the arguments are derived from a polynomial coefficient listing $n = N(s)$ and $d = D(s)$, in descending powers of s. The S-file "pf.s" assumes the polynomial $D(s)$ is monic (i.e., leading coefficient is unity). M-file "residue" can be used to extract the Heaviside coefficients from a rational polynomial. The M-file uses a format $[r,p,k]=\text{residue}(A,B)$ to convert $X(s) = A(s)/B(s)$ into a list of Heaviside coefficients r, pole locations p, and quotient terms k. Evaluate $X(s) = s^2/[(s+1)(s+0.5)] = s^2/[s^2 + 1.5s + 0.5]$, when the roots of $D(s)$ are $s = -1$ and -0.5.

SIGLAB	MATLAB
| ```
>include ''pf.s''
>n=[1,0,0] # N(s)=s^2
>d=[1,1.5,0.5] #D(s)=
s^2+1.5s+0.5
>pfexp(n,d) # partial fraction
Quotient polynomial
1=> a_0
pole @-1+0j with multiplicity
1 and coefficients
-2+0j =>a_1
pole @ -0.5+0j with multiplicity
1 and coefficients
0.5+0j=>a_2
``` | ```
>> a = [1, 0, 0]; b
= [1, 1.5, 0.5];
>> [r, p, k] = residue(a, b)
r = -2.0000   0.5000
p = -1.0000 -0.5000
k =  1
>> % r = Heaviside
coefficients
>> % p = poles
>> % k = Quotient polynomial
``` |

This corresponds to the computed values.

Example 2-8   Repeated roots

If $X(s) = s/(s+1)^2$, then repeated roots arise (i.e., $n(1) = 2$). Here $X(s)$ is a strictly proper polynomial and its partial fraction expansion is given by

$$X(s) = \frac{a_{11}}{(s+1)} + \frac{a_{12}}{(s+1)^2}$$

The partial fraction coefficients are given by

$$a_{12} = \lim_{s\to-1}(s+1)^2 X(s) = \lim_{s\to-1} s = -1$$
$$a_{11} = \lim_{s\to-1} \frac{d}{ds}(s+1)^2 X(s) = \lim_{s\to-1} \frac{d}{ds} s = 1$$

which result in $X(s) = 1/(s+1) - 1/(s+1)^2$. From Table 2-1 (assuming $x(t)$ is causal), it can be seen that $1/(s+1) \overset{\mathcal{L}}{\leftrightarrow} e^{-t}u(t)$ and $1/(s+1)^2 \overset{\mathcal{L}}{\leftrightarrow} te^{-t}u(t)$ or $x(t) = e^{-t}u(t)(1-t)$. Since $X(s)$ is strictly proper, there can be no impulse occurring at time $t = 0$. Therefore, the initial value theorem is applicable and states that $x(0) = \lim_{s\to\infty} sX(s)$, which in this case is

$$x(0) = \lim_{s\to\infty} \frac{s^2}{(s+1)^2} = 1$$

the computed value of $x(0)$. The final value theorem states $\lim_{t\to\infty} x(t) = \lim_{s\to 0} s X(s)$, or

$$\lim_{t\to\infty} x(t) = \lim_{s\to 0} \frac{s^2}{(s+1)^2} = 0$$

which is consistent with the exponential decay of $x(t)$.

Computer Study Compute the partial fraction expansion of $X(S) = N(s)/D(s) = s/(s+1)^2$, having roots of $D(s)$ given by $s = -1$ and -1 where `poly(r)` computes the polynomial $D(s)$ from known roots.

SIGLAB	MATLAB
`>include ''pf.s''`	`>> n = [1 0] % N(s) = s`
`>n=[0,1,0] #N(s)=s^1`	`n =   1   0`
`>d=poly([-1,-1]) #D(s)`	`>> d = poly ([-1 -1])`
`=(s+1)(s+1)`	`% N(s) = (s + 1) (s + 1)`
`>pfexp(n,d) # partial fraction`	`= s^2 + 2s + 1`
`Quotient polynomial`	`d =   1   2   1`
`0=> a_0`	`>> [r, p, k] = residue(n, d)`
`pole @-1+0j with multiplicity`	`r =   1 -1`
`2 and coefficients:`	`p = -1 -1`
`-1-0j 1+0j=>a_12 a_11`	`k =   []`
	`>> % a_11 = 1, a_12 = -1`

This agrees with the computed values.

2-5 SUMMARY

In Chapter 2, the Laplace transform was introduced and shown to be capable of reducing a continuous-time signal to an algebraic expression. The Laplace transforms of common signals have, in general, been cataloged and listed in standard tables. Complicated signals are assumed to be represented as a collection of these primitive signals having known or computable Laplace transforms.

A return to the time-domain from the transform-domain required that a set of algebraic operations be performed. The principal method, called partial-fraction (or Heaviside) expansion, once was a tedious task with a high potential for introducing calculation errors. Now, computers can be used to overcome these problems.

If engineering was simply the study of signals, then the utility of these transform methods would be marginal. However, the real world also contains systems that manipulate and modify signals. In Chapter 3 it will be shown that the signals studied in Chapter 2 also can be represented in the frequency-domain. As such, frequency-selective systems can be designed to enhance certain signal attributes and suppress others. In Chapter 4, it will be shown that the mechanism by which this takes place is convolution. In Chapter 5 the convolution process again will be interpreted in the transform domain. At that time it will become apparent that by transforming both systems and signals, significant analytic efficiencies can be gained. It is because of this that transform methods have gained their reputation as being a fundamental, essential signal and system design and analysis tool.

2-6 COMPUTER FILES

The following computer files were used in Chapter 2.

File Name	Type	Location
	Subdirectory CHAP2	
	Subdirectory SFILES or MATLAB	
pf.s	C	Examples 2-7, 2-8
	Subdirectory SIGNALS	
	Subdirectory SYSTEMS	

Where "s" denotes S-file, "m" a M-file, "x" either S-file or M-file, "T" a tutorial program, "D" a data file and "C" a computer-based-instruction (CBI) program. CBI programs can be altered and parameterized by the user.

2-7 PROBLEMS

2-1 Laplace transform Derive a formula for $X_n(s)$ if $x_n(t) = (d^n x(t)/dt^n)u(t)$ and $x(t) \overset{\mathcal{L}}{\leftrightarrow} X(s)$.

2-2 Laplace transform Compute the Laplace transform of the following signals
  a $x(t) = \exp(-t)u(t)$
  b $x(t) = \exp(-|t|)$
  c $x(t) = \exp(-t)\cos(\omega_o t + \phi)u(t)$

2-3 Laplace transform Determine the Laplace transform of the following causal signals where $x(t) = \sin(t)u(t)$, and $x(t) = \sin(t + \pi/8)u(t)$, using the formal definition of the Laplace integral. Also graph the signals and determine the singular values of $X(s)$.

2-4 Laplace transform In Table 2-1, the Laplace transform of $x(t) = \cos(t)u(t)$ was given to be

$$X(s) = \frac{s}{s^2 + 1}$$

Determine the Laplace transform of the following causal signals (see Problem 2-3), using the properties of the Laplace transform (see Table 2-2)
  a $y(t) = x(t - 1)$
  b $y(t) = dx(t)/dt$
  c $y(t) = x(2t) + x(t/2)$

2-5 Laplace transform Given

$$x_1(t) \overset{\mathcal{L}}{\leftrightarrow} X_1(s) = s^2/(s^2 + 1.5s + 0.5)$$

$$x_2(t) \overset{\mathcal{L}}{\leftrightarrow} X_2(s) = s^2/(s - 1)^2$$

what is $Y(s)$ if

a $y(t) = ax_1(t) + bx_2(t)$
b $y(t) = ax_1(2t) + bx_2(3t)$
c $y(t) = ax_1(2t - 2) + bx_2(3t - 3)$
d $y(t) = t^2(x_1(t) + x_2(t))$

2-6 Laplace transform   Repeat Problem 2-5 for $W(s)$, where

a $w(t) = \int_{-\infty}^{t} y(\tau)d\tau$
b $w(t) = \exp(j\omega_0 t)y(t)$

2-7 Sinusoid formula   Show the Laplace transform of

$$x(t) = \exp(-at)[A\cos(bt) + \frac{(B - Aa)}{b}\sin(bt)]u(t)$$

is

$$X(s) = \frac{(As + B)}{s^2 + 2as + c}$$

where $c = b^2 + a^2$ and $a > 0$.

2-8 Signal generator   A signal generator $y(t)$ is assumed to be given by the homo-geneous equation

$$d^3y(t)/dt^3 + 3d^2y(t)/dt^2 + 3dy(t)/dt + y(t) = 0$$

for $t \geq 0$ and 0 otherwise. The initial conditions are given by $y(0) = y_0, dy(0)/dt = y_1, d^2y(0)/dt^2 = y_2$.

a What is $Y(s)$?
b Verify that your solution meets the given initial conditions using the initial value theorem.

2-9 Region of convergence   Compute the region of convergence for the transforms computed in Problem 2-2.

2-10 Region of convergence   Compute the region of convergence for the transforms computed in Problem 2-4.

2-11 Region of convergence   Given a two-sided signal $x(t) = \exp(-at)u(t) - \exp(at)u(-t)$ where $a > 0$, show that $X(s) = 2s/(s^2 - a^2)$ and that the ROC for $X(s)$ is $-a < \text{Re}(s) < a$. Verify the initial and final values of $x(t)$ using the Initial and Final Value Theorems.

2-12 Even-odd properties   If $x(t)$ is even $(x(t) = x(-t))$, prove that $X(s) = X(-s)$. If $x(t)$ is odd $(x(t) = -x(-t))$, prove that $X(s) = -X(-s)$ and $X(s) = 0$ at $s = 0$.

2-13 RLC circuit   The RLC circuit shown in Figure 2-4 is used to generate signals $y(t)$.

a What is the ODE that defines $y(t)$ (a voltage)?
b Given knowledge of $X(s)$, what is $Y(s)$ if all initial conditions are zero?

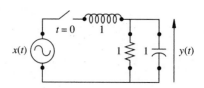

FIGURE 2-4   Circuit for Problem 2-13.

2-14 Inverse Laplace transform Compute the inverse Laplace transform of the following transforms (assume $x(t)$ is causal).

a $X(s) = \dfrac{s+1}{s^2 + 3s - 4}$

b $X(s) = \dfrac{s^2 + 1}{s^3 - 4s^2 + 5s - 2}$

c $X(s) = \dfrac{s+1}{s^3 + s^2 + s + 1}$

d $X(s) = \dfrac{s+1}{s^3 - s^2 + s - 1}$

2-15 Inverse Laplace transform Compute the inverse Laplace transform of the following transforms

a $X(s) = \dfrac{(s-1)}{(s^2+1)^2}$

b $X(s) = \dfrac{(s-1)^4}{(s+1)^5}$

c $X(s) = \dfrac{(s+1)}{(s^2-1)^2}$

d $X(s) = \dfrac{(s+1)^4}{(s-1)^5}$

2-16 Inverse Laplace transform Invert the following Laplace transforms and compute their initial and final values (if possible).

a $X(s) = \dfrac{1}{s^3}$

b $X(s) = \dfrac{1}{s^6 + 21s^5 + 175s^4 + 735s^3 + 1624s^2 + 1764s + 720}$

c $X(s) = \dfrac{s^3 + 3s^2 + 3s + 1}{s^4 + 4s^3 + 6s^2 + 4s + 1}$

d $X(s) = \dfrac{s^2 + 2s + 1}{s^5 + 5s^4 + 11s^3 + 13s^2 + 8s + 2}$

Verify your results by taking the Laplace transform of $x(t)$ to reconstruct $X(s)$.

2-17 Signal analysis It is desired to evaluate a casual signal's behavior at far distant times (i.e., $t \to \infty$), which is sometimes referred to as *steady-state*. A causal signal having the form

$$x(t) = \left(\sum_{i=1}^{L} a_i e^{(\lambda_i)t} \right) u(t) = \left(\sum_{i=1}^{L} a_i e^{(\sigma_i + j\omega_i)t} \right) u(t)$$

will remain bounded for all time if the real part of the eigenvalues λ_i are nonrepeated and nonpositive, namely $\sigma_i \leq 0$. At steady-state, the envelope on which $x(t)$ resides will be dominated by $a_i e^{(\sigma_i + j\omega_i)t} u(t)$ if $\sigma_i > \sigma_j$, for all $j \neq i$. Here $\lambda_i = \sigma_i + j\omega_i$ is called the *dominant eigenvalue*. Devise a scheme based on determining the pole locations of $X(s)$, which define the dominant eigenvalue solution. Apply your discovery to the signals studied in Problem 2-14. What can be said about the case where the eigenvalues may be repeated?

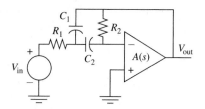

FIGURE 2-5   Electronic circuit for Computer Project 2-2.

2-18 Signal analysis   Explain if the final value theorem can be used to detect if there is a DC component to a causal signal whose Laplace transform is strictly proper and is given by $X(s) = N(s)/D(s)$ where

$$D(s) = \prod_{i=0}^{K} (s - \lambda_i)$$

and $\text{Re}(\lambda_i) < 0$ and $\lambda_i \neq \lambda_j$? How about a sinusoidal component?

2-8 COMPUTER PROJECTS

2-1 Heaviside expansion   Write an S-file or M-file, that is a modification of `pf.s` or `residue.m`, that will accept a transform polynomial $X(s) = N(s)/D(s)$, where $\deg(N(s)) \leq 8$ and $\deg(D(s)) \leq 8$, and produce the corresponding time-domain signal $x(t)$.

2-2 KRC circuit analysis   The electronic KRC circuit shown in Figure 2-5 is to be used as a signal generator. Its principal application will be to change the phase of the the reference input denoted $v_{in}(t)$. The Laplace transform of the output signal, denoted $v_{out}(t)$, is given by

$$V_{out}(s) = \left[G_0 \frac{s}{s^2 + G_1 s + G_2} \right] V_{in}(s)$$

where $G_0 = \frac{1}{R_1 C_1}$, $G_1 = \frac{1}{R_2}(\frac{1}{C_1} + \frac{1}{C_2})$, and $G_2 = \frac{1}{R_1 R_2 C_1 C_2}$ and $v_{in}(t) = \sin(\omega_0 t)u(t)$. For $C_1 = C_2 = 10^{-6}$, $R_1 = R_2 = 10^3$, and $\omega_0 = 2\pi \times 10^3$, use the program developed in Project 2-1 to compute $v_{out}(t)$. Graph $v_{in}(t)$ and $v_{out}(t)$ locally at $t = 0, 10^{-3}, 10^{-1}$ and $t \to \infty$. Measure their phase difference and verify analytically.

3

FREQUENCY DOMAIN REPRESENTATION OF CONTINUOUS-TIME SIGNALS

. . . where you can measure what you are speaking about, and express it in numbers, you know something about it; but when you cannot measure it, when you cannot express it in numbers, your knowledge is of meagre and unsatisfactory kind; it may be the beginning of knowledge, but you have scarcely, in your thoughts, advanced to the stage of Science whatever the matter may be.

—Lord Kelvin

3-1 INTRODUCTION

In the first chapter the time-domain properties of signals were presented. Time, however, is but one possible space in which signals can be represented. In the second chapter signals were developed in the context of the Laplace transform. In Chapter 3 signal information will be mapped into the *frequency-domain*, where signal energy is displayed and analyzed on the basis of a signal's frequency content. In many cases, the frequency-domain signal representation of a signal is preferred over a time-domain paradigm. For example, striking a guitar string will produce a sinusoidal acoustic wave. Ideally, the tone produced by the guitar would have the form $x(t) = A \cos(\omega_0 t + \phi)$ and persist forever. In this model the frequency ω_0 would be referred to as the natural frequency of the string being struck. A more compact means of quantifying the information found in the ideal acoustic wave $x(t)$ involves representing the entire signal as the three-tuple $[A, \omega, \phi]$, which represents signal amplitude, frequency, and phase. From only these three parameters, can the ideal signal be reconstructed precisely. In this case, and in many others, a frequency- or phase-domain signal representation can be shown to be both more efficient and more physically meaningful than a time-domain model.

Expressing a continuous-time waveform as a set of parameters in the frequency-domain has been an important element of science and engineering for centuries. Representing a periodic signal as a trigonometric series can be traced back to Leonhard Euler [1707–1783]. Euler considered a cosine-only series that inspired the work of d'Alembert, Lagrange, and Bernoulli. Lagrange's interest was orbital mechanics, while Bernoulli's work focused on a vibrating string. Jean Baptiste Fourier, whose memory is honored with the *Fourier series*, made a major contribution to this field with his inspired studies of heat and diffusion equations using a trigonometric series. Carl F. Gauss extended this early work and developed what is now called a *discrete Fourier transform* (DFT) in 1805. From this rich legacy has evolved an important set of tools for engineers and scientists. Collectively known as *Fourier analysis,* this branch of applied mathematics has become the foundation for the study of electrical, optical, acoustical, electromagnetic, and other signals.

Because determining the frequency-domain representation of a signal is important to modern science and engineering, much effort has been devoted to its computation. A number of clever mechanical systems, using wheels, pulleys, and cams, have been developed over the past two centuries to compute the harmonic motion of planets, analyze acoustic waves, and study vibratory signals. It is also known that lenses and prisms can separate light into its frequency components, called a color spectrum, or simply *spectrum*. Analog electronic devices, called wave analyzers, use frequency selective filters to measure the frequency content of an arbitrary signal over a range of frequencies. Digital technology has brought with it a wealth of instruments to accurately measure the frequency content of a signal and alter it using digital filters. This rich history has resulted in a technology used in every facet of science and engineering.

This collective experience has established that there are a number of advantages to representing a signal's attributes in the frequency-domain. The more salient advantages are

1 Physical interpretability—Natural and synthetic signals are often produced by a single or finite collection of harmonic oscillations.

2 Filtering—The modification of a signal is often accomplished using a frequency-selective filter.

3 Information compression—Long time-domain records can be replaced often by a few parameters that capture the frequency-domain properties of a signal. These parameters can be archived or transmitted to a remote location and used to reconstruct a facsimile of the original signal.

For these reasons and others, frequency-domain analysis and synthesis techniques have become important engineering tools.

3-2 SPECTRAL ANALYSIS

The study and measurement of the frequency content of a signal is called *spectral analysis*. Analysis and measurement procedures abound in the fields of communication, control, defense, medicine, geophysics, entertainment, and so forth. Methods and instruments have been developed to measure high-frequency signal behavior beyond optical frequencies, as well as low-frequency processes having the periods of the planets. While some of this equipment is highly specialized and complicated, other frequency measurement instruments are in everyday use. Owners of high-tech sound reproduction equipment, for example, can easily alter the frequency response provided by sophisticated studio sound-recording equipment and skilled audio engineers by sliding the bass adjustment on their equalizers to its maximum.

Spectral analysis techniques are broadly classified on the basis of assumed signal attributes. The frequency content of a signal can be *stationary* or *nonstationary* in time. A held musical note, which does not change its natural frequency over time, is an example of a stationary spectral process. Speech, however, is a highly dynamic nonstationary spectral process. Rarely is a real-world signal truly stationary. However, if it is essentially stationary over an observation interval, it is called *quasi-stationary*. Stationary and quasi-stationary signals can often be modeled by deterministic signals. Methods have been developed to mathematically map the information found within a continuous-time deterministic signal into the frequency-domain. The principal method of performing this mapping is called the *Fourier transform*.

Intuitively, signals that exhibit strong, sustained repetitive behavior should be the easiest to analyze in the frequency-domain. An example is a periodic signal $x(t)$, with period T, which satisfies the defining equation

$$x(t) = x(t \pm kT) \tag{3-1}$$

A signal that is not periodic is said to be *aperiodic*.

A periodic signal having a period T has a fundamental frequency of oscillation given by $f_0 = 1/T$ Hz. Multiplies of f_0 are called *harmonics* and often accompany a periodic signal. The *first harmonic* is located at the fundamental frequency f_0. The *zeroth harmonic* is a DC signal and is located at a frequency of 0 Hz (i.e., *DC component*). The *k*th harmonic is found at frequency of $k f_0$ Hz. The study of periodic signals in terms of their harmonic content is called *harmonic analysis*. The primary analysis tool used in the study of periodic continuous-time signals is called the *Fourier series* (FS).

3-3 FOURIER SERIES

A periodic signal with period $T > 0$ can be expressed in terms of a set of periodic approximation functions $\{\phi_k\}$, which are also called a set of *basis* functions. In such a case

$$x(t) = \sum_{k=-\infty}^{\infty} c_k \phi_k(t) \tag{3-2}$$

Equation 3-2 defines the *exponential Fourier series* of $x(t)$ if $\phi_k(t) = \exp(jk\omega_0 t)$, where $\omega_0 = 2\pi/T$. The problem, of course, is to efficiently determine the *Fourier coefficients* $\{c_k\}$. The production rule for c_k can be derived as in the following manner. Consider first multiplying both sides of Equation 3-2 by $\exp(-jn\omega_0 t)$, and then integrating over any period ($t \in [0, T]$, for instance). That is, for the exponential Fourier series

$$\int_0^T x(t) \exp(-jn\omega_0 t)\, dt = \sum_{k=-\infty}^{\infty} c_k \int_0^T \exp(j(k-n)\omega_0 t)\, dt \qquad (3\text{-}3)$$

where the integral on the right-hand side of Equation 3-3 simplifies to

$$\int_0^T \exp(j(k-n)\omega_0 t)\, dt = \begin{cases} T & \text{if } n = k \\ 0 & \text{otherwise} \end{cases} \qquad (3\text{-}4)$$

It immediately follows that the exponential Fourier series, denoted $x(t) \overset{\mathcal{F}}{\leftrightarrow} \{c_k\}$, satisfies

$$x(t) = \sum_{k=-\infty}^{\infty} c_k \exp(jk\omega_0 t) \qquad \text{Synthesis equation} \qquad (3\text{-}5)$$

$$c_k = \frac{1}{T} \int_C^{C+T} x(t) \exp(-jk\omega_0 t)\, dt \qquad \text{Analysis equation} \qquad (3\text{-}6)$$

where C is an arbitrary constant, and c_k is the kth complex harmonic coefficient. The production of the Fourier coefficients is graphically interpreted in Figure 3-1.

FIGURE 3-1   Production of a Fourier series of a periodic signal $x(t)$.

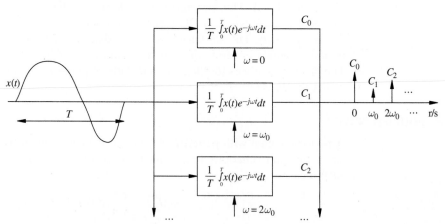

Since the Fourier coefficients are, in general, complex numbers they can be interpreted in cartesian or polar form. In cartesian form, $c_k = \text{Re}\,(c_k) + j\text{Im}(c_k)$. Furthermore

$$|c_k| = \sqrt{\text{Re}\,(c_k)^2 + j\text{Im}(c_k)^2} \tag{3-7}$$

$$\arg[c_k] = \tan^{-1}(\text{Im}(c_k)/\,\text{Re}(c_k))$$

The display of $|c_k|$ versus frequency (or harmonic) is called the *magnitude spectrum*. The display of $\arg[c_k]$ versus frequency (or harmonic) is called the *phase spectrum*. The coefficients in polar form are

$$c_k = |c_k|\exp(j\arg[c_k]) \tag{3-8}$$

Example 3-1   Fourier series

The continuous-time periodic signal $x(t) = 1.0 + \sin(\omega_0 t)$ has period $T = 2\pi/\omega_0$. Using Euler's equation, $x(t)$ can be expressed as

$$x(t) = 1.0 + \sin(\omega_0 t) = 1.0 + \frac{[\exp(j\omega_0 t) - \exp(-j\omega_0 t)]}{2j}$$

The Fourier coefficients, defined by Equation 3-6, are obviously $c_0 = 1.0$, $c_1 = -0.5j$ and $c_{-1} = 0.5j$. The spectrum of $x(t)$ is graphically interpreted in magnitude, phase, and polar forms in Figure 3-2.

Example 3-1 raises an interesting observation. Note that one of the non-zero harmonics of $x(t) = 1.0 + \sin(\omega_0 t)$ is c_{-1}. If c_1 is the Fourier coefficient corresponding to a frequency $\omega_0 = 2\pi/T = 2\pi f_0$, then logically c_{-1} corresponds

FIGURE 3-2   Spectrum of $x(t) = 1.0 + \sin(\omega_0 t)$ in magnitude, phase, and polar forms.

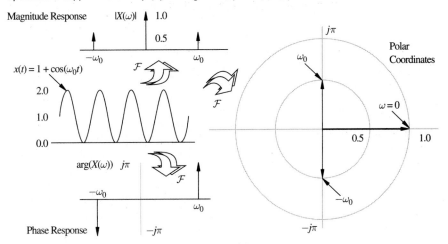

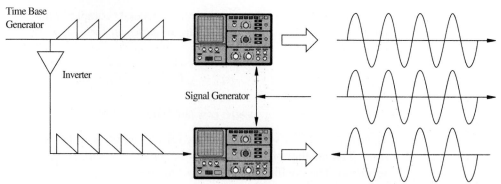

FIGURE 3-3   Negative frequency experiment.

to a frequency $\omega_0 = -2\pi/T = -2\pi f_0$, which is called a *negative frequency*. Negative frequencies may seem to be an abstract concept until it is recognized that the *instantaneous frequency* of the signal $y(t) = \sin(\phi(t))$ is formally defined to be $\omega(t) = d\phi(t)/dt$. From calculus it is known that the sign of a derivative can be changed if time is reversed (i.e., $\Delta_t^+ = t_i - t_{i-1}$ versus $\Delta_t^- = t_{i-1} - t_i$) where $t_i > t_{i-1}$. Therefore, upon reversing time, the instantaneous frequency of the sinusoid becomes $d\phi(t)/dt = -\omega(t)$. To illustrate this point, suppose that two students conduct an oscilloscope experiment in the laboratory. Both send $v(t) = \sin(\omega_0 t)$ to the vertical input of an oscilloscope. One student attaches a fast ramp $h(t) = t \bmod (T)$[1] to the horizontal input while the other mistakenly uses the complement of $h(t)$, namely $h'(t) = 1 - h(t)$, as shown in Figure 3-3. Could the laboratory instructor tell the difference between the two by looking at the scope displays? Probably not, since both would trace a sine wave on the screen. The only perceptible difference would be that one trace would begin at the left of the screen and the other at the right. Mathematically, however, one would be a display of a positive frequency cosine and the other a negative.

The exponential Fourier series for a real signal $x(t)$ can also be expressed as a *trigonometric Fourier series* using Euler's identity. In this form

$$x(t) = a_0 + \sum_{k=1}^{\infty} a_k \cos(k\omega_0 t) + b_k \sin(k\omega_0 t) \qquad (3\text{-}9)$$

where

$$\begin{cases} a_o = c_o \\ a_k = 2\operatorname{Re}[c_k] \text{ for } k > o \\ b_k = 2\operatorname{Im}[c_k] \text{ for } k > o \end{cases} \qquad (3\text{-}10)$$

The spectrum of the periodic signal $x(t)$, whether exponential or trigonometric, is seen to exist only at specific frequencies that correspond to the harmonics of

---

[1] Definition: $c = a$ modulo(b) if $a = kb + c$, where k is an integer constant and $0 \le c < b$.

a periodic signal. For a signal with period T, the harmonics are on $\Delta f = 1/T$ Hz centers. The kth harmonic, for example, physically resides at a frequency, $k f_0$, where $f_0 = 1/T$. A discrete frequency spectrum is also called a *line spectrum*.

Example 3-2   Sinusoidal representation

An A chord played on a guitar consists of the following notes and corresponding fundamental frequencies:

1 A = 110.00 Hz

2 E = 164.81 Hz

3 A = 220 Hz

4 C# = 277.18 Hz

5 E = 329.63 Hz

If all the strings are struck simultaneously with equal force, the resulting acoustic waveform would have the shape

$$x(t) = \cos(2\pi 110t) + \cos(2\pi 164.81t) + \cos(2\pi 220t)$$

$$+ \cos(2\pi 277.18t)) + \cos(2\pi 329.63t)$$

If they are struck at different times, another waveform results, namely

$$y(t) = \cos(2\pi 110t + \phi_1) + \cos(2\pi 164.81t + \phi_2) + \cos(2\pi 220t + \phi_3)$$

$$+ \cos(2\pi 277.18t + \phi_4) + \cos(2\pi 329.63t + \phi_5)$$

It is important to note that, in both cases, the time-domain signal can be completely quantified in terms of the amplitude, frequency, and phase three-tuple of the five generating sinusoids. It should be apparent that the phase spectra of the two signals differ but have identical magnitude spectra. In addition, psychoacoustic tests have shown that when the notes were all struck within a 10 ms interval, most humans could not discern the relative phasing of the individual tones.

Computer Study   S-file `ch3_2.s` and M-file `ch3_2.m` can be used to simulate the A chord, where all strings are struck simultaneously and randomly.

SIGLAB	MATLAB
`>include "ch3_2.s"`	`>>ch3_2`

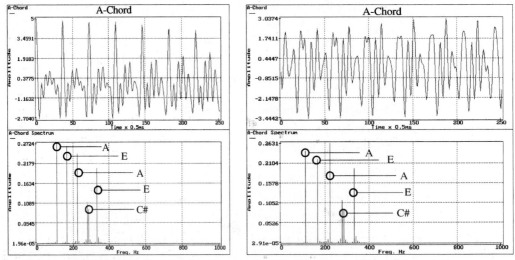

FIGURE 3-4   Time-domain composite of the tones used to construct the A chord on a guitar (left) simultaneously struck, and (right) randomly struck, along with their magnitude squared spectra. (Results may vary due to the use of a random signal.)

The results are presented in Figure 3-4[2].

The existence of a Fourier series of a periodic signal $x(t)$ with period T is established by *Dirichlet conditions* for periodic signals. The conditions are:

1 $x(t)$ is absolutely integrable over an interval T (i.e., $\int_T |x(t)| < \infty$);

2 $x(t)$ has at most a finite number of maxima and minima in a period T;

3 The discontinuities of $x(t)$ exist only at isolated points and, over a period T, are finite in number.

Example 3-3   Fourier series

A continuous-time periodic signal, of period $T = 1$, is given by $x(t) = 1.0 + 1.0\cos(2\pi t) + 1.0\sin(4\pi t + \pi/4)$. The signal $x(t)$ is everywhere bounded and continuously differentiable, therefore $x(t)$ satisfies all the Dirichlet conditions. The computation of $X(j\omega)$ at $\omega = i2\pi$ (see Exercise 3-1) would obviously produce values of $c_0 = 1$, $c_1 = c_{-1} = 1/2$, $c_2 = c_{-2}^* = (1 + j)/2\sqrt{2}$, and $c_i = c_{-i} = 0$ for $|i| > 2$.

Computer Study   S-file `ch3_3.s` can be used to evaluate $X(j\omega)$ at ω. The format is given by `coef(x,k)`, where x is the signal and k is the harmonic index and $k = 2$.

---

[2]A discrete Fourier transform (DFT) was used to compute the spectra displayed in Figure 3-4. In Chapter 8 the properties of the DFT will be developed. At that time the fact that the C# note seems to be "split" between two harmonics will be attributed to a leakage problem.

SIGLAB	MATLAB
```>include "ch3_3.s"```	```>>x = ones(128, 1) +```
```>x=ones(128)+mksin(1,1/128,```	```sin(2*pi/128*(0:127))' +```
```0,128)+```	```sin(2*pi*2/128*(0:127) +```
```mksin(1,2/128,pi/4,128)```	```pi/4)';```
```>coef(x,2)```	```>> c = coef(x, 2)```
```Fourier coefficient 0.353553+0```	```c = 0.3536 + 0.3536i```
```.353553j```	```>> abs(c)```
```Magnitude 0.5```	```ans = 0.5000```
```Phase Angle - Degrees 45```	```>> angle(c)/pi*180```
	```ans = 45```

The Fourier series belongs to a class of mappings called *orthogonal transforms.* Specifically, the Fourier series approximates an arbitrary periodic signal $x(t)$ by expressing it in terms of a collection (possibly infinite) of orthogonal basis functions given by $\phi_i(t) = \exp(j\omega_i t)$. Two signals $\phi_i(t)$ and $\phi_j(t)$ are said to be *orthogonal,* over the interval $[a, b]$, if their *inner product* is zero, or

$$\langle \phi_i(t), \phi_j(t) \rangle = \int_a^b \phi_i(\tau)\phi_j^*(\tau)\, d\tau = 0 \quad \text{for } i \neq j \qquad (3\text{-}11)$$

where * denotes complex conjugation. Since integration is an averaging operation, Equation 3-11 is a measure of the average similarity shared by $\phi_i(t)$ and $\phi_j(t)$ over $t \in [a, b]$. This is an important concept because any approximation scheme, Fourier or otherwise, should be as efficient as possible. If $\phi_i(t)$ and $\phi_j(t)$ have much in common, they would provide redundant information.

Modern engineering makes use of many different sets of orthogonal basis functions. The complex exponential or trigonometric functions used in the definition of a Fourier series are only examples. One of the principal advantages of this particular set is physical interpretability. Intuitively, we know what is meant by a 60 Hz signal. In the case of spatial signal, the frequency would be measured in meters/second or some equivalent unit of measure. If, driving down a road at 60 mph, we count the elapsed time between mile-markers, the spatial frequency would be 1 mile/minute.

Example 3-4   Orthogonal basis

The set of N-trigonometric basis functions $\{\phi_i(t)\}$, where $\phi_0(t) = 1$, $\phi_{2i-1}(t) = \cos(2\pi i \omega_0 t)$, and $\phi_{2i}(t) = \sin(2\pi i \omega_0 t)$ for $i = 1, 2, \ldots, N-1$, are claimed to form an orthogonal set. Using an N-sample discrete-time signal $\Phi_i[k]$ to represent a continuous signal $\phi_i(t)$, the value of $\langle \phi_i(t), \phi_j(t) \rangle$ can be approximated and saved in the ij location of an $N \times N$ matrix Φ (i.e., $\Phi_{ij} = \sum_{k=0}^{N-1} \phi_i[k]\phi_j[k] \approx \langle \phi_i(t), \phi_j(t) \rangle$ for large N). If the bases are indeed orthogonal, then the ijth element of $\Phi\Phi^T$ will be finite if $i = j$ and (essentially) zero otherwise.

Computer Study   S-file `ch3_4.s` and M-file `ch3_4.m` compute the discrete-time approximation to $\phi_i(t)$ and $\phi_j(t)$ and approximates $\langle \phi_i(t), \phi_j(t) \rangle$.

SIGLAB	MATLAB
`>include "ch3_4.s"`	`>> ch3_4`
`16.0 ~0.0 ~0.0 ~0.0 ~0.0 ~0.0 ~0.0`	`ans =`
`~0.0  8.0 ~0.0 ~0.0 ~0.0 ~0.0 ~0.0`	`16. -0. -0. -0. -0. -0.  0.`
`~0.0 ~0.0  8.0 ~0.0 ~0.0 ~0.0 ~0.0`	`-0.  8. -0. -0. -0.  0.  0.`
`~0.0 ~0.0 ~0.0  8.0 ~0.0 ~0.0 ~0.0`	`-0. -0.  8. -0.  0. -0. -0.`
`~0.0 ~0.0 ~0.0 ~0.0 ~0.0 ~0.0 ~0.0`	`-0. -0. -0.  8.  0. -0.  0.`
`~0.0 ~0.0 ~0.0 ~0.0 ~0.0  8.0 ~0.0`	`-0. -0.  0.  0.  8. -0.  0.`
`~0.0 ~0.0 ~0.0 ~0.0 ~0.0 ~0.0  8.0`	`-0.  0. -0. -0. -0.  8. -0.`
	`0.  0. -0.  0.  0. -0.  8.`

Observe that the off-diagonal elements in the matrix $\Phi\Phi^T$ are essentially zero, which indicates that $\langle \phi_i(t), \phi_j(t) \rangle = 0$ (for $i \neq j$) or, equivalently, $\phi_i(t)$ and $\phi_j(t)$ are considered orthogonal.

The trigonometric Fourier series uses a collection of orthogonal sine and cosine waves of period T/k to represent a periodic signal of period T. Periodic signals can also possess odd and even symmetry. Knowledge of a signal's symmetry can be used to simplify the production of a Fourier series. Recall that

1 Even symmetry, $x(t) = x(-t)$;

2 Odd symmetry, $x(t) = -x(-t)$;

3 Half-wave odd symmetry, $x(t) = -x(t + T/2)$.

Based on these symmetry conditions, the Fourier series representation of a symmetric signal simplifies in the manner shown in Table 3-1. The symmetry conditions provide a means of identifying which coefficients need to be computed and which can be ignored. For example, Table 3-1 states that if $x(t)$ is an even function, then the sine coefficients b_i need not be computed. These properties, when properly applied, can reduce the computational burden associated with the production of a Fourier series of a periodic signal with symmetry. However, like Laplace transforms, a Fourier series is rarely computed. Instead, a signal is usually recognized to be constructed from elementary functions whose Fourier series are known, such as those reported in Table 3-2. Here it is assumed that $x(t) = x(t + T)$.

TABLE 3-1   FOURIER SERIES SYMMETRY CONDITIONS

Type of Symmetry	DC-term, a_0	Cosine terms, a_n	Sine terms, b_n
Even	a_0	a_n	$b_n = 0$
Odd	$a_0 = 0$	$a_n = 0$	b_n
Half-wave odd	$a_0 = 0$	$a_{2n} = 0$	$b_{2n} = 0$
		a_{2n+1}	b_{2n+1}

TABLE 3-2   FOURIER SERIES OF PRIMITIVE SIGNALS OF PERIOD T

Periodic Signal	Fourier Series Coefficients	Remark
$x(t) = \text{rect}\left(\dfrac{t}{\tau}\right), t \in \left[-\dfrac{T}{2}, \dfrac{T}{2}\right]$	$c_k = \dfrac{\tau}{T}\text{sinc}\left(\dfrac{k\pi\tau}{T}\right)$ where $\text{sinc}(x) = \dfrac{\sin(x)}{x}$	pulse train
$x(t) = 1 - \left(\dfrac{2t}{T}\right), t \in \left[-\dfrac{T}{2}, \dfrac{T}{2}\right]$	$c_0 = \dfrac{1}{2},\ c_k = \dfrac{2}{(k\pi)^2}$	triangle train
$x(t) = \dfrac{t}{T},\ t \in [0, T]$	$c_0 = \dfrac{1}{2},\ c_k = j\dfrac{1}{2k\pi}$	ramp
$\cos\left(2\pi t\,\dfrac{K}{T}\right)$	$c_{\pm k} = \dfrac{1}{2}\delta(k \pm K)$	cosine
$\sin\left(2\pi t\,\dfrac{K}{T}\right)$	$c_k = \dfrac{j}{2}\delta(k - K)$ if $k > 0$ $c_k = \dfrac{-j}{2}\delta(k + K)$ if $k < 0$	sine

Example 3-5   Periodic pulse

A classic test signal is the even periodic pulse-train $x(t)$ shown in Figure 3-5. The complex exponential Fourier coefficients are given by

$$c_0 = \frac{1}{2}\int_{-1}^{1} x(t)\,dt = \frac{1}{2}\int_{-1/2}^{1/2} 1\,dt = \frac{1}{2}$$

$$c_n = \frac{1}{2}\int_{-1}^{1} \exp(-jn\pi t)\,dt = \frac{1}{2}\int_{-1/2}^{1/2}\exp(-jn\pi t)\,dt$$

$$= \frac{-1}{j2n\pi}\exp(-jn\pi t)\Big|_{t=-0.5}^{t=0.5} = \frac{\sin(n\pi/2)}{n\pi} = \frac{1}{2}\text{sinc}(n\pi/2)$$

The magnitude and phase spectra, therefore, become

$$c_0 = 1/2$$

$$|c_n| = \begin{cases} \dfrac{1}{n\pi} & \text{for odd } n \\ 0 & \text{for } n \text{ even, } n \neq 0 \end{cases}$$

$$\arg[c_n] = \begin{cases} \pi & \text{for odd } n = \pm\{3, 7, 11, \ldots\} \\ 0 & \text{otherwise} \end{cases}$$

where $\arg[c_n] = \tan^{-1}(\text{Im}(c_n)/\text{Re}(c_n))$. Therefore

$$x(t) = \frac{1}{2} + \sum_{\substack{n=1 \\ n \text{ odd}}}^{\infty} c_n \exp(jn\pi t) = \frac{1}{2} + \sum_{\substack{n \text{ odd}}}^{\infty} \frac{\sin(n\pi/2)}{n\pi} \exp(jn\pi t)$$

Recalling that $\cos(n\pi t) = \cos(-n\pi t)$ and $\sin(n\pi t) = -\sin(-n\pi t)$, it follows

$$x(t) = \frac{1}{2} + \sum_{\substack{n=1 \\ n \text{ odd}}}^{\infty} \frac{2}{n\pi} \cos(n\pi t + \arg[c_n])$$

$$= \frac{1}{2} + \frac{2}{\pi}\cos(\pi t) - \frac{2}{3\pi}\cos(3\pi t) + \frac{2}{5\pi}\cos(5\pi t) - \dots$$

Note that the Fourier coefficients are zero for all even harmonics except the DC-term (i.e., c_0). It is also worthwhile noting that the Fourier series representation of the signal $x(t)$ could also be derived by breaking $x(t)$ into the two signals shown in Figure 3-5. The constant signal $x_1(t) = c_0$ would possess a one-term Fourier series expansion, namely $x_1(t) \overset{\mathcal{F}}{\leftrightarrow} c_0$. The second term, namely $x_2(t)$, has half-wave odd symmetry and therefore, from Table 3-1, would contain only odd harmonics in its expansion.

Computer Study   Using a 256-second discrete-time approximation to a continuous-time signal, S-file ch3_5.s and M-file ch3_5.m can be used to compute and compare the Fourier series representation of a periodic pulse-train having a period of 32 seconds, with a duty cycle of 50%.

SIGLAB	MATLAB
>include "ch3_5.s"	>> ch3_5
># data reported	

The magnitude spectrum is partially displayed in Figure 3-6.

FIGURE 3-5   Pulse-train with a 50% duty cycle.

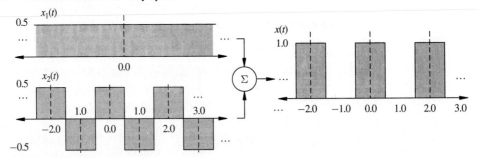

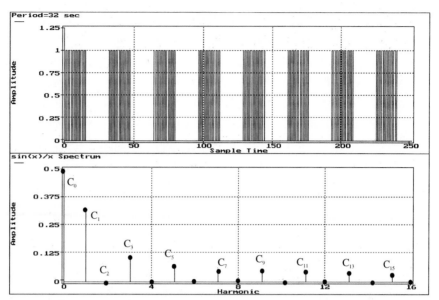

FIGURE 3-6   Periodic pulse-train and magnitude spectrum.

3-4   PROPERTIES OF FOURIER SERIES

The Fourier series possess a number of important properties that provide a mechanism by which a collection of precomputed mappings of the form $x_i(t) \overset{\mathcal{F}}{\leftrightarrow} c_{i,k}$, where $x_i(t)$ is a continuous-time periodic signal of period T and $c_{i,k}$ is the kth harmonic of $x_i(t)$, can be used to construct the Fourier series of more complicated signals. The more important properties of the Fourier series are summarized in Table 3-3. These properties will be developed in greater detail later in this chapter (see Section 3-6). However, linearity, Parseval's theorem, and a condition called Gibbs phenomenon will be studied in this section.

Linearity   The linearity property is useful in deriving the Fourier series of a periodic signal $x(t) = \sum x_i(t)$, where each signal $x_i(t)$ is periodic with period T. Of equal importance is the observation that the linear combination of periodic signals, of period T, can create no new frequencies. That is, if a Fourier coefficient c_m does not exist in the spectrum of $x_i(t) \overset{\mathcal{F}}{\leftrightarrow} c_{i,k}$, then it cannot appear in the spectrum of $x(t)$. This is not the case if nonlinear operations are present. Consider the case where $x_1(t) = x_2(t) = \cos(\omega t)$ and $x(t) = x_1(t)x_2(t) = \cos(\omega t)^2 = (1 + \cos(2\omega t))/2$. Whereas the spectrum of $x_1(t) = x_2(t)$ exhibits spectral components located at $f = 2\pi\omega$, the spectrum of the nonlinear $x(t)$ has components at DC, as well as $f = 4\pi\omega$.

Example 3-6   Nonlinear (rectified) signals

A 60 Hz 120 V line voltage signal is to be converted into a DC voltage source using a half-wave rectifier shown in Figure 3-7. The actual input is assumed to

TABLE 3-3   FOURIER SERIES PROPERTIES

Continuous-time function $x(t)$	Fourier Series c_k	Property
$\sum_i \alpha_i x_i(t)$	$\sum_i \alpha_i c_{i,k}$	Linearity
$x^*(t)$	c^*_{-k}	Conjugation
$x(-t)$	c_{-k}	Time-reversal
$x(t - t_0)$	$c_k \exp\left(\dfrac{-jk2\pi t_0}{T}\right)$	Time-delay
$x(\alpha t), \alpha > 0$	c_k if $x(\alpha t)$ has period $\dfrac{T}{\alpha}$	Time-scaling
$\int_{-\infty}^{t} x(\tau)\,d\tau < \infty$	$\dfrac{1}{jk\left(\dfrac{2\pi}{T}\right)} c_k$ if $c_0 = 0$	Integration
$\dfrac{dx(t)}{dt}$	$\dfrac{jk2\pi}{T} c_k$	Differentiation
$x(t) \exp\left(\dfrac{jK2\pi t}{T}\right)$	c_{k-K}	Modulation
$x(t)$ real	$c_k = c^*_{-k}$	Real

Parseval's theorem: $\dfrac{1}{T} \displaystyle\int_T |x(t)|^2\, dt = \sum_{-\infty}^{\infty} |c_k|^2$

FIGURE 3-7   Half-wave and full-wave rectified sine wave.

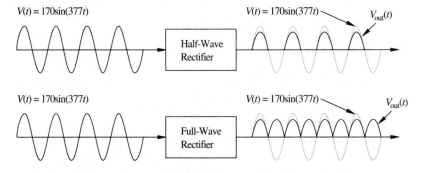

be $E = 120\sqrt{2} \sin(\omega_0 t)$, where $\omega_0 = 120\pi$ radians per second. The half-wave rectified waveform is given by

$$V(t) = \begin{cases} 0 & \text{if } -\pi/\omega_0 < t < 0 \\ V_0 \sin(\omega_0 t) & \text{if } 0 < t < \pi/\omega_0 \end{cases}$$

where $V(t) = V(t + T)$, $T = 2\pi/\omega_0$, and $V_0 = 120\sqrt{2}(R_{\text{load}}/(R_{\text{load}} + R_{\text{source}}))$. The Fourier coefficients are, for $T = 2\pi/\omega_0$,

$$c_n = \frac{1}{T} \int_0^T V_0 \sin(\omega_0 t) e^{-j2n\pi t/T}\, dt$$

$$= \frac{\omega_0 V_0}{2\pi} \int_0^{\pi/\omega_0} \frac{1}{2j} \left[e^{j\omega_0 t} - e^{-j\omega_0 t} \right] e^{-jn\omega_0 t} \, dt$$

$$= \begin{cases} \dfrac{V_0}{\pi(1 - n^2)} & \text{for } n \text{ even} \\[2ex] 0 & \text{if } n \text{ odd where } n \neq \pm 1 \\[2ex] \pm \dfrac{V_0}{4j} & \text{for } n = \pm 1 \text{ respectively} \end{cases}$$

Observe that only the even harmonics are present along with the fundamental. The zeroth harmonic is the DC value of the signal, an important parameter in the conversion of AC to DC.

By adding a second diode, a full-wave rectifier results. A full-wave rectified signal is given by

$$V(t) = |V_0 \sin(\omega_0 t)|$$

where $V(t) = V(t + \pi/\omega_0)$. The Fourier series of $V(t)$ can now be shown to be

$$V(t) = \frac{2V_0}{\pi} - \frac{2V_0}{\pi} \sum_{\substack{i=-\infty \\ i \text{ even} \\ i \neq 0}}^{\infty} \frac{e^{ji\omega_0 t}}{(1 - i^2)}$$

$$= \frac{2V_0}{\pi} - \frac{4V_0}{\pi} \sum_{\substack{i=2 \\ i \text{ even}}}^{\infty} \frac{\cos(i\omega_0 t)}{(1 - i^2)}$$

For the full-wave case observe that only the even harmonics are present and the fundamental frequency is now missing. Again, the zeroth harmonic is the DC value of the signal, which is important to the design of AC to DC converters. Observe also that the zeroth harmonic, or DC term, found at the output of the full-wave rectifier, is twice that of the half-wave rectifier.

Computer Study   Half- and full-wave rectified signals have been previously defined. Their Fourier series representation can be verified experimentally. S-file ch3_6.s and M-file ch3_6 create a finite length discrete-time approximation to a 60 Hz signal as well as its half- and full-wave rectified signals and their amplitude spectra.

SIGLAB	MATLAB
>include "ch3_6.s"	>> ch3_6

Observe that the full-wave rectified signal spectrum, shown in Figure 3-8, contains only even harmonics while the half-wave spectrum has an additional odd harmonic, located at the fundamental frequency (i.e., first harmonic). If the rectifier circuits are to be used as an AC to DC converter, then ideally only

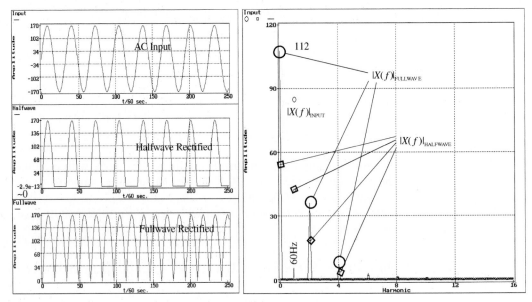

FIGURE 3-8   Half-wave and full-wave rectification example.

the DC term would be retained. Filters can be used to suppress all but the DC component.

Parseval's Theorem   Power, it may be recalled, satisfies the equation $P = E/T$ where E is the energy in a signal computed over a T second interval. Since a Fourier series is defined for a periodic signal of period T, it should also be able to provide a direct power measure. *Parseval's theorem* provides a mechanism by which the power of a periodic signal can be computed either in the time- or frequency-domain. Since the frequency content of a signal can be experimentally measured by instruments called spectral analyzers or wave analyzers, being able to compute power in the frequency-domain is an asset.

Parseval's Theorem (Periodic Signals)   Given $x(t) \overset{\mathcal{F}}{\leftrightarrow} c_k$, or $x(t) = \sum c_n \exp (jn\omega_0 t)$, for $x(t)$ a periodic signal of period T, the power in $x(t)$ over an interval T can be computed using

$$P = \frac{1}{T} \int_{-T/2}^{T/2} x(t)^2 \, dt = \sum_{k-\infty}^{\infty} |c_k|^2 \tag{3-12}$$

It can be seen that calculations can be expressed in terms of time- or frequency-domain signal representations. If $\{c_k\}$ denotes the Fourier series coefficients of $x(t)$, then $c_k c_k^* = |c_k|^2$ is the power in the kth harmonic of $x(t)$. The display of $|c_k|^2$, over all k, is called the *power spectrum*.

Example 3-7   Parseval's theorem

The power in $x(t)$, $d(x(t))/dt$, and $\int x(\tau)\,d\tau$, where $x(t)$ is an EKG record of duration $T = 1.89$ seconds, is to be computed using Parseval's theorem. The EKG signal is assumed to be periodic with period T.

Computer Study   S-file `ch3_7.s` accepts a signal $x(t)$, produces $d(x(t))/dt$, $\int x(\tau)\,d\tau$, and their respective power measures based on Equation 3-12, using the format `parsev(x)`. The results are graphically displayed and tabulated as a ratio of power measures. The discrete-time EKG signal is used to model a periodic continuous-time signal, where the EKG is also assumed to be periodic.

SIGLAB	MATLAB
`>include "ch3_7.s"`	`>> load ekg1.imp;`
`>x=rf("ekg1.imp")`	`>> [px, pfx, pxd, pxfd, pxi,`
`>parsev(x)`	`pxfi] = parsev(ekg1)`
`Ratio = Power(t)/Power(f)`	`Hit any key for plot`
`Power(t)=sum(x(t)^2)`	`px = 0.0965`
`Power(f)=sum(|X(f)|^2)`	`pfx = 0.0965`
`Ratio - x(t) = 1.0`	`pxd = 0.0213`
`Ratio - dx(t)/dt = 1.0`	`pxfd = 0.0213`
`Ratio - int(x(t)dt) = 1.0`	`pxi = 0.0152`
	`pxfi = 0.0152`

The ratio of time-domain and frequency-domain signal power calculations is seen to be equal as predicted by Equation 3-12. The power spectrum, shown in Figure 3-9, indicates how the power is distributed as a function of frequency. Note that relative to the EKG power spectrum, the derivative signal concentrates its power at high frequencies, while the integrated signal is dominated by powerful low-frequency components.

Gibbs Phenomenon   Since the Fourier series represents a continuous-time signal as a linear combination of continuous functions, therefore it should be expected that the Fourier series is well-suited for modeling smooth signals. Recall that the Fourier series is the infinite sum of weighted complex exponentials given by

$$x(t) = \sum_{k=-\infty}^{\infty} c_k \exp(jk\omega_0 t) \tag{3-13}$$

In practice, it may be more practical to consider using only a finite sum to approximate $x(t)$. Using $2N + 1$ coefficients, the reconstructed approximation of $x(t)$ shall be defined to be

$$x_N(t) = \sum_{k=-N}^{N} c_k \exp(jk\omega_0 t) \tag{3-14}$$

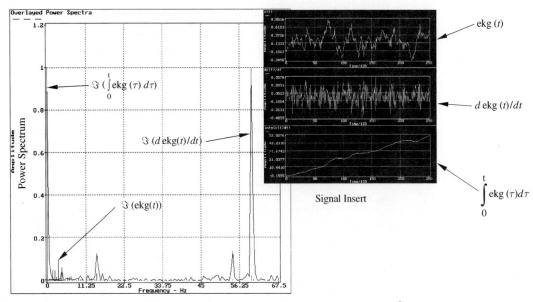

FIGURE 3-9   Parseval's theorem example showing, in the insert, $x(t)$, $dx(t)/dt$, $\int x(\tau)\,d\tau$, and the power spectra of $x(t)$, $dx(t)/dt$, and $\int x(\tau)\,d\tau$ for $x(t) = \mathrm{ekg}(t)$.

Any difference between $x_N(t)$ and $x(t)$ is attributed to the use of a finite number of terms (i.e., harmonics) to reconstruct $x(t)$. Defining the approximation error to be

$$e_N(t) = x(t) - x_N(t) = x(t) - \sum_{k=-N}^{N} c_k \exp(jk\omega_0 t) \qquad (3\text{-}15)$$

the approximation error power can be computed to be

$$P_{e_N} = \int_0^{T_0} |e_N(t)|^2\,dt = \int_0^{T_0} e_N(t) e_N^*(t)\,dt \qquad (3\text{-}16)$$

The number of harmonics required to produce an error that does not exceed a given mean error or error power bound, is signal-dependent. It has been noted, however, that at points of discontinuity of a signal $x(t)$, the error signal $e_N(t)$ has a tendency to oscillate. This is called the Gibbs phenomenon.

Example 3-8   Gibbs phenomenon

The Gibbs phenomenon can be observed using a pulse-train and its reconstruction, using a finite number of Fourier coefficients.

Computer Study   S-file `ch3_8.s` and M-file `ch3_8.m` create a pulse and compute its reconstruction using $N = 254, 236, 56$, and 14 harmonics. The format for the S-file is given by `gibbs(x,M)` where $N = 256 - 2M$. The data is plotted and the error is displayed and analyzed in terms of the mean (average) error and the error variance (error power).

SIGLAB	MATLAB
>include "ch3_8.s"	>> ch3_8
>xs=cshift(sq(256,32/256,256),	Hit any key for plot
128-16)	
>x=(1+xs)/2 # centered pulse	
>gibbs(x,100) # discard 2(100)	
harmonics	Number of harmonic | MSE
Mean Error 6.86434e-18	ans =
Variance 0.003424	1.0000 0.1094
>gibbs(x,121) # discard 2(121)	10.0000 0.0117
harmonics	100.0000 0.0005
Mean Error -2.6563e-18	121.0000 0.0001
Variance 0.0122571	

Gibbs phenomenon is graphically interpreted in Figure 3-10. The error can be seen to be oscillatory and directly related to the number of terms kept in the reconstruction formula. For large N, the error is as small as expected. For N small, the errors are significant, and largest in the vicinity of the points of discontinuity of $x(t)$.

The requirement that a signal be periodic is restrictive since many important real-world elementary signals, such as $x(t) = \cos(\omega_0 t)u(t)$, are not periodic. Periodicity, as defined by Equation 3-1, requires that a signal be noncausal since it must persist forever. Therefore, many real-world signals of finite duration (e.g., causal) technically would not qualify for Fourier series analysis. Instead, a more general mapping, called the Fourier transform, must be used.

3-5 THE FOURIER TRANSFORM

The previous section introduced the Fourier series, which can be used to analyze periodic signals in the frequency-domain. Unfortunately, signals such as the unit step $u(t)$, or causal sinusoid $x(t) = \cos(\omega_0 t)u(t)$, would not possess a Fourier series representation since they are *aperiodic*. Therefore the frequency-domain representation of such signals must be produced by other means. Generally the frequency spectra of such signals are defined in terms of the Fourier transform. The *continuous-time Fourier transform* (CTFT) of a complex signal $x(t)$ is given by

$$X(j\omega) = \int_{-\infty}^{\infty} x(t)e^{-j\omega t}\, dt \quad \text{Analysis equation} \qquad (3\text{-}17)$$

where $X(j\omega)$ is complex and is called the *frequency spectrum* of $x(t)$, or simply the *spectrum* of $x(t)$. The CTFT of $x(t)$ is denoted $x(t) \overset{\mathcal{F}}{\leftrightarrow} X(j\omega)$. The complex

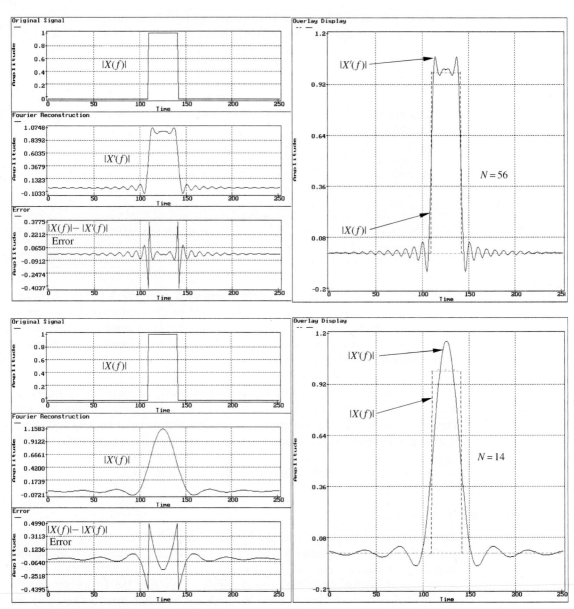

FIGURE 3-10   Examples of Gibbs phenomenon showing original, reconstructed, and error time-series for $N = 56$ (top) and 14 (bottom), and original, reconstructed, error time series, and an overlay of original and reconstructed time-series.

spectrum can be expressed in cartesian coordinates as $X(j\omega) = \text{Re}[X(j\omega)] + j\,\text{Im}[X(j\omega)]$, and in polar form as $X(j\omega) = |X(j\omega)| \exp(j\phi(j\omega))$. The term $|X(j\omega)| = \sqrt{\text{Re}[X(j\omega)]^2 + \text{Im}[X(j\omega)]^2}$ is called the magnitude spectrum and $\phi(j\omega) = \tan^{-1}(\text{Im}[X(j\omega)]/\text{Re}[X(j\omega)])$ is called the phase spectrum. The values of $X(j\omega)$ for $\omega \geq 0$ are called the *positive frequency spectrum* and the values of $X(j\omega)$ for $\omega < 0$ are called the *negative frequency spectrum*. These concepts are displayed in Figure 3-11, where the EKG signal studied in Chapter 1 is interpreted in the frequency-domain.

The *inverse Fourier transform* is given by

$$x(t) = \frac{1}{2\pi} \int_{-\infty}^{\infty} X(j\omega)e^{j\omega t}\, d\omega \quad \text{Synthesis equation} \qquad (3\text{-}18)$$

The inverse Fourier transform is seen to have a form similar to the forward transform (Equation 3-17), with the principal difference being the sign of the complex exponential's exponent. In both cases, the transform pair represent a mapping of a complex signal into another complex signal.

Another observation that can be drawn is that the definition of the Fourier transform closely resembles the two-sided Laplace transform, which is given by

$$X(s) = \int_{-\infty}^{\infty} x(t)e^{-st}\, dt \qquad (3\text{-}19)$$

Substituting $s = j\omega$ into Equation 3-19 produces

$$X(j\omega) = \int_{-\infty}^{\infty} x(t)e^{-j\omega t}\, dt \qquad (3\text{-}20)$$

which is seen to be the definition of the Fourier transform of $x(t)$, as given by Equation 3-17. Therefore, if Equation 3-19 converges along the contour $s = j\omega$, then

$$X(j\omega) = X(s)|_{s=j\omega} \qquad (3\text{-}21)$$

it will produce the Fourier transform of $x(t)$. Since much is documented about the Laplace transforms of elementary signals, it may seem, at first glance, that Equation 3-21 may be the only tool needed to compute the Fourier transform of a Laplace transformable signal. Unfortunately, the Laplace transform exists for many important signals for which the Fourier transform does not. For example, the simple unit step function $u(t)$ has a Laplace transform given by $U(s) = 1/s$, which has a region of convergence (ROC) given by $\text{Re}(s) > 0$. However, the contour $s = j\omega$ does not belong to this region of convergence. Therefore, the Laplace transform of $u(t)$ cannot be used to obtain the Fourier transform of $u(t)$. As a result, the conditions relating to the existence of a Fourier transform will require special attention.

The Fourier transform of $x(t)$ will exist if $x(t)$ has finite energy. That is

$$\int_{-\infty}^{\infty} |x(t)|^2\, dt < \infty \qquad (3\text{-}22)$$

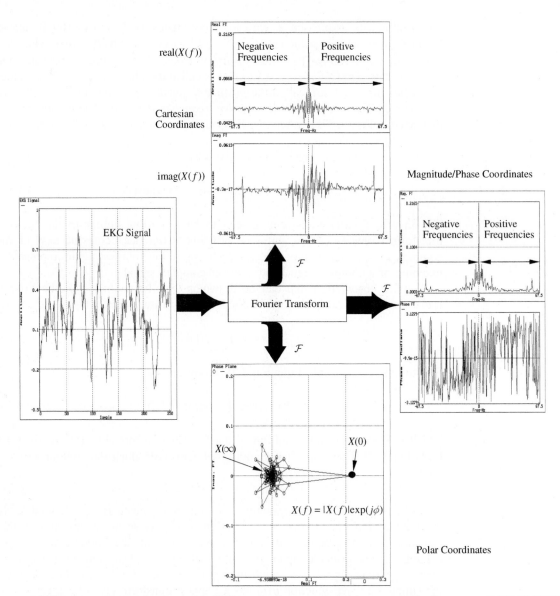

FIGURE 3-11   Fourier transform of an EKG in cartesian, magnitude/phase, and polar forms.

However, the fact that $x(t)$ does not satisfy Equation 3-22 does not preclude the existence of the Fourier transform of $x(t)$. Another existence test for Fourier transforms is the *Dirichlet conditions*. These are:

1 The signal $x(t)$ has a finite number of discontinuities;
2 The signal $x(t)$ has a finite number of minima and maxima;
3 The signal is absolutely integrable.

$$\int_{-\infty}^{\infty} |x(t)|dt < \infty \qquad (3\text{-}23)$$

Dirichlet conditions are also sufficient, but not necessary conditions for the existance of the Fourier transform of $x(t)$.

Example 3-9   Existence

The noncausal signal $x(t) = \sin(\omega_0 t)/\pi t = \omega_0 \text{sinc}(\omega_0 t)/\pi$ (see Table 3-2) is of finite energy, but not absolutely integrable. Therefore the Fourier transform exists by virtue of Equation 3-22, and can be shown to equal

$$X(j\omega) = \begin{cases} 1 & \text{for } |\omega| \le \omega_0 \\ 0 & \text{otherwise} \end{cases} = \text{rect}\left(\frac{\omega}{2\omega_0}\right)$$

which has a value of unity for all $\omega \in [-\omega_0, \omega_0]$. Here ω_0 is called the *break frequency*. The frequency response is seen to have a rectangular shape, which will later be called an ideal lowpass filter. The width of the baseband[3] spectrum is seen to be parameterized by ω_0 and is inversely related to the width of the main lobe of $x(t)$, as shown in Figure 3-12. Therefore, if $X(j\omega)$ exhibits a broad spectrum, the corresponding $x(t)$ will have a narrowly shaped main lobe. In fact, as $\omega_0 \to \infty$, $x(t) \to \delta(t)$ and the resulting spectrum $X(j\omega)$ is constant over $\omega \in [-\infty, \infty]$. This spectrum is studied in greater detail in Example 3-10. *Computer Study*   A signal $x(t) = \sin(x)/x$ can be approximated by a discrete time-series, which is generated by S-file ch3_9.s and M-file ch3_9.m, using

---

[3]A spectrum beginning at 0 Hz (DC) and extending contiguously over an increasing frequency range is called a *baseband spectrum*.

FIGURE 3-12   Relationship of a sinc function and its Fourier transform.

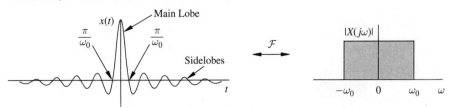

a format `shannon(p,n)` where p sets the break frequency and n is the number of samples used to represent $x(t)$. The program is used to study the relationship between $x(t)$ and $X(j\omega)$, for $p = \{3, 6, 12\}$.

SIGLAB	MATLAB
>include "ch3_9.s"	>> ch3_9
>shannon(3,256); shannon(6,256)	
>shannon(12,256) # p=3,6,12, n=256	

The results are presented in Figure 3-13, where $|X(j\omega)|$ is displayed as a two-sided approximate Fourier transform, with the negative frequencies appearing on the left-hand side, the positive frequencies appearing on the right-hand side, and 0 Hz (DC) in the middle. Note that the width of the baseband spectrum is inversely related to the value of p. The ringing that appears in the spectra is due to Gibbs phenomenon.

Many signal spectra have easily recognizable shapes called *signatures*. Signature analysis is an important element of automatic recognition systems (i.e., speech, image, etc.). Common signatures are the *line spectra* of a pure tone (sine or cosine wave), $\sin(\omega)/\omega$ spectra, which are produced by rectangular pulses, and a white spectrum, which is developed in Example 3-10.

Example 3-10   White spectrum

An impulse distribution $x(t) = \delta(t)$ is absolutely integrable and meets the other Dirichlet conditions. Therefore, its Fourier transform exists and is given

FIGURE 3-13   Shannon window and its magnitude spectrum for $p = 3$ (left), Shannon window and its spectrum for $p = 6$ (middle), Shannon window and its spectrum for $p = 12$ (right).

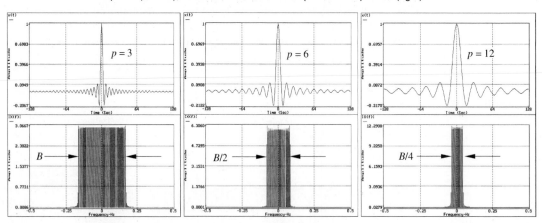

by

$$X(j\omega) = \int_{-\infty}^{\infty} \delta(t)e^{-j\omega t}\,dt = 1$$

This result is graphically interpreted in Figure 3-14 and is seen to be constant for all frequencies. Such a spectrum is said to be *white*. The term is derived from physics, where a light source containing equal amounts of energy over the visible spectrum is called white light.

The impulse distribution $x(t) = \delta(t)$ is a mathematical ideal, which does not exist in the real world. The physically realizable signal $y(t) = K < \infty$ for $t = 0$ and 0 otherwise, may seem, at first glance, to be a fair approximation of an impulse. The Fourier transform of the absolutely integrable $y(t)$ exists and is given by

$$Y(j\omega) = \int_{-\infty}^{\infty} y(t)e^{-j\omega t}\,dt = \int_{0}^{0} Ke^{-j\omega t}\,dt = 0$$

This, in fact, should have been anticipated since an integral will ignore the contribution of an isolated point if it has a bounded value. However, a value of $Y(j\omega) = 0$ obviously does not qualify as a white process. Therefore, by this criteria, $y(t)$ does not qualify as a viable approximation to $\delta(t)$.

The signal $x(t) = \text{rect}(t/\tau)/\tau$ approximates a Dirac impulse distribution as $\tau \to 0$. If the width (i.e., τ) is finite, the resulting spectrum is no longer white (flat), but is called *colored*, which refers to a nonconstant broadband spectrum.

Computer Study S-file `ch3_10.s` and M-file `ch3_10.m` can be used to approximate the spectrum $x(t) = \text{rect}(t/\tau)/\tau$, using a discrete-time signal as shown in Figure 3-15 for ascending values of τ beginning with $\tau = 0$.

SIGLAB	MATLAB
>include "ch3_10.s"	>> ch3_10

The results are displayed over a finite baseband of frequencies in Figure 3-15. The spectrum for the discrete-time approximation of a continuous-time impulse is seen to go from flat (white) to nonflat (colored) spectrum as τ increases.

FIGURE 3-14   Fourier transform of an impulse distribution.

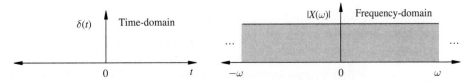

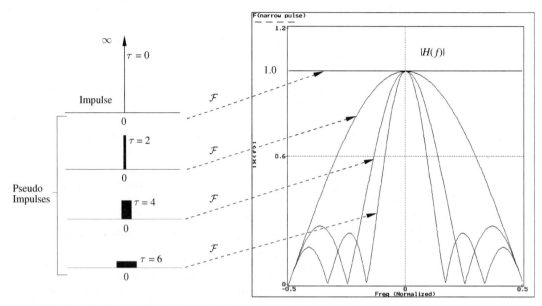

FIGURE 3-15   Fourier transform of an approximate impulse ($\tau = 0,2,4,6$).

Assuming that the Fourier transform of a signal $x(t)$ exists, the next logical question to ask is how accurate is it? The Fourier transform and its inverse are defined by an integral formula (Equations 3-17 and 3-18). Therefore, it is possible that these operations could introduce isolated errors. In Example 3-10, the Fourier transform for $y(t) = K < \infty$ for $t = 0$ and 0 otherwise, was shown to be $Y(j\omega) = 0$. It follows immediately from the definition that the inverse Fourier transform of $Y(j\omega)$ would produce a continuous-time signal that is zero for all time and, therefore, would not perfectly reconstruct $y(t)$. It can also be seen that an error has occurred and it is located at $t = 0$. If the error measured between the actual signal $x(t)$ and that produced by reconstructing a signal from its Fourier transform, namely $x'(t) = \mathcal{F}^{-1}[\mathcal{F}[x(t)]]$, is defined to be $e(t) = x(t) - x'(t)$, then it is known that

$$\int_{-\infty}^{\infty} |e(t)|^2 \, dt = 0 \qquad (3\text{-}24)$$

In other words, Equation 3-24 states that the error energy in the Fourier reconstruction of a signal is zero. Again, this does not mean that the error itself is zero. Errors of finite value can exist at isolated points and still satisfy Equation 3-24.

3-6 PROPERTIES OF FOURIER TRANSFORMS

As a signal becomes more complicated, the production of a Fourier transform from the integral definition (Equation 3-17) can become difficult. Fortunately, the Fourier transforms of the most important signals, called *basic* or *primitive signals*,

TABLE 3-4   FOURIER TRANSFORM PROPERTIES

Continuous-time function $x(t)$	Fourier Transform $X(j\omega)$	Property
$\sum_i \alpha_i x_i(t)$	$\sum_i \alpha_i X_i(j\omega)$	Linearity
$x^*(t)$	$X^*(-j\omega)$	Conjugation
$x(-t)$	$X(-j\omega)$	Time-reversal
$x(t - t_0)$	$\exp(-j\omega t_0)X(j\omega)$	Time-delay
$x(\alpha t)$	$\dfrac{1}{|\alpha|}X\left(\dfrac{j\omega}{\alpha}\right)$	Time-scaling
$\int_{-\infty}^{t} x(\tau)\, d\tau$	$\dfrac{X(j\omega)}{j\omega} + \pi X(0)\delta(\omega)$	Integration
$\dfrac{d^n x(t)}{dt^n}$	$(j\omega)^n X(j\omega)$	Differentiation
$x(t)\exp(j\omega_0 t)$	$X(j(\omega - \omega_0))$	Modulation
$X(t)$	$2\pi x(-j\omega)$	Duality

Parseval's theorem: $\displaystyle\int_{-\infty}^{\infty} |x(t)|^2 \, dt = \frac{1}{2\pi} \int_{-\infty}^{\infty} |X(j\omega)|^2 \, d\omega$

have been precomputed and cataloged. A complicated signal can often be defined by a combination of simple primitive signals with known Fourier transforms. The mappings, modifiers, and operators summarized in Table 3-4 are often used to represent a complicated signal as a collection of primitive signals.

Linearity   The Fourier transform is a linear operator in that if $x(t) \overset{\mathcal{F}}{\leftrightarrow} X(j\omega)$ and $y(t) \overset{\mathcal{F}}{\leftrightarrow} Y(j\omega)$, then

$$\alpha x(t) + \beta y(t) \overset{\mathcal{F}}{\leftrightarrow} \alpha X(j\omega) + \beta Y(j\omega) \tag{3-25}$$

It is important to note that the frequency content of the sum-of-signals is the sum of their respective Fourier transforms. Therefore, no new frequencies are produced by a linear operation. It has been previously pointed out that such is not the case if a nonlinear operation is encountered. Under such conditions, frequency components that are not in the original spectrum can appear in the output spectrum. A classic example is the squaring of a causal cosine wave. Here, if $x(t) = \cos(\omega_0 t)u(t)$, then $y(t) = x^2(t) = \cos^2(\omega_0 t)u(t) = (0.5 + 0.5\cos(2\omega_0 t))u(t)$. It can be seen that the only frequency contained in the original signal is ω_0, whereas the output spectrum contains 0 (DC) and $2\omega_0$ with nothing remaining at ω_0.

Example 3-11   Linearity

A pedestal signal $p(t)$, shown in Figure 3-16, is given by

$$p(t) = \text{rect}(t/2) + \text{rect}(t/4) + \text{rect}(t/8) + \text{rect}(t/16)$$

$$= x_2(t) + x_4(t) + x_8(t) + x_{16}(t)$$

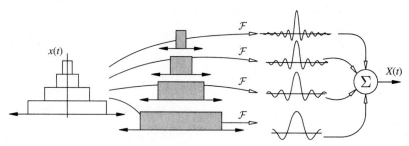

FIGURE 3-16   Pedestal signal and its Fourier transform.

and has a Fourier transform given by

$$P(j\omega) = X_2(j\omega) + X_4(j\omega) + X_8(j\omega) + X_{16}(j\omega)$$

Computer Study   S-file `ch3_11.s` and M-file `ch3_11.m` create the elements of the pedestal signal, form the pedestal signal, and compute the Fourier transform.

SIGLAB	MATLAB
`>include "ch3_11.s"`	`>> ch3_11`

The results are displayed in Figure 3-17. The linearity property is seen to be experimentally verified.

Parseval's Theorem   The total energy in a signal $x(t)$ can also be expressed in terms of a Fourier transform using *Parseval's Theorem*.

Parseval's Theorem   If $x(t) \overset{\mathcal{F}}{\leftrightarrow} X(j\omega)$, then the energy in the signal $x(t)$, denoted E_x, can be computed using

$$E_x = \int_{-\infty}^{\infty} |x(t)|^2 \, dt = \int_{-\infty}^{\infty} x(t)x^*(t) \, dt = \frac{1}{2\pi} \int_{-\infty}^{\infty} |X(j\omega)|^2 \, d\omega \qquad (3\text{-}26)$$

The integrand of the right-most term in Equation 3-26 is referred to as the *energy spectral density* of $x(t)$ and can alternatively be represented as $|X(j\omega)|^2 = X(j\omega)X^*(j\omega)$. Generally, power is a more important measurement than energy. Power is defined by measuring the energy in a signal over an interval of time T and forming the ratio $P_x = E_x/T$. If a signal $x(t)$ is of finite energy $E_x < \infty$, over $T \in [-\infty, \infty]$, then the power calculation produces a meaningless value of zero. A signal's energy at a specific frequency, however, can be an important measurement. The Fourier transform of a signal $x(t)$, at a specific frequency ω_0, can be selectively computed using

$$X(j\omega_0) = \int_{-\infty}^{\infty} x(t)e^{-j\omega_0 t} \, dt \qquad (3\text{-}27)$$

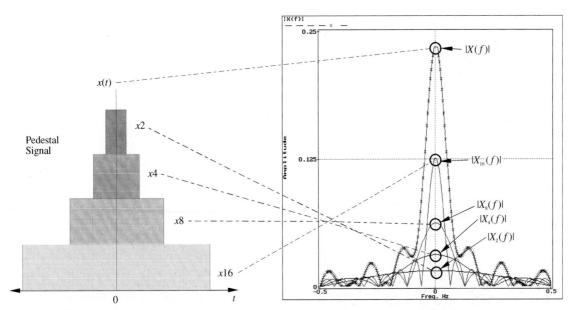

FIGURE 3-17 Linear property (left) pedestal signal and (right) spectral representation.

Here $X(j\omega_0)$ is the called *Fourier image* of $x(t)$ at $\omega = \omega_0$. Such a calculation can be useful in detecting and verifying the existence of a particular signal event, such as the arrival of a telephone dial tone. It can also be used to measure signal- and noise-energy levels. For example, to verify whether or not a 60 Hz component is embedded in a signal, $X(j2\pi 60)$ would be computed. Its magnitude would indicate the presence or absence of a 60 Hz signal component.

Example 3-12 Parseval's theorem

A voltage $v(t) = e^{-at}u(t)$, for $a > 0$, is developed across a 1 ohm resistor. The resistor dissipates energy through heat loss (in joules), which is numerically given by

$$E_v = \int_{-\infty}^{\infty} v^2(t)dt = \int_0^{\infty} e^{-2at}dt = \frac{1}{2a}$$

The Fourier transform of $v(t)$ can be shown to be given by (see Table 3-5)

$$V(j\omega) = \frac{1}{j\omega + a}$$

Therefore,

$$|V(j\omega)|^2 = V(j\omega)V^*(j\omega) = \frac{1}{(j\omega + a)}\frac{1}{(-j\omega + a)} = \frac{1}{(\omega^2 + a^2)}$$

and, from Parseval's theorem, it follows that

$$E_v = \frac{1}{2\pi} \int_{-\infty}^{\infty} \frac{d\omega}{(\omega^2 + a^2)} = \frac{1}{2a\pi} \tan^{-1}(\omega/a) \Big|_{-\infty}^{\infty}$$

$$= \frac{1}{2a\pi} \left[\frac{\pi}{2} - \frac{-\pi}{2} \right] = \frac{1}{2a}$$

which is the expected result.

Differentiation and Integration   If $x(t) \overset{\mathcal{F}}{\leftrightarrow} X(j\omega)$, then for any positive real integer m

$$\frac{d^m x(t)}{dt^m} \overset{\mathcal{F}}{\leftrightarrow} (j\omega)^m X(j\omega) \tag{3-28}$$

and

$$\int_{-\infty}^{t} x(\tau)\, d\tau \overset{\mathcal{F}}{\leftrightarrow} \frac{1}{j\omega} X(j\omega) + \pi X(0)\delta(\omega) \tag{3-29}$$

Differentiators are seen to amplify all frequency components of $X(j\omega)$ by a factor $j\omega$. This can be undesirable in some cases, especially if high frequency noise is present. In this case, differentiating can then be seen to accentuate the noise to a point where it can dominate a desired signal. Integration has just the opposite effect, in that it scales the spectrum by $1/j\omega$. Integration, therefore, emphasizes low frequencies, attenuates high frequencies, and therefore is consistent with the notion that an integrator is intrinsically an averaging (DC) device.

Example 3-13   Calculus

Using a raised cosine waveform, $x(t) = 0.5(1 - \cos(\omega_0 t))$ for $t \in [-\pi/\omega_0, \pi/\omega_0]$ and 0 elsewhere, and an EKG test signal, the scaling of $X(j\omega)$ under the rules $dx(t)/dt$ and $\int x(\tau)\, d\tau$, can be experimentally verified.
Computer Study S-file ch3_13.s and M-file ch3_13.m accept a signal $x(t)$ and produce its derivative, its integral, and then compute and display their magnitude spectra as shown in Figure 3-18. A raised cosine is also known as a Hann window.

SIGLAB	MATLAB
```>include "ch3_13.s"```	```>> ch3_13```
```>x=cshift(han(32) & zeros(250-31),125-16)```	```>> load ekg1.imp```
```>diffint(x) #raised cosine```	```>> diffint(ekg1)```
```>x=rf("ekg3.imp");x=x[0:250] #prune```	
```>diffint(x) # read ekg signal```	

Compared to the original spectra, the differentiated signals are seen to exhibit more high frequency characteristics and the integrated signals are seen to be low-frequency dominated.

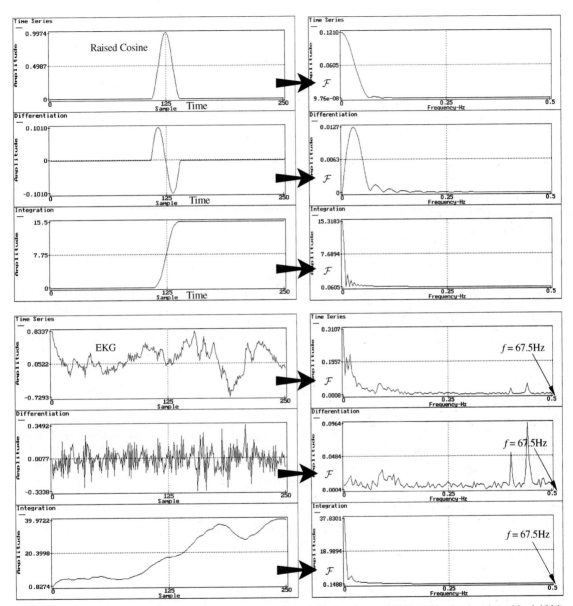

FIGURE 3-18   Effects of differentiating and integrating a raised cosine and EKG: (left) raised cosine $x(t)$, $dx(t)/dt$, and $\int x(t)\, dt$ and their spectra, and (right) EKG $= x(t)$, $dx(t)/dt$, and $\int x(t)\, dt$ and their spectra.

Modulation   If $x(t) \overset{\mathcal{F}}{\leftrightarrow} X(\omega)$, then for any real w_0,

$$x(t) \exp(j\omega_0 t) \overset{\mathcal{F}}{\leftrightarrow} X(\omega - \omega_0) \tag{3-30}$$

Equation 3-30 has a special case based on Euler's identity, namely

$$x(t) \cos(\omega_0 t) \overset{\mathcal{F}}{\leftrightarrow} 0.5[X(j(\omega - \omega_0)) + X(j(\omega + \omega_0))] \tag{3-31}$$

$$x(t) \sin(\omega_0 t) \overset{\mathcal{F}}{\leftrightarrow} j0.5[X(j(\omega + \omega_0)) - X(j(\omega - \omega_0))] \tag{3-32}$$

The terms $X(j(\omega + \omega_0))$ and $X(j(\omega - \omega_0))$ are called the sum and difference frequency terms, or *upper* and *lower sidebands*, respectively.

Example 3-14   Modulation

The spectrum given by Equations 3-31 and 3-32 exhibits distinct sideband activity about the specific frequency ω_0. For example, let $x(t) = \cos(\omega_1 t)$, then $X(j\omega) = \pi(\delta(\omega - \omega_1) + \delta(\omega + \omega_1))$. Then if $y(t) = x(t)\cos(j\omega_0 t)$, the modulated spectrum is given by Equation 3-31, which reduces to $Y(j\omega) = 0.5[X(j(\omega - \omega_0)) + X(j(\omega + \omega_0))] = 0.5\pi[\delta(\omega + \omega_0 - \omega_1) + \delta(\omega + \omega_0 + \omega_1) + \delta(\omega - \omega_0 - \omega_1) + \delta(\omega - \omega_0 + \omega_1)]$. The resulting spectrum is seen to consist of four distinct sidebands existing adjacent to the modulation frequency $\pm\omega_0$. For the case where $\omega_0 > \omega_1$, $0.5\pi\delta(\omega - \omega_0 + \omega_1)$ is called the lower sideband and $0.5\pi\delta(\omega - \omega_0 - \omega_1)$ is the upper sideband. A similar relationship exists for negative frequencies.

Computer Study   S-file `ch3_14.s` and M-file `modulate.m` modulate two cosine waves $x_1(t)$ and $x_2(t)$, located at two distinct frequencies, and display the modulated signal and its two-sided spectrum as shown in Figure 3-19. The format is `modulate(f_1, f_2)` where $f_1, f_2 \in [0, 0.25]$ and $x_1(t) = \cos(2\pi f_1 t)$ and $x_2(t) = \cos(2\pi f_2 t)$.

SIGLAB	MATLAB
`>include "ch3_14.s"` `>modulate(0.05, 0.15)`	`>> modulate(0.05, 0.15)`

Observe that the spectrum consists of upper and lower sidebands.

Duality   If $x(t) \overset{\mathcal{F}}{\leftrightarrow} X(j\omega)$ then

$$X(t) \overset{\mathcal{F}}{\leftrightarrow} 2\pi x(-j\omega) \tag{3-33}$$

That is, if the shapes of $x(t)$ and $X(j\omega)$ are known, then this information can be used to interpret the relationship between $X(t)$ and $x(j\omega)$. In short, time and frequency properties can be exchanged.

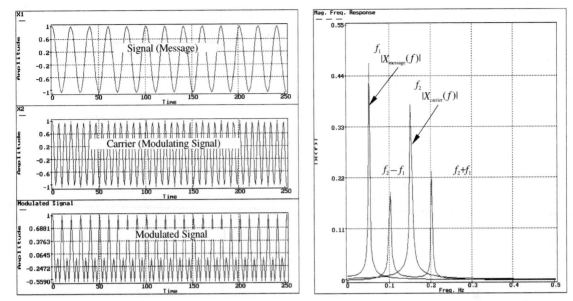

FIGURE 3-19   Modulation example with (left) $x_1(t)$, $x_2(t)$, and $x_1(t)x_2(t)$, and (right) $\mathcal{F}[x_1(t)]$, $\mathcal{F}[x_2(t)]$ and $\mathcal{F}[x_1(t)x_2(t)]$.

Example 3-15   Duality

The Fourier transform of a rectangular pulse $\mathrm{rect}(t/\tau)$ will be derived in Section 3-7 and shown to be (see Table 3-5)

$$\left[\mathrm{rect}\left(\frac{t}{\tau}\right)\right] \overset{\mathcal{F}}{\leftrightarrow} \tau\,\mathrm{sinc}\left(\frac{\omega\tau}{2}\right) = \frac{2\sin(\omega\tau/2)}{\omega}$$

The duality theorem states that if a time-domain signal has a shape given by $x(t) = \tau\,\mathrm{sinc}(t\tau/2)$, then its Fourier representation has the shape of a rectangular pulse. In particular (see Example 3-9)

$$\tau\,\mathrm{sinc}\left(\frac{t\tau}{2}\right) \overset{\mathcal{F}}{\leftrightarrow} 2\pi\,\mathrm{rect}\left(\frac{-\omega}{\tau}\right) = 2\pi\,\mathrm{rect}\left(\frac{\omega}{\tau}\right)$$

Computer Study   S-file `ch3_15.s` and M-file `ch3_15.m` create a discrete-time sinc signal that approximates a continuous-time sinc signal and computes its Fourier spectrum. The format is `width(tau)` and the sinc-function is implemented as $\tau\,\mathrm{sinc}(\tau t/2)$ and $\tau = $ `tau /128`.

SIGLAB	MATLAB
`>include "ch3_15.s"` `>width(16); width(64)`	`>> ch3_15`

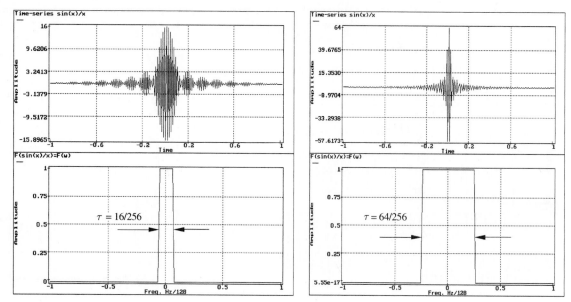

FIGURE 3-20 τ sinc($\tau t/2\pi$) time-domain signal and its Fourier transform for (left) $\tau = 16/256$, and (right) $\tau = 64/256$.

The results are plotted in Figure 3-20 and it is seen that a discrete-time approximation to a continuous $\sin(x)/x$ function has a Fourier spectrum, which has the shape of a rectangular pulse as predicted by the duality property.

Left/Right Shifts, Time Scaling, Time Reversal, and Conjugation   Shifts, scalings, reversals, and conjugations can be developed in the context of the previously stated properties. In particular, if $x(t) \overset{\mathcal{F}}{\leftrightarrow} X(j\omega)$, then

$$x(t \pm \tau) \overset{\mathcal{F}}{\leftrightarrow} X(j\omega)\exp(\mp j\omega\tau) \qquad (3\text{-}34)$$

If $x(t) \overset{\mathcal{F}}{\leftrightarrow} X(j\omega)$, then

$$x(-t) \overset{\mathcal{F}}{\leftrightarrow} X^*(j\omega) \qquad (3\text{-}35)$$

where * denotes complex conjugation. Observe, in all cases, that the magnitude spectra of the left-hand sides of the above equations all equal $|X(j\omega)|$. The differences between the spectra, therefore, must be attributed to the phase. This points to an important fact; that is, phase shift and time delays are equivalent. This can be exemplified by noting that if $x(t) = \cos(\omega_0 t)$, then $x(t - \pi/2\omega_0) = \cos(\omega_0(t - \pi/2\omega_0)) = \cos(\omega_0 t - \pi/2) = \sin(\omega_0 t)$. Finally, if $x(t) \overset{\mathcal{F}}{\leftrightarrow} X(j\omega)$, then $x^*(t) \overset{\mathcal{F}}{\leftrightarrow} X^*(-j\omega)$. This property follows immediately from the fact that $\exp(-j\omega t)^* = \exp(j\omega t)$.

Example 3-16   Signal delay and time reversal

The spectrum of a simple bipolar square wave $x(t)$ can be modified by imparting a simple signal delay or time reversal. In both cases the magnitude frequency response is that of the original signal. The only difference (if any) is in the phase spectrum.

Computer Study   S-file `ch3_16.s` and `ch3_16.m` produce a pulse, then delay and time reverse it and display the spectra as shown in Figure 3-21. The phase display represents phase differences between $x(t)$, $x(t-32)$, $x(t+32)$ and $x(-t)$ over $f \in [0, 20/256]$Hz. From this data it can be seen that the magnitude frequency responses agree, but differ in phase.

SIGLAB	MATLAB
>include "ch3_16.s"	>> ch3_16

Important elementary signals, such as $x(t) = \sin(\omega_0 t)$, are not absolutely integrable nor do they satisfy Dirichlet's conditions for $t \in [-\infty, \infty]$. This, as previously stated, does not mean that their Fourier transforms do not exist. To compute their Fourier transforms, the *generalized Fourier transform* is sometimes be used. It is based on the selective use of an impulse distribution and the duality theorem of Table 3-4. More specifically, recall that if $x(t) = \delta(t)$, then $X(j\omega) = \mathcal{F}(\delta(t)) = 1$. From the duality theorem, it also follows that if $x(t) = 1$ for $t \in [-\infty, \infty]$, then $X(j\omega) = 2\pi\delta(-\omega) = 2\pi\delta(\omega)$. This is called the generalized Fourier transform of $x(t)$ and is seen to be an *indirect* rather than a *direct* computation. By using this technique, the Fourier transforms of many important signals can be derived. Fortunately, the Fourier transforms of many primitive signals have been precomputed and tabled. Using the properties found in Table 3-4, the Fourier transform of complex signals can be synthesized from elementary signals.

3-7 BASIC SIGNALS

As noted, computing a Fourier transform can be a tedious process. It was also stated that a number of basic signals have well-known Fourier transforms (see Table 3-5). A number of important basic signals, or primitives, now will be more fully explored.

Rectangular Pulse   A rectangular pulse (see Example 3-15) $x(t) = \text{rect}(t/\tau)$, of width τ, as displayed in Figure 3-22, has a Fourier transform given by

$$X(j\omega) = \int_{-\infty}^{\infty} \text{rect}(t/\tau) \exp(-j\omega t)\, dt = \int_{-\tau/2}^{\tau/2} \exp(-j\omega t)\, dt$$

$$= \frac{1}{-j\omega} \left[\exp\left(\frac{-j\omega\tau}{2}\right) - \exp\left(\frac{j\omega\tau}{2}\right) \right] \tag{3-36}$$

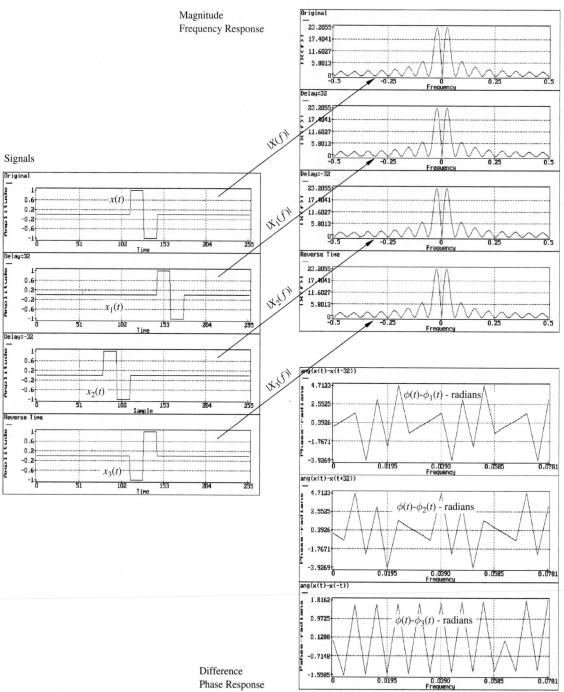

FIGURE 3-21   Effect of time delays and time reversal: (left) time-series, (top) magnitude frequency spectra, and (bottom) phase spectra and phase difference between $x(t)$, $x(t-32)$, $x(t+32)$, and $x(-t)$.

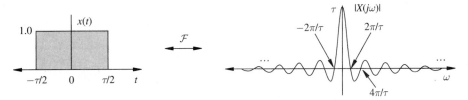

FIGURE 3-22   Fourier transform of a pulse of width τ.

which can be reduced to

$$X(j\omega) = \frac{2}{\omega} \sin \frac{\omega \tau}{2} = \tau \operatorname{sinc} \left(\frac{\omega \tau}{2} \right) \tag{3-37}$$

also shown in Figure 3-22.

The Fourier transform of a pulse is seen to be parameterized by τ. For a large value of τ, rect(t/τ) becomes DC-like and, as a result, the spectrum narrows to a concentration of low frequencies. Another special case is where $\tau \to \infty$, $x(t) = 1$, and $X(j\omega) = 2\pi \delta(j\omega)$. Alternatively, as $\tau \to 0$, the spectrum $X(j\omega)$ whitens.

Example 3-17   Rectangular pulse processes

The magnitude spectrum of $x(t) = \operatorname{rect}(t/\tau)$ is graphically interpreted in Figure 3-22 for a fixed value of τ. The nulls (zeros) of $|X(j\omega)|$ occur at $\omega = 2\pi k/\tau$. Therefore, doubling τ will also double the number of nulls that occur over a given range of frequencies.

Computer Study   S-file ch3_17.s and M-file 3_17.m compute and display (see Figure 3-23) the magnitude frequency response for a discrete-time approximation of rect(t/τ) for $\tau = 2$, 4, 8, and 16 and observed for $0 \le t \le 256$ seconds. The spectra are displayed over a finite baseband range of frequencies.

SIGLAB	MATLAB
>include "ch3_17.s"	>> ch3_17

Observe that the width of the pulse is inversely related to the width of the produced spectrum that has the predicted $\sin(x)/x$ envelope.

Triangular Wave   A triangular wave, for $t \in [-\tau, \tau]$, given by

$$x(t) = \begin{cases} 1 - \frac{|t|}{\tau} & \text{for } |t| \le \tau \\ 0 & \text{for } |t| > \tau \end{cases} \tag{3-38}$$

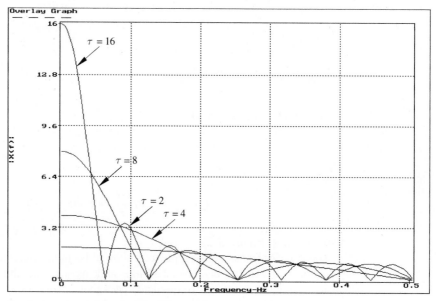

FIGURE 3-23   Fourier transform of a pulse of width τ.

has a Fourier transform given by

$$X(j\omega) = \int_{-\infty}^{\infty} x(t) \exp(-j\omega t)\, dt$$

$$= \int_{-\tau}^{0} \left(1 + \frac{t}{\tau}\right) \exp(-j\omega t)\, dt + \int_{0}^{\tau} \left(1 - \frac{t}{\tau}\right) \exp(-j\omega t)\, dt$$

$$= 2 \int_{0}^{\tau} \left(1 - \frac{t}{\tau}\right) \cos(\omega t)\, dt \qquad (3\text{-}39)$$

which reduces to

$$X(j\omega) = \tau \operatorname{sinc}^2\left(\frac{\omega\tau}{2}\right) \qquad (3\text{-}40)$$

Note again that small values of τ tend to produce broadband spectra, and vice versa, as found in the study of the rectangular pulse.

Example 3-18   Triangle wave

The locations of the nulls of the sinc^2 function are the same as those of the sinc function. However, by squaring, the amplitude of the sidelobes (for $|\omega| > 2\pi/\tau$) of the sinc^2 function are much smaller than those of the sinc envelope. *Computer Study*   S-file ch3_18.s and M-file trix.m create a discrete-time 256-second triangle pulse and displays its two-sided spectra (i.e., positive and

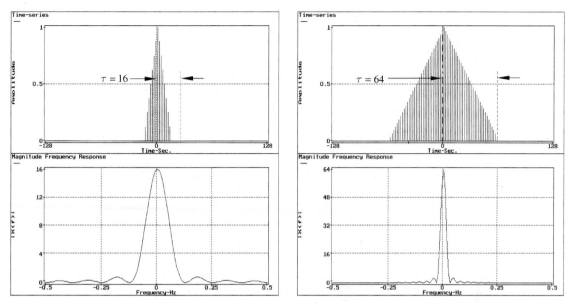

FIGURE 3-24   Fourier transform of triangular wave of various widths: (left) $\tau = 16$, and (right) $\tau = 64$.

negative frequencies) for τ even, using the format `trix(tau)`. For $\tau = \{16, 64\}$, the data shown in Figure 3-24 results.

SIGLAB	MATLAB
`>include "ch3_18.s"`	`>> trix(16)`
`>trix(16); trix(64)`	`>> trix(64)`

Observe that, as in Example 3-9, the width of the signal (measured by τ) is inversely related to the effective width of the produced spectrum.

Complex Exponential   One of the most important signals that can be studied analytically is the noncasual complex exponential given by

$$x(t) = \exp(j\omega_0 t) \tag{3-41}$$

The generalized Fourier transform of a complex exponential is based on knowledge of the transform of a pure DC (0 Hz) process (i.e., $1 \overset{\mathcal{F}}{\leftrightarrow} 2\pi\delta(\omega)$), and the modulation theorem (i.e., $x(t)\exp(j\omega_0 t) \overset{\mathcal{F}}{\leftrightarrow} X(j(\omega - \omega_0))$). From this, it follows that

$$X(j\omega) = 2\pi\delta(\omega - \omega_0) \tag{3-42}$$

which states that all the energy in $x(t)$ is concentrated at one positive frequency $\omega = \omega_0$. An extension of this result is based on the use of Euler's equation, which yields

$$\cos(\omega_0 t) = \frac{\exp(j\omega_0 t) + \exp(-j\omega_0 t)}{2} \overset{\mathcal{F}}{\leftrightarrow} \pi[\delta(\omega - \omega_0) + \delta(\omega + \omega_0)] \quad (3\text{-}43)$$

$$\sin(\omega_0 t) = \frac{\exp(j\omega_0 t) - \exp(-j\omega_0 t)}{2j} \overset{\mathcal{F}}{\leftrightarrow} \frac{\pi}{j}[\delta(\omega - \omega_0) - \delta(\omega + \omega_0)] \quad (3\text{-}44)$$

The Fourier transforms of these signals are seen to possess both positive and negative frequency components. It can also be seen that the magnitude spectra of the sine and cosine waves are identical. Their phase spectra differ, as expected. In addition, the cosine spectrum is seen to be purely real, while the sine spectrum is purely imaginary.

Example 3-19   Harmonic oscillations

The Fourier transform of a complex exponential, cosine wave, and sine wave show that their energy is concentrated at specific frequencies. The complex exponential's energy is concentrated only at $\omega = \omega_0$, whereas the sine and cosine waves have line spectra located at $\omega = \pm\omega_0$. Because of this redistribution of energy, the amplitude of the sine and cosine Fourier transforms is half that of the complex exponential at $\omega = \omega_0$.

Computer Study   S-file ch3_19.s and M-file comexp.m compute and display the signal and the two-sided spectra (magnitude, real, imaginary) of a complex exponential, sine wave, and cosine wave at a frequency f_0 for a T second-time record, using the format comexp(f0/T).

SIGLAB	MATLAB
>include "ch3_19.s" >> comexp(50/256)	
>comexp(50/256)	

The results are presented in Figure 3-25. Note the relationship between the sine and cosine spectra, and the fact that the complex exponential spectrum has but one positive harmonic.

Decaying Exponential   A two-sided symmetric (noncausal) decaying exponential is given by

$$x(t) = \exp(-a|t|) \quad (3\text{-}45)$$

where $a > 0$ and real is called the decay parameter and is shown as signal $x_1(t)$

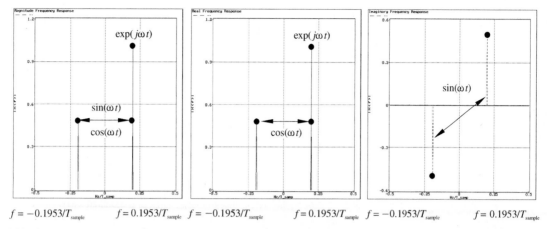

$f = -0.1953/T_{\text{sample}}$ $f = 0.1953/T_{\text{sample}}$ $f = -0.1953/T_{\text{sample}}$ $f = 0.1953/T_{\text{sample}}$ $f = -0.1953/T_{\text{sample}}$ $f = 0.1953/T_{\text{sample}}$

FIGURE 3-25   Fourier transforms of a complex exponential, sine, and cosine showing time-series, magnitude frequency responses, and real and imaginary spectra.

in Figure 3-26. Note that $x(t)$ is absolutely integrable for $a > 0$ since

$$\int_{-\infty}^{\infty} |x(t)| \, dt = 2 \int_{0}^{\infty} \exp(-at) \, dt = \frac{2}{a} < \infty \tag{3-46}$$

and $x(t)$ satisfies the other Dirichlet conditions. Thus, $x(t)$ has a Fourier transform. The Fourier transform of $x(t)$ is given by

$$X(j\omega) = \int_{-\infty}^{0} \exp(at) \exp(j\omega t) \, dt + \int_{0}^{\infty} \exp(-at) \exp(-j\omega t) \, dt$$

$$= \frac{1}{a - j\omega} + \frac{1}{a + j\omega} = \frac{2a}{a^2 + \omega^2} \tag{3-47}$$

which is seen to be real for all ω. Observe that for small a, $x(t)$ decays slowly and takes on the appearance of DC signals. As a increases, the time-domain response changes rapidly from its maximal value to zero, similar to an impulse. For the DC-like signal, the spectrum is narrowly concentrated about 0 Hz. A rapidly changing signal, however, produces a spectrum rich in high frequencies.

The two-sided antisymmetric (noncausal) decaying exponential, shown in Figure 3-26, is given by

$$x_2(t) = \begin{cases} \text{sgn}(t) \exp(-a|t|) & \text{if } t \neq 0 \\ 0 & \text{if } t = 0 \end{cases} \tag{3-48}$$

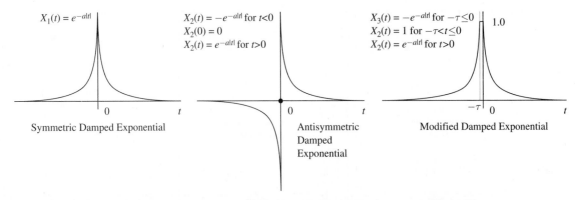

FIGURE 3-26 Symmetric decaying exponential $x_1(t)$, antisymmetric $x_2(t)$, and modified $x_3(t)$.

Note again that $x(t)$ is absolutely integrable for $a > 0$ and a Fourier transform. The Fourier transform of $x(t)$ is given by

$$X(j\omega) = -\int_{-\infty}^{0} \exp(at)\exp(-j\omega t)\,dt + \int_{0}^{\infty} \exp(-at)\exp(-j\omega t)\,dt$$

$$= \frac{-1}{a - j\omega} + \frac{1}{a + j\omega} = \frac{-2j\omega}{a^2 + \omega^2} \tag{3-49}$$

which is seen to be complex for all ω.

It should be appreciated that symmetry can often be a fragile concept. Even a slight alteration in a symmetric signal (e.g., $x_3(t)$ shown in Figure 3-26) can create a new signal with a Fourier image distinctly different from the original.

Example 3-20   Decaying exponential

The Fourier transforms of the noncausal exponential signals shown in Figure 3-26 can be displayed in cartesian and polar coordinates.
Computer Study   S-file `ch3_20.s` and M-file `expat.m` capture a 20-second record of $x(t)$ and display the real and imaginary spectra of $x_i(t)$. For $x_3(t)$, $\tau = 1$, the spectrum $X_3(f)$ is also displayed in polar form. For $a = 50$ the format used is `expat(a)`.

SIGLAB	MATLAB
>include "ch3_20.s" >expat(50) #a=50	>> expat(50)

The results are displayed in Figure 3-27. The Fourier transform of the signal $x_1(t)$ is seen to be real and nearly flat over the displayed frequency range. This is due to the fact that the signal closely resembles a Dirac delta distribution

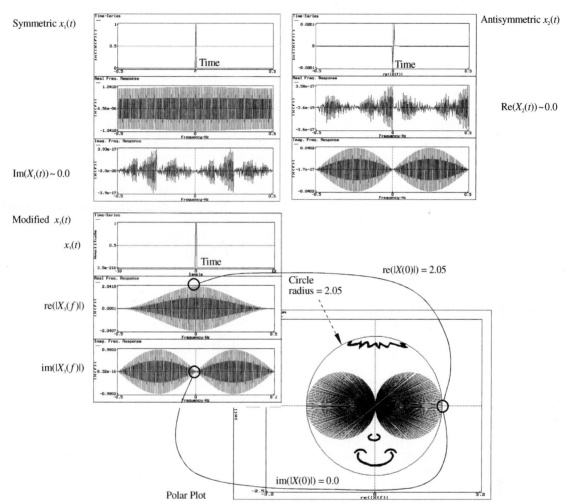

FIGURE 3-27   Fourier transform of a decaying exponential frequency response of non-causal exponentials showing the real and imaginary spectra, plus a polar plot of $X(f)$.

for the given choice of "a". Other choices of "a" having values nearer to zero would have produced a slowly decaying exponential having a Fourier image dominated by low frequency components. The Fourier transform of $x_2(t)$ is seen to be imaginary. Finally the Fourier transform of $x_3(t)$ is observed to be truly complex. In the polar-plane, each ray represents a magnitude and phase projection of $X(j\omega)$ at a specific frequency ω.

The Fourier transforms of many elementary signals have been tabled. If an arbitrary signal can be represented as a combination of elementary signals, then its Fourier transform can be produced using table look-ups. A summary of the

Fourier transforms of a number of elementary signals are provided in Table 3-5. The signals in Table 3-5 are also classified as being causal (i.e., $x(t), t \in (0, \infty)$), or non-causal (i.e., $x(t), t \in (-\infty, \infty)$).

3-8   UNCERTAINTY PRINCIPLE

The Fourier transform has another limitation with an analogy in the uncertainty principle studied in physics. In physics, the uncertainty principle states that simultaneous position and velocity measurements cannot, in general, be accurately made. There is a similar relationship that exists between frequency-domain and time-domain measurements. A short time duration signal, like an impulse, has a broadband spectrum. Therefore, the closer we attempt to resolve the width of a narrow signal (e.g., $\text{rect}(t/\tau)$ for $\tau \to 0$), the wider the bandwidth requirement of the frequency-domain measurement instrument becomes. Similarly, verifying that a line spectrum is produced by a DC signal (e.g., $X(j\omega) = 2\pi \delta(\omega)$), would require that the signal $x(t)$ be observed for all time (i.e., for $t \in (-\infty, \infty)$).

One measure of the *effective duration* of a signal $x(t)$, distributed about $t = 0$, is given by

$$ T = \frac{1}{x(0)} \int_{-\infty}^{\infty} |x(t)| \, dt \tag{3-50} $$

which has a companion *effective bandwidth* measure

$$ B = \frac{1}{X(0)} \int_{-\infty}^{\infty} |X(j\omega)| \, d\omega \tag{3-51} $$

Other measures of effective duration and bandwidth will, of course, give different results. Nevertheless, based on this pair of definitions, the time-bandwidth product can be shown to be bounded by

$$ BT \geq 2\pi \tag{3-52} $$

Equation 3-52 states that bandwidth and time duration cannot be independently specified. This result is called the *uncertainty principle*. For example, if $x(t) = 1$ for all time ($T \to \infty$), it follows that $X(j\omega) = 2\pi \delta(\omega)$ ($B \to 0$), and $TB = 2\pi$. The duality theorem can be used to show that juxtaposing the time- and frequency-domain signal distributions (i.e., $x(t) = \delta(t)$), produces the same result. In short, the uncertainty principle states that it is impossible to measure both the time and frequency-domain properties of a signal to an arbitrarily high precision.

3-9   BANDWIDTH

A spectrum beginning at 0 Hz (DC) and extending contiguously over an increasing frequency range is called a baseband spectrum. Signals with spectra distributed over a wide range of frequencies are called *broadband*. Alternatively, if the spectrum is localized about a point in the frequency-domain, the signal is called a *narrowband* process. Any signal that has a definable maximal frequency is called a *bandlimited* process. The range of frequencies over which a signal distributes its spectrum is called the signal's *bandwidth*, and is denoted BW. If a signal possesses a spectrum only over a contiguous frequency range of B Hz, and is zero

TABLE 3-5   FOURIER TRANSFORMS OF ELEMENTARY FUNCTIONS

Continuous Time Function $x(t)$	Fourier Transform $X(j\omega)$	Remark
1	$2\pi\delta(\omega)$	Constant, noncausal.
$u(t)$	$\pi\delta(\omega) + \dfrac{1}{j\omega}$	Unit-step function, causal.
$\delta(t)$	1	Delta distribution, noncausal.
$\delta(t - t_0)$	$\exp(-j\omega t_0)$	Delayed delta distribution, noncausal.
$\sum_{n=-\infty}^{\infty}\delta(t - nT)$	$\dfrac{2\pi}{T}\sum_{n=-\infty}^{\infty}\delta\left(\omega - \dfrac{2n\pi}{T}\right)$	Impulse train.
$\mathrm{rect}(t/\tau)$	$\dfrac{2\sin(\omega\tau/2)}{\omega} = \tau\,\mathrm{sinc}(\omega\tau/2)$	Rectangular pulse, noncausal.
$\dfrac{\sin(\omega_0 t)}{\pi t} = \dfrac{\omega_0}{\pi}\,\mathrm{sinc}\left(\omega_0 t\right)$	$\mathrm{rect}\left(\dfrac{\omega}{2\omega_0}\right)$	Noncausal.
$\exp(j\omega_0 t)$	$2\pi\delta(\omega - \omega_0)$	Complex exponential, noncausal.
$\cos(\omega_0 t)$	$\pi[\delta(\omega - \omega_0) + \delta(\omega + \omega_0)]$	Noncausal.
$\sin(\omega_0 t)$	$\dfrac{\pi}{j}[\delta(\omega - \omega_0) - \delta(\omega + \omega_0)]$	Noncausal.
$\cos(\omega_0 t)u(t)$	$\dfrac{\pi}{2}[\delta(\omega - \omega_0) + \delta(\omega + \omega_0)] + \dfrac{j\omega}{\omega_0^2 - \omega^2}$	Causal.
$\sin(\omega_0 t)u(t)$	$\dfrac{\pi}{2j}[\delta(\omega - \omega_0) - \delta(\omega + \omega_0)] + \dfrac{\omega_0}{\omega_0^2 - \omega^2}$	Causal.
$\exp(-at)u(t)$	$\dfrac{1}{a + j\omega}$	Re[a]>0, causal.
$t\exp(-at)u(t)$	$\dfrac{1}{(a + j\omega)^2}$	Re[a]>0, causal.
$\exp(-a|t|)$	$\dfrac{2a}{a^2 + \omega^2}$	Re[a]>0, noncausal.
$|t|\exp(-a|t|)$	$\dfrac{2(a^2 - \omega^2)}{a^2 + \omega^2}$	Noncausal.

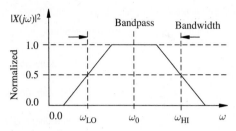
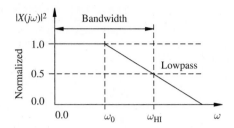

FIGURE 3-28   Bandlimited bandwidth models.

elsewhere, it is said to have an *absolute bandwidth* BW = B. The problem with this definition is that many real signals technically possess a measurable (but possibly low-level) amount of activity over a wide range of frequencies. As a result, an intrinsically narrowband process could be misclassified as being broadband. A more realistic definition is called the 3 dB, or *halfpower bandwidth*. The boundaries of the 3 dB bandwidth are interpreted in Figure 3-28. The maximal gain is assumed to occur at a frequency ω_0. Bandwidth is defined to be BW = $\omega_{HI} - \omega_{LO}$ for the general case and BW= ω_{HI} for a baseband signal. The edge frequencies correspond to the 1/2 power frequencies defined by Equation 3-53.

$$\begin{cases} \dfrac{|X(j\omega_{HI})_{max}|^2}{|X(j\omega_0)|^2} = \dfrac{1}{2} \\[4mm] \dfrac{|X(j\omega_{LO})_{max}|^2}{|X(j\omega_0)|^2} = \dfrac{1}{2} \end{cases} \quad \text{General case, if } \omega_{HI} \text{ and } \omega_{LO} > 0 \qquad (3\text{-}53)$$

$$\frac{|X(j\omega_{HI})_{max}|^2}{|X(j0)|^2} = \frac{1}{2} \quad \text{Baseband case, if } \omega_{HI} > 0 \text{ and } \omega_{LO} = 0 \qquad (3\text{-}54)$$

As a simple example, suppose a baseband frequency spectrum of a signal is modeled by $X(j\omega) = 1/(1 + j\omega)$. Then the 3 dB bandwidth BW = 1 since $|X(j1)|^2/|X(j0)|^2 = 1/2$ (halfpower) in this case. The 3 dB nomenclature comes from the fact that $10\log_{10}(1/2) \sim -3$ dB. The 3 dB condition is also often expressed as an amplitude ratio given by $|X(j1)|/|X(j0)| \sim 1/\sqrt{2} = 0.707$.

Example 3-21   Bandwidth measure

The frequency spectrum of a narrow baseband EKG signal is shown in Figure 3-29 out to 67.5 Hz. It is seen that the concept of absolute bandwidth is meaningless in this case, since the spectrum persists at high frequencies. The 3 dB bandwidth can, however, be directly measured.

Computer Study   To produce the data shown in Figure 3-29, run the S-file.

SIGLAB	MATLAB
>include "ch3_21.s"	Not available in Matlab

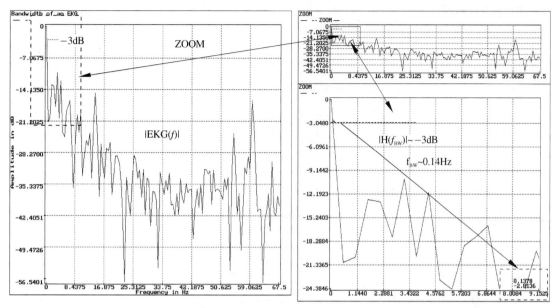

FIGURE 3-29   Baseband bandwidth measurement.

Based on the measured frequency response in Figure 3-29, the maximum normalized component is seen to be 0 dB at 0 Hz with the -3 dB frequency located at $f_{\mathrm{HI}} \approx 0.1$ Hz. The baseband 3 dB bandwidth is, therefore, BW $= 0.1$ Hz.

3-10 SUMMARY

In Chapter 3 the concept of the frequency-domain representation of a signal was introduced. Since many important and real-world signals are produced by oscillators and periodic signal generators, the interpretation of a signal on the base of its frequency content is natural.

The principal spectral analysis tool for continuous-time signals is the Fourier transform (CTFT). If the continuous-time signal under study is periodic, then a simpler form of the Fourier transform, called the Fourier series (FS), can be used. Both of these techniques are extremely important to the study of signals and systems. A signal's spectrum provides considerable information about the signal's structure and how it may interact with other signals and systems.

From all this, the field of spectral analysis has steadily developed. Spectral analysis is both rich in theory and application. To many, the study of signals and systems is an exercise in applied spectral analysis. In Chapter 5 it will be shown how the Laplace and Fourier transforms can be applied to the study of systems, and to the interaction between systems and signals.

3-11 COMPUTER FILES

These computer files were used in Chapter 3.

File Name	Type	Location
Subdirectory CHAP3		
ch3_2.x	C	Example 3-2
ch3_3.s	T	Example 3-3
coef.m	C	Example 3-3
ch3_4.x	T	Example 3-4
ch3_5.x	T	Example 3-5
ch3_6.x	T	Example 3-6
ch3_7.s	C	Example 3-7
ch3_8.x	C	Example 3-8
gibbs.m	C	Example 3-8
ch3_9.x	T	Example 3-9
ch3_10.x	C	Example 3-10
cshift.m	C	Example 3-10
white.m	C	Example 3-10
ch3_11.x	C	Example 3-11
ch3_13.x	C	Example 3-13
diffint.m	C	Example 3-13
ch3_14.s	C	Example 3-14
modulate.m	C	Example 3-14
ch3_15.x	C	Example 3-15
ch3_16.x	T	Example 3-16
ch3_17.x	T	Example 3-17
ch3_18.s	C	Example 3-18
trix.m	C	Example 3-18
triang.m	C	Example 3-18
ch3_19.s	C	Example 3-19
comexp.m	C	Example 3-19
ch3_20.s	C	Example 3-20
expat.m	C	Example 3-20
ch3_21.s	C	Example 3-21
Subdirectory SFILES or MATLAB		
fcoef.s	C	Example 3-5
shannon.s	C	Example 3-9
Subdirectory SIGNALS		
EKG1.imp	D	Examples 3-7, 3-21
EKG3.imp	D	Example 3-13
Subdirectory SYSTEMS		

Where "s" denotes S-file, "m" an M-file, "x" either S-file or M-file, "T" a tutorial program, "D" a data file and "C" denotes a computer-based-instruction (CBI) program. CBI programs can be altered and parameterized by the user.

3-12 PROBLEMS

3-1 Symmetry Prove the Fourier series symmetry results shown in Table 3-1.

3-2 Fourier series Verify the values of c_i found in Table 3-2.

3-3 Fourier series properties Prove the Fourier series properties summarized in Table 3-3.

3-4 Full-wave rectifier Derive the Fourier series representation of a full-wave rectified signal $V(t) = |V_o \sin(\omega_0 t)|$, as defined in Example 3-6.

3-5 Fourier series Compute the Fourier series of the signal shown in Figure 3-30 (only one period shown) and $|x_i(t)| \le 1$. Sketch and compare their magnitude and phase spectra.

3-6 Fourier series Repeat Problem 3-5 for $y_i(t) = x_i(8-t)$ for $t \in [0,8]$.

3-7 Fourier series Repeat Problem 3-5 for $y_i(t) = 1 - x_i(t)$.

3-8 Fourier series Compute the Fourier series of the signal shown in Figure 3-31 (only one period shown) and $|x_i(t)| \le 1$. Sketch and compare their magnitude and phase spectra.

FIGURE 3-30 Waveforms used in Problems 3-5 through 3-7.

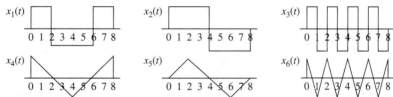

FIGURE 3-31 Waveforms used in Problems 3-8 through 3-10.

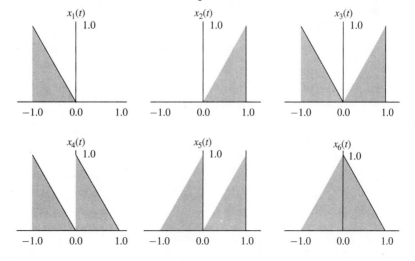

3-9 Fourier series Repeat Problem 3-8 for $y_i(t) = x_i(-t)$ over the period shown.

3-10 Fourier series Repeat Problem 3-8 for $y_i(t) = 1 - x_i(t)$.

3-11 Existence Which of the following signals have a Fourier series guaranteed by the Dirichelt conditions? All signals have a period $T = 1$ and are shown over $t \in [0, 1]$.

a $x_1(t) = \sin(2\pi/t)$

b $x_2(t) = 1/t$

c $x_3(t) = t \exp(-10t)$

3-12 Orthogonality Prove that the Fourier series exponential basis function $\phi_m(t) = \exp(jm\omega_0 t)$, $t\varepsilon[0, T]$, from an orthogonal set for all m (m and integer).

3-13 Orthogonality Determine which pairs or sets of the signals (assumed to be periodic with period 8), shown in Figure 3-30, are orthogonal.

3-14 Properties of Fourier series Let $x(t)$ be a periodic signal with period T. Show that if $x(t) = x_{\text{even}}(t) + x_{\text{odd}}(t)$, x_{even} and x_{odd} are orthogonal over a period T.

3-15 A chord In Example 3-2 a musical A chord was studied. The Fourier representation of the A chord was produced using a computer that produced a spectrum having harmonics spaced $\Delta(f) = f_{\text{sample}}/N$ Hz apart (see Figure 3-4) . Here f_{sample} is the rate at which the A chord is sampled and N is the number of samples of the A chord used to compute an estimate of its Fourier series. Given two frequencies of an A chord, say $f_1 = 110$ Hz and $f_z = 220$ Hz, are there choices of N and f_{sample} that will guarantee that all the notes fall directly on harmonic lines, and if so, what are they?

3-16 Fourier transform of noncausal signals Compute the Fourier transform of the following noncausal signals

a $x(t) = \exp(-a|t|); a > 0$

b $x(t) = 1 - |t|$ if $|t| \le 1$ and 0 otherwise

c $x(t) = sgn(t)$

3-17 Fourier transform Given $x(t) = \exp(-at)u(t)$, $a > 0$, what is the energy spectral density of $x(t)$? Verify that the energy in the signal equals the integral of the energy spectral density (viz, Parseval's equation).

3-18 Fourier transform The signals found in Figure 3-30 are assumed to equal $x_i(t)$, for $t \in [0, 8]$ is zero outside this interval. The signals are modulated by a sinusoid given by $c(t) = \sin(\omega_0 t)$. Sketch the modulated waveforms. Compute and sketch the Fourier transform of the modulated signals.

3-19 Modulation A waveform $y(t) = \sin(t)(u(t) - u(t - 2\pi))$ is created by modulating a waveform $s(t) = \sin(t)$, for $t\in[0, 2\pi]$, with $w(t)$, as shown in Figure 3-32. What is the Fourier transform of $s(t)$, $w(t)$, and $y(t)$? Experimentally verify.

FIGURE 3-32 Modulation of a sinewave waveform.

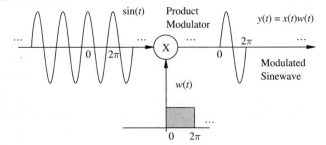

3-20 Energy and power spectral density   What is the energy spectral density of the signal $x(t) = \cos(2\pi t/8)$ for all $t \in [0, 8]$ and zero elsewhere. Sketch the energy spectral density. Suppose $x(t)$ is periodically extended for all time with period $T = 8$. What is the power spectral density of $x(t)$. Sketch the power spectral densities.

3-21 Gaussian signals   A Gaussian shaped signal is given by $x(t) = (a/\pi)^{0.5} \exp(-at^2)$ for $a > 0$. Compute $X(\omega)$. Verify your results experimentally for $a = 1/10, 1, 10,$ and 100.

3-22 EKG signal   Given an EKG signal, conduct an experiment to determine the compression abilities of the Fourier transform. Assume that the EKG signal is periodic. Saving only Fourier coefficients which are 10%, 20%, and 30% of the maximum coefficient, reconstruct the EKG from the compressed database. Graph the reconstruction error power as a function of the compression ratio.

3-23 Walsh functions   The Walsh functions are claimed to be an orthogonal set of basis functions. A set of $N = 2^n$ Walsh functions of length $N = 2^n$ are computed recursively beginning with $W_1 = 1$ and

$$W_2 = \begin{pmatrix} W_1 & W_1 \\ W_1 & -W_1 \end{pmatrix}$$

$$W_3 = \begin{pmatrix} W_2 & W_2 \\ W_2 & -W_2 \end{pmatrix}$$

$$\cdots$$

$$W_n = \begin{pmatrix} W_{n-1} & W_{n-1} \\ W_{n-1} & -W_{n-1} \end{pmatrix}$$

It can been seen that the Walsh functions are binary valued (i.e., ±1). Display the Walsh functions for $N = 8$. Show that they are orthogonal.

Recall a Fourier series used an orthogonal set of functions which were periodic over the observation interval. Are the Walsh functions of constant frequencies?

3-24 Bandwidth   Find the approximate time-bandwidth product of a signal given by $x(t) = \exp(-a|t|)$ for $a > 0$.

3-25 Two-dimensional Fourier transform   A two-dimensional Fourier transform is given by

$$X(j\omega_1, j\omega_2) = \int_{-\infty}^{\infty} \int_{-\infty}^{\infty} x(t_1, t_2) \exp(-(j\omega_1 t_1 + j\omega_2 t_2)) dt_1 dt_2$$

What is $X(j\omega_1, j\omega_2)$ if
a $x(t_1, t_2) = \exp(-|t_1| - |t_2|)$ for $|t_1| \le 1$, $|t_2| \le 1$, and 0 otherwise;
b $x(t_1, t_2) = 1$ for $|t_1| \le 1$, $|t_2| \le 1$, and 0 otherwise.
Sketch $x(t_1, t_2)$ and $X(j\omega_1, j\omega_2)$.

3-26 Shannon interpolation   In light of the results presented in Example 3-9, can you provide an intuitive explanation of why the ideal Shannon interpolator (see Equation 1-15) can reconstruct a signal from its sample values.

3-27 DC rectifier   Fullwave and halfwave rectification was studied in Example 3-6 from a frequency domain viewpoint. Suppose a filter was available that could pass DC without attenuation, signal energy located at 120 Hz (i.e., second harmonic) with an attenuation of "a," and attenuate all other higher harmonics to zero. What would the DC output of the filter be if the input is a half- and fullwave rectified AC line-power signal? What must "a" be so that the second harmonic distortion in the output is less that 0.1% of the DC value?

3-13   COMPUTER PROJECTS

3-1 Signature analysis   Signals are often identified by their frequency response elements called signatures. The signature of an unknown signal is often compared to a set of pre-computed spectral templates in order to classify it. Many defense and communication signal processing systems are based on this concept. Using an ekg1 as a template, develop a frequency-domain scheme by which ekg2 and ekg3 can be classified as being similar to the template. Continue to add noise to determine the robustness of your scheme.

3-2 Doppler measurements   A pulse Doppler radar system is shown in Figure 3-33. It has the ability to track moving targets by using the pulse to estimate range, and the Doppler frequency shift in the transmitted carrier frequency to estimate velocity. The spectral representation of the transmitted pulse $x(t)$ with a pulse repetition frequency (PRF) of 640 pulses per second and a pulse width $\tau = 1\mu s$ is a sinc function. At a transmitted frequency of $f_0 = 5650$ MHz, a shift in frequency by $\delta f = 1$ Hz corresponds to a velocity change of 0.029 yd/s.

Derive the value that calibrates the distance between adjacent harmonics, in terms of a velocity change (yd/s). Also compute the number of spectral lines residing between $[f_0, f_0 + 1/\tau]$. Comment on the practicality of this method in simultaneously measuring range (pulse echo delay) and velocity (Doppler shift), in terms of the instrument precision requirements.

For the purpose of simulation, assume that the PRF $= 100$ and $\tau = 0.5$ ms. A $\delta f = 1$ Hz will be assumed to correspond to a velocity change of 0.029 yd/s. Assume that the target is moving with a constant velocity of 0.1 yd/sec. Simulate the radar's response to the target. Develop a means of estimating the target velocity from the measured spectrum. Conduct experiments to determine how well the system will perform in the presence of white noise (place your results in signal to noise ratio parameters).

FIGURE 3-33   Pulse Doppler radar system.

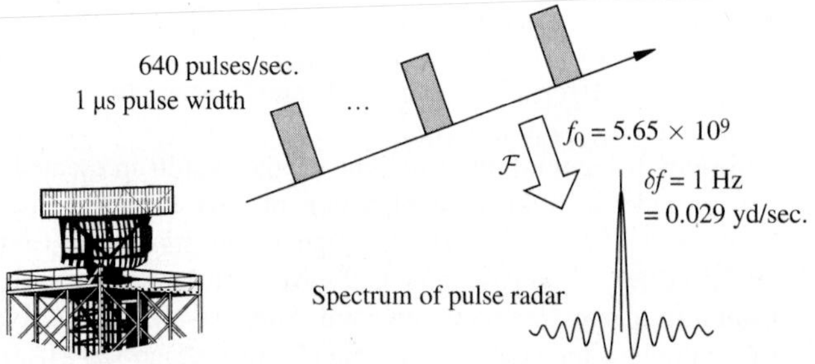

TIME DOMAIN REPRESENTATION OF CONTINUOUS-TIME SYSTEMS

The specialist knows more about less and less until he knows everything about nothing, while the generalist learns less and less about more and more until he knows nothing about everything.

—Dosen

4-1 INTRODUCTION

A *system* alters a signal in some manner. Our bodies, for example, are systems that convert inputs (e.g., energy, sound, light) from one form to another. When systems are designed to achieve a specific effect over a class of signals, they are sometimes referred to as *filters*. The ozone layer, for example, is a system that filters harmful ultraviolet radiation from a signal source, the sun. Many systems have physical, or hardware, manifestations. Continuous-time (analog) systems, for example, can be fabricated using energy storage and dissipation devices. For electrical systems, these devices can take the form of capacitors, inductors, and resistors. Mechanical systems can be created using dashpots (dampers), springs, and masses. Acoustical systems are formed using baffles and resonators, while optical systems are designed using polarizers, lenses, and color-selective plates. Indeed nature is abundant with systems which are governed by the principles studied in physics, biology, and chemistry.

113

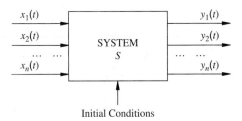

FIGURE 4-1   A MIMO system.

4-2   SYSTEMS

A *system* is a set of physical or mathematical components that respond to an input stimulus with a result called an output. If a system's input and output are scalar processes, then the system is called a *single-input single-output* (SISO) system. If the input and output signals are vectors, then the system is a said to be a *multiple-input multiple-output* (MIMO) system. A MIMO system is shown in Figure 4-1. The system, denoted S, may be a physical collection of electrical, mechanical, acoustical, chemical, biological, or optical devices. If the system is a combination of these technologies, then it is called a *hybrid system*. Systems may also be a collection of mathematical operations that may or may not model a physical process or system.

A mathematical *model* reflects how information is managed, manipulated, and modified by a system. The model can be an accurate representation of a physical system if the system is simple and its physical principles are well-known. At other times this is not the case, and the model is a "best guess" of reality. It is not uncommon for some systems to have several models, each reflecting different system behavior based on a unique set of circumstances called *modes of operation*.

The system model defines how inputs interact with elements of that system and initial conditions to produce an output. If the initial conditions are zero, then the system is said to be *at rest*. The output of an at-rest system, with respect to an input $x(t)$, is abstractly represented as

$$y(t) = (Sx)(t) \tag{4-1}$$

Here S is called the system *input-output operator*. Equation 4-1 reflects the fact that the output may be dependent on $x(t)$ over a time-interval $t \in [a, b]$. The mapping, $y(t) = (Sx)(t) = \int x(\tau)d\tau$, for $t \in [-\infty, t]$, is an example of such a dependency. If $y(t)$ depends on $x(t)$ only at time t, for example a memoryless amplifier that produces an output $y(t) = Kx(t)$, then notationally $y(t) = (Sx)(t) = S(x(t))$.

4-3   SYSTEM CLASSIFICATION

Systems can be grouped into broadly defined classes. By classifying systems by their properties and attributes, they can be more readily compared, and efficiencies are gained by insuring that the correct mathematical tool is applied to a specific problem.

4-3-1   Superposition

The most obvious classification scheme is to label a system linear or nonlinear. Superposition is a fundamentally important property possessed by a linear system.

Superposition Principle   The *superposition principle* states: if the output of a system to a given input $x_i(t)$ is $y_i(t)$, denoted $x_i(t) \rightarrow y_i(t)$, then the output of a linear system to an input $x(t)$, where $x(t)$ is a linear combination of the $x_i(t)$, namely

$$x(t) = \sum_{i=0}^{N} \alpha_i x_i(t) \tag{4-2}$$

is given by

$$y(t) = \sum_{i=0}^{N} \alpha_i y_i(t) \tag{4-3}$$

which can be written in a more compact form as

$$x(t) = \sum \alpha_i x_i(t) \rightarrow \sum \alpha_i y_i(t) = y(t) \tag{4-4}$$

That is, if S is a linear system and is presented with an input $x(t) = a_0 x_0(t) + a_1 x_1(t)$, then the output is given by $y(t) = S(a_0 x_0 + a_1 x_1)(t) = (S a_0 x_0)(t) + (S a_1 x_1)(t) = a_0(S x_0)(t) + a_1(S x_1)(t)$.

The superposition principle is graphically interpreted in Figure 4-2. Observe that the output to a null (zero) input must likewise be null (zero). If a system does not satisfy the superposition principal for any α_i or $x_i(t)$ found in Equation 4-4, then the system is said to be *nonlinear*. It has been previously established that linear operations cannot create new frequencies (see Section 3-4). Using the same argument, it can be shown that a fundamental property of linear systems is that they cannot create *new* frequencies. However, this is not the case for nonlinear systems. The output of a linear system to an input $x(t) = \sin(\omega_0 t)$ can only be of

FIGURE 4-2   Superposition.

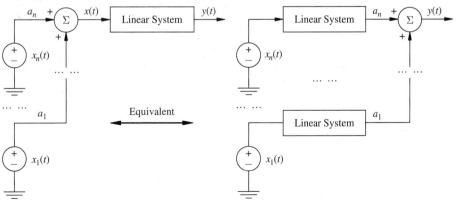

the form $y(t) = A\sin(\omega_0 t + \phi)$. The nonlinear system, defined by the input-output relationship $y(t) = x^2(t)$, maps $x(t) = \cos(\omega_0 t)$ into $y(t) = 1/2 + 1/2\cos(2\omega_0 t)$. Here the output obviously contains frequencies that were not present at the input. The quadratic system can also be seen to violate the superposition principle in that

$$y(t) = (\mathcal{S}(ax_1 + bx_2))(t) = (ax_1(t) + bx_2(t))^2 = a^2 x_1(t)^2 + 2abx_1(t)x_2(t)$$

$$+ b^2 x_2(t)^2$$

$$\neq a^2 x_1(t)^2 + b^2 x_2(t)^2 = (\mathcal{S}(ax_1))(t) + (\mathcal{S}(bx_2))(t) \tag{4-5}$$

for all $x_1(t)$ and $x_2(t) \neq 0$.

Some physical systems may naturally add a DC component, a bias, to the output. Suppose the measured response of a system to $x_1(t)$ is $y_1(t) = \mathcal{S}x_1(t) = Kx_1(t) + b$, where K is a constant and $b \neq 0$ (i.e., a bias). The response to $x_2(t)$ can be similarly computed to be $y_2(t) = Kx_2(t) + b$. The output to an additive input $x_3(t) = x_1(t) + x_2(t)$ is given by $y_3(t) = \mathcal{S}(x_1(t) + x_2(t)) = Kx_1(t) + Kx_2(t) + b \neq y_1(t) + y_2(t) = Kx_1(t) + Kx_2(t) + 2b$. The superposition principle is then seen to fail and, as a result, a biased system is nonlinear. This technical problem could, of course, be corrected by subtracting the bias from the system to produce a new unbiased linear system.

To demonstrate another unapparent consequence of superposition, consider a system modeled by the first-order ordinary constant coefficient differential equation $dy(t)/dt + ay(t) = x(t)$, where $y(0) = y_0$, presented later in this chapter. The complete solution to an ODE will be shown to be given by

$$y(t) = e^{-at}y_0 + \int_0^t e^{-a(t-\tau)}x(\tau)d\tau \tag{4-6}$$

If the model represents a linear system, then the superposition principle must apply. Two arbitrary inputs, say $x_1(t)$ and $x_2(t)$ will individually produce outputs $y_1(t)$ and $y_2(t)$, where

$$y_1(t) = e^{-at}y_0 + \int_0^t e^{-a(t-\tau)}x_1(\tau)d\tau \tag{4-7}$$

$$y_2(t) = e^{-at}y_0 + \int_0^t e^{-a(t-\tau)}x_2(\tau)d\tau \tag{4-8}$$

Satisfying the superposition principle would require that $y_3(t) = \mathcal{S}(x_1 + x_2)(t) = y_1(t) + y_2(t)$. Computing $y_3(t)$, it is seen that

$$y_3(t) = e^{-at}y_0 + \int_0^t e^{-a(t-\tau)}(x_1(\tau) + x_2(\tau))\,d\tau$$

$$\neq 2e^{-at}y_0 + \int_0^t e^{-a(t-\tau)}(x_1(\tau) + x_2(\tau))\,d\tau$$

$$= y_1(t) + y_2(t) \tag{4-9}$$

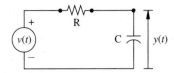

FIGURE 4-3   RC circuit.

If $y_0 \neq 0$, the superposition principle is seen to fail and the system is technically nonlinear. If the initial condition was zero (i.e., $y_0 = 0$), then the superposition would hold and the system would be classified as being linear. Therefore, a system characterized by a linear differential equation is said to represent a system that is linear with respect to zero initial conditions.[1] Assuming that a system is at rest (i.e., initial conditions are zero) is common practice. Later it will be shown that transfer functions and many stability theorems are based on this assumption.

Example 4-1   RC circuit

The RC circuit, shown in Figure 4-3, is at rest if the capacitor's voltage is initially zero. The response to an arbitrary causal $v(t)$ is given by

$$y(t) = \frac{1}{RC} \int_0^t e^{-(t-\tau)/RC} v(\tau) \, d\tau$$

For $v(t) = u(t)$ (a unit step), the output of an at-rest RC circuit can be computed to be

$$y_1(t) = \frac{1}{RC} \int_0^t e^{-(t-\tau)/RC} \, d\tau = \frac{e^{-t/RC}}{RC} \int_0^t e^{\tau/RC} \, d\tau$$

$$= 1 - e^{-t/RC}$$

for $t \geq 0$, and 0 otherwise. The response to a one-second delayed step, namely $u(t-1)$, is given by

$$y_2(t) = \frac{1}{RC} \int_1^t e^{-(t-\tau)/RC} \, d\tau = 1 - e^{-(t-1)/RC}$$

for $t \geq 1$, and 0 otherwise. The response to the combined input $v(t) = u(t) + u(t-1)$, is defined by

$$y_3(t) = \frac{1}{RC} \int_0^t e^{-(t-\tau)/RC} (u(\tau) + u(\tau - 1)) \, d\tau$$

$$= \frac{1}{RC} \int_0^t e^{-(t-\tau)/RC} u(\tau) \, d\tau + \frac{1}{RC} \int_1^t e^{-(t-\tau)/RC} u(\tau - 1) \, d\tau$$

$$= y_1(t) + y_2(t)$$

[1]We shall often refer to any system defined by a linear constant coefficient differential equation as being linear, and thereby make an implicit assumption that the initial conditions are zero.

This calculation shows that the at-rest system behaves linearly to a composite step input. To verify that the system is, in fact, linear would require that it be tested for all possible inputs. Obviously this is unrealistic. Therefore, linearity is generally assumed to be a mathematical, rather than an experimental, test. **Computer Study**   The response of the RC circuit shown in Figure 4-3 to $u(t)$, $u(t-1)$, and $u(t)+u(t-1)$ is simulated and displayed by S-file `ch4_1.s` and M-file `ch4_1.m` for RC = 1.0.

SIGLAB	MATLAB
>include "ch4_1.s"	>> ch4_1

The results, shown in Figure 4-4, demonstrate experimentally that the superposition principle is satisfied.

4-3-2   Time-Invariant and Time-Varying Systems

Another classification technique is based on a system being either *time-invariant* or *time-varying*. A continuous-time system is said to be time-invariant if a time shift, or delay, at the input produces an identical time shift, or delay, at the output. That is, a continuous-time system is time-invariant if $x(t) \rightarrow y(t)$ implies $x(t+\tau) \rightarrow y(t+\tau)$. An example of a time-varying system is one in which the coefficients may vary in time, due to aging or temperature.

FIGURE 4-4   Superposition showing $y_1(t)$, $y_2(t)$, and $y_3(t)$ (top), and an overlay comparison of $y_1(t)+y_2(t)$ and $y_3(t)$ (bottom).

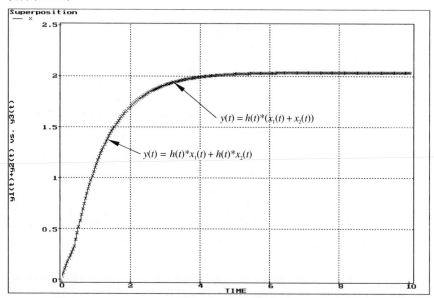

4-3-3 Memory and Memoryless Systems

The simplest of all systems is a device that scales the input by some prespecified constant, say α, and is represented by $y(t) = \alpha x(t)$. An example is a resistive voltage divider network. Such systems are called *memoryless* since they have no knowledge of the past. They have representation $y(t) = \mathcal{S}x(t)$.

Systems may also contain memory. For continuous-time systems, energy storage devices (e.g., capacitors and inductors) are the memory subsystems. These devices can store information (e.g., initial conditions) and, therefore, archive the history of the system. Any linear system, whether electrical, mechanical, biological, mathematical, or other, modeled by a linear differential equation has memory, since its solution is predicated on an a priori knowledge of the system's initial conditions.

4-3-4 Causal and Noncausal Systems

A system is said to be causal, or *nonanticipatory*, if it cannot produce an output until an input is present and the system is turned on. If a causal system is turned on at or before time t_0 and an input $x(t) = 0$, for all $t < t_0$, then the response of a causal linear system $y(t) = (\mathcal{S}x)(t)$ must also be zero for $t < t_0$. For systems with memory, the initial conditions are assumed to be zero, or, equivalently, the system is at rest. If this were not the case, then the output could have a nonzero value prior to $t = t_0$. If a system is not causal, it is said to be noncausal or *anticipatory*. We may be hard pressed to find a noncausal engineered system. However, in the biological world we all know people who give advice (output) long before they are asked for it (input). Such biological systems would be classified as noncausal.

4-4 CONTINUOUS-TIME LTI SYSTEMS

A linear system with fixed coefficients is called a *linear time-invariant (LTI)* system. Many physical systems are exactly, or at least approximately, LTI systems. Continuous-time LTI systems are normally characterized by linear ODEs with fixed coefficients and provide the basis for modeling many important continuous-time engineering systems. For example, the LTI system given by[2]

$$\sum_{k=0}^{N} a_k \frac{d^k y(t)}{dt^k} = \sum_{k=0}^{M} b_k \frac{d^k x(t)}{dt^k} \tag{4-10}$$

for $M < N$, has a homogeneous solution y(t), which is the solution to the ODE

$$\sum_{k=0}^{N} a_k \frac{d^k y(t)}{dt^k} = 0 \tag{4-11}$$

$$y(0) = y_0, \quad \frac{dy(0)}{dt} = y_1, \dots, \quad \frac{d^{(N-1)}y(0)}{dt^{N-1}} = y_{(N-1)}$$

---

[2]It shall be assumed, unless otherwise specified, that LTI systems, as defined in Equation 4-10, are strictly proper. That is, $M < N$. This condition will be relaxed in later chapters.

where y_i is an initial condition. The homogeneous solution is known to be a linear combination of basis, or eigenfunctions, of the form

$$\phi_j(t) = t^{n[j]}e^{s_j t} \tag{4-12}$$

Here s_j is called an eigenvalue and is a solution to the characteristic equation $\sum a_k s_j^k = 0$ for $k \in [0, N]$. Eigenvalues can appear with multiplicity up to $n[j]$, where $0 \le n[j] < N$, $1 \le j \le N$.

Example 4-2   Repeated eigenvalues

The second-order ODE, given by

$$\frac{d^2 y(t)}{dt^2} + 2a\frac{dy(t)}{dt} + a^2 y(t) = 0$$

has eigenvalues that satisfy the characteristic equation

$$s_i^2 + 2as_i + a^2 = 0$$

or $s_1 = s_2 = -a$ and $n[1] = 0$ with $n[2] = 1$ (i.e., multiplicity = 2).

The eigenfunctions $\phi_j(t)$ are also known to satisfy the system's characteristic equation, in that

$$\sum_{k=0}^{N} a_k \frac{d^k \phi_j(t)}{dt^k} = 0 \qquad \text{for } j = 1, 2, \ldots, N \tag{4-13}$$

Example 4-3   Eigenfunctions

The ODE studied in Example 4-2 has as eigenfunctions (after Equation 4-12)

$$\phi_1(t) = t^0 e^{-at}$$

$$\phi_2(t) = t^1 e^{-at}$$

Substituting the candidate eigenfunctions into the ODE $d^2 y(t)/dt^2 + 2ady(t)/dt + a^2 y(t) = 0$, and evaluating, we obtain

$$\frac{d^2 \phi_1(t)}{dt^2} + 2a\frac{d\phi_1(t)}{dt} + a^2\phi_1(t) = a^2\phi_1(t) - 2a^2\phi_1(t) + a^2\phi_1(t) = 0$$

$$\frac{d^2 \phi_2(t)}{dt^2} + 2a\frac{d\phi_2(t)}{dt} + a^2\phi_2(t) = [a^2 t\phi_1(t) - 2a\phi_1(t)]$$

$$+ 2a[-at\phi_1(t) + \phi_1(t)] + a^2 t\phi_1(t) = 0$$

Finally, the homogeneous solution can be expressed as a linear combination of these eigenfunctions and has the form

$$y(t) = \sum_{j=1}^{N} \gamma_j \phi_j(t) \tag{4-14}$$

where $y(t)$ and the first $N-1$ derivatives of $y(t)$, along with the specified initial conditions (see Equation 4-11), result in N equations used to solve for the N unknown γ_i.

Example 4-4   RLC circuit

The behavior of the RLC circuit shown in Figure 4-5, for example, can be defined by an ODE expressed in terms of the loop current $i(t)$. The circuit equations are given by

$$\frac{1}{C} \int_0^t i(\tau)\, d\tau + Ri(t) + L\frac{di(t)}{dt} = v(t)$$

or

$$i(t) + RC\frac{di(t)}{dt} + LC\frac{d^2i(t)}{dt^2} = C\frac{dv(t)}{dt}$$

and

$$v_c(t) = \frac{1}{C} \int_0^t i(\tau)\, d\tau \tag{4-15}$$

The output is defined by the voltage developed across the capacitor and is therefore related to the input voltage $v(t)$ by the equation

$$v_c(t) + RC\frac{dv_c(t)}{dt} + LC\frac{d^2v_c(t)}{dt^2} = v(t) \tag{4-16}$$

The homogeneous equation is, therefore, given by

$$v_c(t) + RC\frac{dv_c(t)}{dt} + LC\frac{d^2v_c(t)}{dt^2} = 0 \tag{4-17}$$

The eigenvalues are the solution to the characteristic equation

$$1 + RCs + LCs^2 = 0 \tag{4-18}$$

which shall be denoted s_1 and s_2. From the eigenvalues, the eigenfunctions can be generated. For the case where $s_1 \neq s_2$, they are

$$\phi_1(t) = e^{s_1 t}, \text{ and } \phi_2(t) = e^{s_2 t} \tag{4-19}$$

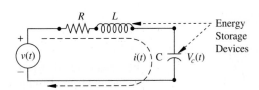

FIGURE 4-5   RLC network.

Substituting this result into Equation 4-17 yields

$$\phi_i(t) + RC\frac{d\phi_i(t)}{dt} + LC\frac{d^2\phi_i(t)}{dt^2} = \phi_i(t) + RCs_i\phi_i(t) + LCs_i^2\phi_i(t)$$

$$= (1 + RCs_i + LCs_i^2)\phi_i(t) = 0 \quad (4\text{-}20)$$

as required. Finally, for a given set of initial conditions y_0 and y_1, Equation 4-14 produces

$$y(0) = y_0 = \gamma_1\phi_1(0) + \gamma_2\phi_2(0) = \gamma_1 + \gamma_2$$

$$\frac{dy(0)}{dt} = y_1 = s_1\gamma_1\phi_1(0) + s_2\gamma_2\phi_2(0) = s_1\gamma_1 + s_2\gamma_2 \quad (4\text{-}21)$$

which can be solved for γ_1 and γ_2 using linear algebra (see Appendix A).

It is important to note that the homogeneous response of a linear continuous-time constant coefficient system is established by its eigenfunctions and initial conditions. Initial conditions can be externally controlled, but the eigenfunctions are defined in terms of the eigenvalues that are, in turn, specified by the system coefficients. Therefore, knowledge of a system's eigenvalues can play a major role in predicting the performance of a system. It will be shown in this chapter that eigenvalues and their attendant eigenfunctions are critical elements in specifying a system's *inhomogeneous*, or forced, response.

4-5   CONTINUOUS-TIME CONVOLUTION

In general, an input signal $x(t)$ will mathematically undergo both amplitude and phase modification before it emerges from the system as $y(t)$. The process by which a continuous-time linear system maps an input $x(t)$ into a continuous-time output $y(t)$ is called *convolution*. Convolution is a fundamental operation that ex-plains why a point source of light from a distant planet appears as a scintillating blur when viewed from earth. Convolution is at work when an amplified elec-trical signal is passed to a speaker and converted into an audible acoustic wave. Whenever a linear system exists between a signal source (origin of energy) and a signal sink (termination of energy), convolution takes place. Convolution, also referred to as *linear filtering*, is an extremely important concept pervasive in the study of signals and systems.

An abstract continuous-time linear system is shown in Figure 4-6. Suppose the input to the system, namely $x(t)$, is expressible as a linear combination of primitive signals, $\theta_i(t)$, such that

$$x(t) = \sum_{i=1}^{N} c_i\theta_i(t) \quad (4\text{-}22)$$

For example, if $\theta_i(t) = \cos(\omega_i t + \phi_i)$, then $x(t)$ is assumed to be represented by a linear combination of harmonic oscillations, as studied in Chapter 3. Other

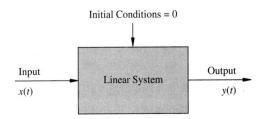

FIGURE 4-6   LTI system.

representations are, of course, possible. Suppose further that the system input and output can be measured by instruments. The output to an input $\theta_i(t)$ is recorded and denoted $y_i(t)$. The measurement process is continued until all $\theta_i(t)$ and $y_i(t)$ are tested and recorded. Using superposition, the output of the linear system to an input $x(t)$ can then be written as

$$y(t) = \sum_{i=1}^{N} c_i y_i(t) \tag{4-23}$$

$$\theta_i(t) \rightarrow y_i(t)$$

Therefore, through the use of superposition, a linear system's response to the complicated signal $x(t)$, can be expressed as a sum of individually recorded responses, as shown in Figure 4-7. Carrying this idea to a higher level, suppose an arbitrary input signal is expressed as an infinite collection of Dirac delta distributions, such that

$$x(t) = \int_{-\infty}^{\infty} x(\tau)\delta(t - \tau)\,d\tau \tag{4-24}$$

If $t = 1$, for example, then

$$x(1) = \int_{-\infty}^{\infty} x(\tau)\delta(1 - \tau)\,d\tau \tag{4-25}$$

FIGURE 4-7   Superposition principle.

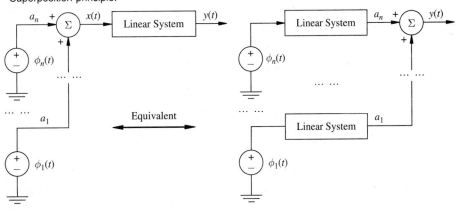

Therefore, $\delta(t-\tau)$ can isolate and capture the value of $x(t)$ at $t = \tau$. By evaluating $x(\tau)\delta(t - \tau)$, over all τ, Equation 4-24 can reproduce $x(t)$ for all t. Suppose that only one impulse, say $\delta(t)$, is applied to the input of the LTI system, and the output response is measured to be $h(t)$. For obvious reasons, the function $h(t)$ is called the system's impulse response to $x(t) = \delta(t)$, or more simply, the impulse response. The impulse response will be shown to be of major importance in the study of linear systems.

The impulse response of an at-rest system is defined to be the input-output mapping $h(t) = (\mathcal{S}\delta)(t)$. In particular, the impulse response for the at-rest LTI system, defined in Equation 4-10, is given by

$$\sum_{k=0}^{N} a_k \frac{d^k h(t)}{dt^k} = \sum_{k=0}^{M} b_k \frac{d^k \delta(t)}{dt^k} \tag{4-26}$$

for $M < N$. The solution to Equation 4-26 can, in turn, be expressed as a linear combination of eigenfunctions, as shown in Equation 4-14. Therefore

$$h(t) = \sum_{k=1}^{N} \gamma_k \phi_k(t) \tag{4-27}$$

where $\phi_j(t) = t^{n[j]} \exp(s_j t)$, $\phi_j(t)$ is an eigenfunction of Equation 4-26 and s_j is a root (eigenvalue) of the system's characteristic equation $\sum a_k s_k^k = 0$. In Chapter 5, efficient methods will be presented to compute the weighting coefficients γ_i found in Equation 4-27. Based on the superposition principle and the signal representation technique given in Equation 4-24, it follows that the output of an LTI system to an input $x(t)$ is given by

$$y(t) = \int_{-\infty}^{\infty} x(\tau)h(t - \tau)\,d\tau \tag{4-28}$$

The integral defined in Equation 4-28 is an important and fundamental equation in the study of linear systems; it is called the *convolution integral*. Because it is so often used, it has been given a special shorthand representation, namely $y(t) = h(t) * x(t)$.

Example 4-5   Impulse response and convolution

The superposition principle states that the response of an LTI system, having an impulse response $h(t)$ to an input $ax_1(t)+bx_2(t)+cx_3(t)$, is $y(t) = ax_1(t)* h(t) + bx_2(t) * h(t) + cx_3(t) * h(t)$.

Computer Study   S-file `ch4_5.s` and M-file `superpos.m` create a signal $x(t) = ax_1(t) + bx_2(t) + cx_3(t)$, where $x_1(t) = u(t)$, $x_2(t) = \cos(2\pi f_1 t)$, and $x_3(t) = \cos(2\pi f_2 t)$, using the format `superpos(a,b,c,d1,d2)`, where $d_i = f_i/250$. It then convolves each $x_i(t)$ individually and collectively as $x(t) = ax_1(t) + bx_2(t) + cx_3(t)$ with an LTI system with an impulse response known to be $h(t)$. The response $y(t) = x(t) * h(t)$ is compared to $z(t) = ax_1(t) * h(t) + bx_2(t) * h(t) + cx_3(t) * h(t)$.

SIGLAB	MATLAB
>include "ch4_5.s" >superpos(1,.5,-.25,10,20)	>> superpos(1, 0.5, -0.25, 10, 20)

The data shown in Figure 4-8 demonstrates experimentally that the two representations are in agreement.

The convolution integral specifies how the output of an LTI system evolves in time due to the presence of an input $x(t)$. As a result, convolution plays a critical role in such fields as filtering, controls, communications, and biological signal processing. Observe also that the output response of an at-rest LTI system is characterized by the system's impulse response $h(t)$ and the input signal $x(t)$. Of these two functions, only $h(t)$ is defined by the system equations. Therefore, the study of a linear system has, in many cases, become a study of impulse responses. The next example uses a simple RLC circuit to study the convolution process.

FIGURE 4-8 Superposition experiment showing (left) individually convolved signals, and (right) a demonstration of the superposition principle.

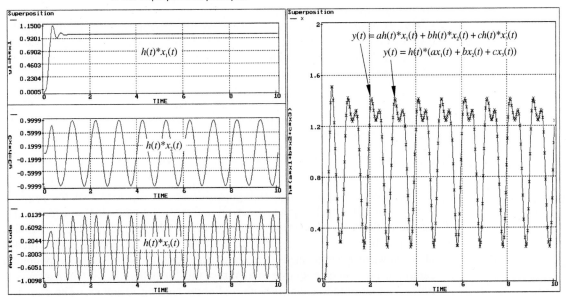

Example 4-6   RLC circuit

The impulse response of the RLC circuit shown in Figure 4-5 is given in terms of the solution to the ODE

$$LC\frac{d^2v_c(t)}{dt^2} + RC\frac{dv_c(t)}{dt} + v_c(t) = v(t)$$

where $v(t) = \delta(t)$. For a specific RLC circuit model

$$\frac{d^2v_c(t)}{dt^2} + 139.1\frac{dv_c(t)}{dt} + 9680.3v_c(t) = 9680.3v(t)$$

the impulse response is given by

$$h(t) = h_1(t) + h_2(t) = \lambda_1 e^{s_1 t} + \lambda_2 e^{s_2 t}$$

where $\lambda_1 = j69.55$, $\lambda_2^* = \lambda_1$, $s_1 = -69.55 - j69.59$, and $s_2^* = s_1$, are the roots to the characteristic equation $s^2 + 139.1s + 9680.3 = 0$. Again, methods will be developed in Chapter 5 to efficiently compute λ_1 and λ_2. Based on these parameters, the impulse response can be defined as

$$h(t) = j69.55e^{-69.55t}(e^{-j69.59t} - e^{j69.55t}) = 139.1e^{-69.55t}\sin(69.59t)$$

The impulse response can be seen to be a damped sinewave. Therefore, if the RLC circuit is presented with an input that can be mathematically represented as a collection of impulses, the output will take on the characteristics of a linear combination of damped sinusoids.

Computer Study   S-file ch4_6.s and M-file ch4_6.m produce the impulse response $h(t)$ for t ranging from 0 out to 0.185, 0.37, 0.741, 1.852 seconds. The convolution of $h(t)$ with a pulse train period of 0.474 seconds, and an EKG signal is also computed and displayed.

SIGLAB	MATLAB
>include "ch4_6.s"	>> ch4_6

The results are displayed in Figure 4-9. It can be seen that the impulse response is effectively spread over an interval of about 0.09 seconds, with the major build-up concentrated over the first 0.045 seconds. This interval also corresponds to the first half-period of $\sin(69.55t)$. In fact, all effective evidence of a sinewave has vanished after about one period, due to severe exponential damping. The forced response to a pulse train shows that the system attempts to follow the amplitude changes with a slightly lagging response. The response

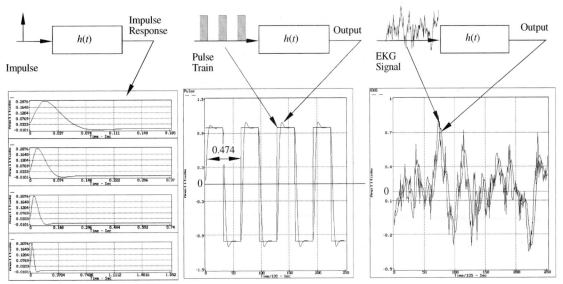

FIGURE 4-9   RLC response to (left) an impulse over an interval of 0.185, 0.37, 0.74, 1.852, (middle) convolution with a pulse train, and (right) and convolution with an EKG signal.

to a pulse train also indicates that the solution has periods of rapid change, followed by an interval in which the solution closely follows the input. The period of dynamic change is called the *transient response*, which is followed by the *steady-state response*. The EKG signal, after being convolved, is seen to lag behind the input. In addition, the output is also seen to have smoothed EKG signal peaks and edges. It will be shown in later chapters that the RLC circuit, when properly parameterized, has the ability to suppress high-frequencies (i.e., lowpass or bandpass filter) in the output. It is the suppression of the high-frequency components that results in a smoothed output.

Convolution can also be graphically interpreted, as shown in Figure 4-10. By investigating the mathematical process defined by Equation 4-28 in a graphical format, the act of convolution can be better understood. Refer to Equation 4-28 and observe that for a given time t, the integrand of the convolution integral consists of two terms, namely $x(\tau)$ and $h(t-\tau)$. These two signals can be plotted as a function of τ, where it can also be noted that $h(t - \tau)$ is a time-reversed signal that is reflected, or "folded," about the point $\tau = t$. This folding effect gives rise to an alternative name to the convolution integral, the *folding integral*. Integrating $g(t, \tau) = x(\tau)h(t - \tau)$ over time, produces $y(t) = h(t) * x(t)$ as shown in Figure 4-10.

A graphical interpretation of convolution can also be used to explain why the output found in Figure 4-10 exists for a longer interval of time than either $x(t)$ or $h(t)$. Refer again to Figure 4-10 and assume that $x(t)$ and $h(t)$ are both causal and of finite duration over the time interval $t \in [0, T]$. It can be seen that the first instant of time that $x(\tau)$ and $h(t - \tau)$ can possibly overlap is at $t = 0$. Therefore, the first possible nonzero value of $g(t, \tau)$ will also occur at $t = 0$. The signal overlap will continue to time $t = 2T$. In general, the convolution of two causal signals of duration T_1 and T_2 respectively, will be of length $T_1 + T_2$. Because of this, the output of a linear filter, driven by a narrow pulse, may persist for a considerably long interval of time. In fact, another pulse may arrive before the effects of the first have diminished. Therefore, in real applications, the effects of isolated input events may become intermixed so that they can no longer be individually identified.

FIGURE 4-10   Graphical convolution of $x(t)$ and $h(t)$, showing the evolution of the integrand and convolution integral as a function of t and τ.

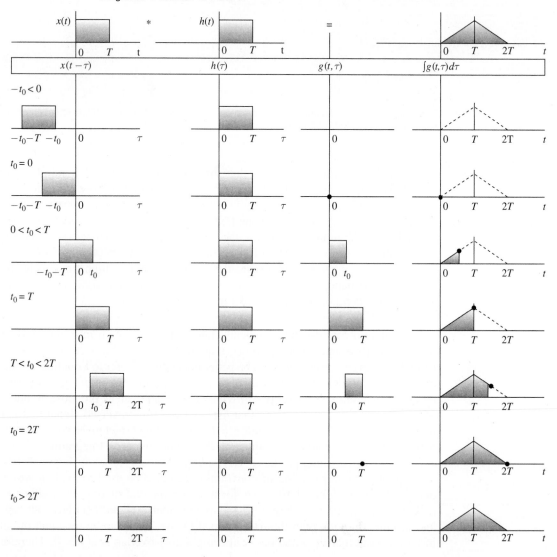

Example 4-7 Graphical convolution

The convolution of a signal $x(t)$ of 32 msec duration by an LTI system having an impulse response $h(t) = e^{-0.004t}u(t)$ for $t \in [0,\ 0.032]$ and 0 otherwise, can be graphically produced. The convolution outcome should have a duration of 64 msec and have a value 0 outside this range.

Computer Study S-file ch4_7.s and M-file gconv.m can be used to convolve a 32 ms signal of finite duration with $h(t)$. Both $h(t)$ and $x(t)$ will be approximated by discrete-time signals of sample length 32. The format used is gconv(x,ΔT) where x is a signal of length 32-samples and ΔT defines the incremental change in time between convolution calculations (i.e., $t = i\Delta T$). For $x(t)$, a triangular signal, and $\Delta T = 16$ ms, the following experiment is conducted.

SIGLAB	MATLAB
>include "ch4_7.s" >x=tri(4*32,32) >#triangle wave, slope 1, period 32 >gconv(x,16)	>> x = triang(32); >> gconv(x, 16)

The results are shown in Figure 4-11. They represent five distinct evolutions of the convolution computation. It can be seen that convolution builds up during the first 32 ms. During this period the folded input $x(t - \tau)$ is shifted to the right and has increasing overlap with $h(t)$. Thereafter, the overlap decreases in size until at time 64 ms, it is completely vanished.

4-6 CONVOLUTION PROPERTIES

The convolution mapping possesses a number of important properties. Among these are

Commutative Property: If $x(t)$ is a signal and $h(t)$ an impulse response, then $x(t) * h(t) = h(t) * x(t)$;

Associative Property: If $x(t)$ is a signal and $h_1(t)$ and $h_2(t)$ are impulse responses, then $[x(t) * h_1(t)] * h_2(t) = x(t) * [h_1(t) * h_2(t)]$;

Distributive Property: If $x(t)$ is a signal and $h_1(t)$ and $h_2(t)$ are impulse responses, then $x(t) * [h_1(t) + h_2(t)] = x(t) * h_1(t) + x(t) * h_2(t)$.

Example 4-8 Convolution properties

The convolution properties of the RLC circuit studied in Example 4-6, will be examined using the first 0.93 seconds of the system's damped sinusoid impulse response $h(t)$, and a 1.9-second section of an EKG signal.

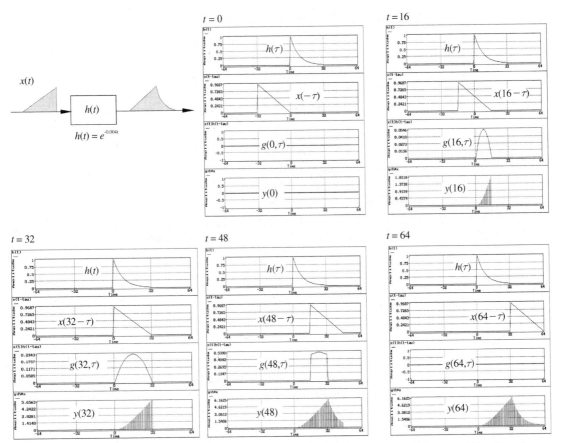

FIGURE 4-11   Graphical convolution example for $t = [0,16,32,48,64]$. From top to bottom, $h(t)$, $x(t - \tau)$, $g(t,\tau) = h(t)x(t - \tau)$, and $y(t) = h(t) * x(t)$.

Computer Study   S-file `ch4_8.s` M-file `convprop.m` captures 0.93 seconds of the impulse responses $h(t) = h_1(t) = h_2(t)$ (see Convolution Properties), along with a 1.9 second record from a signal $x(t)$. It then displays the first 0.23 seconds of $h(t)$ and all of $x(t)$, using the S-file format `convprop(x)` or M-file format `convprop(x, h1, h2)`. The program then implements the commutative, associative, and distributive equations for the case $h(t) = h_1(t) = h_2(t)$, and displays the results.

SIGLAB	MATLAB
`>include "ch4_8.s"`	`>> load ekg1.imp`
`>x=rf("ekg1.imp")`	`>> h1 = rand(31, 1); h2 = rand(31, 1);`
`>convprop(x)`	`>> convprop(ekg1, h1, h2)`

The results are displayed in Figure 4-12. The properties are seen to experimentally agree with the predicted results. Note that the convolution of a 0.93-second impulse response and 1.9 second signal is a 2.83 second signal (commutative and distributive graph), and the convolution of the 1.9-second EKG and two 0.93-second impulse responses is equal to their sum, namely 3.86 seconds (associative graph).

4-7   STABILITY OF CONTINUOUS SYSTEMS

Real systems designed with physical components can perform unacceptably on occasion. For example, a public address system often emits a loud shriek when the amplifier gain is turned too high. The system, in this case, has entered a regime of run-away large amplitude oscillation. Knowledge of signal divergence is of major importance, since divergence can lead to self-destruction or violent system behavior. These are all questions of *stability*. Stability, in the simplest sense, means that the system's response to a bounded input is likewise bounded. This is called the *bounded-input bounded-output principle* (BIBO).

Bounded-Input Bounded-Output (BIBO) Stability   If $x(t)$ is a bounded input, such that $|x(t)| < \infty$, then the system is BIBO stable if the output $|y(t)| < \infty$ for all t and all $x(t)$.

Since the output response to an arbitrary input can be described by a complicated system of equations (possibly nonlinear), establishing the stability of a system can be a challenging problem. For memoryless systems, however, the issue of stability can be established simply by evaluating the input-output equation. For example, if $y(t) = \exp(x(t))$, then $y(t)$ is maximized when $x(t) = M < \infty$ and produces $y(t)_{max} = |\exp(M)| < \infty$. Therefore the system is stable. However, if $y(t) = \log(|x(t)|)$, then the worst (maximizing) case value of $y(t)$ occurs when $x(t) = 0$. Here, $y(t)_{max} = |\log(0)| \to \infty$ results in the conclusion that the system is unstable in a BIBO sense.

Since stability can be defined in the time domain, it would stand to reason that the stability of an LTI system can be analyzed by investigating the properties of the convolution integral. Assume that an Nth order LTI system is characterized by the ODE

$$a_N \frac{d^N y(t)}{dt^N} + \cdots + a_1 \frac{dy(t)}{dt} + a_0 y(t) = x(t) \tag{4-29}$$

The impulse response of the system described by Equation 4-29 can be written as a linear combination of its weighted eigenfunctions, namely

$$h(t) = \sum_{i=1}^{N} h_i(t) = \sum_{i=1}^{N} \gamma_i t^{n[i]} e^{(\sigma_i + j\omega_i)t} \tag{4-30}$$

where $0 \le n[i] < N$ and $h_i(t) = \gamma_i t^{n[i]} \exp(\sigma_i + j\omega_i)$ where $\text{Re}\{s_i\} = \sigma_1$ and $\text{Im}\{s_i\} = \omega_i$ and s_i is an eigenvalue of Equation 4-29. Substituting Equation 4-30

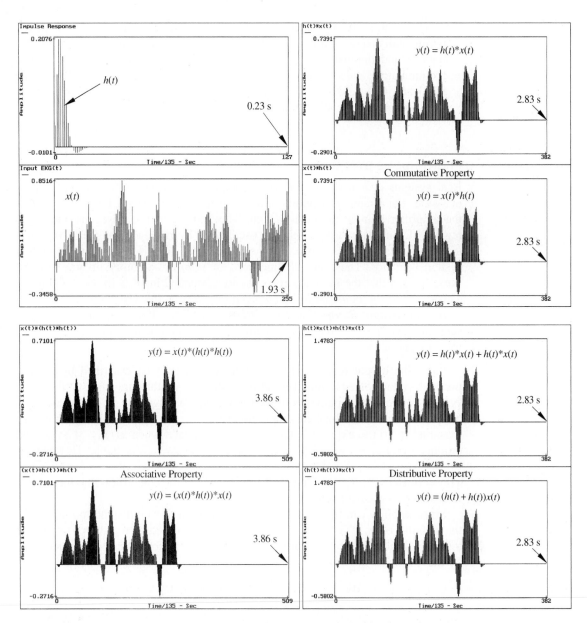

FIGURE 4-12   RLC circuit-impulse response, an EKG signal, and their convolution, demonstrating the commutative, associative, and distributive property.

into the convolution integral, it follows that if $x(t)$ is a bounded input, such that $|x(t)| < M$, then

$$y(t) = \int_{-\infty}^{\infty} x(t - \tau)h(\tau)\,d\tau \le \left| \int_{-\infty}^{t} Mh(\tau)\,d\tau \right|$$

$$= M \left| \int_{-\infty}^{t} h(\tau)\,d\tau \right| = M \left| \int_{-\infty}^{t} \sum_{i=1}^{N} h_i(\tau)d\tau \right|$$

$$\le M \sum_{i=1}^{N} \left| \int_{-\infty}^{t} h_i(\tau)\,d\tau \right| = M \sum_{i=1}^{N} \left| \int_{t=-\infty}^{t} \gamma_i t^{n[i]} e^{(\sigma_i + j\omega_i)\tau}\,d\tau \right| \quad (4\text{-}31)$$

Equation 4-31 can be seen to remain bounded if

$$\left| \int_{\tau=-\infty}^{t} h_i(\tau)\,d\tau \right| = \left| \int_{\tau=-\infty}^{t} \gamma_i t^{n[i]} e^{(\sigma_i + j\omega_i)\tau}\,d\tau \right| < \infty \quad (4\text{-}32)$$

If the system is causal, then $h_i(t)$ is causal and Equation 4-31 is finite if $\sigma_i < 0$. Whenever this condition occurs, each $y_i(t) = h_i(t) * x(t)$ will remain bounded; therefore their sum will also be bounded. This analysis presumes that the system's initial conditions are zero (i.e, the system is at rest). If initial conditions are considered, then the homogeneous solution's contribution to $y(t)$ must also be included. For a causal input, the *complete solution* to the LTI system given by the ODE shown in Equation 4-29, would have the form

$$y(t) = \sum_{i=1}^{N} c_i t^{n[i]} e^{(\sigma_i + j\omega_i)} + \int_{0}^{\infty} x(t - \tau)h(\tau)\,d\tau \quad (4\text{-}33)$$

where the c_i's are chosen so that $dy^i(0)/dt^i = y_i$ (the initial condition on the ith derivative of $y(t)$, where $i \in [0, N - 1]$). Again it is concluded that the system is BIBO stable if $\sigma_i < 0$. Therefore, if the real part of the eigenvalue's s_i, which were computed from the system's characteristic equation $\sum_{i=1}^{N} a_i s^i = 0$, are negative (i.e., $\sigma_i = \mathrm{Re}(s_i) < 0$), then BIBO system stability is assured.

There is also a body of knowledge that relates to the stability of systems being forced only by initial conditions. For arbitrary initial conditions and a zero (null) input, the homogeneous solution of a stable system will converge to what is called the *equilibrium state*[3], or *equilibrium solution*, of an LTI system. The rate at which the homogeneous solution will converge is directly influenced by the value of σ_i and the nonnegative integer $n[i]$, which is determined by the multiplicity of the eigenvalues. Nevertheless, as long as $\sigma_i < 0$, convergence of the homogeneous solution to zero is guaranteed.

---

[3]Technically, an equilibrium solution $x(t)$ has an expected value of $dx(t)/dt$ of zero.

Therefore, having eigenvalues with a negative real part will guarantee BIBO operation. If σ_i is a large negative number (i.e., $\sigma \to -\infty$), the exponential term $e^{s_i t}$ will almost immediately converge to zero. This action has the effect of rapidly dissipating energy from the system. The response of such system to $x(t) = \delta(t)$ would approach a null value in a brief period of time. If, however, $\sigma_i \to 0^-$, the exponential term $e^{s_i t}$ will require a long time to converge to zero. The energy supplied by a single impulse is now retained over a long period of time. If the eigenvalues are ordered, such that $\sigma_1 \geq \sigma_2 \geq ... \geq \sigma_n$, where $\sigma_i < 0$ for all i, then it follows that $|e^{\sigma_1 t}| \geq |e^{\sigma_2 t}| \geq ... \geq |e^{\sigma_n t}|$. The exponential $e^{\sigma_1 t}$ is said to dominate all others and σ_1 is called the *dominant eigenvalue*. Dominant eigenvalues are sometimes used to model those elements of a system response that persist for a long period of time.

Example 4-9   BIBO stability

A first-order system is assumed to be modeled as $dy(t)/dt + ay(t) = u(t)$, $y(0) = y_0$. The characteristic equation is given by $s + a = 0$ or $s = \sigma = -a$, in this case. The homogeneous solution is given by $y(t) = \exp(-at)y_0$ converges to zero as $t \to \infty$ if $a > 0$, which establishes the equilibrium state to be the origin. The system's impulse response is given by $h(t) = \exp(-at)u(t)$. The system is initially at rest and its response to a unit step function (called step response) is given by

$$y(t) = \int_{\tau=0}^{t} e^{-a(t-\tau)} \, d\tau = e^{-at} \int_{\tau=0}^{t} e^{a\tau} \, d\tau = e^{-at} \left[\frac{e^{a\tau}}{a} \right]_0^t = \frac{[1 - e^{-at}]}{a}$$

provided $a \neq 0$. If $a < 0$, the steady state solution $y(t)$ will become unbounded (unstable) and if $a > 0$, $y(t)$ will converge to $1/a$ as $t \to \infty$ (stable). If $a = 0$, then $dy(t)/dt = u(t) = 1$ and $y(t) = \int_0^t 1 d\tau \to \infty$ as $t \to \infty$ (unstable).

Computer Study   S-file `ch4_9.s` and M-file `firstode.m` produce the solutions to $dy(t)/dt + ay(t) = u(t)$, $y(0) = 0$ using the S-file format `exp(a)` or M-file format `firstode(a)`, where $|a| \leq 20$. To produce the data shown in Figure 4-13, choose $a = 10$.

SIGLAB	MATLAB
>include "ch4_9.s"	>> firstode(-10)
>exp(10) # select |a|>=20.0	>> firstode(10)

The data shows the stable and unstable solutions obtained for $a = \pm 10$. The stable solution is seen to converge to a steady-state value of $1/10$. For $a = -10$, the solution is seen to be unstable and diverges toward infinity.

In the real world, unstable systems never actually reach infinity due to the dynamic range limitation of physical components. Devices will either saturate or fail if they are stressed too much. Nonlinear effects may only appear when the system attempts to produce arbitrarily large amplitude results. An amplifier, for example, attached to a ±15 volt power supply may appear to operate linearly over a portion of this voltage range. As the output approaches ±15 volts, nonlinear effects would be noticed as saturation begins to take place. Note that independent of the range of the input, the output will always be bounded by ±15 volts. Predicting the response of a nonlinear system to an arbitrary input is, in general, a difficult problem. In many cases simulation is used to mitigate this problem. While it does not produce a mathematically acceptable proof of stability or instability, it can nevertheless provide important insights into system performance.

Nonlinear system elements appear in a wide variety of forms. There are, however, several standard nonlinear operators that appear with high repetition in practice. Assume that an analog system has a finite dynamic range $v \in [-V, V]$ (i.e., ±V). Over this range the system may behave in a linear manner, but once the range limits are exceeded, the output is "clamped" to the nearest ±V value. Mathematically, this is defined by the *saturation function* and is denoted sat(x), where

$$\text{sat}(x) = \begin{cases} 1 & \text{if } x > 1 \\ x & \text{if } x \in [-1, 1] \\ -1 & \text{if } x < -1 \end{cases} \tag{4-34}$$

FIGURE 4-13   Stable and unstable step responses for $a = 10$ and $a = -10$.

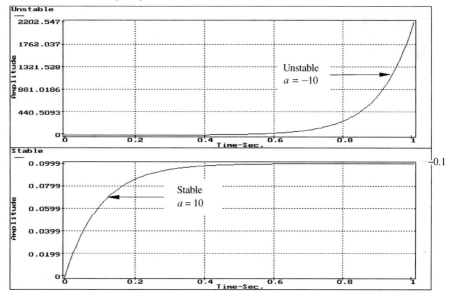

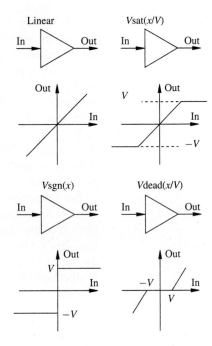

FIGURE 4-14   Elementary memoryless nonlinear devices.

The $\pm V$ volt saturating unit gain amplifier would have a model given by $y(t) = V\,\mathrm{sat}(x(t)/V)$.

Another typical nonlinear function is the *sgn, signum,* or *severely clipped* function which, is given by

$$\mathrm{sgn}(x) = \begin{cases} 1 & \text{if } x > 0 \\ 0 & \text{if } x = 0 \\ -1 & \text{if } x < 0 \end{cases} \qquad (4\text{-}35)$$

A high-gain amplifier (i.e., operational amplifier) can approximate a sgn function. A $\pm V$ volt high gain amplifier would be modeled as $y(t) = V\,\mathrm{sgn}(x(t))$.

Systems can also exhibit what is commonly called *deadband* behavior. Many mechanical systems must be physically moved away from a static resting point before they can begin to exhibit dynamic motion. A deadband is a range of values, say $v \in [-V, V]$, which produce a zero output. Outside this range the system behaves in a linear manner. The mathematical model for a deadband is given by

$$\mathrm{dead}(x) = \begin{cases} x - 1 & \text{if } x > 1 \\ x = 0 & \text{if } x \in [-1, 1] \\ x + 1 & \text{if } x < -1 \end{cases} \qquad (4\text{-}36)$$

A $\pm V$ volt deadband device would be modeled as $y(t) = V\,\mathrm{dead}(x(t)/V)$.

Ideal saturation, sgn, and deadband devices are graphically interpreted in Figure 4-14.

Example 4-10   Nonlinear effects

The linear ODE equation

$$\frac{dy(t)}{dt} = 0.0005y(t) + u(t)$$

$$y(0) = 0$$

is unstable. Simulation can be used to compare the solution $y(t)$ with the solutions of the nonlinear differential equations

$$\frac{dy_c(t)}{dt} = 0.0005(\text{sgn}(y_c(t)) + u(t))$$

(± 1 volt sgn function)

$$y(0) = 0$$

and

$$\frac{dy_s(t)}{dt} = 0.0005(\text{sat}(y_s(t)) + u(t)$$

(± 1 volt sat function)

$$y(0) = 0$$

Computer Study   S-file `ch4_10.s` and M-file `clip.m` simulate the step responses $y(t)$, $y_c(t)$, and $y_s(t)$, using the format `clip(z)`, where z is an input of length < 16 seconds. The test signals are a pulse train and a zero-mean triangular wave with additive zero-mean gaussian noise of length 16 seconds (corresponds to 510 samples of its discrete-time approximation).

SIGLAB	MATLAB
`>include "ch4_10.s"`	`>> x1 = 1 + square((0:511)`
`># pulse train 0<=x(t)<=2`	`*2*pi/32)';`
`>x1=1+sq(64,32/64,510)`	`>> rand('normal');`
`>#triangle and gaussian noise`	`>> x2 = sawtooth((0:511)`
`>x2=tri(510/5,510)+.25*gn(510)`	`*2*pi/32)' +`
`>clip(x1); clip(x2)`	`0.25*rand(512,1);`
	`>> clip(x1);`
	`>> clip(x2);`

The results are shown in Figure 4-15, which display, from top to bottom, the linear, severely clipped, and saturated system outputs. The pulse train pumps energy into the system at regular intervals, thus causing the system to continuously accumulate power and resulting in instability. The clipped and saturated systems produce similar bounded outputs. This is due to the fact that when

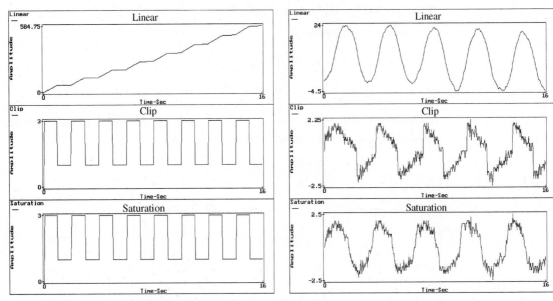

FIGURE 4-15   Simulated responses of linear (top), clipped (middle), and saturated (bottom) nonlinear differential equations, with a pulse train (left), and triangle plus gaussian noise input (right) (may vary due to random signal generator).

an unstable system receives a pulse, its reponse is to immediately produce a high-gain output. During these periods of large amplitude swings, the sat and sgn functions behave in a similar manner.

When the input is numerically generated by a triangular signal with gaussian noise, the output of the unstable system tends to wander about amplifying small differences (biases) from an ideal mean of zero. The saturation response is affected little by the nonlinearity, because the signal generally remains within the linear range of the system. The clipped response, however, is significantly affected by the nonlinearity. In both nonlinear cases, the output will remain bounded at all times.

In the next chapter, the issue of stability will be further investigated. At that time, stability will be discussed as it relates to transforms.

4-8   CONTINUOUS TRAJECTORIES

The response of a system to an arbitrary input stimulus can be analyzed in a graphical format, as well. The traditional method of displaying the solution to $y(t) = (\mathcal{S}x)(t)$ (called *trajectories*) is to graph the solution $y(t)$ versus time. To provide a common means of comparison, system trajectories are generally produced by a commonly agreed upon class of test signals. Next to the impulse response, the step response is by far the most often used to study LTI systems.

An LTI system can also be classified on the basis of the order of its ODE model. The analysis of high-order LTI systems can be mathematically challenging. Instead, systems are often reduced to collections of low-order *subsystems* that can be more easily analyzed and studied. In addition, the trajectories of these low-order subsystems are generally well known. Therefore, predicting the behavior of a high-order system can often be based on understanding the impulse and step responses of its low-order subsystems.

4-8-1   First-Order LTI Systems

The simplest continuous-time system is modeled by a first-order linear differential equation that takes the form

$$\frac{dy(t)}{dt} + a(t)y(t) = b(t)x(t) \tag{4-37}$$

The first-order system defined by Equation 4-37 is sometimes represented as a two-tuple $T_2 = \{a(t), b(t)\}$. A *block diagram* representation of the ODE Equation 4-37 is shown in Figure 4-16. The diagram, also called a *simulation diagram*, displays input, output, and all other relevant information in a block-diagram format.

The homogeneous response to a first-order time-varying system with a system description given by Equation 4-37, is the solution to

$$\frac{dy(t)}{dt} = -a(t)y(t)$$

$$y(0) = y_0 \tag{4-38}$$

which is $y(t)$, and is given by

$$y(t) = y_0 \exp\left(\int_0^t -a(\sigma)\,d\sigma\right) \tag{4-39}$$

FIGURE 4-16   Simulation diagram for a first-order system.

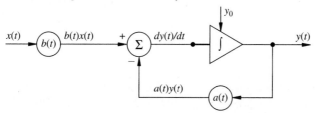

where the term $\exp(\int a(\sigma)\,d\sigma)$ reduces to $\exp(at)$ if $a(t) = a$ (i.e., a constant). The solution, given by Equation 4-39 can be verified by substituting $y(t)$ into Equation 4-38 and showing that it satisfies the differential equation.

The inhomogeneous solution is defined by the convolution integral. The impulse response of a first-order LTI system, where $a(t) = a$ and $b(t) = b$, is

$$h(t) = be^{-at}u(t) \tag{4-40}$$

The impulse response can be seen to be that of a stable BIBO system if $a > 0$. The step response of the same at-rest, first-order LTI system is given by $y(t) = h(t) * u(t)$, which equals

$$y(t) = b \int_{\tau=0}^{t} e^{-a(t-\tau)}d\tau = \frac{b(1 - e^{-at})}{a}u(t) \tag{4-41}$$

where $y(t)$ begins with $y(0) = 0$ and exponentially converges to $y(\infty) \to b/a$, the steady state value of $y(t)$, if $a > 0$.

Example 4-11   First-order system

A first-order LTI system of the form $dy(t)/dt + ay(t) = bx(t)$, for $a > 0$, will be called a *lowpass* filter in Chapter 13. A lowpass filter, having a cut-off frequency of f_0, will pass the signal components over the baseband frequency range $f \in [0, f_0]$ with little or no attenuation, and heavily attenuate signal components outside this range.

Computer Study   The impulse response of a first-order LTI is given by $h(t) = \exp(-at)$, for $b = 1$ and $a = \{0.1k, 0.25k, 0.5k, k, 2k\}$, is to be convolved with a signal $x(t)$. S-file ch4_11.s and M-file first.m perform the convolution of $h(t)$ with a signal, and display the results using the format first(k,x), where x is a maximum length 250 samples and is sampled at rate of $f_s = 25.6$ Hz.

SIGLAB	MATLAB
>include "ch4_11.s"	>> x1 = [1; zeros(100,1)];
>x1=impulse(101); x2=ones(101)	>> first(-1/10, x1)
>first(-1/10,x1);first(-1/10,x2)	>> x2 = ones(100, 1);
	>> first(-1/10, x2)

The results are reported in Figure 4-17. It can be seen that the impulse and step responses have the shapes predicted by Equations 4-40 and 4-41.

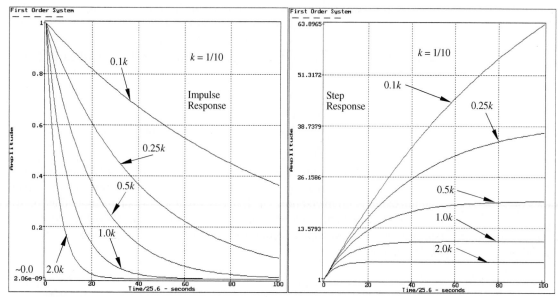

FIGURE 4-17   First-order system impulse and step response.

4-8-2   Second-Order LTI Systems

The next level of sophistication is a linear second-order system, having the form

$$\frac{d^2y(t)}{dt^2} + a(t)\frac{dy(t)}{dt} + b(t)y(t) = c(t)x(t) \tag{4-42}$$

This second-order LTI system is graphically interpreted in Figure 4-18. The system can be represented in three-tuple form as $T_3 = \{a(t), b(t), c(t)\}$. Solving system equations containing time-varying coefficients is generally a difficult problem. Numerical simulations are sometimes used to mitigate this problem. Fortunately, a large number of important physical systems can be satisfactorily represented by a constant coefficient second-order system $T_3 = \{a, b, c\}$. This is indeed important, since many systems are intentionally designed to have two (or more) energy-

FIGURE 4-18   Simulation diagram of a second-order system.

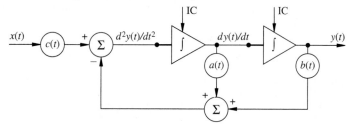

storage subsystems that exchange energy levels in some oscillating manner. These systems can be successfully modeled by second-order ODEs.

A typical step response of a stable, causal second-order system is displayed in Figure 4-19. The step response (i.e., $x(t) = u(t)$) can be quantified in terms of the following parameters:

1 Final (steady state) value $y(100\%) \neq 0$
2 Overshoot $= y_{max} - y(100\%)$
3 Percent overshoot $= (\text{Overshoot}/y(100\%)) \times 100\%$
4 Rise time $= t_2 - t_1$

An *underdamped* solution exhibits a finite overshoot of its steady-state value, then converges to its final value in a damped oscillatory manner. The *overdamped* solution exhibits no overshoot. The trajectory that resides on the boundary between being overdamped and underdamped is *critically damped*. Overdamped, critically damped, and underdamped second-order systems share a common mathematical model, namely

$$\frac{d^2y(t)}{dt^2} + 2\delta\omega_0\frac{dy(t)}{dt} + \omega_0^2 y(t) = \omega_0^2 x(t) \qquad (4\text{-}43)$$

where δ is called the *damping factor* and $\omega_0 > 0$ is called the system's *natural resonant frequency*. The roots of the system's characteristic equation

$$s^2 + 2\delta\omega_0 s + \omega_0^2 = 0 \qquad (4\text{-}44)$$

are

$$s_1, s_2 = -\delta\omega_0 \mp \omega_0\sqrt{\delta^2 - 1} = \omega_0(-\delta \mp \sqrt{\delta^2 - 1}) \qquad (4\text{-}45)$$

The solution to the second-order ODE described in Equation 4-43 is defined by the values of δ and ω_0. Of these two, it is δ that determines if the solution is stable or unstable, and if the impulse response is exponential or oscillatory.

FIGURE 4-19   Typical LTI step response.

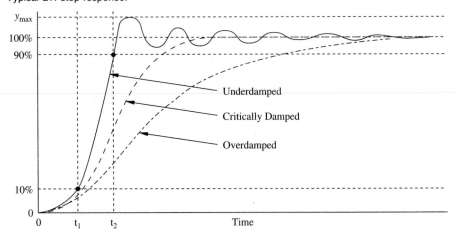

Example 4-12   Second-order systems

The influence of δ on the step response of the basic second-order system, defined by Equation 4-43, can be isolated by setting ω_0 to a fixed value, say $\omega_0 = 1$.

Computer Study   S-file `ch4_12.s` and M-file `second.m` compute and display the impulse response $h(t)$ for $t \in [0, 100]$ of a second-order system, using the format `second(d,x)` for $d = \delta$ and $x(t)$, an input of duration 39 to 78 seconds (100 to 200 samples), using discrete-time approximations. The program also displays the output $y(t) = h(t) * x(t)$ for $t \in [0, 39]$. The following procedure can be used to generate and display the impulse and step responses for $\delta = \{0.25, 0.5, 0.75, 1.0, 1.5, 2.5, 5.0\}$.

SIGLAB	MATLAB
`>include "ch4_12.s"`	`>> u = ones(100,1);`
`>u=ones(100) #39 second`	`>> y25 = second(0.25, u)`
`step`	`>> y50 = second(0.5, u)`
`>y25=second(.25,u);y50`	`>> y75 = second(0.75, u);`
`=second(.5,u)`	`>> y100 = second(1, u)`
`>y75=second(.75,u);y100`	`>> y150 = second(1.5, u)`
`=second(1,u)`	`>> y250 = second (2.5, u);`
`>y150=second(1.5,u);y250`	`>> y500 = second(5, u);`
`=second(2.5,u)`	`>> plot([y25 y50 y75 y100 y150`
`>y500=second(5,u)`	`y250 y500])`
`>ograph(y25,y50,y75,y100,`	`>> grid`
`y150,y250,y500)`	

The results are displayed in Figure 4-20. The impulse response of the underdamped impulse response is seen to exhibit a large amount of oscillation along with some damping. The step responses are seen to range from strongly underdamped to strongly overdamped.

The second-order system defined by Equation 4-43 is BIBO stable if $\delta > 0$ and $\omega > 0$, since $\text{Re}(s_i) < 0$ (from Equation 4-45). If $\delta < 1$, the system is underdamped and the roots s_i appear as complex conjugate pairs. The system's impulse response is given by

$$h(t) = He^{-\sigma t}\sin(\omega t)u(t) \tag{4-46}$$

$$H = \frac{\omega_0}{\sqrt{(1 - \delta^2)}}$$

$$\sigma = \delta\omega_0$$

$$\omega = \omega_0\sqrt{(1 - \delta^2)}$$

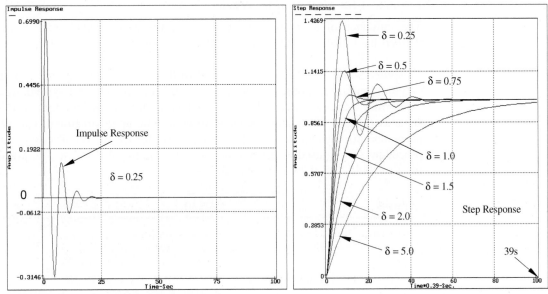

FIGURE 4-20   Simulated responses of a second-order system for $\delta = 0.25$ and an overlay display of the system's step responses.

The solution can be seen to be that of a damped sinusoid. If $\delta = 0$, then $h(t)$ reduces to a pure sinusoid that oscillates at the natural resonant frequency ω_0.

The step response (i.e., $x(t) = u(t)$) of the underdamped system is given by $y(t) = h(t) * u(t)$, where

$$y(t) = u(t) - e^{-\sigma t}\cos(\omega t)u(t) \qquad (4\text{-}47)$$

The step response starts out at $y(0) = 0$ and follows an underdamped trajectory shown in Figure 4-19 (i.e., $0 < \delta < 1.0$ in Figure 4-20), reaching a steady-state value of $y(\infty) = 1.0$.

If $\delta = 1.0$, the system is said to be critically damped and the roots of the characteristic equation are given by $s_1 = s_2 = -\omega_0$ (real) and $\sigma = \omega_0$. The impulse response, in this case, is given by

$$h(t) = \omega_0^2 t e^{-\sigma t} u(t) \qquad (4\text{-}48)$$

and the step response satisfies

$$y(t) = u(t) - e^{-\sigma t}u(t) - \omega_0 t e^{-\sigma t}u(t) \qquad (4\text{-}49)$$

which follows the critically damped (i.e., $\delta = 1.0$) trajectory in Figure 4-20. A direct calculation shows that $y(0) = 0.0$ and $y(\infty) = 1.0$. The critically-damped step response is seen to approach its steady value from below.

If $\delta > 1$, the system is said to be overdamped and the roots of the characteristic equation, s_1 and s_2 (see Equation 4-45) are real, and $s_1 < s_2$. The impulse response is given by

$$h(t) = ae^{s_2t} - ae^{s_1t} \tag{4-50}$$

$$a = \frac{\omega_0^2}{s_2 - s_1} = \frac{\omega_0}{2\sqrt{\delta^2 - 1}} > 0$$

Finally, the step response is given by

$$y(t) = u(t) + \frac{a}{s_1}e^{s_2t}u(t) - \frac{a}{s_2}e^{s_1t}u(t) \tag{4-51}$$

The overdamped step response is displayed in Figure 4-20 for $\delta > 1.0$, where it can be seen that $y(0) = 0$ and $y(\infty) \to 1.0$. Like the critically damped solution, the overdamped response is seen to approach its steady-state value from below. However, it takes longer to achieve a near-final value than the critically damped solution.

Example 4-13   Trajectory analysis

A second-order RLC system studied in Example 4-4, for $L = C = 1$, is defined by the input-output equation $d^2v_c(t)/dt^2 + Rdv_c(t)/dt + v_c(t) = v(t)$. The impulse response is the solution to the at-rest system for $v(t) = \delta(t)$. The impulse response can have one of three possible forms depending on how the characteristic equation $s^2 + Rs + 1 = 0$ factors. This corresponds to the general second-order ODE (Equation 4-43) for $\delta = R/2$ and $\omega_0^2 = 1.0$. The possible roots to the characteristic equation are

Real distinct if $R > 2$

$$s_1 = -\frac{R}{2} - \sqrt{\left(\frac{R}{2}\right)^2 - 1}$$

$$s_2 = -\frac{R}{2} + \sqrt{\left(\frac{R}{2}\right)^2 - 1}$$

Real repeated if $R = 2$

$$s_1 = s_2 = -\frac{R}{2} = -1$$

Complex if $R < 2$

$$s_1 = -\frac{R}{2} - j\sqrt{\left|\left(\frac{R}{2}\right)^2 - 1\right|} = -\sigma - j\omega$$

$$s_2 = -\frac{R}{2} + j\sqrt{\left|\left(\frac{R}{2}\right)^2 - 1\right|} = -\sigma + j\omega$$

which yield the following impulse response models

$$
\begin{cases}
h(t) = (ae^{s_2 t} - ae^{s_1 t})u(t) & \text{if } R > 2 \\
h(t) = te^{s_1 t}u(t) & \text{if } R = 2 \\
h(t) = (be^{s_2 t} - be^{s_1 t})u(t) = be^{-\sigma t}(e^{j\omega t} - e^{-j\omega t})u(t) & \\
\quad = 2jbe^{-\sigma t}\sin(\omega t)u(t) = \left(e^{-\sigma t}\sin(\omega t)\right)\dfrac{u(t)}{\omega} & \text{if } R < 2
\end{cases}
$$

where $a = 1/(s_2 - s_1)$ and $b = 1/j2\mathrm{Im}(s_2) = 1/j2\omega$. The unit step response is given by the convolution integral

$$
y(t) = \int_0^t u(\tau)h(t - \tau)\,d\tau = \int_0^t h(t - \tau)\,d\tau
$$

The three cases translate to the following instances:

Overdamped if $R > 2$

$$
y(t) = \int_0^t (ae^{s_2(t-\tau)} - ae^{s_1(t-\tau)})\,d\tau
$$

$$
= ae^{s_2 t}\int_0^t e^{-s_2 \tau}\,d\tau - ae^{s_1 t}\int_0^t e^{-s_1 \tau}\,d\tau
$$

$$
= u(t) - a\left(\frac{-e^{s_2 t}}{s_1} + \frac{e^{s_1 t}}{s_2}\right)u(t)
$$

$$
= \left(1 + a\left(\frac{e^{s_2 t}}{s_2} - \frac{e^{s_1 t}}{s_1}\right)\right)u(t)
$$

Critically damped if $R = 2$

$$
y(t) = \int_0^t (t - \tau)e^{s_1(t-\tau)}\,d\tau
$$

$$
= u(t) - e^{-t}u(t) - te^{-t}u(t)
$$

Underdamped if $R < 2$

$$
y(t) = \int_0^t \frac{e^{-\sigma(t-2)}}{\omega}\sin(\omega(t - \tau))\,d\tau
$$

$$
= u(t) - \frac{e^{-\sigma t}}{\omega}(\omega\cos(\omega t) + \sigma\sin(\omega t))u(t)
$$

By continuously adjusting R, the solution can be manipulated through all these regimes.

Computer Study   S-file `ch4_13.s` produces overdamped, critically damped, slightly underdamped, and underdamped solutions, computes their overshoot, and displays the output step responses. The format used is `overshoot(n)` where n is the number of samples used in the discrete-time signal approximations ($n \leq 510$). The values of s_1 and s_2 are the roots of the equation $s^2 + Rs + 1$. The values of R to be tested are $R = \{4.0, 2.0, 1.5, 1.0\}$. The first is an overdamped case, the second is critically damped, and the last two are underdamped.

SIGLAB	MATLAB
| ```
>include "ch4_13.s"
>overshoot(250)
y1 steady state value_1.9999
y1 overshoot %_0
y2 steady state value_2
y2 overshoot %_0
y3 steady state value_2
y3 overshoot %_45.8792
y4 steady state value_2
y4 overshoot %_55.3942
``` | Not available in Matlab |

The responses are graphically interpreted in Figure 4-21. It can be seen that the solutions range from being overdamped to underdamped.

FIGURE 4-21   Simulated step response of a second-order LTI.

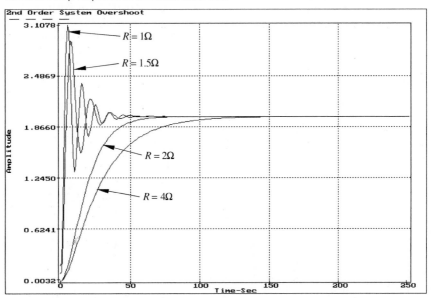
```

Phase-plane Trajectories (Optional)   Another useful trajectory representation technique is called a *phase-plane* display. In the phase-plane, one system variable is plotted against another using time as an implicit variable. The stability of a system can be deduced from a system's trajectory in the phase-plane. Referring to Figure 4-22, a stable LTI system is shown to exhibit a phase-plane trajectory that spirals inward toward the equilibrium solution (assumed to be the origin). If the system is unstable, the trajectories will spiral outward toward infinity or, in some cases, systems will enter a regime of stable oscillations called *limit cycling*. For example, a pure harmonic oscillator would trace out a circular pattern in the phase-plane.

Example 4-14   Phase-plane analysis

The second-order ODE, $dy^2(t)/dt^2 - 2a(dy(t)/dt) + (a^2 + 1)y(t) = 0$, can be expressed as a system of coupled first-order ODEs. By defining $y_1(t) = y(t)$, it follows that

$$\frac{dy_1(t)}{dt} = ay_1(t) + y_2(t)$$

$$\frac{dy_2(t)}{dt} = -y_1(t) + ay_2(t)$$

with initial conditions

$$y_2(0) = y_{20}$$

$$y_1(0) = y_{10}$$

The homogeneous solutions are given by

$$y_1(t) = \exp(at)(\cos(t)y_{10} - \sin(t)y_{20})$$

$$y_2(t) = \exp(at)(-\sin(t)y_{10} - \cos(t)y_{20})$$

FIGURE 4-22   Phase-plane trajectories.

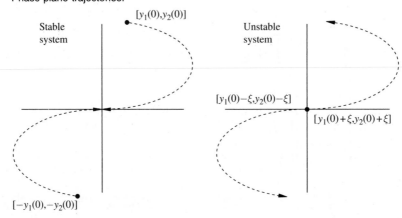

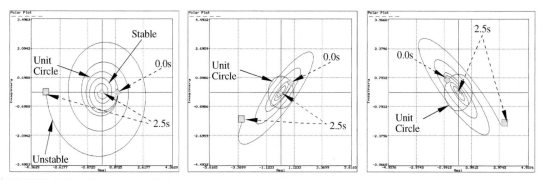

FIGURE 4-23   Phase-plane display of a second-order ODE.

which can be verified by substituting $y_1(t) = y(t)$ into the original ODE. The phase-plane plot is given by $y_1(t)$ versus $y_2(t)$, using time as an implicit variable.

Computer Study   S-file `ch4_14.s` and M-file `phasepla.m` produce the phase-plane plot for both $\pm a$ ($a > 0$ (unstable), $a < 0$ (stable)), using the formats `phaseplane(a,y10,y20)` and `phasepla(a,y10,y20)` respectively, where $y10$ and $y20$ are the initial conditions. The initial conditions are chosen to reside on the periphery of the unit circle at angles of $\{0, \pi/6, \pi - \pi/6\}$ radians.

SIGLAB	MATLAB
```>include "ch4_14.s"```	```>> phasepla(.5,cos(0),```
```>phaseplane(.5,cos(0),```	```sin(0))```
```sin(0))```	```>> phasepla(.5,cos(pi/6),```
```>phaseplane(.5,cos(pi/6),```	```sin(pi/6))```
```sin(pi/6))```	```>> phasepla(.5,-cos(pi/6),```
```>phaseplane(.5,cos(-pi/6),```	```sin(pi/6))```
```ttcode [-1pt] sin(pi/6))```	

The results are displayed in Figure 4-23 for $a = 0.5$ and the specified initial conditions. The stable solution is seen to spiral inward toward an equilibrium solution located at the origin. The response of the unstable system is seen to spiral outwards.

4-9   CONTINUOUS-TIME ANALOGS

Second-order ODEs often occur in the study of electrical, mechanical, hydraulic, and thermodynamic systems. Mathematical models, called *analogs* or *analogies*, can be used to establish a link between a system studied in one discipline and that studied in another. The two systems shown in Figure 4-24, for example,

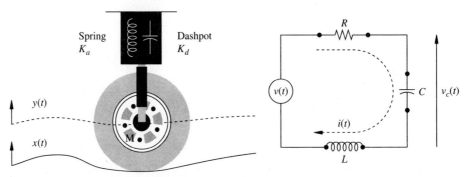

FIGURE 4-24   Automobile suspension system and electrical analog.

are mathematically equivalent at the differential equation level, but are distinctly different physical manifestations of that equation. The analog can establish a relationship between the mechanical and electrical parameters using the substitution rule $M \sim L$, $K_d \sim R$, and $K_s \sim 1/C$. The implication is that a mechanical circuit can be used to model an electrical circuit and vice versa. These observations were very important several decades ago when analog computers were routinely used to simulate systems based on electrical analogies. Since then, the digital computer has become the principal simulation tool and analog computers and analogies have decreased in importance.

Example 4-15   Mechanical analogs

The RLC circuit studied in Example 4-6 can also be used to model a mechanical automotive suspension system shown in Figure 4-24, since their differential equations are the same (except for coefficients). The RLC circuit is called an analog of the mechanical system. The electrical system has an input that is a voltage $v(t)$, whereas the mechanical system's input is the road surface. The electrical system's output is measured across the capacitor while the mechanical system's output is the movement of the axle relative to the car's body.

Computer Study   A car is tested using simulation on several road surfaces. The car is first tested using a simple curb or block obstacle. The second test is conducted on the streets of San Francisco, producing interesting oscillations when driving at extremely high speeds. The third test is in Cincinnati, reported to have the the most exciting potholes in North America. The results of the tests presented using S-file ch4_15.s are displayed in Figure 4-25. The test results show that when the block obstacle is met, the wheel follows with a lagging, but rapid response. It exhibits some overshoot of the top of the curb before coming to rest. The San Francisco experiment produces a lagging sinusoidal

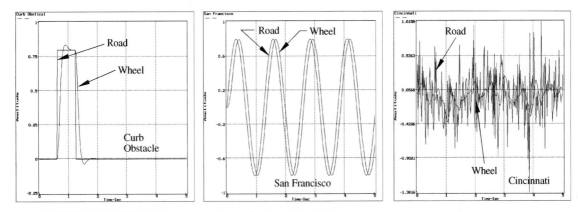

FIGURE 4-25   Automotive suspension system field tests using a curb obstacle (left), streets of San Francisco (middle), and the potholes of Cincinnati (right).

ride. In Cincinnati, the car attempts to do its best, but is seen to be hopelessly outclassed by the potholes.[4]

SIGLAB	MATLAB
>include "ch4-15.s"	Not available in Matlab

4-10   SUMMARY

In Chapter 4 systems were introduced. The study focused on linear time invariant (LTI) models, which form the foundation for the study of most engineering systems. LTI at-rest systems were shown to interact with signals through a process called convolution. Convolution was defined in terms of a given input signal and the system's impulse response. In the next chapter, transform-domain methods will be used to convert the complicated time-domain implementation of a convolution integral as a set of simple algebraic operations.

Stability was also studied in the context of a bounded-input bounded-output (BIBO) criteria. The stability of an LTI system was seen to be established by the eigenvalues of the system's characteristic equation. In the next chapter, this concept will be extended using transform-domain methods.

The fundamental building blocks of many LTI systems were defined to be first- and second-order subsystems. Manufactured systems, such as filters, are often defined in terms of Nth order linear differential equations. These high-order systems are often factored into lower order systems, which are then collected together and used to reconstruct the original Nth order system. The factoring process will be developed in detail in the next two chapters. At that time, the factors of the system

[4]For users of MONARCH, the model of an automotive suspension system has been saved under the name "car".

will be shown to be the roots of the system's characteristic equation. It has been previously noted that these roots are either real or complex.

4-11   COMPUTER FILES

These computer files were used in Chapter 4.

File Name	Type	Location
Subdirectory CHAP4		
ch4_1.x	C	Example 4-1
ch4_5.s	C	Example 4-5
superpos.m	C	Example 4-5
ch4_6.x	C	Example 4-6
square.m	C	Example 4-6
ch4_7.s	C	Example 4-7
gconv.m	C	Example 4-7
triang.m	C	Example 4-7
ch4_8.s	C	Example 4-8
convprop.m	C	Example 4-8
ch4_9.s	C	Example 4-9
firstode.m	C	Example 4-9
ch4_10.s	C	Example 4-10
clip.m	C	Example 4-5
ch4_11.s	C	Example 4-11
first.m	C	Example 4-11
ch4_12.s	C	Example 4-12
second.m	C	Example 4-12
ch4_13.s	C	Example 4-13
ch4_14.s	C	Example 4-14
phasepla.m	C	Example 4-14
ch4_15.s	T	Example 4-15
Subdirectory SFILES or MATLAB		
Subdirectory SIGNALS		
EKG1.imp	D	Examples 4-6, 4-8
Subdirectory SYSTEMS		
system.arc	D	Example 4-5
system-5.arc	D	Examples 4-6, 4-8
car.arc	D	Example 4-15

Here, "s" denotes S-file, "m" an M-file, "x" either a S-file or M-file, "T" a tutorial program, "D" a data file, and "C" denotes a computer-based-instruction (CBI) program. CBI programs can be altered and parameterized by the user.

4-12 PROBLEMS

4-1 Superposition principle Prove that an at-rest linear time-varying system, given by

$$\frac{d^n y(t)}{dt^n} = \sum_{k=0}^{N-1} a_k(t) \frac{d^k y(t)}{dt^k} + \sum_{k=0}^{M} b_k(t) \frac{d^k x(t)}{dt^k}$$

satisfies the superposition principle.

4-2 Eigenfunctions Equation 4-13 established a relationship between an LTI system's characteristic equation and the eigenfunctions. Suppose $d^3 y(t)/dt^3 = x(t)$, $y(0) = y_0 = a$, $dy(0)/dt = y_1 = b$, and $d^2 y(0)/dt^2 = y_2 = c$. Compute the eigenfunctions and show they satisfy the system's characteristic equation.

4-3 Eigenfunctions Suppose $d^2 y(t)/dt^2 + 2dy(t)/dt + y(t) = x(t)$, $y(0) = y_0 = a$ and $dy(0)/dt = y_1 = b$. Compute the eigenfunctions and show they satisfy the system's characteristic equation. What is the system's at-rest impulse response in terms of the computed eigenfunctions?

4-4 Eigenfunctions Suppose $d^2 y(t)/dt^2 + a(dy(t)/dt) + y(t) = x(t) + dx(t)/dt$. Compute the eigenfunctions as a function of the real parameter "a" where $a > 0$. What is the range of "a" that will guarantee that all the eigenfunctions will monotonically decrease to zero as $t \to \infty$? What is the range of "a" that will guarantee that the magnitude of the eigenfunctions will monotonically decrease to zero as $t \to \infty$?

4-5 LTIs Show that if S is an at-rest LTI system after Equation 4-10, then if inputs $x(t)$ and $s(t)$ produce outputs $y(t)$ and $r(t)$ respectively, and if $x(t) = s(t)$, it follows that $y(t) = r(t)$.

4-6 Graphical convolution Pairwise graphically convolve the signals shown in Figure 4-26.

4-7 Graphical convolution Graphically convolve $h(t)$ and $x(t)$ to form $y(t) = h(t) * x(t)$, where

$$h(t) = \cos(t) \quad \text{for } t \in [0, 2\pi], 0 \text{ otherwise}$$

$$x(t) = \sin(t) \quad \text{for } t \in [0, 2\pi], 0 \text{ otherwise}$$

4-8 Convolution Suppose $x(t) = \exp(-t)u(t)$, and $h(t) = (t/T)u(t)$ for $t \in [0, T]$ and 0 otherwise. Show that for $t \leq T$

$$y(t) = x(t) * h(t) = (1/T)(\exp(-t) + t - 1)$$

and for $t > T$

$$y(t) = x(t) * h(t) = (1/T)(\exp(-t) + \exp(-T)(t - T - 1))$$

Sketch the solution for $T = 1$. How does the choice of T affect the final value of $y(\infty)$?

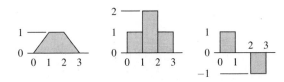

FIGURE 4-26 Signals for Problem 4-6.

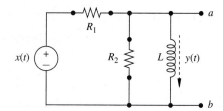

FIGURE 4-27   RL circuit for Problem 4-10 where $y(t)$ is the inductor current.

4-9 RC circuit   An RC circuit was studied in Example 4-1. Suppose $RC = 10^{-2}$ and the system is at-rest. What is the output $y(t)$ if $v(t) = u(t) - u(t - 10^{-2})$?

4-10 RL circuit   Verify that the RL circuit shown in Figure 4-27 can be modeled by

$$dy(t)/dt + (R_1 R_2)/(L(R_1 + R_2))y(t) = R_2 x(t)/(L(R_1 + R_2))$$

If $y(0) = 1$ and $x(t) = 0$, show that the system does not satisfy the superposition principle. In this case, what is $y(t)$? If $y(0) = 0$, show that the superposition principle is satisfied. Compute $h(t)$ and $y(t)$.

4-11 Step response   Using the LTI system defined in Example 4-6, compute and experimentally verify the system's step response. Measure the rise-time and overshoot of the system's step response. Derive a formula to determine the time at which the maximal overshoot will occur, and verify.

4-12 Second-order system   A second-order system has a step response given by $y(t) = (e^{-t} + te^{-t} - 1)u(t)$. Verify that the system's impulse response is $h(t) = -te^{-t}u(t)$. What would the system response be to $x(t) = \sin(\omega_0 t)u(t)$? What is the steady value of $y(t)$ as a function of ω_0?

4-13 Full-wave rectification   In Example 3-6, a full-wave AC rectifier circuit was presented. Assume an ideal rectification network (i.e., passes only the DC component) is connected between points (a) and (b) in Figure 4-27 and that it presents no load to the RL circuit. If $x(t) = \sin(\omega_0 t)$, $\omega_0 = 69.59$ r/s, what is the response of the RL circuit measured as $v_{ab}(t)$? What is the response of the system measured at the output of the rectifier as $t \to \infty$?

4-14 Stability, causality, and inhomogeneous response   Determine the stability, causality, and output response for all signal and impulse-response pairs (assume systems are at rest):

$$h_1(t) = u(t) - u(t - 1)$$

$$h_2(t) = \exp(at)u(t), a < 0$$

$$h_3(t) = \exp(at)u(-t), a < 0$$

$$x_1(t) = \delta(t)$$

$$x_2(t) = u(t)$$

$$x_3(t) = \cos(\omega_0 t)$$

$$x_4(t) = \cos(\omega_0 t)u(t)$$

4-15 Stability   Suppose $d^2y(t)/dt^2 + a(dy(t)/dt) + 2y(t) = x(t) + dx(t)/dt$ models an at-rest system. Compute the range of "a" that guarantees the system to be stable. What effect would nonzero initial conditions have on your answer?

4-16 Dominant eigenvalues   If the second-order system given by Equation 4-43 was evaluated at $\delta = 0.9, 1.0, 1.1$, and $\omega_0 = 1.0$, what is the dominant eigenvalue and eigenfunction? Experimentally show dominance.

4-17 Phase-plane trajectories   Using the study conducted in Example 4-14, initialize the system to $y_{10} = \pm y_{20}$ and produce the system's phase-plane trajectories for $a = -0.9$ and -0.5. Mathematically explain what you see. Initialize the system so that $y_{10} = y_{20} \ne 0$ and let $a \to 0$. What effect are you simulating?

4-18 Inverse systems   Two LTI systems, denoted S_1 and S_2, are said to be inverses of each other if

$$h_1(t) * h_2(t) = \delta(t)$$

If so, system S_1 is said to be invertible.

Suppose S_1 is given by $dy(t)/dt + ay(t) = dx(t)/dt + bx(t)$. What is S_2?

4-19 Inverse systems   Invertibility is often stated in terms of input-output behavior. Suppose a system S_1, having an input $x(t)$ and output $y(t)$ is invertible. Then there exists an inverse system S_2, such that S_2 uniquely (perfectly) reconstructs $x(t)$ from $y(t)$. Which of the following systems have inverses, and if so, what are they?

$$y(t) = \exp(jx(t))$$

$$y(t) = \int_{-\infty}^{t} x(\tau)d\tau, \ y(-\infty) = 0$$

$$y(t) = x(t+1)$$

$$y(t) = 1/x(t)$$

4-20 Automobile analog   Return to the suspension system shown in Figure 4-24. Using the RLC analog for $C = 1.0$, $L = 2.0$, and $R = 3.0$, derive the axle's step response. Verify your claim using simulation.

4-21 Echo cancellation   A signal $x(t)$ is transmitted along a telephone channel with $AL = 2.0$, and $R = 3.0$, an impulse response $h(t) = \exp(-10^3 t)u(t)$. The channel also carries an echo of $x(t)$ along a path having an impulse response $h_{echo}(t) = 0.2\exp(-10^3(t - 10^{-3}))u(t - 10^{-3})$ as shown in Figure 4.28. Sketch a simulation diagram for the system. What is the output $y(t) = (h(t) + h_{echo}(t)) * x(t)$? Suppose $x(t) = \cos(2\pi 10^3 t)$. What is $y(t)$? Sketch $y(t)$. Can a system be found to remove the echo at the telephone and, if so, what is it?

FIGURE 4-28   Telephone echo cancellation.

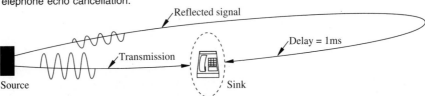

4-13 COMPUTER PROJECTS

4-1 Space rendezvous   Orbit transfers are often required to rendezvous with several spacebourne vehicles. In the terminal maneuver, the differences between two space-craft can be on the order of 1 to 10 nautical miles, with velocity differences on the order of 100 ft/sec. Assuming circular orbits, the equations of motion are assumed to be

$$\frac{d^2x(t)}{dt^2} - 2w\frac{dy(t)}{dt} = F_x$$

$$\frac{d^2y(t)}{dt^2} + 2w\frac{dx(t)}{dt} - 3w^2y(t) = F_y$$

$$\frac{d^2z(t)}{dt^2} + w^2\frac{dz(t)}{dt} = F_z$$

where x, y, and z are the cartesian coordinates of the controlled spacecraft in a target-centered coordinated system, $w^2 = g/r^3$, g is the Earth's gravitational constant, r is the radius of the earth-centered circular rotation of the target, and F_x, F_y, and F_z are relative thrust accelerations between the two space vehicles.

Consider using the first two equations of motion to study the rendezvous problem in a two-dimensional space (the z-axis will be assumed to be out-of-plane). Assume, for the sake of simplification, that $w = 0.707$ and that the spacecraft are aligned in the y-axis with a unit x-axis separation. The differential velocities in the x- and y-axis directions will be assumed to be initially zero. Assuming that you can use one second unit-strength thrusting bursts with nine-second separation (in any or all directions), find an initial thrust policy for closing on the target and reaching it with minimum differential velocity. What would you do if you had 10 or 50 seconds to actively maneuver?

4-2 Inverted pendulum   An inverted pendulum is shown in Figure 4-29 (also see Computer Project 6-2). It consists of a stick with mass m, supported by a cart on mass M through a hinge. The stick motion is constrained to be in one plane (i.e., y direction). Applying Newton's laws, we obtain

$$I\frac{d^2\phi(t)}{dt^2} = VL\sin(\phi(t)) - HL\cos(\phi(t))$$

$$m\frac{d^2(L\cos(\phi(t)))}{dt^2} = -mg + V$$

$$m\frac{d^2(y + L\sin(\phi(t)))}{dt^2} = H$$

$$M\frac{d^2y}{dt^2} = u - H$$

where $I = mL^2/3$, the moment of inertia of the stick about its center, L is one-half the length of the stick, V is the vertical force in the upward direction exerted by the cart on the stick, and H is the horizontal force.

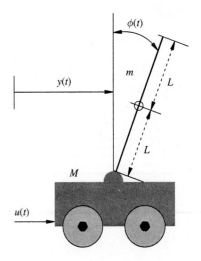

FIGURE 4-29   Inverted pendulum.

Assuming only small angle excursions for ϕ, such that $\sin(\phi) \approx \phi$ and $\cos(\phi) \approx 1$, the equations of motion can be approximated by

$$\frac{dx_1(t)}{dt} = x_2(t)$$

$$\frac{dx_2(t)}{dt} = a_{21}x_1(t) + b_2u(t)$$

$$\frac{dx_3(t)}{dt} = x_4(t)$$

$$\frac{dx_4(t)}{dt} = a_{41}x_1(t) + b_4u(t)$$

where $x_1(t) = \phi(t)$, $x_2(t) = d\phi(t)/dt$, $x_3(t) = y(t)$, and $x_4(t) = dy(t)/dt$. Also $a_{21} = 3g(m+M)/L)(m+4M)$, $a_{41} = -3mg/(m+4M)$, $b_2 = -3/L(m+4M)$, and $b_4 = 4/(m+4M)$. Using $g = M = 4m = 4L = 32.2$, is the at-rest system stable? If the system is at $x_1(0) = 0.1$ initially, and all other initial conditions are zero, what would be the pendulum's response to $u(t) = 0.1\sin(\omega_0 t)$ for $\omega_0 = 0.1$, 1 and 10? Devise a control policy using $u(t) = \pm 1$, which will tend to keep the stick vertical. Verify your results experimentally.

5

TRANSFORM DOMAIN REPRESENTATION OF CONTINUOUS-TIME SYSTEMS

Everything should be made as simple as possible, but not simpler.

—Albert Einstein

5-1 INTRODUCTION

In the previous chapter, continuous-time LTI systems in the time-domain were studied. The response of an LTI system was defined in terms of a homogeneous solution to an ODE, impulse response, and convolution integral. BIBO stability was defined on the basis of time-domain computations with predictions based on knowledge of the system impulse response. Unfortunately, this type of analysis can become computationally intensive whenever the system under study is of high order. LTI systems, however, can also be analyzed efficiently in a transform-domain. This is a logical extension of the Laplace transform representation methods developed in Chapter 2, and the Fourier transform techniques introduced in Chapter 3. By using transform methods, the computational problems introduced by performing complicated time-domain calculations can be replaced by simpler algebraic operations.

5-2 CONTINUOUS-TIME SYSTEMS

Linear constant-coefficient, ordinary differential equations (ODE) are often used to model continuous-time LTI systems (e.g., electronic, electromagnetic, acoustic, mechanical, and optical systems). The response of an at-rest, causal LTI system having an impulse response $h(t)$ to a causal input signal $x(t)$, is defined by the

158

convolution process $y(t) = h(t) * x(t)$. As previously noted, the production of a system's impulse response and its convolution with an arbitrary input can be a complicated mathematical exercise. It has also been previously established that continuous-time signals can be analyzed and manipulated in a transform domain. Since both $x(t)$ and $h(t)$ can be considered signals, it is logical to interpret convolution in a transform-domain. Assume that the system's impulse response $h(t)$ is Laplace transformable and is $H(s)$. Assume also that a causal input signal $x(t)$ is also Laplace transformable and is $X(s)$. Formally taking the Laplace transform of the convolution $y(t) = x(t) * h(t)$, yields

$$Y(s) = \mathcal{L}(y(t)) = \mathcal{L}(x(t) * h(t)) = \mathcal{L}\left(\int_0^\infty x(\tau)h(t - \tau)\,d\tau\right)$$

$$= \int_0^\infty \left[\int_0^\infty x(\tau)h(t - \tau)\,d\tau\right]e^{-st}\,dt$$

$$= \int_0^\infty x(\tau)\left[\int_0^\infty h(t - \tau)e^{-st}\,dt\right]d\tau \qquad (5\text{-}1)$$

By making a substitution of variables $\lambda = t - \tau$, it follows that

$$Y(s) = \int_0^\infty x(\tau)e^{-s\tau}\left[\int_0^\infty h(\lambda)e^{-s\lambda}\,d\lambda\right]d\tau = \int_0^\infty x(\tau)e^{-s\tau}H(s)\,d\tau$$

$$= H(s)\int_0^\infty x(\tau)e^{-s\tau}\,d\tau = H(s)X(s) \qquad (5\text{-}2)$$

That is, the convolution of $h(t)$ and $x(t)$ is mathematically equivalent to the multiplication of their respective Laplace transforms in the s-domain. Finally, to produce the desired output response $y(t)$, simply compute

$$y(t) = \mathcal{L}^{-1}[Y(s)] \qquad (5\text{-}3)$$

Equations 5-2 and 5-3 define the celebrated *convolution theorem of Laplace transforms*, or simply the *convolution theorem*.

Convolution Theorem   The output to a continuous-time at-rest LTI system having an impulse response $h(t)$, to an input signal $x(t)$, having Laplace transforms $H(s)$ and $X(s)$, respectively, is $y(t)$, where

$$Y(s) = X(s)H(s) \qquad (5\text{-}4)$$

If the region of convergence for $X(s)$ and $H(s)$ is $\mathcal{R}_x$ and $\mathcal{R}_h$ respectively, then the *region of convergence* for $Y(s)$ is given by $\mathcal{R}_y$ where

$$\mathcal{R}_y \supset \mathcal{R}_x \bigcap \mathcal{R}_h \qquad (5\text{-}5)$$

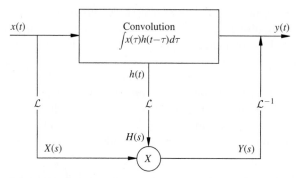

FIGURE 5-1 Convolution theorem.

The convolution theorem is graphically interpreted in Figure 5-1. The principal justification for analyzing a system in the transform domain is a mathematical simplification of the convolution process. It may seem, at first glance, that replacing the convolution integral with a collection of Laplace transforms may be counter-productive. However, remember that a potentially complicated integration is being exchanged for a set of usually straightforward algebraic operations. Since algebra is generally considered to be the simpler of these tasks, the Laplace transform path is considered more efficient for manual computations. The exception would be the case where $h(t)$ or $x(t)$ do not have a directly recognizable or computable Laplace transforms.

Referring again to Equation 5-2, note that $Y(s) = H(s)X(s)$. It therefore follows that if $x(t) = \delta(t)$, then $X(s) = 1$ and $Y(s) = H(s)$. Thus $H(s)$ is the Laplace transform of the system's impulse response and, of course, the inverse Laplace transform of $H(s)$ is the system's impulse response. From another point of view, $H(s) = Y(s)/X(s)$ describes how the transform of input and output are related to each other. As a result, the function $H(s)$ is also referred to by its popular name, *transfer function*. As the name suggests, the transfer function represents how the transform of the input is transferred to the output.

Consider the procedure required to compute the output of a continuous-time linear system using manual calculations. Suppose a continuous-time, at-rest LTI system is modeled by the ODE

$$\frac{d^N y(t)}{dt^N} + a_{N-1}\frac{d^{N-1}y(t)}{dt^{N-1}} + \cdots + a_1\frac{dy(t)}{dt} + a_0 y(t)$$

$$= b_N\frac{d^N x(t)}{dt^N} + b_{N-1}\frac{d^{N-1}x(t)}{dt^{N-1}} + \cdots + b_1\frac{dx(t)}{dt} + b_0 x(t) \tag{5-6}$$

Implementing the top path of Figure 5-1 formally requires that the ordinary differential equation (Equation 5-6) be solved to produce the impulse response $h(t)$. Then $h(t)$ is convolved (integrated) with $x(t)$ to create $y(t)$. The bottom path is assumed to be implemented using tables to produce $X(s)$ and $H(s)$. The transfer function $H(s)$ can be defined in terms of the Laplace transform of Equation 5-6 and is

$$s^N Y(s) + a_{N-1}s^{N-1}Y(s) + \cdots + a_1 s Y(s) + a_0 Y(s)$$
$$= b_N s^N X(s) + b_{N-1}s^{N-1}X(s) + \cdots + b_1 s X(s) + b_0 X(s) \qquad (5\text{-}7)$$

which yields

$$H(s) = \frac{Y(s)}{X(s)} = \frac{b_N s^N + b_{N-1}s^{N-1} + \cdots + b_1 s + b_0}{s^N + a_{N-1}s^{N-1} + \cdots + a_1 s + a_0} \qquad (5\text{-}8)$$

The multiplication of the two polynomials $X(s)$ and $H(s)$ is straightforward and leaves $Y(s)$. If $Y(s)$ can be mapped back into $y(t)$ efficiently, then again it should be apparent that the bottom path is computationally more efficient. The study of Laplace transforms and their inverse, presented in Chapter 2, provides encouragement that the inversion of $Y(s)$ can be efficiently implemented. This, in fact, will be shown to be the case later in this chapter.

Example 5-1   RLC circuit

The response of a series RLC circuit studied in Example 4-6 satisfies the defining continuous-time ODE equation $d^2 v_c(t)/dt^2 + (R/L)dv_c(t)/dt + (1/LC)v_c(t) = (1/LC)v(t)$. If the system is at rest, the Laplace transform representation of the RLC system output is given by

$$s^2 V_c(s) + \frac{R}{L}s V_c(s) + \frac{1}{LC}V_c(s) = \frac{1}{LC}V(s)$$

or

$$V_c(s) = \frac{1}{LC}\frac{1}{s^2 + (R/L)s + (1/LC)}V(s) = H(s)V(s)$$

The zeros of the denominator of $H(s) = 1/[LC(s^2 + (R/L)s + (1/LC))]$, namely $LC(s^2 + (R/L)s + (1/LC))$, are the same as the roots of the characteristic equation of the original ODE. The values of s that satisfy $s^2 + (R/L)s + (1/LC) = LCs^2 + RCs + 1.0 = 0.0$ are

$$s_1 = -\frac{R}{2L} - \sqrt{\left(\frac{R}{2L}\right)^2 - \frac{1}{LC}}$$

$$s_2 = -\frac{R}{2L} + \sqrt{\left(\frac{R}{2L}\right)^2 - \frac{1}{LC}}$$

If $(R/2L)^2 - (1/LC) > 0$, then both roots are real. If $(R/2L)^2 - (1/LC) = 0$, both roots are real, but repeated. Finally, if $(R/2L)^2 - (1/LC) < 0$, then the roots appear as complex conjugate pairs.

For the purpose of numerical analysis, suppose the transfer function $H(s) = b/((s+a)(s+2a))$, where a and b are real. Using techniques developed in

Chapter 2 to invert $H(s)$, it can be shown that $h(t) = [(b/a)\exp(-at) - (b/a)\exp(-2at)]u(t)$. If the system's input is a step function, then $V(s) = U(s) = 1/s$ and, from the convolution theorem, it follows that the Laplace transform of the output is given by $V_c(s) = H(s)V(s) = b/(s(s+a)(s+2a))$. Upon inverse transforming $V_c(s)$ (using methods developed later in this chapter), $v_c(t) = \mathcal{L}^{-1}[V_c(s)] = [(b/2a^2) - (b/a^2)\exp(-at) + (b/2a^2)\exp(-2at)]u(t) = [(b/a^2)[1/2 - \exp(-at) + 1/2\exp(-2at)]]u(t)$ is obtained. Notice that the solution consists of a constant plus two decaying exponential terms. In addition, it can be seen that $y(0) = 0.0$ and $y(\infty) \to (b/2a^2)$ if $a > 0$.

Computer Study   Assume $a = b = 1.0$. Simulation will be used to verify that convolution of $h(t) = \exp(-t) - \exp(-2t)$, with a unit step, produces the result predicted by the inverse Laplace transform of $Y(s)$, namely $y(t) = 0.5 - \exp(-t) + 0.5\exp(-2t)$, simulation will be used. The study is performed using S-file `ch5_1.s` or M-file `ch5_1.m` over 10 seconds of output history. The program compares the system responses computed, using the convolution integral and convolution theorem.

SIGLAB	MATLAB
>include "ch5_1.s"	>> ch5_1

The results are shown in Figure 5-2. The step-response solution consists of two components. The initial phase is the *transient response* and represents that

FIGURE 5-2   RLC response derived from time and frequency-domain methods and their overlay.

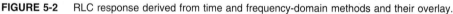

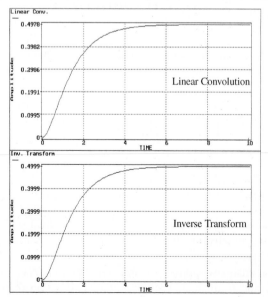
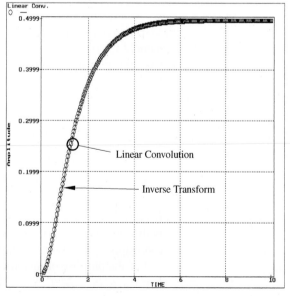

part of the solution history undergoing rapid dynamic change. The second part of the solution is the *steady-state response* and represents an epoch in which the initial transient response has died away, leaving only a periodic pattern (in this case, a constant).

5-3 TRANSFER FUNCTION

The transfer function is a concept and mathematical tool of fundamental importance to the study of linear systems. A transfer function encapsulates sufficient information about an at-rest LTI system to predict its behavior under forced-signal conditions.

A transfer function of a continuous-time LTI system will be assumed to be the rational function given by (also see Equation 5-8)

$$H(s) = \frac{b_N s^N + b_{N-1} s^{N-1} + \cdots + b_1 s + b_0}{s^N + a_{N-1} s^{N-1} + \cdots + a_1 s + a_0} = \frac{\sum_{i=0}^{M} b_i s^i}{\sum_{i=0}^{N} a_i s^i} \qquad (5-9)$$

A polynomial $\sum_{i=0}^{N} c_i p^i$ is said to be *monic* if its leading coefficient $c_N = 1$. If $a_N = 1$ in Equation 5-9, then the transfer function $H(s)$ is said to be monic. The transfer function, given by Equation 5-9, can be interpreted as a simulation diagram, as well. One simulation diagram interpretation of $H(s)$ consists of an interconnected network of integrators (i.e., $1/s$), multipliers, and summers as shown in Figure 5-3. Other simulation diagrams are possible and will be developed in the next chapter. The coefficients $\{b_i\}$ found in Figure 5-3 are said to populate the *feedforward* paths while coefficients $\{a_i\}$ belong to *feedback* paths. Feedforward paths are nonclosed, or unidirectional pathways between the input and output, along which information flows. Feedback paths, however, are closed loops and continue to recirculate information within that path.

FIGURE 5-3   Simulation diagram representation of $H(s)$.

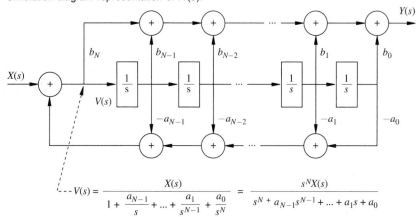

The model, defined by Equation 5-9, presumes that $M < N$. If this is so, $H(s)$ is said to be strictly proper. If $M = N$, the system is said to be proper. If $M \geq N$, $H(s)$ will be assumed to be factored into two distinct polynomials, namely

$$H(s) = \frac{N(s)}{D(s)} = Q(s) + \frac{R(s)}{D(s)} \qquad (5\text{-}10)$$

where $Q(s)$, called the *quotient polynomial*, is of degree $(M - N)$. The polynomial $R(s)$ is called the *remainder polynomial* and is of degree (K), $0 \leq K \leq (N - 1)$, while $D(s)$ is of degree N. Therefore, $R(s)/D(s)$ is strictly proper. The quotient polynomial can be computed using long division and has the form

$$Q(s) = \alpha_k s^k + \cdots + \alpha_1 s + \alpha_0 \quad \text{where } k = M - N \qquad (5\text{-}11)$$

Referring to Tables 2-1 and 2-2, it follows that the inverse Laplace transform of $H(s) = Q(s) + R(s)/D(s)$ should be

$$h(t) = \alpha_k \frac{d^n \delta(t)}{dt^n} + \cdots + \alpha_1 \frac{d\delta_D(t)}{dt} + \alpha_0 \delta(t) + \mathcal{L}^{-1} \left[\frac{R(s)}{D(s)} \right] \qquad (5\text{-}12)$$

Proper systems (i.e., $N \leq M$), it should be noted, possibly have a nonzero value for a_0. Such systems often exist in the real-world as evidenced by a system consisting of a piece of wire that simply connects the input to the output (i.e., $H(s) = 1$). For a proper system, $H(s) = \alpha_0 + R(s)/D(s)$ and since the $\deg(R(s)) < \deg(D(s))$, it follows that

$$\alpha_0 = \lim_{s \to \infty} H(s) \qquad (5\text{-}13)$$

Finally, if $M > N$, then the output will contain a linear combination of differentiated Dirac impulse distributions.

5-4   PARTIAL FRACTION EXPANSION

Recall that Equation 5-9 represented a transfer function expressed as the ratio of two polynomials. The numerator and denominator polynomials, namely $N(s)$ and $D(s)$, can be individually factored. The factors of $N(s) = 0$ are the system *zeros* and the factors that satisfy $D(s) = 0$ are the system *poles*. In particular

$$H(s) = \frac{Y(s)}{X(s)} = \frac{N(s)}{D(s)} = K \frac{\prod_{i=1}^{M} (s - z_i)}{\prod_{i=1}^{N} (s - p_i)} \qquad (5\text{-}14)$$

where z_i are the zeros and p_i the poles of $H(s)$. Observe that the polynomial

$D(s)$ can be represented in a more compact form as

$$D(s) = \prod_{i=1}^{L}(s - p_i)^{n(i)} \qquad \text{where } N = \sum_{i=1}^{L} n(i) \qquad (5\text{-}15)$$

where the pole p_i appears with *multiplicity* $n(i)$. If $n(i) > 1$, then p_i is said to be a *repeated* pole, otherwise if $n(i) = 1$, p_i is said to be *distinct*.

The transfer function $H(s)$ can also be expressed in *partial fraction expansion,* or *Heaviside expansion* form as introduced in Chapter 2. A strictly proper $H(s)$ can be represented as

$$H(s) = \sum_{j=1}^{L} \sum_{i=1}^{n(j)} \frac{\alpha_{j,i}}{(s - p_j)^i} \qquad (5\text{-}16)$$

The production rule for the coefficients $\alpha_{j,i}$ was derived in Chapter 2 and is

$$\alpha_{j,n(j)} = \lim_{s \to p_j} H(s)(s - p_j)^{n(j)} \qquad (5\text{-}17)$$

If p_j is a distinct pole of $D(s)$ (i.e., $n(j) = 1$), then upon computing $\alpha_{j,1}$, one would proceed to compute the next Heaviside coefficient $a_{j+1,1}$, and so forth. If p_j is a repeated pole, the coefficient $a_{j,n(j)-k}$ is computed using the rule

$$\alpha_{j,n(j)-1} = \lim_{s \to p_j} \frac{d}{ds}(s - p_j)^{n(j)} H(s) \qquad (5\text{-}18)$$

$$\alpha_{j,n(j)-2} = \lim_{s \to p_j} \frac{d^2}{ds^2}\frac{1}{2}(s - p_j)^{n(j)} H(s) \qquad (5\text{-}19)$$

and, in general

$$\alpha_{j,k} = \lim_{s \to p_j} \frac{d^{n(j)-k}}{ds^{n(j)-k}} \frac{1}{(n(j) - k)!}(s - p_j)^{n(j)} H(s) \qquad (5\text{-}20)$$

Once the Heaviside coefficients are known, the impulse response can be produced by computing the inverse Laplace of $H(s)$. That is

$$h(t) = \mathcal{L}^{-1}[H(s)] = \mathcal{L}^{-1}\left[\sum_{j=1}^{L} \sum_{i=1}^{n(j)} \frac{\alpha_{j,i}}{(s - p_j)^i}\right] = \sum_{j=1}^{L} \sum_{i=1}^{n(j)} \mathcal{L}^{-1}\left[\frac{\alpha_{j,i}}{(s - p_j)^i}\right]$$

$$(5\text{-}21)$$

which can be expanded to read as

$$h(t) = \cdots + \mathcal{L}^{-1}\left[\frac{\alpha_{j,1}}{(s - p_j)}\right] + \mathcal{L}^{-1}\left[\frac{\alpha_{j,2}}{(s - p_j)^2}\right] + \cdots + \mathcal{L}^{-1}\left[\frac{\alpha_{j,n(j)}}{(s - p_j)^{n(j)}}\right] + \cdots$$

$$(5\text{-}22)$$

which produces

$$y(t) = \cdots + \alpha_{j,1}e^{p_j t}u(t) + \alpha_{j,2}te^{p_j t}u(t) + \cdots$$

$$+ \frac{\alpha_{j,n(j)}}{(n-1)!}t^{n(j)-1}e^{p_j t}u(t) + \cdots \qquad (5\text{-}23)$$

where the inverse Laplace transform of the individual terms having the form $1/(s + p_j)^k$, are found in tables.

Example 5-2   First-order continuous-time LTI

A monic second-order transfer function is given by $H(s) = s^2/(s^2 + 1.5s + 0.5) = s^2/[(s + 1)(s + .5)]$ and the simulation diagram shown in Figure 5-4. In Chapter 6, this particular model will be called a *companion form* implementation of $H(s)$. The roots of $H(s)$ are seen to be nonrepeated and the partial fraction expansion of $H(s)$, where $N = M = 2$, is given by

$$H(s) = \alpha_0 + \alpha_1 \frac{1}{(s+1)} + \alpha_2 \frac{1}{(s+0.5)}$$

where α_0 correspond to the weighting of a delta distribution and $\alpha_1/(s + 1)$ and $\alpha_2/(s + 0.5)$ corresponds to the weighting of exponentials. The Heaviside coefficients are computed to be

$$\alpha_0 = \lim_{s \to \infty} H(s) = 1$$

$$\alpha_1 = \lim_{s \to -1}(s + 1)H(s) = \lim_{s \to -1}\frac{s^2}{s + 0.5} = \frac{1}{-1 + 0.5} = -2$$

$$\alpha_2 = \lim_{s \to -0.5}(s + 0.5)H(s) = \lim_{s \to -0.5}\frac{s^2}{s - 1} = \frac{(-0.5)^2}{-0.5 + 1} = \frac{0.25}{0.5} = 0.5$$

From Table 2.1 the inverse Laplace transform of $H(s)$ can now be explicitly written as $h(t) = \delta(t) - 2e^{-t}u(t) + 0.5e^{-0.5t}u(t)$. Since $h(t)$ is known to contain the term $\delta(t)$, the initial value is, in fact, infinite. The final value theorem states

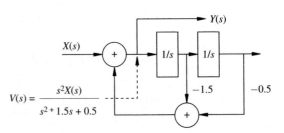

FIGURE 5-4   Block diagram representation of a second order $H(s)$.

$h(\infty) = \lim s H(s)$ as $s \to 0$, or $h(\infty) = \lim s^3/[(s+1)(s+0.5)] \to 0$ as $s \to 0$. This is consistent with the exponential decay of $h(t)$.

Computer Study S-file pf.s and M-file residue.m, introduced in Chapter 2, can be used to perform a partial fraction expansion of $H(s) = s^2/[(s+1)(s+0.5)]$. This corresponds to the derived values.

SIGLAB	MATLAB
>include "pf.s"	>> n = [1 0 0];
># N(s)=s^2; D(s)=(s+1)(s+0.5)	>> d = poly([-1, -0.5])
>n=[1,0,0]; d=poly([-1,-.5])	d =
>d	1.0000 1.5000 0.5000
1 1.5 0.5	>> % r = [a1 a2]
>pfexp(n,d) # partial fraction	>> % p = [p1 p2]
Quotient polynomial: 1 => a0	>> % k = a0
Pole @ -1+0j with multiplicity 1	>> [r p k] = residue(n, d)
and coefficients:-2-0j => a1	r =
Pole @ -0.5+0j with multiplicity	-2.0000
1 and coefficients:	0.5000
0.5-0j => a1	p =
	-1.0000
	-0.5000
	k =
	1

Systems can often contain poles of multiplicity greater than one. A system constructed using a double integrator of the form $d^2y(t)/dt^2 = x(t)$, for example, would have two poles at the origin and have a transfer-function representation given by $H(s) = 1/s^2$. Another example is a system constructed by cascading k identical electronic sections together to form $H(s) = 1/(s+\alpha)^k$. This would introduce a pole at $s = -\alpha$ of multiplicity k.

Example 5-3   Repeated poles

The transfer function $H(s) = s/(s^3 + 3s^2 + 3s + 1) = s/(s+1)^3$ has repeated roots at $s = -1$ of multiplicity 3 (i.e., $n(1) = 3$). A simulation diagram representation of $H(s)$ is shown in Figure 5-5. In Chapter 11, this model will be called a Cascade implementation.

FIGURE 5-5   Block diagram representation of a third-order $H(s)$.

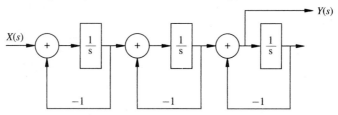

The partial fraction expansion of $H(s)$ is given by

$$H(s) = \alpha_0 + \frac{a_{11}}{(s+1)} + \frac{\alpha_{12}}{(s+1)^2} + \frac{\alpha_{13}}{(s+1)^3}$$

Since $H(s)$ is strictly proper, $\alpha_0 = 0$. The remaining partial fraction coefficients are given by

$$\alpha_{13} = \lim_{s \to -1} (s+1)^3 H(s) = \lim_{s \to -1} s = -1$$

$$\alpha_{12} = \lim_{s \to -1} \frac{d}{ds}(s+1)^3 H(s) = \lim_{s \to -1} \frac{ds}{ds} = 1$$

$$\alpha_{11} = \lim_{s \to -1} \frac{d^2}{ds^2}\frac{(s+1)^3}{2} H(s) = \frac{1}{2}\lim_{s \to -1}\frac{d^2 s}{ds^2} 1 = 0$$

which translates to $H(s) = 1/(s+1)^2 - 1/(s+1)^3$. From Table 2-1, it can be seen that $1/(s+1)^2 \overset{\mathcal{L}}{\leftrightarrow} te^{-t}u(t)$, and $1/(s+1)^3 \overset{\mathcal{L}}{\leftrightarrow} 0.5t^2 e^{-t}u(t)$. Therefore, $h(t) = te^{-t}u(t) - 0.5t^2 e^{-t}u(t)$. From the initial value theorem, $h(0) = \lim(sX(s))$ as $s \to \infty$, which in this case is $h(0) = \lim(s^2/(s+1)^3) \to 0$ as $s \to \infty$. This is also the computed value of $h(0)$ as $t \to 0$. The final value theorem states $h(\infty) = \lim(sH(s))$ as $s \to 0$, or $h(\infty) = \lim(s^2/(s+1)^3) \to 0$ as $s \to 0$. This is consistent with the exponential decay of $h(t)$.

Computer Study   S-file `pf.s` or M-file `residue.m` performs a partial fraction expansion of $H(s) = N(s)/D(s)$. The roots of $D(s)$ are $s = -1, -1$, and -1. This will correspond to the derived values.

SIGLAB	MATLAB
| ```
>include "pf.s"
>n=[0,1,0]; d=poly([-1,-1,-1])
># N(s)=s^1; D(s)=(s+1)(s+1)(s+1)
>d
    1   3   3   1
>pfexp(n,d) # partial fraction
Pole @ -1+0j with multiplicity 3 and
coefficents:
  -1-0j 1+0j 0+0j=>a13, a12, and a11
``` | ```
>> n = [0 1 0];
>> d = poly([-1 -1 -1])
d =
    1   3   3   1
>> % r = [a11 a12 a13]
>> % p = [p1 p1 p1]
>> % k = a0
>> [r p k] = residue(n, d)
r =
         0
    1.0000
   -1.0000
p =
   -1.0000
   -1.0000
   -1.0000
k =
   []
``` |

Transfer functions are particularly useful in developing and analyzing mathematical models of a physical process. Given a transfer function representation of a continuous-time LTI system, a computer simulation can be used to study a system under dynamic conditions. From a blend of mathematical and simulated studies, a system's performance can be adjusted and modified to meet a specific set of objectives. This is particularly important when the performance of a complicated system must be predicted before an expensive prototype is designed.

Example 5-4   Automotive suspension

An automotive suspension system was studied earlier in Example 4-15. The details of this suspension system are described in Figure 5-6 and are defined in terms of

$$M_{cw} = \text{mass of the car over the wheel}$$

$$M_w = \text{mass of the wheel}$$

$$K_s = \text{stiffness of the shock absorber}$$

$$K_t = \text{stiffness of the tire}$$

$$K_d = \text{damping factor of the shock absorber}$$

The system is instrumented with acceleration, velocity, and position measurement transducers. The input to the system is the profile of the road surface, denoted $x(t)$, and the output is the motion of the car body, denoted $y(t)$. The

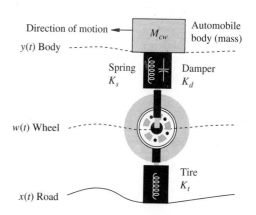

FIGURE 5-6   Automobile suspension system.

equations of dynamic motion are

$$M_{cw}\frac{d^2w(t)}{dt^2} + K_t(w(t) - x(t)) = K_s(y(t) - w(t)) + K_d\left(\frac{dy(t)}{dt} - \frac{dw(t)}{dt}\right)$$

$$M_w\frac{d^2y(t)}{dt^2} + K_s(y(t) - w(t)) + K_d\left(\frac{dy(t)}{dt} - \frac{dw(t)}{dt}\right) = 0$$

The Laplace transform representation of the at-rest system is

$$M_{cw}s^2W(s) + K_t(W(s) - X(s)) = (K_s + K_ds)(Y(s) - W(s))$$

$$M_ws^2Y(s) + (K_s + K_ds)(Y(s) - W(s)) = 0$$

The transfer function $H(s)$ is given by the ratio of the input and output transforms, which is

$$H(s) = \frac{Y(s)}{X(s)} = \frac{(K_s + K_ds)K_t}{(M_{cw}s^2 + K_t)(M_ws^2 + K_s + K_ds) + (K_s + K_ds)M_ws^2}$$

The location of the poles of the transfer function $H(s)$ establishes whether the car rides softly or firmly. Depending on the specific design parameters, the poles of the fourth-order transfer function can be complex and/or real, repeated and/or distinct.

Suppose that the car was placed in a test apparatus that holds the body rigid but allows the wheel to respond to road conditions. That is, $y(t) = 0$ for all time. Then the suspension system would be modeled as

$$M_{cw}s^2W(s) + K_t(W(s) - X(s)) = -(K_s + K_ds)W(s)$$

Again using $x(t)$ as the input (road surface) and $w(t)$ as the output (wheel motion), the transfer function becomes $H(s) = W(s)/X(s)$, where

$$H(s) = \frac{K_t}{M_{cw}s^2 + K_ds + (K_t + K_s)} = \frac{A}{s^2 + Bs + C}$$

where A, B, and C are defined in the obvious manner. For the purpose of a numerical study, assume that the second-order model of the suspension system is given by

$$H(s) = \frac{0.095811}{s^2 + 0.43774s + 0.095811}$$

The system is second-order because there are now only two energy storage subsystems present, namely the wheel mass (M_w) and the two springs acting together ($K_s + K_t$). The vehicle is instrumented to measure the wheel deflection. The suspension system is tested by running the vehicle over a $6'' \times 8'$ raised block at 20 mph. Another pass is made over the obstacle course after it was used to test automobiles designed by graduates of that other, rival university. It is not surprising that one can conclude that *their* engineers completely forgot to add damping to their system, as evidenced by a track strewn with car parts. The results of both tests were analyzed to evaluate the performance of the car under both ideal and hazardous conditions.

Computer Study   The system response is produced using S-file `ch5_4.s`, which makes use of file "car" that contains transfer function information.

SIGLAB	MATLAB
>include "ch5_4.s"	Not available in Matlab

The results, displayed in Figure 5-7, show that the input is initially felt as a step input and responds with a slight lagging movement as the wheel attempts to rise to the top of the obstacle. At the top of the wheel's trajectory, a slight overshoot is noted, which means that the wheel leaves the road for a brief instant. Shortly after the initial mechanical jump discontinuity is encountered, the wheel moves along the top of the block, maintaining contact. When the end of the block is encountered, the trailing edge overshoot found at the bottom of the wheel's trajectory would correspond to the tire's deformation (compression) that occurs when the tire meets the pavement again. After a brief settling-out period, the car again rides evenly along the road. This agrees with the final value of $W(s) = X(s)H(s)$, where $H(s)$ is shown above and $X(s) = (1 - \exp(-st_0))/s$ (i.e, $x(t) = u(t) - u(t - t_0)$). Here, $w(t) = \lim(sX(s)H(s)) = 0$ as $s \to 0$.

FIGURE 5-7   Automobile suspension response: (left) clear test track, and (right) littered test track.

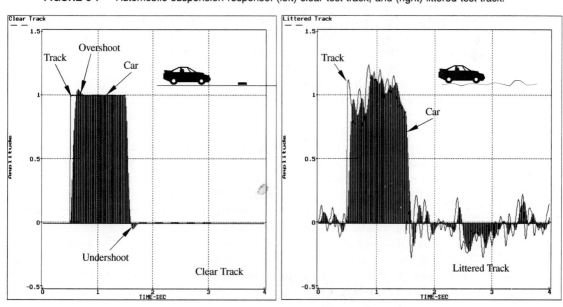

The second test over rough road conditions was conducted using random noise to simulate the altered road conditions. The lagging response is again noted and the resulting ride is seen to be extremely rough.

5-5   STABILITY OF CONTINUOUS-TIME SYSTEMS

When continuous-time systems were studied in the time domain in Chapter 4, the concept of BIBO stability was introduced. There are other stability definitions and criteria that can be used, as well. If, for all possible bounded initial conditions, the system's homogeneous solution converges to $y(t) \rightarrow 0$ as $t \rightarrow \infty$, then the system is said to be *asymptotically stable*. The output of an asymptotically stable system to a bounded input is known to be BIBO stable. If the homogeneous response of a system is simply bounded, such that $|y(t)| < M$ for all $t \geq 0$ (not guaranteeing convergence to zero), then the system is said to be *marginally stable*. Marginally stable systems can behave in a stable manner over a suite of inputs, but may become unstable for a specific input or class of input signals, as demonstrated in Example 5-5. If the homogeneous response to any possible initial condition is unbounded (i.e., $|y(t)| < \infty$ as $t \rightarrow \infty$), then the system is said to be *unstable*. If an unstable system is initialized to a nonzero state, then regardless of what the forced (inhomogeneous) solution is, the complete solution (homogeneous plus inhomogeneous) will become unbounded.

Example 5-5   Marginal stability

A marginally stable system has a tendency to oscillate at a fixed frequency f_0. An example is an LTI system having an undamped oscillatory impulse response $h(t) = \cos(2\pi f_0 t)u(t)$, or $H(s) = s/(s^2 + (2\pi f_0)^2)$. The homogeneous response of this system is simply a harmonic oscillation, which is bounded if the initial conditions are bounded. The forced response of this system, if started at rest and driven by a sinusoid of the form $x(t) = \sin(2\pi f_0 t)$ or $X(s) = 2\pi f_0/(s^2 + (2\pi f_0)^2)$, can be computed using the convolution theorem. Here $Y(s) = H(s)X(s)$, which yields $Y(s) = 2\pi f_0 s/(s^2 + (2\pi f_0)^2)^2$. By manipulating the data found in a table of Laplace transforms (e.g., Table 2-1), the inverse Laplace transform of $Y(s)$ can be recognized as given by $y(t) = 0.5t \sin(2\pi f_0 t)$, which becomes unbounded as $t \rightarrow \infty$. An input having energy at a frequency other than f_0, however, will produce an output that remains bounded, though possibly large, for all time.

This example helps us understand why sometimes a physical system, which seems to be stable, may suddenly become unstable when presented with certain types of inputs. Mechanical structures, such as bridges, can be destroyed by winds that gust at the bridge's natural resonant frequency. The natural resonant frequency of a structure, sometimes called its mode, can often be experimentally determined by striking, shaking, pulsing, or exciting a system with an external source of energy, then measuring the system's frequency response at critical locations. This process is called modal analysis. Such information can be used to reduce the possibility of a system becoming marginally unstable in practice. ***Computer Study*** S-file `ch5_5.s` and M-file `ch5_5.m` create the impulse response $h(t) = \cos(2\pi f_0 t)u(t)$ and three test input signals given by $u(t)$, $x(t) = \cos(2\pi f_0 t)u(t)$ and $y(t) = \cos(2\pi(f_0 + \Delta)t)u(t)$. The forced system response to $x(t)$ is mathematically of the form $z(t) = (c_1 t \cos(2\pi f_0 t) + c_2 \sin(2\pi f_0 t))u(t)$ where c_1 and c_2 are real constants, which is becoming unbounded in time. The system response to the other two inputs are bounded.

SIGLAB	MATLAB
>include "ch5_5.s"	>>ch5_5

The response of the marginally stable system to $u(t)$, $x(t)$, and $y(t)$ is shown in Figure 5-8, for normalized values of $f_0 = 1/8$ and $\Delta f = 1/64$. It can be seen that the response to $x(t)$ is unstable. However, the output produced by an input at any other frequency other than f_0 is bounded.

The analysis provided in Example 5-5 points out that the stability of the system can also be examined in the transform domain. In particular, it can again be noted that there is a correspondence between the roots of the characteristic equation of an ODE and the roots of the Laplace transform of the same ODE. That is, the

FIGURE 5-8 Marginally stable system's response to $x(t)$, $u(t)$, and $y(t)$.

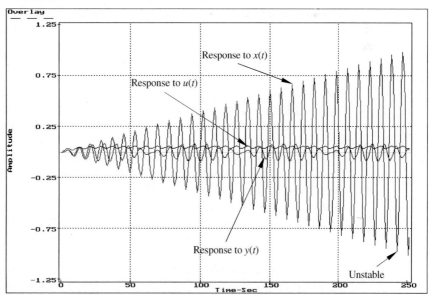

homogeneous ODE, having the form $a_n dy^n(t)/dt^n + a_{n-1} dy^{n-1}(t)/dt^{n-1} + \cdots + a_1 dy(t)/dt + a_0 y(t) = 0$, produces a characteristic equation $a_n s^n + a_{n-1} s^{n-1} + \cdots + a_1 s + a_0 = 0$. This is the same equation used to determine the poles of $H(s)$ (see Equation 5-15). The poles were called eigenvalues when they were studied in the context of a characteristic equation. The eigenvalues, in turn, established the nature of the eigenfunction solution to the ODE. The eigenfunctions can be used as a stability test using the time-domain methods found in Chapter 4. Since the eigenvalues and poles contain the same information, there should also be a correlation between the study of stability in the time-domain with the frequency-domain.

Referring to Equation 5-14, the N poles of the Nth order transfer function $H(s)$ are given by $s = p_i$, where p_i may be real or complex, distinct or repeated. A continuous-time LTI system was said to be *asymptotically stable* if its homogeneous solution converges to zero. This means that each eigenfunction, which has the form $\phi_r(t) = t^j \exp(p_i t)$ (see Equation 4-12), must converge to zero. The Laplace transform of the rth eigenfunction can be found in Table 2-1 and it is $\Phi_r(s) = q(s)/(s - p_i)^j$, where $\deg(q(s)) < j$ (i.e., $\Phi_r(s)$ is strictly proper). A system having a transfer-function representation $H(s)$ and a partial-fraction representation given by

$$H(s) = \frac{\alpha_{j1}}{s - p_1} + \cdots + \frac{\alpha_{j1}}{(s - p_j)} + \frac{\alpha_{j2}}{(s - p_j)^2} + \cdots + \frac{\alpha_{jn(j)}}{(s - p_j)^{n(j)}} + \cdots \quad (5\text{-}24)$$

is seen to have the denominator identical to that of the Laplace transform of the eigenfunctions $\phi_r(t) = t^j \exp(p_i t)$. If the ith pole is complex, say $p_i = \sigma_i + j\omega_i$, then it follows that $\phi_r(t) = t^j \exp(\sigma_i t)\exp(j\omega_i t))$ and $|\phi_r(t)| \leq |t^j \exp(\sigma_i t)||\exp(j\omega_i t)| = |t^j \exp(\sigma_i t)|$, which converges to zero as $t \to \infty$ if $\sigma_i < 0$. If the ith pole is real, namely $s_i = \sigma_i$, then it follows that $\phi_r(t) = t^j \exp(\sigma_i t)$ and $|\phi_r(t)| = |t^j \exp(\sigma_i t)|$, which also converges to zero as $t \to \infty$ if $\sigma_i < 0$. Therefore, an LTI system is *asymptotically stable* if and only if $\text{Re}(p_i) = \sigma_i < 0$ for all i.

The same line of reasoning can be used to determine the marginal stability conditions of a continuous-time LTI system in the transform-domain. An LTI system is *marginally stable* if and only if $\text{Re}(p_i) = \sigma_i \leq 0$ for all nonrepeated roots and $\text{Re}(p_i) < 0$ if p_i is repeated. An LTI system is *unstable* if there exists at least one pole, such that $\text{Re}(p_i) = \sigma_i > 0$, or in the case where p_i is repeated, $\text{Re}(p_i) = \sigma_i \geq 0$. One must be careful, however, to insure that the locations of potentially troublesome poles have not been masked by zeros (called *masking zeros*) located at the same point. If, for example, an LTI system was specified by the ODE $dy(t)/dt - y(t) = dx(t)/dt - x(t)$, then the transfer function (which assumes the system is at rest) is

$$H(s) = \frac{Y(s)}{X(s)} = \frac{(s - 1)}{(s - 1)} = 1 \quad (5\text{-}25)$$

The transfer function $H(s)$, in this case, is equivalent to a memoryless unity-gain amplifier. The system would, therefore, appear to be stable for all bounded inputs.

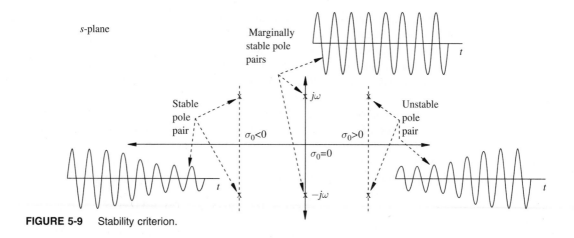

FIGURE 5-9   Stability criterion.

However, for any nonzero initial condition, say $y(0) = y_0$, the homogeneous solution is given by $y(t) = \exp(t)y_0$, which diverges to infinity as $t \rightarrow \infty$. From this it can be concluded that an LTI system, having a transfer function $H(s)$, and possessing no marginally stable or unstable poles covered or canceled by masking zeros, is asymptotically stable.

The transform-domain stability tests are to be based on the value of the real part of the system's poles in the s-plane. The stability of an LTI system can, therefore, be determined by mapping the system's poles into the s-plane and classifying them as suggested in Figure 5-9. Stability is guaranteed if *all* poles reside in the left-hand plane; instability will occur if any pole is in the right-hand plane. Poles located along the $j\omega$-axis are seen to be marginally stable, or unstable, depending on their multiplicity.

There are alternative parametric stability determination methods (e.g., Routh-Hurwitz method). Their advantage is that the transfer function does not require factoring in order to generate the system poles from which stability can be determined. However, with the recent availability of general purpose computer and various polynomial factoring routines, the importance and use of these methods has decreased.

5-6   CONTINUOUS-TIME STEADY STATE RESPONSE

Consider the input to an LTI system to be $x(t) = p(t)u(t)$, where $p(t)$ is a periodic signal of infinite duration and $u(t)$ is a causal step function. The signal $x(t)$ is sometimes referred to as a *one-sided periodic* signal. The forced response of a stable LTI system to a one-sided periodic signal can be divided into two regimes, called the *transient* and *steady-state* solutions. These concepts have already been introduced informally. Assuming that a system is asymptotically stable, the transient response represents the early time-history of a system where the initial conditions have an influence on the output. During the transient period, energy is also being supplied to the system by the input process, giving rise to an inhomogeneous solution component, as well. Since the effects of the initial conditions must be self-extinguishing for an asymptotically stable system, after a period of time the system output must be completely characterized by the input process. Since the input process is a one-side periodic process (e.g., $p(t) = u(t)$, $p(t) = \sin(\omega_0 t)u(t)$), the output of the linear system should begin to take on a periodic character, as well. This claim can be justified by returning to a partial fraction expansion interpretation of the forced response of an at-rest causal LTI system. Equation 5-26 represents the partial fraction expansion of $Y(s) = H(s)X(s)$, where $X(s)$ is assumed to have a Laplace transform $X(s) = N_x(s)/D_x(s)$ and $R(s)$ is a remainder polynomial that is a by-product of the partial fraction expansion of $Y(s)$

$$Y(s) = H(s)X(s) = Y_1(s) + Y_2(s)$$

$$= \underbrace{\frac{\alpha_{11}}{s - p_1} + \cdots + \frac{\alpha_{j1}}{(s - p_j)} + \frac{\alpha_{j2}}{(s - p_j)^2} + \cdots + \frac{\alpha_{jn(j)}}{(s - p_j)^{n(j)}}}_{Y_1(s)} + \underbrace{\frac{R(s)}{D_x(s)}}_{Y_2(s)} \quad (5\text{-}26)$$

The forced response $y(t)$ can be obtained by computing the inverse Laplace transform of Equation 5-26. If $H(s)$ is assumed to be asymptotically stable, then all elements grouped together as $Y_1(s)$ will have an inverse Laplace transform that will converge to zero (i.e., $\text{Re}(s_i) < 0$). If $x(t) = p(t)u(t)$ is a nonzero, one-sided periodic signal of period T, then $p(t) = p(t + kT)$ for all k and $t \geq 0$. Table 2-1, illustrates that the poles of periodic signals, such as $\cos(\omega t)$ and $u(t)$, lie along the $j\omega$-axis in the s-plane. Therefore the inverse Laplace transform of $Y_2(s) = R(s)/D_x(s)$ will produce terms that do not decay to zero, but exhibit periodic behavior at the frequency of the input process. The solution $y_2(t)$ is called the steady-state response.

If the input to an LTI system is a continuous-time, one-sided periodic sinusoidal signal of the form $x(t) = A\cos(\omega_0 t)u(t)$, then the solution for large values of time is called the *steady-state sinusoidal solution*. The steady-state sinusoidal solution has the form $y(t) = G(j\omega_0)(\cos(\omega_0 t + \phi(j\omega_0)))$, where $G(j\omega_0)$ and $\phi(j\omega_0)$ are called the system's steady-state sinusoidal gain and phase shift at $s = j\omega_0$. In particular, if $H(s)$, is given by

$$H(s) = K\frac{\prod_{i=1}^{n}(s - z_i)}{\prod_{i=1}^{n}(s - p_i)} \quad (5\text{-}27)$$

then

$$G(j\omega_0) = |H(s)|_{s=j\omega_0} = K\frac{\prod_{i=1}^{n}|(j\omega_0 - z_i)|}{\prod_{i=1}^{n}|(j\omega_0 - p_i)|} \quad (5\text{-}28)$$

and

$$\phi(j\omega_0) = \arctan\left(\frac{\text{Im}(H(j\omega_0))}{\text{Re}(H(j\omega_0))}\right) \quad (5\text{-}29)$$

Example 5-6   Steady-state frequency response and measurement

The automotive suspension system studied in Example 5-4 was modeled as

$$H(s) = \frac{0.095811}{s^2 + 0.43774s + 0.095811}$$

The system's steady-state sinusoidal frequency response can be directly computed by evaluating $H(s)$ along the path $s = j\omega$. Evaluating the transfer function $H(s)$ at a fixed frequency, say $\omega_0 = 2\pi f_0$, produces $G(j\omega_0)$ and $\phi(j\omega_0)$; these are the system's steady-state gain and phase response to a sinusoidal input at a frequency of f_0. For $f_0 = 0$ Hz, the gain and phase behavior of the automotive system can be calculated to be $G(j0) = |H(j0)| = 1.0$ and $\phi(j0) = \tan^{-1}(0) = 0$ radians, respectively. For $f_0 = 5/128$ Hz, the gain and phase behavior of the automotive system can be calculated to be $G(j2\pi5/128) = |H(j2\pi5/128)| = 0.844147$ and $\phi(j2\pi5/128) = \tan^{-1}(0.4377(2\pi5/128)/(0.095811 - (j2\pi5/128)^2)) = -1.25767$ radians, or $-72°$, respectively.

Computer Study   S-file ch5_6.s computes $|H(j2\pi f)|$ over a range of frequencies using the format magf(n,f0), where n is the number of baseband frequency locations to be plotted over a baseband range of $f \in [0, 0.5]$ and $f_0 < 0.5$. The step response is automatically computed, as is the response to a sinusoidal input at a frequency f_0. Finally, the system gain and phase shift, at frequency f_0, are explicitly computed.

To analyze the automotive suspension system response at $s = j2\pi5/128$, the following program is executed.

SIGLAB	MATLAB
>include "ch5_6.s" >magf(200,5/128) Magnitude: 0.844147; Phase angle in deg.: -72.0591	Not available in Matlab

The results are displayed here and in Figure 5-10. First observe that the predicted values of $G(j2\pi5/128)$ and $\phi(j2\pi5/128)$ are numerically verified. Next note that the forced response to $x(t)=u(t)$ and $x(t)=\cos(2\pi5t/128)u(t)$ consists of a transient and steady-state response. The steady-state step response is seen to be unity, and the steady-state response to $x(t)=\cos(2\pi5t/128)u(t)$ equals $|H(j2\pi5/128)| = 0.844147$. The phase difference between the input $x(t) = \cos(2\pi5t/128)u(t)$ and output y(n) is approximately $-72°$ in steady state.

If the phase angle $\phi(j\omega_0)$, defined by Equation 5-29, is positive, then the system is a *lead system*. If negative, the system is a *lag system*. For example, the system studied in Example 5-6 has a $72°$ lagging phase angle at $s = j\omega_0 = j2\pi5/128$.

Steady-state design methods can be used by experienced designers to closely approximate an arbitrary response $H(j\omega_0) = |H(j\omega_0)| \arg(j\omega_0)$ with another

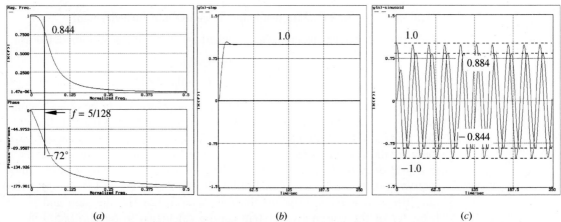

(a)   (b)   (c)

FIGURE 5-10   Automotive suspension system: (a) magnitude frequency and phase response, (b) step response, and (c) sinusoidal response.

system that consists of a collection of well-placed poles and zeros. Referring to Equation 5-27, observe that the magnitude frequency response of $H(s)$ can be expressed as

$$|H(j\omega)| = K \frac{\prod_{i=1}^{n} |(j\omega - z_i)|}{\prod_{i=1}^{n} |(j\omega - p_i)|} = K \frac{\prod_{i=1}^{n} \alpha_i(j\omega)}{\prod_{i=1}^{n} \beta_i(j\omega)} \qquad (5\text{-}30)$$

To produce $|H(j\omega)|$, $H(s)$ is evaluated along the $j\omega$-axis in the s-plane for all values of $s = j\omega$. The information needed to produce $|H(j\omega)|$ can be graphically determined, as shown in Figure 5-11. The distance between a test point $s = j\omega$ and a zero located at z_i is denoted $\alpha_i(j\omega)$, as defined in Equation 5-30. Similarly, the distance between a test point $s = j\omega$ and a pole located at p_i is denoted $\beta_i(j\omega)$, also shown in Figure 5-11.

With practice, the experienced designer will manipulate the terms found in Equation 5-30 to achieve an acceptable design. The manual design procedure can be explained as follows. As a test frequency $j\omega_0$ moves to a point in close proximity to a pole located near the $j\omega$ axis, the distance measure $|j\omega - p_i| = \beta_i(j\omega)$ will rapidly decrease and, correspondingly, the value of $|H(j\omega)|$ will increase. Therefore, placing poles near the $j\omega$ axis will cause the filter to have high localized gain in the frequency-domain. Zeros will behave in an opposite manner. By selectively placing poles and zeros in strategic locations, a system's steady-state sinusoidal frequency response can be shaped.

Example 5-7   Steady-state frequency response

Many real systems are, or contain, second-order sections. The dependence of the pole and zero locations on a second-order polynomial $P(s)$ can also be graphically explored. The loci of roots of the polynomial $P(s) = s^2 + 2\delta\omega_0 s + \omega_0^2$ are displayed in Figure 5-12 as a function of δ. For $\delta < 1/\sqrt{2}$, the frequency at

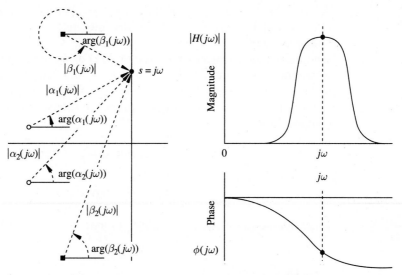

FIGURE 5-11   Graphical relationship between $H(s)$ and the pole-zero distribution.

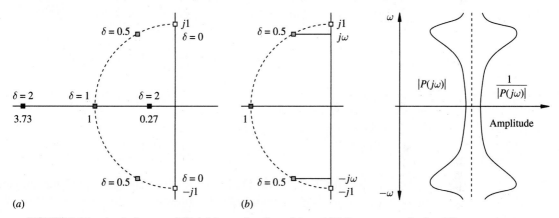

FIGURE 5-12   Loci of zeros of $P(s)$ (a) as a function of δ, and (b) the response for $\delta = 1/2$.

which $|P(j\omega)|$ is minimized can be computed using an unconstrained derivative test, namely

$$\frac{d|P(j\omega)|^2}{d\omega} = \frac{d((\omega_0^2 - \omega^2)^2 + (2\delta\omega_0\omega)^2)}{d\omega}$$

$$= 4\omega^3 + 4\omega\omega_0^2(2\delta^2 - 1) = 0$$

or

$$\omega = \pm\omega_0\sqrt{1 - 2\delta^2}$$

It is important to note that the minimizing frequency is a nonlinear function of δ. For $\delta = 0$, $\omega = \pm\omega_0$; otherwise, the optimizing value of ω along the

$j\omega$-axis satisfies $|\omega| < \omega_0$. Finally, for $\delta = 1/2$, the corresponding values of $|P(j\omega)|$ and $1/|P(j\omega)|$ are graphically interpreted.

Computer Study   S-file ch5_7.s and M-file dist.m compute the steady-state magnitude frequency response of $P(s)$ using the format dist(p,k,n), where the polynomial $P(s)$ is evaluated over the frequency range $f \in [0, 2\pi k]$ radians per second and n is the number of subintervals of the frequency range to be displayed. The magnitude frequency responses of $P(s)$ and $1/P(s)$ are computed for

$$P_1(s) = s^2 + s + 1$$
$$P_2 = s^2 + 0.1s + 1$$
$$P_3 = s^2 + 0.01s + 1$$

which correspond to $\delta = \left[1/2, 1/20, 1/200\right]$.

SIGLAB	MATLAB
>include "ch5-7.s"	>>d1=[1,1,1];
>d1=[1,1,1]; d2=[1,.1,1];	>>d2=[1,0.1,1];
d3=[1,.01,1]	>>d3=[1,0.01,1];
>one=ones(101);e1=dist	>>e1=dist(d1,1,101);
(d1,1,101)	>>e2=dist(d2,1,101);
>e2=dist(d2,1,101); e3=	>>e3=dist(d3,1,101);
dist(d3,1,101)	plot(0:100,[e1',e2',
>ograph(e1,e2,e3)	e3']);
>ograph(one./e1,one./e2,one./e3)	plot(0:100,1./[e1',
	e2',e3']);

The results are displayed in Figure 5-13. The minimizing and maximizing frequency locations can be seen to be a nonlinear function of δ, as predicted.

The term $H(j\omega)$ is, in general, a complex function of frequency and is called the steady-state sinusoidal system gain. The attendant steady-state analysis assumes, of course, that the system is being driven by a sinusoidal input and that all the transients have dissipated. Under such conditions, the system's frequency response can be displayed and analyzed in a number of standard formats, which emphasize both a system's magnitude and phase behavior. The more commonly used frequency-domain system representation formats are:

Format 1 Complex frequency response

$$H(j\omega) = H(s)|_{s=j\omega} = K \frac{\prod_{i=1}^{N}(j\omega - z_i)}{\prod_{i=1}^{N}(j\omega - p_i)} \tag{5-31}$$

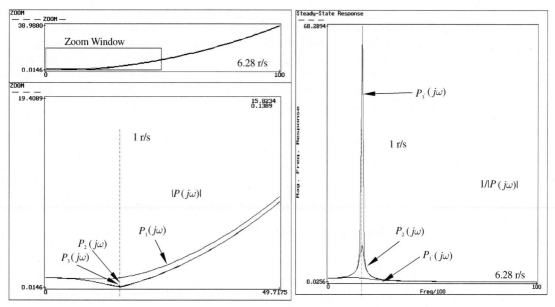

FIGURE 5-13   Steady-state magnitude frequency response of $P(s)$ and $1/P(s)$.

Format 2 Magnitude frequency response

$$G(j\omega) = |H(j\omega)| \qquad (5\text{-}32)$$

Format 3 Logarithmic magnitude frequency response (in decibel units)

$$|H(j\omega)|_{dB} = 20\log_{10}|H(j\omega)| \qquad (5\text{-}33)$$

Format 4 Phase response

$$\phi(j\omega) = \arg(H(j\omega)) = \arctan\left(\frac{\text{Im}[H(j\omega)]}{\text{Re}[H(j\omega)]}\right) \qquad (5\text{-}34)$$

These are all measures of a system's steady-state sinusoidal frequency response and have been extensively used in this chapter. Their interpretations are self-evident. Another commonly-used measure, is called *group delay*.

Format 5 Group delay

$$\tau_g(j\omega) = -\frac{d\phi(j\omega)}{d\omega} \qquad (5\text{-}35)$$

Group delay measures the rate of phase change as a function of frequency. A system that has a constant group delay is called a *linear phase filter*. Linear phase filters are important in some communications applications where the phase of the received signal must be determined in order to decode the transmitted message. In typical communications applications, the instantaneous frequency of the received message would, in general, be constantly changing. Phase detectors often have difficulty tracking signals whose phase and frequency response are rapidly changing in a nonlinear manner. The phase detection subsystem would prefer that the frequency and phase fluctuations of the received signal be small over a wide range. Ideally, for phase detection, the phase versus frequency relationship should be constant, or at least linear. Therefore, group delay can act as a performance predictor for a phase detector. In fact, some systems are intentionally designed to exhibit constant (or near constant) group delay over a range of frequencies. These concepts will be explored in greater detail in Chapter 12.

Example 5-8   Frequency response representation techniques

A typical at-rest system is given by

$$H(s) = \frac{0.000127511(s^4 + 1.639s^2 + 0.545)}{s^5 + 0.20643s^4 + 0.083183s^3 + 0.010672s^2 + 0.001435s + 6.95538 \times 10^{-5}}$$

The various frequency-domain representation schemes, denoted Format 1 through 5, can be directly computed from $H(s)$.
Computer Study $H(s)$ is saved as "fgraph." S-file ch5_8 computes the frequency domain representation of $H(s)$, using the formats $|H(2\pi f)|$, $|H(2\pi f)|_{dB}$, $\arg(H(2\pi f))$ and $\tau_g(2\pi f)$ over the baseband frequency range $f \in [0.0, 0.2502]$.

SIGLAB	MATLAB
>include "ch5_8.s"	Not available in Matlab

The results are displayed in Figure 5-14. The data indicates that the system functions as a lowpass filter in that it possesses signal energy over the frequency range $f \in [0.0, 0.035]$ Hz and severely attenuates signal components above 0.1 Hz. The linear graph of $|H(f)|$ provides a high resolution display of the passband, but compresses the higher frequency response into a flat line of amplitude zero. The logarithmic display sacrifices passband resolution for increased interpretability of the system's high frequency behavior. Based on the logarithmic display, the gain of the system for $f > 0.1$ Hz is seen to be less than -60 dB. The phase profile is observed to be highly nonlinear and is most "ill behaved" over the range of frequencies where the magnitude frequency response is maximally changing (called the *transition band*). This nonlinear behavior carries over into the group delay spectrum. It can be seen that if this particular system was used in a phase detection application, it could create problems if left uncorrected.

In many cases the frequency range of interest may span decades of change. For example, if the system considered in Example 5-8 is to be analyzed over the frequency range $0.001 \leq f \leq 10^6$, the meaningful passband region of a magnitude, or log magnitude frequency response, would be mapped to the extreme left-hand side of the graph. This is obviously undesirable from a visual data interpretation viewpoint. In such cases, the frequency axis can be interpreted in logarithmic

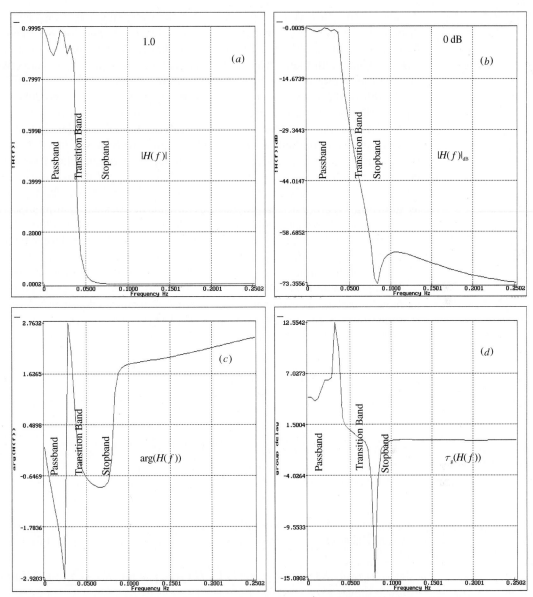

FIGURE 5-14 Steady-state frequency-domain representations (*a*) magnitude frequency response, (*b*) log magnitude (dB) frequency response, (*c*) phase response, and (*d*) group delay.

units that would compress the frequency axis into a manageable linear range. The display of $|H(j\omega)|_{dB}$ versus $\log(f)$, or $\log(\omega)$, is called a *Bode plot*. In a Bode plot display, the frequency axis is subdivided into logarithmically distributed intervals. The two most common frequency-axis display protocols are based on a decade ($\times 10$) or octave ($\times 2$) partitioning of the frequency axis. Consider the data displayed in Figure 5-15, which consist of a log-linear and log-log magnitude-frequency response displays. The log-linear display is implemented by a bank of constant bandwidth filters. This, in fact, is the type of frequency-domain coverage provided by many digital filters and the discrete Fourier transform (DFT). The bandwidth of the filter centered at f_0 is the same as that of its neighbor. If the input to the filter bank is a broadband signal having a constant power density of σ_0^2 W/Hz, then the output power levels of each filter would be identical. However, there are many important signals found in nature with power that falls off at a rate on the order of $1/f$ or $1/f^2$ over a wide range of frequencies. Examples can be found in radio, audio, acoustic, and mechanical applications. Signals of this type can be more easily analyzed in a log-log, or Bode format. The log-log plot, as shown in Figure 5-15, is produced by a bank of constant Q filters (see Exercise 5-21). The frequency axis has been partitioned in octaves and the bandwidth (in Hz) of filters centered at higher frequencies is significantly larger than those centered about lower frequencies. The power output measured at each filter of the filter bank, to an input having a $1/f$ power-density distribution, would be equal.

An approximate Bode plot, based on asymptotic approximations, is often used to rapidly sketch an estimate of a system's frequency response. Consider the transfer function $H(s) = N(s)/D(s) = K \prod H_i(s)$, where $H_i(s) = N_i(s)/D_i(s)$ is a low-order subsystem of $H(s)$ and $K > 0$. Let $H(j\omega)$ be factored as

$$H(j\omega) = K \prod_{i=N}^{L} H_i(j\omega) = \frac{G \prod_{i=1}^{N} N_i(j\omega)}{\prod_{i=1}^{N} D_i(j\omega)} \qquad (5\text{-}36)$$

$$N_i(j\omega) = (1 + j\omega/z_i)$$

$$D_i(j\omega) = (1 + j\omega/p_i)$$

$$G = K \frac{\prod_{i=1}^{N} z_i}{\prod_{i=1}^{N} p_i}$$

where z_i and p_i are the zeros and poles of $H(s)$ respectively. The individual terms found in Equation 5-36 can be analyzed in a piecemeal manner by observing that

$$\log(|H(j\omega)|) = \log(K) + \log\left(|\prod H_i(j\omega)|\right) = \log(K) + \sum \log(|H_i(j\omega)|)$$

$$= \log(G) + \sum \log(|N_i(j\omega)|) - \sum \log(|D_i(j\omega)|) \qquad (5\text{-}37)$$

The log-log magnitude display is seen to be a linear combination of contributions from each low-order subsystems. The phase response can be similarly computed to be (assuming $G > 0$)

$$\arg(H(j\omega)) = \arg\left(\prod H_i(j\omega)\right) = \sum \arg(N_i(j\omega)) - \sum \arg(D_i(j\omega)) \quad (5\text{-}38)$$

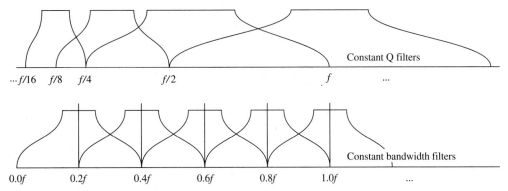

FIGURE 5-15   Frequency spectrum produced by a band of constant Q (top) and constant bandwidth (bottom) filters.

The steady-state log-log frequency response of a simple first-order system, given by $(s - \alpha)$ or $1/(s - \alpha)$, is displayed in Figure 5-16. The gain and phase profile of a simple zero or pole can be asymptotically approximated as follows

$$20 \log_{10} |(1 + j\omega/\alpha)| \rightarrow 20 \log_{10}(\omega/\alpha) \qquad \text{for } \omega \gg \alpha \qquad (5\text{-}39)$$

$$20 \log_{10} |(1 + j\omega/\alpha)| \rightarrow 0 \qquad \text{for } \omega \ll \alpha$$

$$\arg(1 + j\omega/\alpha) \rightarrow \pi/2 \qquad \text{for } \omega \gg \alpha \qquad (5\text{-}40)$$

$$\arg(1 + j\omega/\alpha) \rightarrow 0 \qquad \text{for } \omega \ll \alpha$$

For a pole, the approximate Bode plot approaches

$$20 \log_{10} \left| \frac{1}{1 + j\omega/\alpha} \right| \rightarrow 20 \log_{10}(\alpha/\omega) \qquad \text{for } \omega \gg \alpha \qquad (5\text{-}41)$$

$$20 \log_{10} \left| \frac{1}{1 + j\omega/\alpha} \right| \rightarrow 0 \qquad \text{for } \omega \ll \alpha$$

$$-\arg(1 + j\omega/\alpha) \rightarrow -\pi/2 \qquad \text{for } \omega \gg \alpha \qquad (5\text{-}42)$$

$$-\arg(1 + j\omega/\alpha) \rightarrow 0 \qquad \text{for } \omega \ll \alpha$$

In both cases for $\omega \ll \alpha$, the gain and phase approach 0 dB and 0 radians, respectively. The gain, or attenuation, of a single pole or zero is seen to grow at a rate of ± 20 dB per decade (6 dB per octave) as ω becomes large relative to α, and the phase approaches $\pm \pi/2$ radians. An estimate of the actual log-log magnitude frequency response can be made using a straight line approximation where the low- and high-frequency asymptotes meet at $\omega = \alpha$.

Second-order systems can be approximated by a combination of two first-order subsystems. This approximation is generally acceptable if $N_i(s)$ or

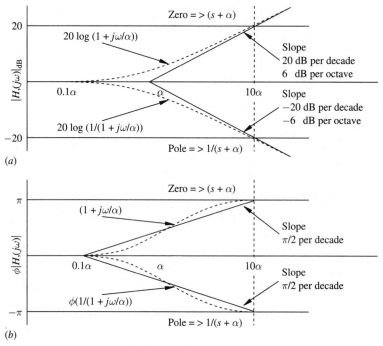

FIGURE 5-16 Bode plot of first-order terms showing (a) magnitude frequency response, and (b) phase response (bottom).

$D_i(s) = s^2 + 2\delta\omega s + \omega^2$ and $\delta \geq 1.0$ (i.e., critically or overdamped). If $\delta < 1.0$ (i.e., underdamped), however, a correction to the straight-line approximation is sometimes needed, as the following example demonstrates.

Example 5-9   Bode plots

A Bode approximation to first- and second-order subsystems is based on a straight-line approximation. Underdamped second-order systems sometimes need a correction in the neighborhood of their natural resonate frequency.

Computer Study   S-file ch5_9i.s and M-file bode.m can compute the log-log magnitude frequency response of a transfer function $H(s)$. The S-file uses the format bode(N(s), D(s),wmax,n) where $H(s) = N(s)/D(s)$, and n is the number of discrete subintervals used to cover $\omega \in [0.0, \omega_{max}]$. The M-file uses a format bbode (N(s),D(s),wmax,n). The log-log plot of $H(s)$ is produced and displayed out to $\omega_{max} = 10^e$ radians per second. The programs also compute the poles of $H(s)$, where

$$H_1(s) = 1/(s+1)$$

$$H_2(s) = 1/(s^2 + 0.5s + 1) \qquad \text{underdamped}$$

$$H_3(s) = 1/(s^2 + 1.0s + 1) \qquad \text{underdamped}$$

$$H_4(s) = 1/(s^2 + 2.0s + 1) \qquad \text{crtically damped}$$

$$H_5(s) = 1/(s^2 + 4.0s + 1) \qquad \text{overdamped}$$

$$H_6(s) = 1/(s^2 + 8.0s + 1) \qquad \text{overdamped}$$

out to frequency 10 radians per second (i.e., 10^1), is performed as follows.

SIGLAB	MATLAB

<pre>>include "ch5_9i.s"
>n=[0,1] # numerator
>h1=bode(n,[1,1],10,100)
# -1+0j---Poles 	
>h2=bode(n,[1,0.5,1],10,100)
# -0.25 +/- 0.968j---Poles 	
>h3=bode(n,[1,1,1],10,100)
# -0.5 +/- 0.886j---Poles 	
>h4=bode(n,[1,2,1],10,100)
# -1+0j -1+0j---Poles 	
>h5=bode(n,[1,4,1],10,100)
# -3.732+0j -0.268+0j---Poles 	
>h6=bode(n,[1,8,1],10,100)
# -7.873+0j -0.127+0j---Poles 	
>ograph(h1,h2,h3,h4,h5,h6)</pre>
	<pre>>> N = 1;
	>> [w, h1, z, p] = bbode(N,
	[1 1], 10, 100); p
	p =
	  -1
	>> [w, h2, z, p] = bbode(N,
	[1 0.5 1], 10, 100); p
	p =
	  -0.2500 + 0.9682i
	  -0.2500 - 0.9682i
	>> [w, h3, z, p] = bbode(N,
	[1 1 1], 10, 100); p
	p =
	  -0.5000 + 0.8660i
	  -0.5000 - 0.8660i
	>> [w, h4, z, p] = bbode(N,
	[1 2 1], 10, 100); p
	p =
	  -1
	  -1
	>> [w, h5, z, p] = bbode(N,
	[1 4 1], 10, 100); p
	p =
	  -3.7321
	  -0.2679
	>> [w, h6, z, p] = bbode(N,
	[1 8 1], 10, 100); p
	p =
	  -7.8730
	  -0.1270
	>> loglog(w, [h1 h2 h3 h4
	h5 h6])</pre>

The data is displayed in Figure 5-17 and demonstrates the effectiveness of the straight-line approximation to over- and critically-damped low-order

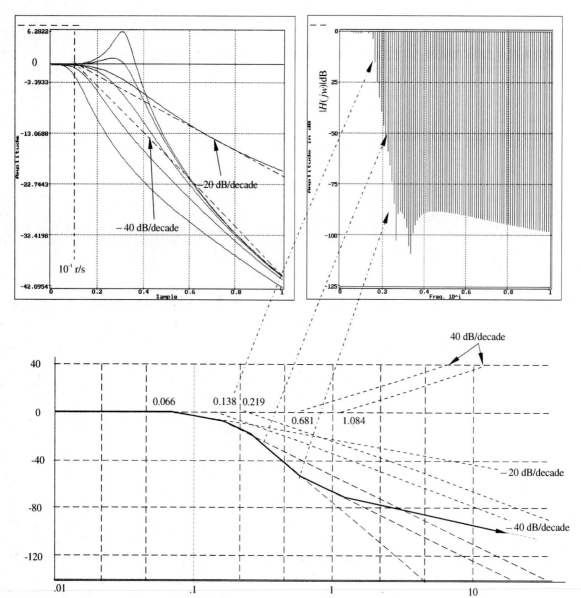

FIGURE 5-17 Bode plot of (left) first- and second-order system, (right) a fifth-order model, and (bottom) an approximate Bode plot of $H(s)$.

subsystems. It can be observed that the straight-line approximation becomes increasingly inaccurate as the system becomes more underdamped.

Approximate Bode plots are often used to rapidly sketch the log-log response of a high-order system. As an example, consider again the system studied in

Example 5-8, given by

$$H(s) = \frac{0.000127511(s^4 + 1.639s^2 + 0.545)}{s^5 + 0.20643s^4 + 0.083183s^3 + 0.010672s^2 + 0.001435s^1 + 6.95538 x 10^{-5}}$$

which can be interpreted as a log-log plot. In Example 5-8, the meaningful portion of the system's frequency response was seen to be concentrated over a baseband range of less than about 1 Hz. To analyze $H(s)$ out to $f_{max} = 10/2\pi$ Hz (i.e., $\omega = 10$ radians per second), use S-file ch5_9ii.s.

SIGLAB	MATLAB
>include "ch5_9ii.s"	Not available in Matlab
Gain in dB -0.00754665	
Zero locations	
0-1.084j 0+1.084j 0-0.681j 0+0.681j	
Pole locations	
-0.066+0 -0.052-0.138j -0.052+0.138j	
-0.019-0.219j -0.019+0.219j	

The gain G, in dB units, is seen to be essentially zero. The numerator polynomial can be seen to be given by $N(s) = (s^2 - 2.175s + 1.175)(s^2 - 1.464s + .464)$ and the denominator polynomial satisfies $D(s) = (s + 0.66)(s^2 - 1.022s + 0.022)(s^2 - 1.048s + 0.048)$. These polynomials establish the breakpoints for an asymptotic approximation to a Bode plot. The log-log plot and the sum of the straight-line approximations are shown in Figure 5-17. The approximate results can be seen to be a good facsimile of the actual log-log display of $H(j\omega)$. However, with the advent of the general purpose, digital computer, fast approximation techniques are now often replaced with computer-generated responses. The utility of these approximation methods can, nevertheless, prove very useful in rapidly verifying the plausibility of a computer-generated solution.

5-7   SUMMARY

In Chapter 5 continuous-time systems were investigated from a transform-domain viewpoint. It was shown that the complicated convolution operations introduced in Chapter 4, based on time-domain methods, are greatly simplified in the transform-domain. This was a direct result of the convolution theorem derived for continuous-time linear systems. Both the homogeneous and inhomogeneous behavior of linear systems can be directly established and computed with Laplace transforms for continuous-time systems. It was shown the inhomogeneous solution can be characterized in terms of a system's transfer function. In general, the qualitative behavior of a linear system was shown to be established by a linear system's pole distribution. In addition, a connection was made between poles and eigenvalues. The stability of linear systems was shown to be definable in terms of the system's pole locations. System stability, based on pole location, was then interpreted in terms of the half-planes in the s-domain for continuous-time systems. If continuous time stable systems are forced by a one-sided periodic process, it was shown that the total solution would consist of a transient and a periodic steady-state solution. Graphical analysis and display techniques were presented to support steady-state system analysis.

In Chapter 6 the material presented in Chapter 5 will be extended to the problem of realizing a system-transfer function in hardware. This study will introduce the concepts of architecture and state variables. This material will provide a foundation upon which systems, having prespecified attributes, can be synthesized and implemented. In Chapter 12 filter synthesis procedures will be presented that will put to practical use the material found in Chapters 4 through 6.

5-8   COMPUTER FILES

These computer files were used in Chapter 5.

File Name	Type	Location
Subdirectory CHAP5		
ch5_1.x	T	Example 5-1
ch5_4.s	T	Example 5-4
ch5_5.x	T	Example 5-5
ch5_6.s	C	Example 5-6
ch5_7.s	C	Example 5-7
dist.m	C	Example 5-7
ch5_8.s	T	Example 5-8
ch5_9i.s	C	Example 5-9
bbode.m	C	Example 5-9
ch5_9ii.s	C	Example 5-9
Subdirectory SFILES or MATLAB		
pf.s	C	Example 5-2, 5-3
directf.s	C	Example 5-6
Subdirectory SIGNALS		
Subdirectory SYSTEMS		
fgraph.arc	D	Example 5-8
car.arc	D	Example 5-4, 5-6
marg.arc	D	Example 5-5

Here "s" denotes S-file, "m" an M-file, "x" either an S-file or M-file, "T" denotes a tutorial program, "D" a data file and "C" denotes a computer-based-instruction (CBI) program. CBI programs can be altered and parameterized by the user.

5-9 PROBLEMS

5-1 Impulse response Find the impulse response of the following systems

$$H_1(s) = \frac{1}{(s+1)(s+2)}$$

$$H_2(s) = \frac{1}{(s^2+4)(s^2+1)}$$

$$H_3(s) = \frac{s^2}{(s+1)^2}$$

$$H_4(s) = \frac{s^2+10s+100}{(s^2+1)}$$

$$H_5(s) = \frac{s^2}{(s^2+1)^3}$$

$$H_6(s) = \frac{1}{(s+j)}$$

and sketch their two-sided steady-state frequency responses in Formats 1 through 5, as illustrated in Section 5-6.

5-2 Inhomogeneous response Determine the forced response of each system defined in Problem 5-1 to

$$x_1(t) = u(t)$$

$$x_2(t) = \exp(-t)u(t)$$

$$x_3(t) = \cos(t)u(t)$$

5-3 Inhomogeneous response Given the at-rest system shown in Figure 5-18, what is the system's transfer function and what is the output if the input is $x(t) = \exp(-t)u(t)$ and $h(t) = \exp(-t)u(t)$?

5-4 Simulation diagram Produce the simulation diagrams for the systems shown in Problem 5-1.

FIGURE 5-18 Linear system.

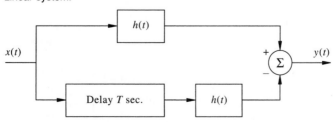

5-5 Convolution   Compute the convolution of $x(t)$ and $h(t)$ using the convolution theorem, where

$$h(t) = e^{-|t|}, t \in (-\infty, \infty) \tag{5-43}$$

and

$$x_1(t) = \delta(t)$$

$$x_2(t) = \delta(t - \tau), \tau \neq 0$$

5-6 Associative property   Derive the associative property of convolution (i.e., $(x(t) * h(t)) * g(t) = x(t) * (h(t) * g(t))$ in the s-domain.

5-7 All-pass system   Show that the polynomial

$$H(s) = \pm \frac{P^*(-s^*)}{P(s)}$$

is an all-pass filter (i.e., $|H(j\omega)| = 1$ for all ω).

5-8 All-pass system   Suppose $P(s)$ in the previous exercise is a seventh-order polynomial given by

$$P(s) = (s - \sigma_0) \prod_{k=1}^{3} (s - (j0.9\pi/k) - \sigma_0) \prod_{k=1}^{3} (s + (j0.9\pi/k) - \sigma_0)$$

Plot the pole-zero distribution of $H(s)$ (see Problem 5-7) for $\sigma_0 > 0$. Compute $|H(j\omega)|$ for $\sigma_0 = 0.1$.

5-9 RLC circuit   Given the circuit shown in Figure 5-19, derive the input-output transfer function if $x(t)$ is the input voltage and the output $y(t)$ is the current through the capacitor.
   Where are the system poles and zeros for $L = C = 10^{-3}$, and $R = 10^3$?

5-10 Inhomogeneous response   Consider again the RLC circuit studied in Problem 5-9. Suppose the system is initially at rest. What is the output $y(t)$, if $x(t) = \cos(2\pi \times 10^3 t)u(t)$?

5-11 Stability   Consider again the RLC circuit studied in Problem 5-9. Suppose the resistor is adjustable. Over what range of $R \in [0, \infty)$ will the circuit remain stable?

5-12 Stability   Determine the stability of each system defined in Problem 5-1 and sketch the pole-zero distribution.

5-13 Stability   If $h(t) = te^{-t}u(t)$, is the system shown in Figure 5-18 stable? What about $h(t) = tu(t)$?

5-14 Stability   A system transfer function is given by

$$H(s) = \frac{s^2 + 10s + 100}{(s^2 + ks + 1)^2}$$

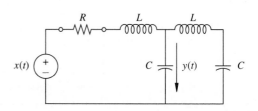

FIGURE 5-19   RLC circuit.

Sketch a simulation diagram for the system. What values of k will ensure that the system is asymptotically stable? Is there a value of k for which the system is marginal stable?

5-15 Pendulum   The simple pendulum shown in Figure 5-20 can be modeled by the equation

$$\frac{I d^2\phi(t)}{dt^2} + MgL\phi(t) = Lx(t)$$

if angle $\phi(t)$ is small. Here, I is the moment of inertia, M is the mass, g is the force of gravity, L is length, and $x(t)$ is the force applied. What is the system's transfer function H(s) if the output is $\phi(t)$? Classify the system's stability.

  If $\phi(0) = \pi/12$ and $d\phi(0)/dt = 0$, what must $x(t)$ be to keep the system at $\phi(t) = \pi/12$ for all $t \geq 0$? If $I = 1$, $MgL = 1/16$, and $L = 1$, what is $\phi(t)$ if $x(t) = u(t) - u(t - 0.1)$? Experimentally verify.

5-16 Two-phase motor   The speed-voltage relationship of the two-phase electric motor given in Figure 5-21 is given by

$$\frac{\Phi(s)}{E(s)} = \frac{K_s K_m K_g}{s(Js + (K_s + B))(T_1 s + 1)}$$

where J is the load inertia, B is the viscous friction, $\phi(t)$ is the angular rotation, $e(t)$ is the applied control voltage, and $J = 0.001$ in.-oz sec^2, $B = 0.01$ in.-oz sec. $T_1 = 0.01$ sec. $K_g = 0.01$ $K_m = 0.0395$.

  The angular rotation is interpreted in radians, where one revolution of the shaft equals 2π radians. The parameter K_s is an adjustable servo-amplifier gain. Classify the stability of the system for $K_s \in [0, \infty)$. For $K_s = 10^{-2}$, compute and display the system's response to $e(t) = \cos(377t)u(t)$. What would the output response be if $e(t) = u(t)$?

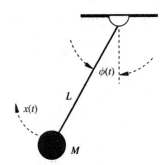

FIGURE 5-20   Pendulum model.

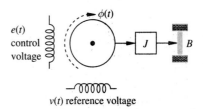

FIGURE 5-21   Two-phase servomotor.

5-17 Graphical design   Graphically compute and display the magnitude and phase-frequency response of

$$H(s) = \frac{s+1}{(s+1)^2 + 4}$$

in lin-lin and Bode form.

5-18 Bode plot   Compute and sketch a straight-line approximate Bode plot of each system defined in Problem 5-1. Verify, using a computer.

5-19 Bode plot   Produce an approximate Bode plot for the RLC circuit shown in Figure 5-19. Compare to the the actual frequency response.

5-20 Feedback   A simple feedback system is shown in Figure 5-22. Show that the input-output transfer function is given by

$$T(s) = \frac{H(s)}{1 - KG(s)H(s)}$$

Does knowledge of the pole locations of $G(s)$ and $H(s)$ provide sufficient information to establish the stability of $T(s)$. Suppose $H(s) = 10/s$ and $G(s) = 1/(s+10)$. Over what range of K can asymptotic stability be guaranteed? What about marginal stability?

Sketch the migration of the pole locations of $T(s)$ for $K \in [0, \infty)$. Historically, this is called a *root-locus* graph. Of what potential importance is the graph?

5-21 Quality Factor   In Chapter 12 frequency-selective filters are studied. Suppose bandpass filter has -3 dB cut-off frequencies of f_{LO} and f_{HI} where $f_{LO} < f_{HI}$, bandwidth $BW = f_{HI} - f_{LO}$, and a geometric center frequency $f_0 = \sqrt{f_{LO}f_{HI}}$. The frequency selectivity of the filter is parameterized in terms of the *quality factor*, or Q, where

$$Q = \frac{f_0}{BW}$$

Suppose that the filter is defined by Equation 4-43, namely

$$\frac{d^2y(t)}{dt^2} + 2\delta\omega_0 \frac{dy(t)}{dt} + \omega_0^2 y(t) = \omega_0^2 x(t)$$

Express the "Q" of the system in terms of δ and ω_0. Sketch the magnitude frequency response for a system having values of $Q = 1, 10$, and 100. What is the phase of the system at frequency ω_0? Physically, what does this represent?

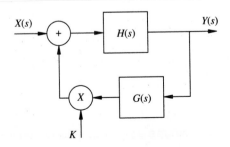

FIGURE 5-22   Simple feedback system.

5-22 Butterworth filter   In Chapter 12, frequency-selective filters will be studied. A second-order Butterworth lowpass filter is given by

$$H(s) = \frac{2\omega_0^2}{(s + \omega_0)^2 + \omega_0^2}$$

What is the filter's steady-state frequency response? What is the filter's impulse response?

5-23 Automotive suspension system   An automotive suspension system was defined in Example 5-4 to be

$$H(s) = \frac{0.095811}{s^2 + 0.43774s + 0.095811}$$

What is the system's magnitude frequency response, phase response, and magnitude and phase Bode plot?

5-10   COMPUTER PROJECTS

5-1 RC amplifier   The system shown in Figure 5-23 represents an active RC-tuned amplifier. The amplifier is assumed to have an infinite gain. The transfer function is given by

$$H(s) = \frac{C_0}{R_0 C^2} \frac{\omega_0^2 Q^2}{1 - (kR/R_1)^2}$$

$$\frac{s}{(s^2 + s(\omega_0/Q) + (\omega_0^2/(1 + k(R/R_1)))(s^2 + s(\omega_0/Q) + (\omega_0^2/(1 - k(R/R_1)))}$$

For $4C_0/R_0 C^2 = 25 * 10^{-5}$ and $Q = 20$ (−3 dB bandwidth), compute a family of magnitude-frequency responses for $kR/R_1 \in [0.01, 0.2]$. Can the system become unstable for $k \geq 0$? Display the results in all presentation formats. Determine the center frequency in all cases. Using both computer and manual techniques, create a

FIGURE 5-23   Active RC circuit.

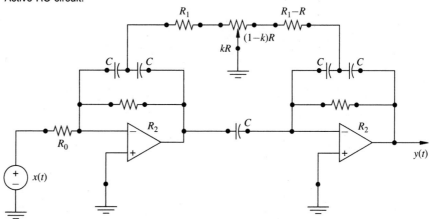

Bode plot of the filter. Add up to two additional zeros and variable gain (if needed), increase the steepness of the system's transition band by 20% without significantly altering the system's center frequency and value of Q.

5-2 AC-DC rectifier   In Example 3-6 and Problems 3-27 and 4-13, a DC rectifier was studied. Upon fully rectifying an AC line power signal $E(t) = 120\sqrt{2}\sin(120\pi)$, $V(t) = 120\sqrt{2}|\sin(120\pi)|$ resulted. The resulting spectrum consisted of a DC term plus even harmonics. Design a third-order system to map $V(t)$ into a signal with a DC value of 5 volts, and the second harmonic distortion less than 0.05.

6

CONTINUOUS-TIME SYSTEM ARCHITECTURE

All architecture is what you do to it when you look upon it.

—Walt Whitman

6-1 SYSTEM ARCHITECTURE

In the previous chapters, continuous-time systems were described in terms of differential equations and transfer functions. These systems, once designed, must often be implemented in hardware. The physical implementation of a system, or filter, is called a *system architecture* or *architecture*. The specification of a system architecture is the next to the last step in the design paradigm presented in Figure 6-1. In this model, system requirements are assumed to be numerically specified. From these parameters, a candidate system-transfer function is derived. If the system is to process actual signals, the candidate transfer function must be converted into an architecture specification representing a system of interconnected hardware entities. However, due to the inaccuracies found in physical systems, the final design must be tested again to ensure that it continues to meet the specified design objectives. If it does not, then either the mathematical model (Step 2) or the architecture (Step 3), or both, must be modified.

A system architecture can be represented in a number of acceptable forms. Three important forms are block diagrams, signal flow graphs, and state variables. Block diagrams and signal-flow graphs provide pictorial representations of a system, while state variables are algebraic models.

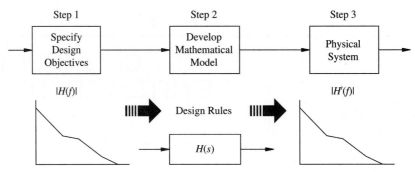

FIGURE 6-1   System design paradigm.

6-2   BLOCK DIAGRAMS AND SIGNAL-FLOW GRAPHS

Continuous-time systems can be represented in block diagram or signal-flow graph form. A block diagram, as the name implies, consists of a system of interconnected blocks or subsystems, each performing a specific function. A signal-flow graph encapsulates this information using a set of nodes connected together by directed line segments called branches. To illustrate, the graphical representation of a continuous-time LTI system, given by the transfer function

$$H(s) = \frac{H_1(s)}{1 - H_1(s)H_2(s)} \tag{6-1}$$

is shown in both block and signal-flow diagram form in Figure 6-2.

A block diagram or signal-flow graph displays how the information is managed by a system and shared among its subsystems and components. Some subsystems manage only input-output data, while others perform specific mathematical operations. Block diagrams and signal-flow graphs can be hierarchical and, as such, are particularly useful in integrating a collection of subsystems into a larger system. The process by which the subsystems are combined to form larger systems is called block or signal-flow diagram *reduction*.

Block diagram reduction techniques are based on two basic operations, *forking* and *joining*. Forking occurs at nodes that are called *pickoff points* or *points-of-*

FIGURE 6-2   Block and signal-flow diagram of $H(s)$.

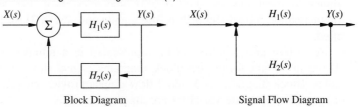

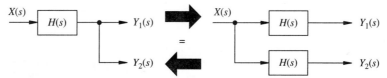

FIGURE 6-3   Pickoff point movement.

bifurcation. The pickoff point can be moved through (or back through) a block, as indicated in Figure 6-3. Externally, the two systems displayed in Figure 6-3 are identical.

Joining operations are performed using adders and subtractors. Joiners can be used to move signals through (or back through) a node, as reported in Figure 6-4. The two systems displayed in Figure 6-4 are functionally identical.

Using these elementary operations, along with Equation 6-1, a highly complex collection of subsystems can be reduced to a system.

Example 6-1   Block diagram reduction

Consider the continuous-time system shown in Figure 6-5. The third-order system consists of a set of simple subsystems interconnected to form what will be called, later in this chapter, an extended companion form. The model can be reduced to a single input-output, third-order system by using the step-by-step process described in Figure 6-5. The example begins the block-diagram reduction process at the output side of the system and then systematically reduces the number of uncombined subsystems at each subsequent step. The reduction process, however, can begin anywhere. Each step of the block-diagram reduction is shown in each shaded area found in Figure 6-5.

FIGURE 6-4   Adder/subtractor movement.

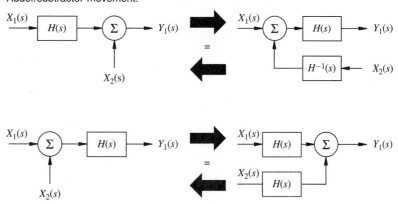

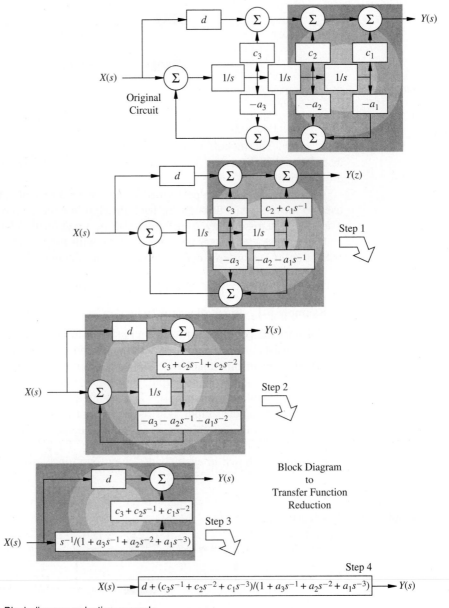

FIGURE 6-5   Block-diagram reduction example.

As a final step, upon simplification, the transfer function can be expressed as

$$H(s) = \frac{d + (c_3 + da_3)s^{-1} + (c_2 + da_2)s^{-2} + (c_1 + da_1)s^{-3}}{1 + a_3s^{-1} + a_2s^{-2} + a_1s^{-3}}$$

or

$$H(s) = \sum_{i=0}^{3} \beta_i s^i \Big/ \sum_{i=0}^{3} \alpha_i s^i$$

where $\alpha_0 = a_1$, $\alpha_1 = a_2$, $\alpha_2 = a_3$, $\alpha_3 = 1$, $\beta_0 = (c_1 + da_1)$, $\beta_1 = (c_2 + da_2)$, $\beta_2 = (c_3 + da_3)$, and $\beta_3 = d$.

6-3 MASON'S GAIN FORMULA

Large systems are sometimes designed in a piecemeal fashion. Reducing a large, complicated block diagram or signal-flow graph into a single input-output model is both a commonplace and necessary operation. Once a system is captured as a transfer function, it can then be studied using any of the previously developed system-analysis techniques. If the system is a complicated collection of many subsystems, then using manual block-diagram reduction techniques can be too cumbersome to be of practical use. It would therefore be desirable to automate this process. Such a procedure is available; it is the *Mason's Gain Formula*. Since it is a formula, it provides a recipe that, if followed, will result in the production of a transfer function.

Mason's Gain Formula is based on graph theoretic methods and the use of signal-flow graphs. To understand how to use the formula, it is important to recall that a signal-flow graph consists of the following list elements and components, as can be seen in Figure 6-6.

Branches—A directed line segment connecting two nodes; [1,2], [3,2]
Input Node [Source]—A node possessing only outgoing branches; [1]
Output Node [Sink]—A node possessing only incoming branches; [4]
Internal Node—Entry and/or exit point for branches; [2], [3]
Path—Any continuous unidirectional collection of branches; [1,2,3,4]
Feedback path—Path that originates and terminates on a common node in which all interconnected nodes are traversed only once; [2,3,2]
Feedforward path—Path connecting an input to an output in which all interconnected nodes are traversed only once; [2,3]
Path gain—Product of all the branch gains along a connected path; B = [2,3]
Loop Gain—Product of all the branch gains along a closed path; BD = [2,3,2]

These parameters are used to define Mason's Gain Formula.

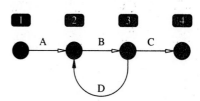

FIGURE 6-6   Sample flow diagram.

Mason's Gain Formula.   Mason's Gain Formula states that the transfer function of an LTI continuous-time case system, having a known flow-diagram representation, can be written as

$$H(s) = \sum_k \frac{M_k(s)\Delta_k(s)}{\Delta(s)} \qquad (6\text{-}2)$$

where the path gain of the kth feedforward path is $M_k(s)$ and $\Delta(s)$ is the *characteristic equation* of $H(s)$, given by

$$\Delta(s) = 1 - \left[\sum \text{ gains from all individual loops } (r = 1)\right]$$

$$+ \left[\sum \text{gains of all possible combinations of two nontouching loops}\right.$$

$$\left.(r = 2)\right] \pm \dots$$

or formally

$$\Delta(s) = 1 - \sum_m P_{m1}(s) + \sum_m P_{m2}(s) - \sum_m P_{m3}(s) + \dots \qquad (6\text{-}3)$$

where $P_{mr}(s)$ = gain of the m possible combinations of r nontouching loops (i.e., no common nodes), and $\Delta_k(s)$ = value of Equation 6-3 for the part of the graph not touching the kth feedforward path.

The rules, while appearing to be complicated in the abstract, are really well-ordered, as demonstrated by the following example.

Example 6-2   Mason's Gain Formula

Consider the system shown at the top of Figure 6-7, containing a "self-loop" attached to node 4. The system is seen to consist of six nodes, with the input assigned to node 1 and the output to node 6. It can be seen that there are three direct input-to-output paths, namely [1, 2, 3, 4, 5, 6], [1, 2, 4, 5, 6], and [1, 2, 5, 6]. There are four feedback loops, namely [2, 4, 3, 2], [3, 4, 3], [2, 3, 2], and [4, 4] where the last two are nontouching.

The feedback paths are isolated in Figure 6-8. From the feedback data the system characteristic equation $\Delta(s)$ can be derived using Equation 6-3. It is given by

$$\Delta(s) = 1 - (P_{11}(s) + P_{21}(s) + P_{31}(s) + P_{41}(s)) + P_{12}(s)$$

$$= 1 - [s^{-1}C_{32} + s^{-1}C_{43} + C_{44} + C_{32}C_{43}C_{24}] + [s^{-1}C_{32}C_{44}]$$

Combining the Mason path gains M_i, shown in Figure 6-7, and $\Delta_i(s)$, the transfer function $H(s)$ can be computed to be

$$H(s) = \frac{s^{-3}C_{12} + s^{-1}C_{12}C_{24} + C_{12}C_{25}[1 - s^{-1}C_{43} - C_{44}]}{1 - [s^{-1}C_{32} + s^{-1}C_{43} + C_{44} + C_{32}C_{43}C_{24}] + [s^{-1}C_{32}C_{44}]}$$

$$= \frac{(C_{12}C_{25}(1 - C_{44}))s^3 + (C_{12}C_{24} - C_{12}C_{25}C_{43})s^2 + C_{12}}{s^2((1 - C_{44} - C_{32}C_{43}C_{24})s - (C_{32} + C_{43} - C_{32}C_{44}))}$$

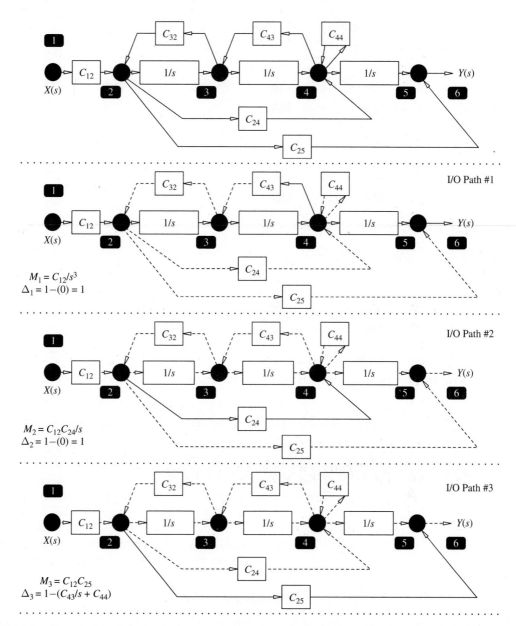

$$M_1 = C_{12}/s^3$$
$$\Delta_1 = 1-(0) = 1$$

$$M_2 = C_{12}C_{24}/s$$
$$\Delta_2 = 1-(0) = 1$$

$$M_3 = C_{12}C_{25}$$
$$\Delta_3 = 1-(C_{43}/s + C_{44})$$

FIGURE 6-7 Mason's Gain Formula example showing original system at the top and the feedforward paths following.

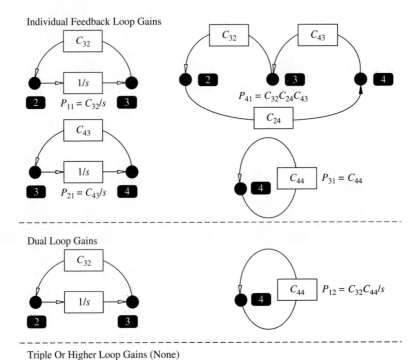

FIGURE 6-8   Mason's Gain Formula for the feedback paths for the system shown in Figure 6-7.

From this example it can be seen that a transfer function can be computed using a formal procedure.

A typical continuous-time Nth order LTI system may contain up to N feedback paths and M feedforward paths. The actual number of feedforward and feedback paths are established by the chosen architecture. The number of paths required to implement a given transfer function can vary significantly. Some architectures reduce the number of paths required to realize a system to a small value in order to maximize the potential bandwidth. Other architectures introduce more paths in order to become less susceptible to the effects of noise or parametric variation.

Example 6-3   Third-order system

A third-order transfer function $H(s)$ is assumed to have a block-diagram representation given in Figure 6-9. (This extended companion-form architecture is discussed later in the chapter.) There are four identifiable feedforward paths that can be used to compute the system's transfer function using Mason's Gain Formula as follows

Step 1: Compute Feedforward Paths (see Figure 6-9)

$$M_1 = d, M_2 = c_3 s^{-1}, M_3 = c_2 s^{-2} \text{ and } M_4 = c_1 s^{-3}$$

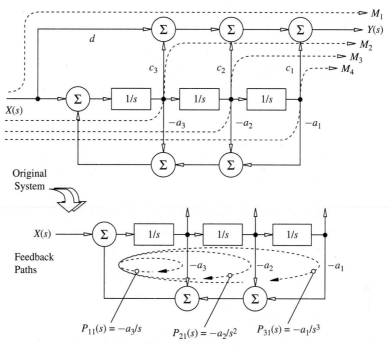

FIGURE 6-9   Original system architecture, feedforward, and feedback loops.

Step 2: Compute characteristic equation $\Delta(s)$. Closed single loops are also shown in Figure 6-9. There are seen to be no nontouching dual loops. Therefore

$$\Delta(s) = 1 - P_{11}(s) - P_{21}(s) - P_{31}(s) = 1 + a_3 s^{-1} + a_2 s^{-2} + a_1 s^{-3}$$

Step 3: Compute $\Delta_i(s)$ for $i \in [1, 4]$.

$$\Delta_1(s) = 1 - P_{11}(s) - P_{21}(s) - P_{21}(s) = 1 + a_3 s^{-1} + a_2 s^{-2} + a_1 s^{-3} = \Delta(s)$$

Note that M_1, shown in Figure 6-9, does not touch the identified feedback paths also shown in Figure 6-9. However, the other feedforward paths touch at least one node of a feedback path. Therefore

$$\Delta_2(s) = \Delta_3(s) = \Delta_4(s) = 1$$

Step 4: Compute $H(s)$.

Upon combining all the terms as defined by Equation 6-2, we obtain

$$H(s) = \frac{d\Delta(s) + c_3 s^{-1} + c_2 s^{-2} + c_1 s^{-3}}{\Delta(s)}$$

$$= d + \frac{c_3 s^2 + c_2 s + c_1}{s^3 + a_3 s^2 + a_2 s + a_1}$$

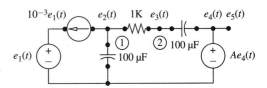

FIGURE 6-10 Electrical circuit with dependent voltage source.

Mason's Gain Formula is an algorithm and therefore amenable to computer implementation. Such a computer program is discussed in detail in Appendix C. Mason's Gain Formula is important because it can be used to reduce a signal-flow graph to a transfer function, as is demonstrated in the next example.

Example 6-4 Computer-generated, continuous-time transfer function

A circuit with a dependent voltage source is shown in Figure 6-10. The Laplace transforms of the circuit equations are given by

Node equation 1

$$10^{-4}sE_2(s) - 10^{-3}E_1(s) + 10^{-3}(E_2(s) - E_3(s)) = 0$$

Node equation 2

$$10^{-3}(E_3(s) - E_2(s)) + 10^{-4}s(E_3(s) - E_4(s)) = 0$$

Voltage equation

$$E_4(s) = AE_3(s)$$

Solving these equations for $E_2(s)$, $E_3(s)$, and $E_4(s)$, the following system of coupled equations is obtained

$$E_2(s) = \frac{10E_1(s)}{(s + 10)} + \frac{10E_3(s)}{(s + 10)}$$

$$E_3(s) = \frac{10E_2(s)}{(s + 10)} + \frac{sE_4(s)}{(s + 10)}$$

$$E_4(s) = AE_3(s)$$

In terms of a signal-flow graph, the electrical circuit shown in Figure 6-10 translates into the diagram shown in Figure 6-11, where node 5 has been added to provide an isolated output node. The signal-flow graph shown in Figure 6-11 is in a form suitable for analysis, using Mason's Gain Formula.

FIGURE 6-11 Signal-flow graph for electrical circuit shown in Figure 6-10.

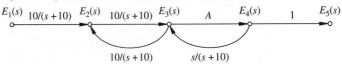

Upon performing this analysis, we obtain

$$H(s, A) = \frac{100A}{(1 - A)s^2 + (20 - 10A)s}$$

Computer Study For $A = 1$ and $A = -1$, the signal-flow graphs are numerically parameterized. The signal-flow graph information is processed by the program `sfg.exe` to produce a transfer-function model from the signal-flow graph shown in Figure 6-11. The format is given by `sfg filename`. The signal-flow graph information is saved in file `ckt1.sfg` for the case where $A = 1$ and `ckt-1.sfg` for $A = -1$. The information has the following structure (also see Section 6-11).

Unit gain paths
$1 = [4, 5]$
Transfer function subsystems
$10/(s + 10) = [1, 2] = [2, 3] = [3, 2], \ s/(s + 10) = [4, 3]$
Dependent gain path
$A = [3, 4]$

The transfer function is produced from a DOS command line using these commands.

SIGLAB	MATLAB
`>! path\sfg path\ckt1.sfg #DOS`	Not available in Matlab

The resulting transfer function is computed and displayed as seen below.

SIGLAB	MATLAB
echo print of branch and node data messages generated by program execution loops and paths in matrix form	Not applicable

```
Signal flow graph transfer function
-----------------------------------
Numerator Polynomial Coefficients
100000; 30000; 3000; 100; 0; 0
Denominator Polynomial Coefficients
-----------------------------------
0; 10000; 3000; 300; 10; 0
```

This establishes the transfer function for the case where $A = 1$ to be

$$H(s) = \frac{100(s^3 + 30s^2 + 300s + 1000)}{10(s^4 + 30s^3 + 300s^2 + 1000s)} = \frac{10}{s}$$

which agrees with $H(s, 1)$.

That is, if $A = 1$, the entire system reduces to an integrator having a transfer function given by $H(s) = 10/s$. If, however, $A = -1$, a completely different result is obtained, as shown here.

SIGLAB	MATLAB
>! path\sfg path\ckt-1.sfg #DOS	Not available in Matlab

The resulting transfer function is computed and displayed as seen below.

SIGLAB	MATLAB
echo print of branch and node data	Not applicable
messages generated by program execution	
loops and paths in matrix form	
Signal flow graph transfer function	

Numerator Polynomial Coefficients	
-100000; -30000; -3000; -100; 0; 0	
Denominator Polynomial Coefficients	

0; 30000; 11000; 1500; 90; 2	

This establishes the transfer function, for the case where $A = -1$, to be

$$H(s) = \frac{-100(s^3 + 30s^2 + 300s + 1000)}{2s^5 + 90s^4 + 1500s^3 + 11000s^2 + 30000s}$$

Using long division, it can be shown that the denominator of $H(s)$ can be factored as $25^5 + 90s^4 + 1500s^3 + 11000s^2 + 30000s = (s^3 + 30s^2 + 300s + 1000)(25^2 + 30s)$. Therefore, $H(s)$ reduces to

$$H(s) = \frac{-100}{2s^2 + 30s} = \frac{-50}{s(s + 15)}$$

which agrees with $H(s, -1)$.

The poles of $H(s)$ can be computed as follows.

SIGLAB	MATLAB
>x=[2,30,0]	>> x = [2 30 0];
>root(x)'	>> roots(x)
-15+0j	ans =
0+0j	
	0
	-15.00

Both systems are seen to have poles at the origin, therefore are conditionally stable.

It can be seen that block and signal-flow diagrams can be used to show how subsystems exchange information. Later in this chapter this concept will be directly related to the concept of system architecture. Before that, another algebraic method of accounting for the interchange of information within a system, called the *state-variable method*, will be presented. Because it is an algebraic method, it can be used as a foundation upon which powerful analysis techniques can be built.

6-4 STATE VARIABLES

A state-variable model of a continuous-time LTI system describes completely how information, both external and internal to the system, is managed. It may be recalled that a similar claim was made in Chapter 4 about the ability of differential equations to model some systems. Furthermore, this claim was extended to transforms in Chapter 5. Why then is there a need for state variables? It will be shown that, when compared to other techniques, state variables can be used to more efficiently model and analyze linear systems.

Loosely stated, the information-bearing variables of a system are called *state variables*. The set of all state variables is called the system's *state*. State variables contain sufficient information so that all future states and outputs can be computed

1 if the system's past history is known (i.e., previous states), and

2 if the system's input-output relationships are known (i.e., system model or equation), and

3 if the system's future inputs are known.

Therefore, knowledge of a system's state is sufficient to specify where the system is, how it got there, and where it is going. The president's State of the Union address, in concept, provides the listener with an assessment of where the nation stands, how it got there, and, assuming that the ecosocial forces can be predicted, predicts where the nation will be at the end of the year.

Engineers are familiar with the basic concept of state variables although they may have not been formally introduced to them as a mathematical study. They are often unaware that they are using state variables to analyze a variety of systems defined by a set of differential equations. However, once mastered, state-variable methods can greatly simplify the study of complicated linear systems. State-variables models, for example, can range from a 50-state rocket booster to a simple one-state RC circuit.

State variables identify locations where information is stored. Therefore, the first task in reducing an arbitrary LTI system to state-variable form is identifying where the system stores information. To represent an RC circuit in state-variable form, we first identify where the system stores information, which in this case is the energy stored in the capacitor. The energy found in the capacitor, in turn, is a function of voltage. It is known that the voltage developed across that capacitor is governed by the first-order ordinary differential equation $RCdv_c(t)/dt + v_c(t) = v(t)$, where $v_c(0) = v_0$. Assume, therefore, that $v_c(t)$ is assigned to be the system's state variable. Then, in the context of the three state requirements,

1 the past state history is the initial condition $v_c(0) = v_0$;

2 the input-output relationship is given by the ODE $RCdv_c(t)/dt + v_c(t) = v(t)$;

3 the future input is $v(t)$;

and the solution $v_c(t)$ follows.

As a general rule, state variables are assumed to be attached to system elements that store information about the system's past history (i.e., initial conditions). It should be pointed out, however, that this assignment rule can sometimes result in a state-variable model of higher order than necessary. For example, two capacitors in parallel would require two state-variables if the stated assignment rule is strictly interpreted. However, the capacitors can be first combined to form a single equivalent device reducing the state requirement to one.

For the purpose of demonstration, assume that the state assignment rule used for a series RC circuit shown in Figure 6-12, being forced by a voltage $v(t)$, is the capacitor voltage $v_c(t) = x(t)$. The system output $y(t)$ is defined to be the voltage developed across the resistor, which is represented by $y(t) = v(t) - v_c(t) = v(t) - x(t)$. The RC circuit equations can be written as a system of two equations. The first defines how the system manages state information and is given by

$$\frac{dx(t)}{dt} = -\frac{1}{RC}x(t) + \frac{1}{RC}v(t)$$

$$x(0) = x_0 \tag{6-4}$$

The second equation defines how state information is presented to the output and is given by

$$y(t) = -1x(t) + 1v(t) \tag{6-5}$$

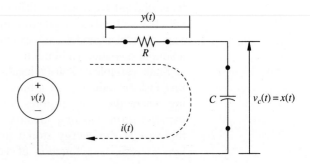

FIGURE 6-12   RC circuit with the output taken across the resistor load.

Such a system of equations can be reduced to an elegant vector-matrix, or state-variable notation, as follows

$$\frac{d\mathbf{x}(t)}{dt} = \mathbf{A}\mathbf{x} - 1t) + \mathbf{b}v(t) \qquad \text{State equation} \qquad (6\text{-}6)$$

$$\mathbf{x}(0) = \mathbf{x}_0$$

$$y(t) = \mathbf{c}^T\mathbf{x}(t) + du(t) \qquad \text{Observation (output) equation} \qquad (6\text{-}7)$$

The coefficients contained in Equations 6-6 and 6-7 can also be expressed as the four-tuple $\mathcal{S} = [\mathbf{A}, \mathbf{b}, \mathbf{c}, d]$. For the RC example, the state vector is a one-dimensional state vector $x(t)$, A is a 1×1 matrix (i.e., $A = -1/RC$), b and c are one-dimensional vectors (i.e., $b = 1/RC$, $c = -1$), and d is a scalar (i.e., $d = 1$). Thus the system's numerical, four-tuple description is given by $\mathcal{S} = [-1/RC, 1/RC, -1, 1]$. The history is specified by x_0 and the input is $u(t)$. This model can be seen to completely characterize the RC network.

Example 6-5 RLC network

A simple RLC network is shown in Figure 6-13. The two energy storage devices in this circuit are the inductor and capacitor. A capacitor stores energy in an electric field while an inductor stores energy in an electric field. Therefore, the state assignments will be assumed to be $x_1(t) = i_L(t)$ and $x_2(t) = v_C(t)$. The system voltage equations are

$$L\frac{di_L(t)}{dt} + Ri_L(t) + v_C(t) = v(t)$$

and

$$\frac{1}{C}\int_0^t i_L(\tau)\,d\tau = v_C(t) \Leftrightarrow \frac{dv_C(t)}{dt} = \frac{i_L(t)}{C}$$

Therefore, the state model can be seen to be specified by the four-tuple $\mathcal{S} = [\mathbf{A}, \mathbf{b}, \mathbf{c}, d]$, which is given by

$$\frac{d\mathbf{x}(t)}{dt} = \begin{pmatrix} -R/L & -1/L \\ 1/C & 0 \end{pmatrix}\mathbf{x}(t) + \begin{pmatrix} 1/L \\ 0 \end{pmatrix}v(t) \qquad \text{State equation}$$

$$y(t) = x_2(t) = (0 \quad 1)\mathbf{x}(t) + 0v(t) \qquad \text{Observation equation}$$

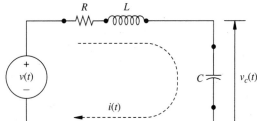

FIGURE 6-13 RLC circuit.

where $\mathbf{x}(t)$ is a two-dimensional state vector and the initial condition $\mathbf{x}(0) = \mathbf{x}_0 = [i_L(0), v_C(0)]^T$, is defined in terms of the initial capacitor voltage and inductor current.

Example 6-6   Mechanical analog

A basic dynamic mechanical system, consisting of an external force, spring, mass, and dashpot (shock absorber), is shown in Figure 6-14. In this system, the energy storage devices are the mass and spring. The energy stored in the spring is a function of displacement, whereas the energy stored by the mass can be expressed in terms of momentum. The dashpot dissipates energy as a function of velocity and f(t) is the force applied to the mass. Therefore, the states should be displacement and momentum. In particular

State assignment
$$x_1(t) = y(t)$$

$$x_2(t) = M\,dy(t)/dt$$

Output assignment
$$y(t) = x_1(t)$$

State equations
$$dx_1(t)/dt = (1/M)x_2(t)$$

$$dx_2(t)/dt = -Kx_1(t) - (D/M)x_2(t) + f(t)$$

State variable model
$$\frac{d\mathbf{x}(t)}{dt} = \begin{pmatrix} 0 & 1/M \\ -K & -D/M \end{pmatrix} \mathbf{x}(t) + \begin{pmatrix} 0 \\ 1 \end{pmatrix} f(t)$$

Observation (output) model
$$y(t) = (\,1 \quad 0\,)\,\mathbf{x}(t) + 0 f(t)$$

Initial conditions would be supplied in terms of initial displacement $x_1(0)$ and momentum $x_2(0)$. Other state assignment rules are equally valid.

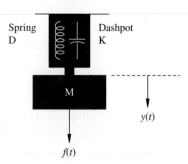

FIGURE 6-14   Mechanical system.

6-5 CONTINUOUS-TIME LTI SYSTEMS

Up to now, most continuous-time LTI systems studied were modeled by a single-input single-output ODE. In general, multiple-input multiple-output continuous-time LTI systems can also be represented using an ODE. A p-input nth-order system, for example, would be modeled as an nth ODE with n initial conditions. In particular

$$\alpha_n \frac{dx^n(t)}{dt^n} + \alpha_{n-1}\frac{d^{n-1}x(t)}{dt^{n-1}} + \cdots + \alpha_1\frac{dx(t)}{dt} + \alpha_0 x(t) = \beta_p u_p(t) + \cdots + \beta_1 u_1(t)$$

(6-8)

$$x(t)|_{t=0} = x_0(0), \frac{dx(t)}{dt}|_{t=0} = x_1(0), \ldots, dx^n(t)/dt^n|_{t=0} = x_n(0)$$

Assume further that there are r-outputs, each having the form

$$y_j(t) = \gamma_{jn}\frac{dx^n(t)}{d^n t} + \gamma_{jn-1}\frac{dx^{n-1}(t)}{d^{n-1}t} + \cdots + \gamma_{j1}\frac{dx(t)}{dt}$$

$$+ \gamma_{j0}x(t) + \lambda_{jp}u_p(t) + \cdots + \lambda_{j1}u_1(t)$$

(6-9)

where $j \in [1, r]$. It can be shown that such a system can be reduced to a state-variable model $\mathcal{S} = [\mathbf{A}, \mathbf{B}, \mathbf{C}, \mathbf{D}]$ given by

$$\frac{d\mathbf{x}(t)}{dt} = \mathbf{A}\mathbf{x}(t) + \mathbf{B}\mathbf{u}(t) \qquad \mathbf{x}(0) = \mathbf{x}_0 \qquad \text{State equation} \qquad (6\text{-}10)$$

$$\mathbf{y}(t) = \mathbf{C}^T\mathbf{x}(t) + \mathbf{D}\mathbf{u}(t) \qquad \text{Observation (output) equation} \qquad (6\text{-}11)$$

where $\mathbf{x}(t)$, $\mathbf{u}(t)$, and $\mathbf{y}(t)$ are n, p, and r-dimensional vectors respectively, and $\mathbf{A}, \mathbf{B}, \mathbf{C}, \mathbf{D}$ are $n \times n$, $n \times p$, $n \times r$, and $r \times p$ matrices respectively.

The state-determined system specified in Equations 6-10 and 6-11 is graphically interpreted in Figure 6-15. The system is seen to consist of two parts, the state and observation (or output) model, consisting of vector data paths of width

FIGURE 6-15 State-variable continuous-time system model.

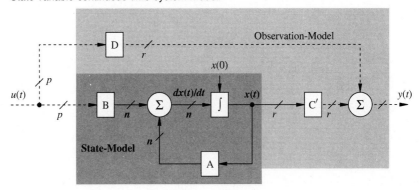

n, p, and r. This partitioning of function is found naturally in systems and system design. The state model, sometimes called a *plant*, represents the dynamic part of the system. The system-feedback structure, and therefore evidence of its eigenfunctions and stability, is completely encapsulated in the plant portion of the model. The observation system describes how the system interfaces to the external world.

6-6   STATE TRANSITION MATRIX

Recall that an ODE possesses both homogeneous and inhomogeneous solutions. The unforced, or homogeneous, response is the solution to the state equation $d\mathbf{x}(t)/dt = \mathbf{A}(t)\mathbf{x}(t)$, $\mathbf{x}(t_0) = \mathbf{x}_0$, for $t \geq t_0$, and is symbolically represented in *state-transition matrix* form as

$$\mathbf{x}(t) = \Phi(t, t_0)\mathbf{x}_0 \qquad (6\text{-}12)$$

Observe that at $t = t_0$, $\mathbf{x}(t_0) = \Phi(t_0, t_0)\mathbf{x}_0$, which means $\Phi(t_0, t_0) = \mathbf{I}$ (the identity matrix). It can also be noted that given an initial condition $\mathbf{x}_0$ at $t = t_0$, the trajectory $\mathbf{x}(t)$ has intermediate values $\mathbf{x}(t_1)$, $\mathbf{x}(t_2)$, ..., $\mathbf{x}(t_n = t)$, where

$$\mathbf{x}(t_i) = \Phi(t_i, t_0)\mathbf{x}(t_0)$$

$$\mathbf{x}(t) = \Phi(t_n, t_{n-1})\Phi(t_{n-1}, t_{n-2}) \dots \Phi(t_1, t_0)\mathbf{x}(t_0) \qquad (6\text{-}13)$$

This is called the *transition property* and is graphically interpreted in Figure 6-16 for a second-order, or two-state, system.

For a first-order equation, $dx(t)/dt = ax(t)$, the matrix exponential $\Phi(t, t_0)$ is a scalar function given by $\Phi(t, t_0) = \exp(a(t - t_0))$. For an nth-order LTI system, the scalar exponential $\Phi(t, t_0) = \exp(a(t - t_0))$ becomes, in form, the *matrix exponential* denoted $\Phi(t) = \exp(\mathbf{A}t)$ if $\mathbf{A}$ is an $n \times n$ constant-coefficient

FIGURE 6-16   State transition property.

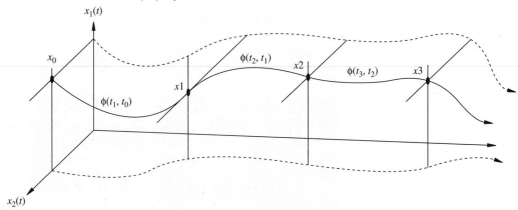

matrix. In such cases, the state-transition matrix and matrix exponential are equivalent. The matrix exponential may be expressed as the following Taylor series expansion

$$\Phi(t) = e^{\mathbf{A}t} = \mathbf{I}_{n \times n} + \mathbf{A}t + \mathbf{A}^2 \frac{t^2}{2!} + \mathbf{A}^3 \frac{t^3}{3!} + \cdots = \sum_{i=0}^{\infty} \mathbf{A}^i \frac{t^i}{i!} \quad \Phi(0) = \mathbf{I}_{n \times n} \quad (6\text{-}14)$$

If Equation 6-14 converges, the infinite series produces the $n \times n$ matrix $\Phi(t)$.

Example 6-7   Matrix exponential

The infinite sum given in Equation 6-14 is impractical from a numerical computation viewpoint. However, if for all i and j, $[\mathbf{A}^k]_{ij} \Rightarrow 0$ rapidly as k increases, then a finite sum can be used to approximate the infinite sum. That is

$$\Phi(t) = e^{\mathbf{A}t} \approx \sum_{i=0}^{K} \mathbf{A}^i \frac{t^i}{i!} \quad (6\text{-}15)$$

Computer Study   S-file `ch6_7.s` and M-file `matexp.m` compute a finite sum approximation of the matrix exponential for $K = 9$ and $t \in [0, 1]$. The format is given by `matexp(A)` where $\mathbf{A}$ is an $n \times n$ matrix. The elements of $\Phi(t)$, namely $\Phi_{ij}(t)$, are computed and displayed along with the computed ratio $r = \max(\mathbf{A})/\max(\mathbf{A}^9/9!)$. While not guaranteeing convergence, a small value of r indicates that the higher-order terms may be negligible, thus allowing a finite sum to be a reasonable approximation of the infinite sum found in Equation 6-14. For example, for

$$A = \begin{pmatrix} -1 & 1 \\ 0 & -2 \end{pmatrix}$$

the following computation is made.

SIGLAB	MATLAB
>include "ch6_7.s"	>> A = [-1 1; 0 -2]
>a={[-1,1],[0,-2]}	A =
>matexp(a)	-1 1
Dimension 2	0 -2
maximum ratio 0.00140818	>> matexp(A)
	maximum ratio = 0.001408

The values $\Phi_{ij}(t)$ are displayed in Figure 6-17. It can be seen that $\Phi_{11}(t)$ and $\Phi_{22}(t)$ are exponential, $\Phi_{12}(t)$ is more complex, and $\Phi_{21}(t)$ is zero for all time.

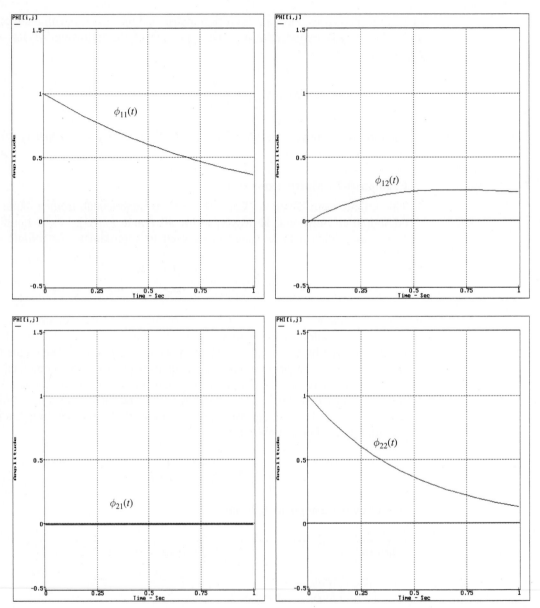

FIGURE 6-17   Trajectories of $\phi(t)$, (top left) $\phi_{11}(t)$, (top right) $\phi_{12}(t)$, (bottom left) $\phi_{21}(t)$, and (bottom right) $\phi_{22}(t)$.

The solution to the homogeneous-state equation $dx(t)/dt = A(t)x(t)$, $x(0) = x_0$ was given in Equation 6-12 as $x(t) = \Phi(t,0)x(0)$. For the case where A is a constant matrix, this reduces to $x(t) = \exp(At)x_0$. To verify that this is the solution to the homogeneous state equation $dx(t)/dt = Ax(t)$, $x(0) = x_0$, we need only show that it satisfies its defining differential equation. Substituting the candidate solution into the homogeneous-state differential equation, one obtains

$$\frac{d\mathbf{x}(t)}{dt} = \frac{de^{\mathbf{A}t}\mathbf{x}_0}{dt} = \left(\mathbf{A} + \mathbf{A}^2 t + \mathbf{A}^3 \frac{t^2}{2!} + \cdots\right)\mathbf{x}_0 = \mathbf{A}\mathbf{x}(t)$$

$$\mathbf{x}(t_0) = e^{\mathbf{A}0}\mathbf{x}(t_0) = \mathbf{x}(t_0) \tag{6-16}$$

Therefore, $\mathbf{x}(t) = \exp(\mathbf{A}t)\mathbf{x}_0$ is the homogeneous solution, and the practical problem reduces to efficiently computing the state-transition matrix $\Phi(t) = \exp(\mathbf{A}t)$. This problem will be resolved later in this chapter. Nevertheless, once $\Phi(t) = \exp(\mathbf{A}t)$ is known, the observed (output) homogeneous solution can be immediately expressed as

$$\mathbf{y}(t) = \mathbf{C}^T\mathbf{x}(t) = \mathbf{C}^T \exp(\mathbf{A}t)\mathbf{x}_0 \tag{6-17}$$

Example 6-8   Homogeneous state solution

A single-input single-output, second-order, continuous-time system is assumed to have a state equation given by

$$\frac{d\mathbf{x}(t)}{dt} = \begin{pmatrix} -1 & 1 \\ 0 & -2 \end{pmatrix}\mathbf{x}(t) + \begin{pmatrix} 0 \\ 1 \end{pmatrix} u(t) \qquad \text{State equation}$$

and an observation equation satisfying

$$y(t) = (1 \quad 1)\,\mathbf{x}(t)$$

The continuous-time LTI system has a simulation diagram given in Figure 6-18. The matrix exponential, in Taylor series form, is given by

$$e^{\mathbf{A}t} = \mathbf{I} + \mathbf{A}t + \mathbf{A}^2\frac{t^2}{2!} + \mathbf{A}^3\frac{t^3}{3!} + \cdots$$

$$= \begin{pmatrix} 1 - t + t^2/2! - t^3/3! + \cdots & 0 + t - 3t^2/2! + 7t^3/3! + \cdots \\ 0 & 1 - 2t + 4t^2/2! - 8t^3/3! + \cdots \end{pmatrix}$$

$$= \begin{pmatrix} e^{-t} & e^{-t} - e^{-2t} \\ 0 & e^{-2t} \end{pmatrix}$$

FIGURE 6-18   Example of second-order system, state-variable model.

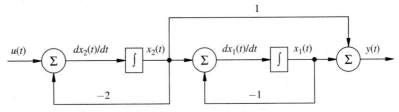

Therefore, for a given $\mathbf{x}_0$, the homogeneous solution satisfies

$$\mathbf{x}(t) = e^{\mathbf{A}(t-t_0)}\mathbf{x}(t_0) = \begin{pmatrix} e^{-t} & e^{-t} - e^{-2t} \\ 0 & e^{-2t} \end{pmatrix} \begin{pmatrix} x_1(0) \\ x_2(0) \end{pmatrix}$$

where

$$x_1(t) = \exp(-t)x_1(0) + (\exp(-t) - \exp(-2t))x_2(0)$$

$$x_2(t) = \exp(-2t)x_2(0)$$

$$y(t) = (1 \quad 1)\mathbf{x}(t) = (x_1(t) + x_2(t)) = \exp(-t)x_1(0) + \exp(-t)x_2(0)$$

Computer Study   The homogeneous trajectories $x_1(t)$ and $x_2(t)$ are defined in terms of $\Phi(t)$ and $\mathbf{x}_0$, and simulated using S-file `ch6_8.s` or M-file `homo.m`. Their format is given by `homo(sec,IC)`, where `sec` is the display period in seconds and `IC` is the initial conditions $x_1(0)$ and $x_2(0)$. The display of the homogeneous solution $\mathbf{x}(t) = \exp(\mathbf{A}t)\mathbf{x}(0)$, for $t \in [0, 5]$, and $\mathbf{x}_0 = (1 \quad 1)^T$ and $\mathbf{x}_0 = (1 \quad -1)^T$ is accomplished as follows.

SIGLAB	MATLAB
`>include "ch6_8.s"`	`>> ic = [1; 1];`
`>IC=[1,1]; homo(5,IC)`	`>> homo(5, ic)`
`>IC=[1,-1]; homo(5,IC)`	`>> ic = [1; -1];`
	`>> homo(5, ic)`

The results are shown in Figure 6-19. The solutions, in all cases, start at the specified initial conditions and exponentially decay toward zero.

As stated, computing the matrix exponential $\Phi(t) = \exp(\mathbf{A}t)$ using a Taylor series "infinite sum" (Equation 6-13), is generally impractical. Fortunately, there are several other methods that can be used to compute the matrix exponential $\Phi(t)$. One technique is referred to as the *Cayley-Hamilton method,* based on the *Cayley-Hamilton theorem* (see Appendix A). The Cayley-Hamilton method belongs to a branch of mathematics called a *function of a square matrix.* The method establishes a connection between the values of a scalar function $f(t)$ and matrix-valued function $f(\mathbf{A})$. For example, suppose $f(t)$ is a scalar function having a Taylor series representation

$$f(t) = \sum_{k=0}^{\infty} \frac{1}{k!}\alpha_k t^k \tag{6-18}$$

If $\mathbf{A}$ is a square matrix, then $f(\mathbf{A})$ is defined by

$$f(\mathbf{A}) = \sum_{k=0}^{\infty} \frac{1}{k!}\alpha_k \mathbf{A}^k \tag{6-19}$$

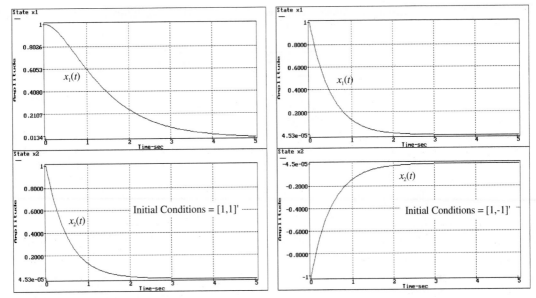

FIGURE 6-19   Homogeneous-state trajectories.

provided that the series converges. An example is the important Cayley-Hamilton theorem, which states that a matrix satisfies its own characteristic equation. That is, if $\mathbf{A}$ is an $n \times n$ matrix, then its characteristic equation is given by

$$f(\lambda) = \det(\mathbf{A} - \lambda\mathbf{I}) = \alpha_0 + \alpha_1\lambda + \alpha_2\lambda^2 + \cdots + \alpha_n\lambda^n \qquad (6\text{-}20)$$

then

$$f(\mathbf{A}) = \alpha_0\mathbf{I} + \alpha_1\mathbf{A} + \alpha_2\mathbf{A}^2 + \cdots + \alpha_n\mathbf{A}^n = \mathbf{O} \qquad (6\text{-}21)$$

This result has a number of uses. For example, assuming $\mathbf{A}$ is nonsingular, upon manipulating Equation 6-21 one obtains

$$\mathbf{A}^{-1} = -(\alpha_0)^{-1}[\alpha_1\mathbf{I} + \alpha_2\mathbf{A} + \cdots + \alpha_n\mathbf{A}^{n-1}] \qquad (6\text{-}22)$$

which provides a useful formula for expressing the inverse of a matrix in terms of powers of $\mathbf{A}$.

Example 6-9   Matrix inverse

Suppose $\mathbf{A}$ is the nonsingular matrix shown below

$$\mathbf{A} = \begin{pmatrix} 0 & 1 \\ -2 & -3 \end{pmatrix}$$

Then $\det(\mathbf{A} - \lambda\mathbf{I}) = \lambda^2 + 3\lambda + 2$. For $\alpha_0 = 2$, $\alpha_1 = 3$, and $\alpha_2 = 1$, then

$$\mathbf{A}^{-1} = -\frac{\alpha_1}{\alpha_0}\mathbf{I} - \frac{\alpha_2}{\alpha_0}\mathbf{A} = -\frac{3}{2}\mathbf{I} - \frac{1}{2}\mathbf{A} = \begin{pmatrix} -3/2 & -1/2 \\ 1 & 0 \end{pmatrix}$$

Computer Study   Suppose **A** is defined above, then Equation 6-22 produces the data, as illustrated.

SIGLAB	MATLAB
># define matrix	>> A = [0 1; -2 -3]
>A={[0,1],[-2,-3]}	A =
>p # characteristic equation	0 1
>p=poly(eig(A))	-2 -3
>a0=p[2]; a1=p[1]; a2=p[0]	>> eig(A)
>(-a1/a0)*eye(2)+(-a2/a0)*A	ans =
># A inverse	-1
-1.5 -0.5	-2
1 0	>> p = poly(eig(A))
	p =
	1 3 2
	>> a0 = p(3);
	>> a1 = p(2);
	>> a2 = p(1);
	>> -a1/a0*eye(2) - a2/a0*A
	ans =
	-1.5000 -0.5000
	1.0000 0
	>> inv(A)
	ans =
	-1.5000 -0.5000
	1.0000 0

Our objective is the efficient production of the matrix exponential $\exp(\mathbf{A}t)$, which is a function of the square matrix **A**. First, suppose that $f(\lambda)$ is a scalar function having a power series expansion given by

$$f(\lambda) = \sum_{k=0}^{\infty} \frac{1}{k!} \beta_k \lambda^k \qquad (6\text{-}23)$$

Then define

$$f(\mathbf{A}) = \sum_{k=0}^{\infty} \frac{1}{k!} \beta_k \mathbf{A}^k \qquad (6\text{-}24)$$

provided the matrix power series converges. If **x** is a eigenvector of the equation $\mathbf{A}\mathbf{x} = \lambda_i \mathbf{I}\mathbf{x}$, where λ_i is an eigenvalue of an $n \times n$ matrix **A**, then for all λ_i

$$[f(\mathbf{A}) - f(\lambda_i)\mathbf{I}]\mathbf{x} = \left[\sum_{k=0}^{\infty} \frac{1}{k!} \beta_k \mathbf{A}^k - \sum_{k=0}^{\infty} \frac{1}{k!} \beta_k \lambda_i^k \mathbf{I} \right] \mathbf{x} = \sum_{k=0}^{\infty} \frac{1}{k!} \beta_k (\mathbf{A}^k - \lambda_i^k \mathbf{I})\mathbf{x} = \mathbf{0}$$

$$(6\text{-}25)$$

Therefore, $f(\mathbf{A})$ exists if $f(\lambda_i)$ exists for all i. This relationship can be exploited to develop a formula that can be used to compute $f(\mathbf{A}) = \exp(\mathbf{A}t)$ in terms of

$f(\lambda) = \exp(\lambda t)$ which can be expanded as an infinite series and expressed in factored form as

$$f(\lambda) = \sum_{i=0}^{\infty} \lambda^i t^i / i! = q(\lambda) p(\lambda) + r(\lambda) \tag{6-26}$$

where $p(\lambda) = \det(\lambda I - A)$ (the characteristic equation), $r(\lambda) = \alpha_0 + \alpha, \lambda + \alpha_L \lambda^L + \ldots + \alpha_{n-1} \lambda^{n-1}$ (the remainder polynomial), and $\deg(r(\lambda)) < \deg(p(\lambda))$. Evaluating Equation 6-26 at $\lambda = \lambda_i$, and noting $p(\lambda) = 0$, it follows that $f(\lambda) = r(\lambda)$. Since $P(A) = 0$ from the Cayley-Hamilton theorem, we conclude

$$f(A) = \alpha_0 I + \alpha, A + \alpha_2 A^2 + \cdots + \alpha_{n-1} A^{n-1} \tag{6-27}$$

Recall λ_i is the eigenvalue of A. What needs to be determined are the coefficients $\{\alpha_i\}$. If the eigenvalues of A are distinct, then a system of n equations in n unknowns can be defined to be

$$\begin{pmatrix} f(\lambda_1) \\ (\lambda_2) \\ \vdots \\ f(\lambda_n) \end{pmatrix} = \begin{pmatrix} 1 & \lambda_1 & \lambda_1^2 & \cdots & \lambda_1^{n-1} \\ 1 & \lambda_2 & \lambda_2^2 & \cdots & \lambda_2^{n-1} \\ \vdots & & & \ddots & \vdots \\ 1 & \lambda_n & \lambda_n^2 & \cdots & \lambda_n^{n-1} \end{pmatrix} \begin{pmatrix} \alpha_0 \\ \alpha_1 \\ \vdots \\ \alpha_{n-1} \end{pmatrix} \tag{6-28}$$

which can be solved for $\{\alpha_i\}$. Therefore, the Cayley-Hamilton method can be used to compute the matrix exponential

$$\Phi(t) = \exp(At) = f(A) = \alpha_0 I + \alpha_1 A + \ldots + \alpha_n A^n \tag{6-29}$$

once the set of n coefficients $\{\alpha\}$ is known. To generate the required coefficients, refer to Equation 6-28 and observe that

$$\exp(\lambda_1 t) = \alpha_0 1 + \alpha_1 \lambda_1 + \cdots + \alpha_{n-1} \lambda_1^{n-1}$$

$$\exp(\lambda_2 t) = \alpha_0 1 + \alpha_1 \lambda_2 + \cdots + \alpha_{n-1} \lambda_2^{n-1}$$

$$\vdots \qquad \vdots$$

$$\exp(\lambda_n t) = \alpha_0 1 + \alpha_1 \lambda_n + \cdots + \alpha_{n-1} \lambda_n^{n-1} \tag{6-30}$$

It can be noted that the method defined by Equation 6-30 presumes that the eigenvalues of A are distinct. If they were not, then two or more of the equations found in Equation 6-29 would be linearly dependent. If the eigenvalues are distinct, then upon solving the system of n equations for $\{\alpha_i\}$, $f(A) = \exp(At) = \sum a_j A^j$ can be computed. It immediately follows that the homogeneous solution becomes

$$\mathbf{y}(t) = \mathbf{C}^T \exp(At) \mathbf{x}_0 \tag{6-31}$$

Example 6-10   RLC circuit

Consider again the RLC circuit of Example 6-5 for $R = 10/3$, $L = 10^{-6}/3$, and $C = 10^{-6}/3$, shown in Figure 6-13 (repeated in Figure 6-20). The state-variable

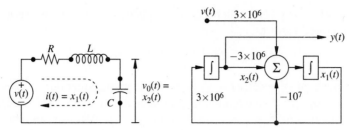

FIGURE 6-20   State variable model of an RLC circuit.

model is given by

$$\mathbf{A} = \begin{pmatrix} -10^7 & -3 \times 10^6 \\ 3 \times 10^6 & 0 \end{pmatrix}$$

$$\mathbf{b} = \begin{pmatrix} 3 \times 10^6 \\ 0 \end{pmatrix}$$

$$\mathbf{c} = \begin{pmatrix} 0 \\ 1 \end{pmatrix}$$

$$d = 0$$

with $\mathbf{x}^0 = (1 \quad 1)^T$. The RLC state-determined system is displayed as a simulation diagram in Figure 6-20. The system's characteristic equation is given by

$$\det(\lambda \mathbf{I} - \mathbf{A}) = \lambda^2 + 10^7 \lambda + 9 \times 10^{12} = (\lambda + 9 \times 10^6)(\lambda + 10^6)$$

which yields distinct stable eigenvalues $\lambda_1 = -9 \times 10^6$ and $\lambda_2 = -10^6$. The Cayley-Hamilton method for computing the matrix exponential is based on solving the system of scalar equations (after Equation 6-30)

$$e^{-9 \times 10^6 t} = \alpha_0 - 9 \times 10^6 \alpha_1$$

$$e^{-10^6 t} = \alpha_0 - 10^6 \alpha_1$$

or in matrix-vector form

$$\begin{pmatrix} e^{-9 \times 10^6 t} \\ e^{-10^6 t} \end{pmatrix} = \begin{pmatrix} 1 & -9 \times 10^6 \\ 1 & -10^6 \end{pmatrix} \begin{pmatrix} \alpha_0 \\ \alpha_1 \end{pmatrix} \Rightarrow \begin{pmatrix} \alpha_0 \\ \alpha_1 \end{pmatrix}$$

$$= \begin{pmatrix} -1/8 & 9/8 \\ -10^{-6}/8 & 10^{-6}/8 \end{pmatrix} \begin{pmatrix} e^{-9 \times 10^6 t} \\ e^{-10^6 t} \end{pmatrix}$$

which defines α_0 and α_1 to be

$$\alpha_0 = -(1/8)e^{-9\times 10^6 t} + (9/8)e^{-10^6 t}$$
$$\alpha_1 = -(10^{-6}/8)e^{-9\times 10^6 t} + (10^{-6}/8)e^{-10^6 t}$$

and yields

$$\exp(\mathbf{A}t) = \alpha_0\mathbf{I} + \alpha_1\mathbf{A} = \begin{pmatrix} \alpha_0 - 10^7\alpha_1 & -3\times 10^6\alpha_1 \\ 3\times 10^6\alpha_1 & \alpha_0 \end{pmatrix}$$

or

$$\exp(\mathbf{A}t) = \frac{1}{8}\begin{pmatrix} 9e^{-9\times 10^6 t} - e^{-10^6 t} & 3e^{-9\times 10^6 t} - 3e^{-10^6 t} \\ -3e^{-9^6 t} + 3e^{-10^6 t} & -e^{-9\times 10^6 t} + 9e^{-10^6 t} \end{pmatrix}$$

Finally, as a check, observe that $\exp(\mathbf{A}0) = \mathbf{I}$.

Computer Study   The production of the matrix exponential $\exp(\mathbf{A}t)$, for $\mathbf{A}$ a square matrix with distinct eigenvalues, consists of two parts. The first is the computation of the coefficients $\{\alpha_i\}$. The second is generating $\exp(\mathbf{A}t) = \sum \alpha_i\mathbf{A}^i$. Restricting the study to 2×2 matrices $\mathbf{A}$ with distinct eigenvalues, S-file ch6-10a.s or M-file alpha.m can be used to compute α_i for $i = 0$ or 1, using the format alpha(A,T,i) where T sets the limits on the range $t \in [0, T]$. S-file ch6_10b.s and M-file yout.m compute and display the four elements of $\exp(\mathbf{A}t)$, then use the matrix exponential to compute $y(t) = \mathbf{c}^T \exp(\mathbf{A}t)\mathbf{x}_0$, using the format yout(A,alpha0,alpha1,c',x0'). For $\mathbf{A}$ given above, the programs compute and display $y(t) = \mathbf{c}^T \exp(\mathbf{A}t)\mathbf{x}_0$. To study the 2×2 matrix $\mathbf{A}$ (defined earlier in this example with $\mathbf{x}_0 = (1 \quad 1)^T$) and display the results out to T $=$ 100 ms, conduct the following experiment.

SIGLAB	MATLAB
>include "ch6_10a.s"	>>A=[-10^7 -3*10^6; 3*10^6 0];
>include "ch6_10b.s"	>>alpha0=alpha(A,10^(-5),1);
>A={[-10^7,-3*10^6],[3*10^6,0]}	>>alpha1=alpha(A,10^(-5),2);
>alpha0=alpha(A,10^(-5),0);	>>yout(A,alpha0,alpha1,
alpha1=alpha(A,10^(-5),1)	[0,1],[1,1]);
># c'=[0,1], x0'=[1,1]	
>yout(A,alpha0,alpha1,	
[0,1],[1,1])	

The results are displayed in Figure 6-21. They display the time-varying behavior of the matrix exponential and the homogeneous output for the stable single-input single-output RLC network.

The previous analysis presumed that the eigenvalues of $\mathbf{A}$ were distinct. This is obviously not always the case. In Chapter 2 it was noted that when repeated

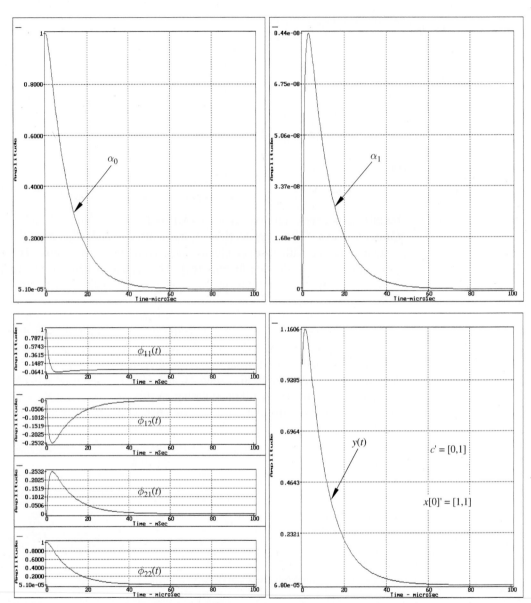

FIGURE 6-21   Homogeneous solution of a RLC network: (top left) α_0; (top right) α_1; (bottom left) $\phi_{11}(t)$, $\phi_{12}(t)$, $\phi_{21}(t)$, and $\phi_{22}(t)$; and (bottom right) $y(t)$ for $\mathbf{x}_0 = (1\ 1)^T$.

eigenvalues were encountered, successive differentiation was used to produce a set of linearly independent equations that could be solved to produce the needed Heaviside coefficients. A similar procedure can be used to modify Equation 6-30 to accept repeated eigenvalues. Refer to that equation and assume that eigenvalue

λ_i appears with multiplicity $n(i)$. Then, simply replacing λ_i for λ, for $i \in [1, n(i)]$, would result in

$$\text{Equation (1)} \qquad \exp(\lambda_i t) = \alpha_0 + \alpha_1 \lambda_i + \cdots + \alpha_{n-1} \lambda_i^{n-1}$$

$$\text{Equation (2)} \qquad \exp(\lambda_i t) = \alpha_0 + \alpha_1 \lambda_i + \cdots + \alpha_{n-1} \lambda_i^{n-1}$$

$$\vdots \quad \vdots$$

$$\text{Equation } (n(i)) \qquad \exp(\lambda_i t) = \alpha_0 + \alpha_1 \lambda_i + \cdots + \alpha_{n-1} \lambda_i^{n-1} \qquad (6\text{-}32)$$

which can be seen to produce $n(i)$ identical (therefore linearly dependent) equations. Using the Heaviside study as a guide, consider successive differentiation of Equation 6-32 to produce

$$\exp(\lambda_i t) = \alpha_0 + \alpha_i \lambda_1 + \cdots + \alpha_{n-1} \lambda_i^{n-1}$$

$$\frac{d}{d\lambda_i} \exp(\lambda_i t) = t \exp(\lambda_i t) = \frac{d}{d\lambda_i}(\alpha_0 + \alpha_1 \lambda_i + \cdots + \alpha_{n-1} \lambda_i^{n-1})$$

$$= \alpha_1 + \cdots + (n-1)\alpha_{n-1} \lambda_i^{n-2}$$

$$\frac{d^2}{d\lambda_i^2} \exp(\lambda_i t) = t^2 \exp(\lambda_i t) = \frac{d^2}{d\lambda_i^2}(\alpha_0 + \alpha_1 \lambda_i + \cdots + \alpha_{n-1} \lambda_i^{n-1})$$

$$= 2a_2 + \cdots + (n-1)(n-2)\alpha_{n-1} \lambda_i^{n-3}$$

$$\vdots \quad \vdots$$

$$\frac{d^{n(i)-1}}{d\lambda_i^{n(i)-1}} \exp(\lambda_i t) = t^{n(i)-1} \exp(\lambda_i t) = \frac{d^{n(i)-1}}{d\lambda_i^{n(i)-1}}(\alpha_0 + \alpha_1 \lambda_i + \cdots + \alpha_{n-1} \lambda_i^{n-1})$$

$$= (n(i) - 1) \cdots (2)(1)\alpha_{(n(i)-1)} + \cdots$$

$$+ (n-1)(n-2) \cdots (n(i) - 1)\alpha_{n-1} \lambda_i^{n-n(i)} \qquad (6\text{-}33)$$

From this, a set of $n(i)$ independent equations results that can be used to compute the required coefficient set $\{\alpha_i\}$.

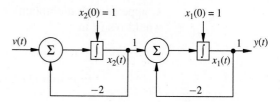

FIGURE 6-22   State variable model of a cascaded second-order system.

Example 6-11   Repeated eigenvalues

Consider the cascaded second-order system shown as a simulation diagram in Figure 6-22. The state-variable four-tuple can be seen to be given by

$$\mathbf{A} = \begin{pmatrix} -2 & 1 \\ 0 & -2 \end{pmatrix}$$

$$\mathbf{b} = \begin{pmatrix} 0 \\ 1 \end{pmatrix}$$

$$\mathbf{c} = \begin{pmatrix} 1 \\ 0 \end{pmatrix}$$

$$d = 0$$

where $\mathbf{x}^0 = (1 \quad 1)^T$. The system's characteristic equation is given by

$$\det(\lambda \mathbf{I} - \mathbf{A}) = \lambda^2 + 4\lambda + 4 = (\lambda + 2)^2$$

which produces repeated eigenvalues $\lambda_1 = -2$ and -2. The Cayley-Hamilton method of computing the matrix exponential for repeated eigenvalues is based on Equation 6-33, or

$$e^{-2t} = \alpha_0 - 2\alpha_1$$

$$te^{-2t} = \alpha_1$$

which reduces to

$$\begin{pmatrix} e^{-2t} \\ te^{-2t} \end{pmatrix} = \begin{pmatrix} 1 & -2 \\ 0 & 1 \end{pmatrix} \begin{pmatrix} \alpha_0 \\ \alpha_1 \end{pmatrix} \Rightarrow \begin{pmatrix} \alpha_0 \\ \alpha_1 \end{pmatrix} = \begin{pmatrix} 1 & 2 \\ 0 & 1 \end{pmatrix} \begin{pmatrix} e^{-2t} \\ te^{-2t} \end{pmatrix}$$

Upon solving for α_0 and α_1, one obtains

$$\alpha_0 = (1 + 2t)e^{-2t}$$

$$\alpha_1 = te^{-2t}$$

and it follows that

$$\exp(\mathbf{A}t) = \alpha_0 \mathbf{I} + \alpha_1 \mathbf{A} = \begin{pmatrix} \alpha_0 - 2\alpha_1 & \alpha_1 \\ 0 & \alpha_0 - 2\alpha_1 \end{pmatrix} = \begin{pmatrix} e^{-2t} & te^{-2t} \\ 0 & e^{2t} \end{pmatrix}$$

Finally, note that $\exp(\mathbf{A}0) = \mathbf{I}$ and that the homogeneous output is given by

$$y(t) = c^T \exp(\mathbf{A}t)\mathbf{x}_0 = \exp(-2t)x_1(0) + t \exp(-2t)x_2(0).$$

Another method of computing the state transition matrix is based on Laplace transforms. Recall that the homogeneous state solution is given by the state differential equation $d\mathbf{x}(t)/dt = \mathbf{A}\mathbf{x}(t)$, $\mathbf{x}(t_0) = \mathbf{x}_0$. The output homogeneous response satisfies the equation $y(t) = c^T\mathbf{x}(t)$. The Laplace transforms of the homogeneous state differential and observation (output) equations can be directly computed to be

$$s\mathbf{X}(s) = \mathbf{A}\mathbf{X}(s) - \mathbf{x}_0 \tag{6-34}$$

and

$$\mathbf{Y}(s) = \mathbf{c}^T\mathbf{X}(s) \tag{6-35}$$

respectively. Simplifying, they define a set of state and output transform equations given by

$$\mathbf{X}(s) = (s\mathbf{I} - \mathbf{A})^{-1}\mathbf{x}_0 \quad \text{State equation} \tag{6-36}$$

$$\mathbf{Y}(s) = \mathbf{c}^T(s\mathbf{I} - \mathbf{A})^{-1}\mathbf{x}_0 \quad \text{Observation equation} \tag{6-37}$$

Realizing that $\mathbf{x}(t) = e^{(\mathbf{A}t)}\mathbf{x}(0)$, it can be seen that Equation 6-36 also establishes that $\Phi(t) = \exp(\mathbf{A}t) = \mathcal{L}^{-1}((s\mathbf{I} - \mathbf{A})^{-1})$. Therefore, a Laplace transform can also be used to define the state transition matrix. This method is usually preferred in those cases where the elements of the matrix $[(s\mathbf{I} - \mathbf{A})^{-1}]$ have a well known inverse Laplace transform.

Example 6-12   RLC example using transforms

Consider once again the RLC circuit studied in Example 4-4. For the given $\mathbf{A}$ matrix, it follows that

$$(s\mathbf{I} - \mathbf{A}) = \begin{pmatrix} s + 10^7 & 3 \times 10^6 \\ -3 \times 10^6 & s \end{pmatrix} \Rightarrow (s\mathbf{I} - \mathbf{A})^{-1}$$

$$= \begin{pmatrix} s & -3 \times 10^6 \\ 3 \times 10^6 & s + 10^7 \end{pmatrix} \frac{1}{(s + 10^6)(s + 9 \times 10^6)}$$

The matrix exponential $\exp(\mathbf{A}t)$ is then defined by $\Phi(t) = \mathcal{L}^{-1}(s\mathbf{I} - \mathbf{A})^{-1}$, where each of the four elements in $(s\mathbf{I} - \mathbf{A})^{-1}$ can be converted to their time-domain equivalents using the inverse Laplace transform techniques introduced in Chapter 2. The four elements in the 2×2 matrix are given by

$$\phi_{11}(t) = \mathcal{L}^{-1}\left[\frac{s}{(s+10^6)(s+9x10^6)} = \frac{A_{11}}{s+10^6} + \frac{B_{11}}{s+9x10^6}\right]$$

$$\phi_{12}(t) = \mathcal{L}^{-1}\left[\frac{-3x10^6}{(s+10^6)(s+9x10^6)} = \frac{A_{12}}{s+10^6} + \frac{B_{12}}{s+9x10^6}\right]$$

$$\phi_{21}(t) = \mathcal{L}^{-1}\left[\frac{3x10^6}{(s+10^6)(s+9x10^6)} = \frac{A_{21}}{s+10^6} + \frac{B_{21}}{s+9x10^6}\right]$$

$$\phi_{22}(t) = \mathcal{L}^{-1}\left[\frac{(s+10^7)}{(s+10^6)(s+9x10^6)} = \frac{A_{22}}{s+10^6} + \frac{B_{22}}{s+9x10^6}\right]$$

$$A_{11} = -0.125, \quad B_{11} = 1.125$$
$$A_{12} = -0.375, \quad B_{12} = 0.375$$
$$A_{21} = 0.375, \quad B_{21} = -0.375$$
$$A_{22} = 1.125, \quad B_{22} = -0.125$$

As an intermediate test, compute $\phi_{ij}(0)$ using the initial value theorem, which states that $\phi_{ij}(0) = s\Phi_{ij}(s)$ as $s \to \infty$. It can be immediately seen that all $\phi_{ij}(0) = 0$ if $i \neq j$ and 1 if $i = j$, as required.

Computer Study   The S-file `pf.s` or M-file `residue.m`, can be used to compute the partial fraction coefficient of the polynomial $N(s)/D(s)$. The data is entered and analyzed below and is seen to agree with the previously computed results.

ϕ_{ij}	$N(s)$	$D(s)$	Root 1	B	Root 2	A
ϕ_{11}	[0,1,0]	[1,10^7,9 × 10^{12}]	$-9e+06$	1.125	$-1e+06$	-0.125
ϕ_{12}	[0,0,-3×10^6]	[1,10^7,9 × 10^{12}]	$-9e+06$	0.375	$-1e+06$	-0.375
ϕ_{21}	[0,0,3 × 10^6]	[1,10^7,9 × 10^{12}]	$-9e+06$	-0.375	$-1e+06$	0.375
ϕ_{22}	[0,0,10^7]	[1,10^7,9 × 10^{12}]	$-9e+06$	-0.125	$-1e+06$	1.125

The previous example demonstrates that once the matrix $(s\mathbf{I} - \mathbf{A})^{-1}$ is known, a computer can be used to perform the partial fraction expansion of the elements of the matrix terms $\Phi(s)$.

6-7   INHOMOGENEOUS STATE SOLUTIONS

To this point only the homogeneous-state solution to a continuous-time state-determined LTI system has been developed. In general, the inhomogeneous-state solution is also required. Consider a system represented by the at-rest, single-input single-output state differential equation

$$\frac{d}{dt}\mathbf{x}(t) = \mathbf{A}\mathbf{x}(t) + \mathbf{b}u(t) \qquad \mathbf{x}(t_0) = \mathbf{0} \tag{6-38}$$

for $t \geq t_0$. When continuous-time inhomogeneous solutions were previously introduced, they were expressed in terms of a convolution integral or the convolution theorem of Laplace transforms. It will be shown that these concepts apply to state-determined continuous-time LTI systems, as well.

Initially assume that the solution to $d\mathbf{x}(t)/dt = \mathbf{A}\mathbf{x}(t) + \mathbf{b}u(t)$, $\mathbf{x}(t_0) = \mathbf{0}$ is given by

$$\mathbf{x}(t) = \int_{t_0}^{t} e^{\mathbf{A}(\tau - t_0)} \mathbf{b}u(\tau)\, d\tau \qquad \text{Inhomogeneous-state solution} \qquad (6\text{-}39)$$

To verify that Equation 6-39 is the inhomogeneous state solution, simply show that it also satisfies Equation 6-38. Taking the total derivative of Equation 6-39, we obtain

$$\frac{d}{dt}\mathbf{x}(t) = \int_{t_0}^{t} \frac{d}{dt}(e^{\mathbf{A}(\tau - t_0)} \mathbf{b}u(\tau))\, d\tau + e^{\mathbf{A}(t - \tau)} \mathbf{b}u(\tau)|_{\tau = t}$$

$$= \mathbf{A} \int_{t_0}^{t} e^{\mathbf{A}(\tau - t_0)} \mathbf{b}u(\tau)\, d\tau + e^{\mathbf{A}0} \mathbf{b}u(t)$$

$$= \mathbf{A}\mathbf{x}(t) + \mathbf{b}u(t) \qquad (6\text{-}40)$$

Therefore, Equation 6-39 is seen to be the inhomogeneous-state solution. Therefore, it logically follows that the output (i.e., inhomogeneous solution) is given by

$$y(t) = \mathbf{d}u(t) + \mathbf{c}^{T} \int_{t_0}^{t} e^{\mathbf{A}(\tau - t_0)} \mathbf{b}u(\tau)\, d\tau \qquad \text{Inhomogeneous-state solution} \quad (6\text{-}41)$$

Example 6-13   Inhomogeneous-state solution

Consider again the RLC example presented in Example 6-5, with the input $v(t)$ equal to a unit-step function. For $v(t) = u(t) = 1$ for $t \geq 0$,

$$\frac{d\mathbf{x}(t)}{dt} = \begin{pmatrix} -R/L & -1/L \\ 1/C & 0 \end{pmatrix} \mathbf{x}(t) + \begin{pmatrix} 1/L \\ 0 \end{pmatrix} v(t) \qquad \text{State equation}$$

$$y(t) = x_2(t) = (\,0\quad 1\,)\,\mathbf{x}(t) + 0v(t) \qquad \text{Output equation}$$

The inhomogeneous solution is defined in terms of the state-transition matrix $\exp(\mathbf{A}t)$. It was shown in Example 6-10 that if $R = 10/3$, $L = 10^{-6}/3$, and $C = 10^{-6}/3$, that

$$\exp(\mathbf{A}t) = \frac{1}{8} \begin{pmatrix} 9e^{-9\times10^6 t} - e^{-10^6 t} & 3e^{-9\times10^6 t} - 3e^{-10^6 t} \\ -3e^{-9\times10^6 t} + 3e^{-10^6 t} & -e^{-9\times10^6 t} + 9e^{-10^6 t} \end{pmatrix}$$

In addition, since $\mathbf{b}^T = (3 \times 10^6, 0)$, $\mathbf{c}^T = (0, 1)$, and $\mathbf{d} = 0$, it follows that

$$y(t) = (0 \quad 1) \int_0^t e^{\mathbf{A}(t-\tau)} \begin{pmatrix} 3 \times 10^6 \\ 0 \end{pmatrix} u(\tau)\, d\tau$$

$$= 3 \times 10^6 \int_0^t (0.375 e^{-10^6(t-\tau)} - 0.375 e^{-9 \times 10^6(t-\tau)})\, d\tau$$

$$= 3 \times 10^6 \left((0.375/(-10^6)) e^{-10^6(t-\tau)} \Big|_0^t \right.$$

$$\left. -(0.375/(-9 \times 10^6)) e^{-9 \times 10^6(t-\tau)} \Big|_0^t \right)$$

$$= 1.125(1 - e^{-10^6 t}) - 0.125(1 - e^{-9 \times 10^6 t})$$

$$= 1.0 + 0.125 e^{-9 \times 10^6 t} - 1.125 e^{-10^6 t}$$

The total-state solution is simply the sum of the homogeneous and inhomogeneous responses. Observe that both the homogeneous and inhomogeneous solutions are defined in term of the state transition matrix. Once computed, $\exp(\mathbf{A}t)$ can be used to quantify the system's response to initial conditions and external forcing functions. This represents a great computational economy when compared to traditional methods that rely on computing a system's homogeneous solution and impulse response, which generally are different functions.

The inhomogeneous solution can also be expressed in terms of a Laplace transform. For an at-rest LTI system, the Laplace transform of the state equation $d\mathbf{x}(t)/dt = \mathbf{A}\mathbf{x}(t) + \mathbf{b}u(t)$ is $s\mathbf{X}(s) = \mathbf{A}\mathbf{X}(s) + \mathbf{b}U(s)$. The output equation given by $y(t) = \mathbf{c}^T\mathbf{x}(t) + \mathbf{d}u(t)$ has a transform representation given by $Y(s) = \mathbf{c}^T\mathbf{X}(s) + \mathbf{d}U(s)$. Simplifying, the following transformed state and observation equations result

$$\mathbf{X}(s) = (s\mathbf{I}-\mathbf{A})^{-1}\mathbf{b}U(s) \qquad \text{State equation} \qquad (6\text{-}42)$$

$$Y(s) = (\mathbf{d}+\mathbf{c}^T(s\mathbf{I}-\mathbf{A})^{-1}\mathbf{b})U(s) \qquad \text{Output equation} \qquad (6\text{-}43)$$

where $\mathbf{x}(t) = \mathcal{L}^{-1}[(s\mathbf{I} - \mathbf{A})^{-1}\mathbf{b}U(s)]$ and $y(t) = \mathcal{L}^{-1}[(\mathbf{d} + \mathbf{c}^T(s\mathbf{I} - \mathbf{A})^{-1}\mathbf{b})U(s)]$. The last equation can also be used to define the system's transfer function, which implicitly assumes that the system is at-rest. The transfer function for a single-input single-output system is formally given by

$$H(s) = \frac{Y(s)}{U(s)} = (\mathbf{d}+\mathbf{c}^T(s\mathbf{I}-\mathbf{A})^{-1}\mathbf{b}) \qquad (6\text{-}44)$$

It is worthwhile to note that the matrix $(s\mathbf{I} - \mathbf{A})^{-1}$, for $\mathbf{A}$ an $n \times n$ constant-coefficient matrix, can be expressed as the following matrix polynomial (called *adjoint form*)

$$(s\mathbf{I}-\mathbf{A})^{-1} = \frac{1}{\Delta(s)}\mathbf{K}(s) = \frac{1}{\Delta(s)}[s^{n-1}\mathbf{K}_1 + s^{n-2}K_2 + \cdots + s\mathbf{K}_{n-1} + \mathbf{K}_n] \quad (6\text{-}45)$$

where $\Delta(s)$ is the determinant of $(s\mathbf{I} - \mathbf{A})$, $\mathbf{K}(s)$ is the adjoint of the matrix $(s\mathbf{I} - \mathbf{A})$ (see Appendix A), and $\mathbf{K}_i$ is an $n \times n$ matrix. Substituting this result into the state-determined transfer-function equation results in the following transfer function representation

$$H(s) = \frac{Y(s)}{U(s)} = (\mathbf{d} + \mathbf{c}^T(s\mathbf{I} - \mathbf{A})^{-1}\mathbf{b}) = \mathbf{d} + \frac{\mathbf{c}^T\mathbf{K}(s)\mathbf{b}}{\Delta} = \frac{\mathbf{d}\Delta(s) + \mathbf{c}^T\mathbf{K}(s)\mathbf{b}}{\Delta(s)}$$

$$(6\text{-}46)$$

It is seen that the denominator of the transfer function is completely specified by $\Delta(s)$. The stability of a linear time-invariant system, when introduced in Chapter 4, was classified by the locations of the roots of the polynomial $\Delta(s) = (s - p_1)\cdots(s - p_n)$, where p_i is a pole of the system. It was noted that stability is guaranteed if the system's poles are located in the left-hand side of the s-plane, and marginal stability may result if the poles lie along the $j\omega$-axis in the s-plane. The poles are also known to be the eigenvalues of the matrix $\mathbf{A}$. In particular, if $\mathbf{A}$ is an $n \times n$ constant-coefficient matrix with a characteristic polynomial $\Delta(s) = \det(s\mathbf{I} - \mathbf{A}) = (s - \lambda_1)(s - \lambda_2)\cdots(s - \lambda_n)$, then the eigenvalues λ_i correspond to the system poles. Furthermore, it is important to note that stability is only a property of the matrix $\mathbf{A}$ and is independent of $\mathbf{b}$, $\mathbf{c}$, and $\mathbf{d}$. Referring to Figure 6-15, it can be seen that it is only the matrix $\mathbf{A}$ that defines the system's feedback structure and, as such, the system's eigenfunctions and pole locations.

Finally, the system output can be derived from knowledge of $H(s)$, using the inverse Laplace transform equation

$$y(t) = \mathcal{L}^{-1}(\mathbf{d}+\mathbf{c}^T(s\mathbf{I}-\mathbf{A})^{-1}\mathbf{b})U(s) \quad (6\text{-}47)$$

Example 6-14   Transfer function

Consider again the RLC example studied in Example 6-10. Here

$$\mathbf{A} = \begin{pmatrix} -10^7 & -3 \times 10^6 \\ 3 \times 10^6 & 0 \end{pmatrix}$$

$$\mathbf{b} = \begin{pmatrix} 3 \times 10^6 \\ 0 \end{pmatrix}$$

$$\mathbf{c} = \begin{pmatrix} 0 \\ 1 \end{pmatrix}$$

$$d = 0 \quad (6\text{-}48)$$

The system poles are given by

$$\det(s\mathbf{I} - \mathbf{A}) = (s + 10^6)(s + 9 \times 10^6)$$

and are equal to $s = -10^6$ and $s = -9 \times 10^6$, which are seen to be well inside the left-hand s-plane. Therefore, the system is stable. The system transfer function is given by

$$H(s) = d + \mathbf{c}^T(s\mathbf{I} - \mathbf{A})^{-1}\mathbf{b}$$

which computes, in this case, to be

$$H(s) = 0 + (0, 1)\frac{\begin{pmatrix} s + 10^7 & 3 \times 10^6 \\ -3 \times 10^6 & s + 10^7 \end{pmatrix}}{(s + 10^6)(s + 9 \times 10^6)}\begin{pmatrix} 3 \times 10^6 \\ 0 \end{pmatrix}$$

which simplifies to

$$H(s) = \frac{9 \times 10^{12}}{(s + 10^6)(s + 9 \times 10^6)}$$

If the input to the system is assumed to be a unit step, then $U(s) = 1/s$ and the Laplace transform of the output satisfies the equation

$$Y(s) = (d + \mathbf{c}^T(s\mathbf{I} - \mathbf{A})^{-1}\mathbf{b})U(s)$$

which equals

$$Y(s) = \frac{9 \times 10^{12}}{s(s + 10^6)(s + 9 \times 10^6)}$$

The output trajectory $y(t)$ of the at-rest RLC circuit is therefore $y(t) = \mathcal{L}^{-1}Y(s)$.

Computer Study   The unit step response of the RLC circuit is given by the inverse Laplace transform of

$$Y(s) = \frac{9 \times 10^{12}}{s(s + 10^6)(s + 9 \times 10^6)}$$

Using S-file pf.s or M-file residue.m, $Y(s)$ can be expressed as a partial fraction expansion.

SIGLAB	MATLAB

``
>include "pf.s"
>n=[0,0,0,9*10^12]
>d=[1,10^7,9*10^12,0]
>pfexp(n,d)
pole @ -9e+06+0j with
multiplicity 1 and coefficients
0.125+0j
pole @ -1e+06+0j with
multiplicity 1 and coefficients
-1.125+0j
pole @ 0+0j with
multiplicity 1 and coefficients 1+0j
``
	``
	>> n = [0 0 0 9*10^12];
	>> d = [1 10^7 9*10^12 0];
	>> % r = [a1 a2 a3],
	>>p = [p1 p2 p3], k = a0
	>> [r p k] = residue(n, d)
	r =
	    0.1250
	   -1.1250
	    1.0000
	p =
	   -9000000
	   -1000000
	          0
	k =
	   []
	``

This yields

$$Y(s) = \frac{1}{s} + \frac{0.125}{(s + 9 \times 10^6)} - \frac{1.125}{(s + 10^6)}$$

or

$$y(t) = (1 + 0.125e^{-9 \times 10^6 t} - 1.125e^{-10^6 t})u(t)$$

which agrees with the results computed in Example 6-14.

The step response, over a 5 microsecond display interval, can be displayed using S-file ch6_14.s or the Matlab program shown below.

SIGLAB	MATLAB
``` >include "ch6_14.s" > a = 1 > b = 0.125 > c = -1.125 > lambda1 = -9*10^6 > lambda2 = -10^6 > rlc(a,b,c,lambda1, lambda2) ```	``` >> a = 1; >> b = 0.125; >> c = -1.125; >> lambda1 = -9*10^6; >> lambda2 = -10^6; >> t = 10^(-4) * (0:255)/255; >> plot(t, a + b * exp(lambda1 * t) + c * exp(lambda2 * t)) ```

The step response of the RLC circuit is shown in Figure 6-23 over a 1/10 millisecond interval. It can be seen that the response has a classic overdamped profile.

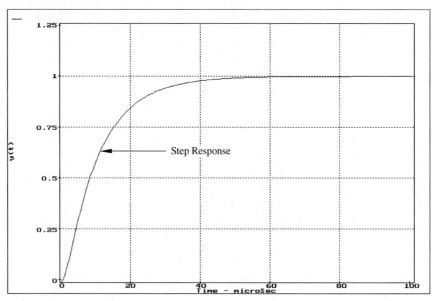

FIGURE 6-23   RLC circuit step response.

6-8   CONTINUOUS-TIME SYSTEM ARCHITECTURE

One of the most useful and important properties of a state-variable model is its ability to model the way a system manages information internally. This nth-order ODE, for example

$$a_n \frac{d^n}{dt^n} y(t) + a_{n-1} \frac{d^{n-1}}{dt^{n-1}} y(t) + \cdots + a_1 \frac{d}{dt} y(t) + a_0 y(t) = v(t) \qquad (6\text{-}49)$$

is typical of many systems. If $a_n = 1$, then the system is also said to be monic. Using the following state assignment rule

$$x_1(t) = y(t)$$

$$x_2(t) = \frac{d}{dt} y(t) = \frac{d}{dt} x_1(t)$$

$$x_3(t) = \frac{d^2}{dt^2} y(t) = \frac{d}{dt} x_2(t) \qquad (6\text{-}50)$$

$$\vdots = \vdots$$

$$x_n(t) = \frac{d^{n-1}}{dt^{n-1}} y(t) = \frac{d}{dt} x_{n-1}(t)$$

and

$$\frac{dx_n(t)}{dt} = -\frac{a_0}{a_n}x_1(t) - \frac{a_1}{a_n}x_2(t) - \cdots - \frac{a_{n-1}}{a_n}x_n(t) + \frac{1}{a_n}v(t) \qquad (6\text{-}51)$$

the nth-order ODE can be placed in state-variable form $dx(t)/dt = \mathbf{A}x(t)+\mathbf{b}v(t)$, $y(t) = \mathbf{c}^T\mathbf{x}(t)+dv(t)$ where $\mathbf{x}(t)$ is an n-dimensional state vector. The state four-tuple $S = [\mathbf{A}, \mathbf{b}, \mathbf{c}, d]$ would, in this case, satisfy

$$\mathbf{A} = \begin{pmatrix} 0 & 1 & 0 & \cdots & 0 \\ 0 & 0 & 1 & & 0 \\ \vdots & \vdots & & \ddots & \vdots \\ 0 & 0 & & & 1 \\ -a_0/a_n & -a_1/a_n & \cdots & -a_{n-2}/a_n & -a_{n-1}/a_n \end{pmatrix}$$

$$\mathbf{b}^T = (0 \quad \cdots \quad 0 \quad 1/a_n)$$

$$\mathbf{c}^T = (1 \quad 0 \quad \cdots \quad 0)$$

$$d = (0) \qquad (6\text{-}52)$$

This is called the *companion form* or *companion architecture* and is graphically interpreted in Figure 6-24.

The output is extracted from the output of the last integrator, which corresponds to the state x_1 location. The input is presented to the system at the other end. Initial conditions are attached to each integrator. What is most important to observe is that each a_{ij} element of $\mathbf{A}$ defines the unidirectional path gain of the path connecting the jth state to the ith state. If, for example, there is no direct path between states x_3 and x_1, then $a_{13} = 0$. Each distinct architecture will produce a unique 4-tuple $S = [\mathbf{A}, \mathbf{b}, \mathbf{c}, d]$. For each unique $\mathbf{A}$, a new, state transition matrix and architecture will result.

FIGURE 6-24   Simulation diagram for an nth-order state-variable model in companion form.

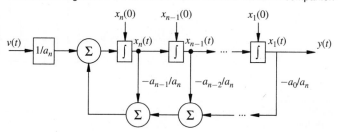

Example 6-15   Companion architecture

In Example 6-14, the transfer function of a specific RLC circuit was shown to be equal to

$$H(s) = \frac{9 \times 10^{12}}{(s^2 + 10^7 s + 9 \times 10^{12})}$$

This has a companion architecture given by

$$\mathbf{A} = \begin{pmatrix} 0 & 1 \\ -9 \times 10^{12} & -10^7 \end{pmatrix}$$

$$\mathbf{b}^T = (0 \quad 1)$$

$$\mathbf{c}^T = (9 \times 10^{12} \quad 0)$$

$$\mathbf{d} = (0)$$

which is graphically interpreted as a simulation diagram in Figure 6-25.

Refer to Equation 6-49 and note that the forcing function is simply the scalar $u(t)$. In general, the forcing function is more complicated and would have the form

$$\text{input } (t) = \sum_{i=0}^{m} b_i \frac{d^i v(t)}{dt^i} \tag{6-53}$$

The modified system can be expressed in *extended companion form,* which is given by

$$\frac{d\mathbf{x}(t)}{dt} = \mathbf{A}\mathbf{x}(t) + \mathbf{b}v(t) \tag{6-54}$$

and

$$\mathbf{y}(t) = \mathbf{c}^T \mathbf{x}(t) + dv(t) \tag{6-55}$$

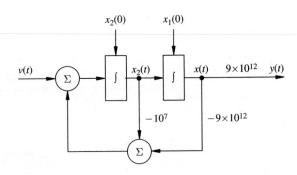

FIGURE 6-25   Simulation diagram of a second-order RLC circuit as a companion architecture.

where

$$A = \begin{pmatrix} 0 & 1 & 0 & \cdots & 0 \\ 0 & 0 & 1 & & 0 \\ \vdots & \vdots & & \ddots & \vdots \\ 0 & 0 & & & 1 \\ -a_0/a_n & -a_1/a_n & -a_2/a_n & \cdots & -a_{n-1}/a_n \end{pmatrix}$$

$$\mathbf{b}^T = (0 \quad \cdots \quad 0 \quad 1/a_n)$$

$$\mathbf{c}^T = (b_0 - (a_0 b_n/a_n) \quad b_1 - (a_1 b_n/a_n) \quad \cdots \quad b_{n-1} - (a_{n-1}b_n/a_n))$$

$$d = (b_n/a_n) \tag{6-56}$$

(details left as a chapter exercise. See Problem 6-10.) The system is graphically interpreted as a simulation diagram in Figure 6-26.

Example 6-16   Fifth-order LTI system

A specific fifth-order LTI system, studied in Example 6-4, has a transfer function given by

$$H(s) = K \frac{(b_3 s^3 + b_2 s^2 + b_1 s + b_0)}{(a_5 s^5 + a_4 s^4 + a_3 s^3 + a_2 s^2 + a_1 s)}$$

$$= \frac{-50(s^3 + 30s^2 + 300s + 1000)}{s^5 + 45s^4 + 750s^3 + 5500s^2 + 15000s}$$

if the dependent voltage gain is $A = -1$. The transfer function is monic and

FIGURE 6-26   Extended form companion architecture.

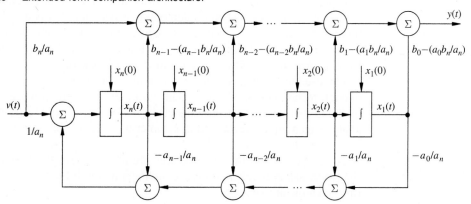

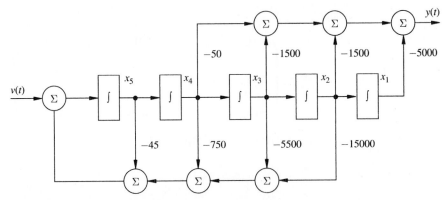

FIGURE 6-27   Simulation diagram of a fifth-order system as a companion architecture.

the companion form architecture is given by

$$A = \begin{pmatrix} 0 & 1 & 0 & 0 & 0 \\ 0 & 0 & 1 & 0 & 0 \\ 0 & 0 & 0 & 1 & 0 \\ 0 & 0 & 0 & 0 & 1 \\ 0 & -15000 & -5500 & -750 & -45 \end{pmatrix}$$

$$\mathbf{b}^T = (0 \quad 0 \quad 0 \quad 0 \quad 1)$$

$$\mathbf{c}^T = (-50000 \quad -15000 \quad -1500 \quad -50 \quad -0)$$

$$\mathbf{d} = (0)$$

which is graphically interpreted as a simulation diagram in Figure 6-27.

6-9   SUMMARY

In Chapter 6 the issue of architecture was introduced. It is generally considered to be the last design step when implementing a system in hardware. The physical embodiment of a system is called a system architecture. Architecture, it was shown, can be expressed as a block diagram, a signal-flow graph, or in state-variable form. In this form, the information contained in a block diagram or signal-flow graph can be used to produce a transfer-function description of a system with manual reduction techniques or Mason's Gain Formula. Also presented were state-variable methods that define and are defined by an architecture. Using the state-variable analysis techniques developed in Chapter 6, systems often can be more efficiently and elegantly analyzed.

A number of specific architectures are available for the design of continuous-time systems. A number of classic architectures will be developed when discrete-time systems are studied.

6-10 COMPUTER FILES

These computer files were used in Chapter 6.

File Name	Type	Location
Subdirectory CHAP6		
ch6_7.s	C	Example 6-7
matexp.m	C	Example 6-7
ch6_8.s	C	Example 6-8
homo.m	C	Example 6-8
ch6_10a.s	C	Example 6-10
ch6_10b.s	C	Example 6-10
alpha.m	C	Example 6-10
yout.m	C	Example 6-10
ch6_14.s	T	Example 6-14
Subdirectory SFILES or MATLAB		
pf.s	C	Example 6-12, 6-14
Subdirectory SIGNALS		
Subdirectory SYSTEMS		
sfg.exe	C	Example 6-4
ckt1.sfg	C	Example 6-4
ckt-1.sfg	C	Example 6-4

Here "s" denotes an S-file, "m" an M-file, "x" either an S-file or M-file, "T" denotes a tutorial program, "D" a data file and "C" a computer-based instruction (CBI) program. CBI programs can be altered and parameterized by the user.

6-11 CHAPTER PROBLEMS

6-1 Block diagram reduction   Convert the system described in the flow diagram found in Figure 6-28 to a block diagram. Using manual reduction techniques, derive the system's transfer function.

6-2 Mason's Gain Formula   Use Mason's Gain Formula to compute a transfer function for the system shown in Figure 6-28.

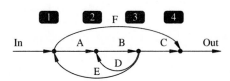

FIGURE 6-28   Four node signal-flow graph.

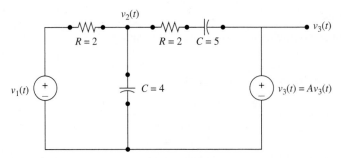

FIGURE 6-29   Electrical network with a dependent source.

6-3 RLC circuit   Compute the transfer function for the RLC circuit studied in Example 6-5 from the simulation diagram shown in Figure 6-13, using Mason's Gain Formula.

6-4 Dependent sources   A network containing a dependent voltage source is shown in Figure 6-29. Refer to Example 6-4 and derive the equations that describe the circuit's electrical behavior. Describe the circuit in flow-diagram form. Compute the system's input-output transfer function $H(s)$. For $A = \pm 1$, where do the system poles and zeros reside?

6-5 Mason's Gain Formula   The computerized version of Mason's Gain Formula is a parametric method requiring that all nodes be indexed and path gains be defined. The RLC circuit shown in Figure 6-13 can be translated into the signal-flow graph shown in Figure 6-30. Use Mason's Gain Formula to compute the transfer function $H(s)$. Also convert the signal-flow graph to a block-diagram representation of a single-input single-output system. Reduce the block diagram to a transfer function. Compare.

6-6 Multiple-input multiple-output systems   A multiple-input multiple-output system is shown in Figure 6-31. What are all four input-output transfer functions? Is the system stable?

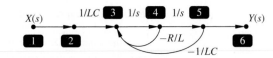

FIGURE 6-30   Signal-flow graph for an RLC circuit.

FIGURE 6-31   Multiple-input multiple-output system.

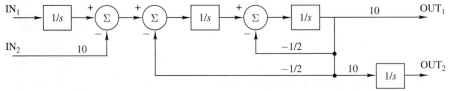

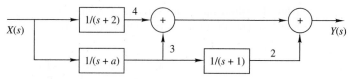

FIGURE 6-32   Signal-flow graph.

6-7 Mason's Gain Formula   Use Mason's Gain Formula to compute a transfer function for the system shown in Figure 6-32. For $a = 0, 1, 2$, what is the stability classification of each system?

6-8 Mason's Gain Formula   Use Mason's Gain Formula to compute a transfer function for the system shown in Figure 6-33 for an output taken to be $Y_1(s)$, $Y_2(s)$, or $Y_3(s)$. What are the values of a and b that will guarantee stability?

6-9 State-transition matrix   Using Equation 6-14, derive a formula for

$$\int_0^t \exp(\mathbf{A}\tau)d\tau$$

6-10 State equations   Verify that Equation 6-56 represents the system defined by Equations 6-53 through 6-55 for $m \le n$.

6-11 State-transition matrix   What is the state-transition matrix for the system defined in Problem 6-4 for $A = \pm 1$. What is the homogeneous response to a set of nonzero initial conditions. Explain the physical meaning of $\mathbf{x}[0]$ in terms of circuit parameters or physics.

6-12 Cayley-Hamilton theorem   Use the Cayley-Hamilton theorem to compute $\mathbf{A}^{-1}$ where

$$\mathbf{A} = \begin{pmatrix} 2 & -4 \\ t & 2 \end{pmatrix}$$

Will $\mathbf{A}^{-1}$ always exist?

FIGURE 6-33   Signal-flow graph.

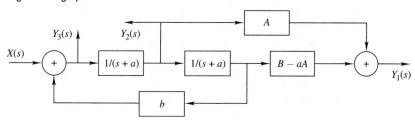

6-13 Cayley-Hamilton method   Suppose **A** is the set of nonsingular matrices shown below. Compute the matrix exponential $\exp(\mathbf{A}t)$ using the Cayley-Hamilton method.

$$\mathbf{A}_1 = \begin{pmatrix} -1 & 1 & 0 \\ 0 & -1 & 1 \\ 0 & 0 & 1 \end{pmatrix}$$

$$\mathbf{A}_2 = \begin{pmatrix} 1 & 2 & 0 \\ 2 & -1 & 2 \\ 1 & 2 & 2 \end{pmatrix}$$

$$\mathbf{A}_3 = \begin{pmatrix} 1 & 1 & 0 \\ 0 & 1 & 1 \\ 0 & 0 & 1 \end{pmatrix}$$

$$\mathbf{A}_4 = \begin{pmatrix} 0 & 1 & 0 \\ 0 & 0 & 1 \\ 1 & 2 & 1 \end{pmatrix}$$

Are these systems stable?

6-14 State transition matrix   For the plant matrix **A** defined in Problem 6-13, compute $\exp(\mathbf{A}t)$ using the Laplace transform method.

6-15 Matrix exponential   A state variable model is given by

$$\frac{dx(t)}{dt} = \begin{pmatrix} 0 & 1 \\ -6 & -5 \end{pmatrix} x(t) + \begin{pmatrix} 0 \\ 1 \end{pmatrix} u(t)$$

Verify that the matrix exponential is given by

$$\Phi_{11}(t) = 3\exp(-2t) - 2\exp(-3t)$$

$$\Phi_{12}(t) = \exp(-2t) - \exp(-3t)$$

$$\Phi_{21}(t) = -6\exp(-2t) + 6\exp(-3t)$$

$$\Phi_{22}(t) = -2\exp(-2t) + 3\exp(-3t)$$

What is $\Phi(t)$ for the transpose system (i.e., replace **A** with the $\mathbf{A}^T$ transpose)?

6-16 State transition matrix   Suppose **A** is the nonsingular matrix shown below. Compute $\exp(\mathbf{A}t)$ using either the Cayley-Hamilton method or the Laplace transform method.

$$\mathbf{A} = \begin{pmatrix} \cos(\theta) & \sin(\theta) \\ -\sin(\theta) & \cos(\theta) \end{pmatrix}$$

What is the physical significance of the **A** matrix?

6-17 Similarity transform   A state-determined system is specified by the four-tuple $S = [\mathbf{A}, \mathbf{b}, \mathbf{c}, d]$. The state-variable vector $\mathbf{x}(t)$ is to be replace by another vector $\mathbf{r}(t)$ where $\mathbf{r}(t) = \mathbf{T}\mathbf{x}(t)$ for **T** a nonsingular square matrix. What is the new system four-tuple? What is the new system's state-transition matrix in terms of $\exp(\mathbf{A}t)$?

6-18 Multiple-input multiple-output system   A multiple-input multiple-output system is shown in Figure 6-34. Using $v_1(t)$ and $v_2(t)$ as inputs, and $y_1(t)$ and $y_2(t)$ as out-

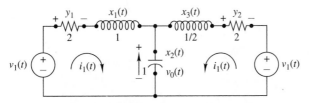

FIGURE 6-34   Multiple-input multiple-output system.

puts, define the state four-tuple $\mathcal{S} = [\mathbf{A}, \mathbf{B}, \mathbf{C}, \mathbf{D}]$ in terms of the three state variables x_1, x_2, and x_3. What is the output if $v_1(t) = v_2(t) = u(t)$ (a step function)?

6-19 Block-diagram reduction   Convert the block diagram of the state-variable system model displayed in Figure 6-15 to a transfer function, using manual reduction techniques (remember that you are working with matrix equations).

6-20 Operational amplifier   An *operational amplifier*, or *op amp*, is a high-gain amplifier. They are ubiquitous and readily available in IC packages. In a typical configuration the input to the op amp are signals with very small values and can be thought of as being at ground potential (i.e., $v_{in} \approx 0$). This is called a *virtual ground*. The input-output transfer function is known to be given by

$$H(s) = -\frac{Z_{out}(s)}{Z_{in}(s)}$$

where Z denotes the impedance if indicated path. If the path, for example, indicated by Z_{out} is a capacitor, then $Z_{out} = 1/Cs$. Using the electronic circuit shown in Figure 6-35, what is $H(s)$. What purpose might this circuit serve?

6-21 Airplane controller   An aircraft's direction is defined in terms of roll, attitude, and yaw. The roll angle of an aircraft can be controlled with the use of ailerons as shown in Figure 6-36 (see page 244). The equations that govern the interaction between the ailerons and roll angle is given by

$$J\frac{d^2\phi(t)}{dt^2} = k\theta(t) - D\frac{d\phi(t)}{dt}$$

where $\phi(t)$ is the roll angle, $\theta(t)$ is the aileron position, J is a moment of inertia, D is the drag coefficient, and k is a constant. This equation is sometimes expressed

FIGURE 6-35   Operational amplifier.

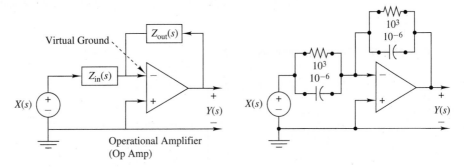

FIGURE 6-36   Roll-axis control system.

in terms of roll-rate, where the roll-rate variable is given by $\omega = d\phi(t)/dt$ and results in

$$J\frac{d\omega(t)}{dt} = k\theta(t) - D\omega(t)$$

Suppose $J = k = \lambda D$. What is the range of $\lambda \geq 0$ that will guarantee stability? Implement a roll-angle and roll-rate controller in companion form.

Suppose that at time $t = 0$, the roll-angle is $\pi/8$ radians and the roll-rate is 0. Devise an aileron control strategy to bring the roll-angle to 0 radians at $t = 1/\lambda$.

6-12   COMPUTER PROJECTS

6-1 Extended companion-form state equations   Refer to the RC circuit shown in Figure 6-29. How many state-variables would be needed to model the circuit in state-variable form? Derive a state-variable model, define the location, and specify the units of measure of each state. What is the state-transition matrix?

6-2 Third-order electrical circuit   The three-state electrical system shown in Figure 6-37 is to be analyzed. Define the system output to be $y(t) = x_1(t) - x_2(t)$. Represent the system in signal-flow graph form and use Mason's Gain Formula to derive the system transfer function (remember there are two independent voltage sources). Represent the circuit in state-variable form and derive the transfer function again using the transform method. Compare.

Suppose that the system is initially at rest and that $v_1(t) = \sin(2\pi 10^5 t)$. What unit amplitude input $v_2(t)$ can be used to produce a steady-state output, measured by $y(t)$, which has a minimum absolute value? Verify your conjecture experimentally.

6-3 Inverted pendulum   An inverted pendulum is described in Project 4-2 of Chapter 4 (see Figure 6-38). It consists of a stick with mass m and moment of inertia

FIGURE 6-37   Electrical network for Project 6-1.

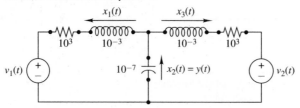

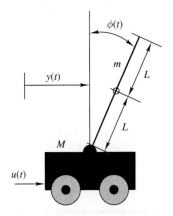

FIGURE 6-38   Inverted pendulum after Problem 4-2, Chapter 4.

given by $I = mL^2/3$, a cart with mass M, and a hinge. The cart and stick can only move in the y direction. A force $u(t)$ is applied to the cart. The defining equations, for small angles ϕ are

$$\frac{dx_1(t)}{dt} = x_2(t)$$

$$\frac{dx_2(t)}{dt} = a_{21}x_1(t) + b_2 u(t)$$

$$\frac{dx_3(t)}{dt} = x_4(t)$$

$$\frac{dx_4(t)}{dt} = a_{41}x_1(t) + b_4 u(t)$$

$$x_1(t) = \phi$$

$$x_2(t) = \frac{d\phi(t)}{dt}$$

$$x_3(t) = y(t)$$

$$x_4(t) = \frac{dy(t)}{dt}$$

$$a_{21} = \frac{3g(m + M)}{L(m + 4M)}$$

$$a_{41} = \frac{-3mg}{L(m + 4M)}$$

$$b_2 = \frac{-3g}{L(m + 4M)}$$

$$a_4 = \frac{4}{L(m + 4M)}$$

Here g is the force of gravity. Represent the system in state-variable form for $M = 2.22$, $m = 1.0$, $L = 1$, and $g = 32.2$. Determine the stability of the system.

Show that if $u(t) = 0$, the at-rest position is unstable in that any perturbation from the vertical will cause the stick to fall.

Determine (intuitively) a control policy for $u(t)$, which will tend to keep the stick vertical. Conduct simulation trials to determine the efficacy of your scheme.

DISCRETE-TIME SIGNALS AND SYSTEMS

PART THREE

DISCRETE-TIME
SIGNALS AND
SYSTEMS

MATHEMATICAL REPRESENTATION OF DISCRETE-TIME SIGNALS

"My dear Watson, try a little analysis yourself" said he, with a touch of impatience. "You know my methods. Apply them and it will be instructive to compare results."

—Sherlock Holmes, in "The Sign of Four" by Arthur Conan Doyle

7-1 INTRODUCTION

In previous chapters, continuous-time signals were studied. Deterministic continuous-time signals were often modeled as a solution to an ordinary differential equation and was expressed and manipulated using Laplace and Fourier transform methods. The study of discrete-time signals can follow a similar path, where deterministic signals are expressed as solutions of discrete-time difference equations and transforms. In this chapter, transform methods that can convert a discrete-time signal into a transform-domain will be introduced. The use of transforms can, as seen in the continuous-time case, simplify the analysis of a signal and how it interacts with systems.

7-2 DIFFERENCE EQUATIONS

A linear constant-coefficient *difference equation*, or simply DE, can be used to model many deterministic discrete-time signal processes. The discrete-time signals, or time-series, can be the result of sampling a continuous-time or analog signal. At other times, discrete-time signals are obtained directly from an algorithm that implements a difference equation. For example, a discrete-time exponential can be modeled by the first-order DE, $x[k+1] - ax[k] = 0$, with initial conditions

249

given by $x[k]|_{k=0} = x_0$. If $x_0 \neq 0$, then $x[k] = a^k x[0]$, for $k \geq 0$. This result can be verified by direct substitution. Extending this concept, higher order signals can be modeled as the homogeneous (unforced) solution to a nth order DE

$$c_n x[k-n] + c_{n-1} x[k-(n-1)] + \cdots + c_1 x[k-1] + c_0 x[k] = 0 \qquad (7\text{-}1)$$

or

$$\sum_{i=0}^{n} c_i x[k-i] = 0 \qquad (7\text{-}2)$$

with initial conditions

$$x[-i] = x_{-i} \qquad \text{for } i = 1, \ldots, n-1$$

Linear difference and differential equations share a considerable amount of common terminology and behavior. Recall that the eigenfunctions of an ODE were assumed to have the form $\phi_n(t) = \exp(s_n t)$. They appear in difference equations as $\Phi_n[k] = \lambda_n^k$. The solution to the DE homogeneous equation (Equation 7-2) can be expressed in terms of eigenfunctions $\Phi_n[k]$, as shown in Equation 7-3

$$x[k] = \sum_{i=1}^{n} a_i \Phi_i[k] \qquad (7\text{-}3)$$

where the coefficients a_i are computed using the method of undetermined coefficients.

7-3   Z-TRANSFORM

Up until the 1950s, signals and systems were generally assumed to be either continuous-time or analog. Historically, Laplace transforms were used to study signals defined by the solutions to linear ordinary differential equations. Beginning in the 1950s, discrete-time signals began to appear. More recently, the descendants of discretely sampled signals, namely digital signals, have become commonplace. Unfortunately, the Laplace transform is not well-suited for the study of discrete-time or digital signals. Instead, another transform, called the *Z-transform*, is used.

Recall that the delay property of the Laplace transform was given by

$$x(t - T_s) \overset{\mathcal{L}}{\leftrightarrow} e^{-sT_s} X(s) \qquad (7\text{-}4)$$

where $x(t) \overset{\mathcal{L}}{\leftrightarrow} X(s)$. A shorthand notation has evolved to represent the term $\exp(-sT_s)$. It is called the *z-operator* and is defined by the equation

$$z = \exp(sT_s) \quad \text{or} \quad z^{-1} = \exp(-sT_s) \qquad (7\text{-}5)$$

The z-operator provides the foundation for the Z-transform, which is fundamental to the study of discrete-time and digital signals and systems.

The Z-transform is known to belong to a class of discrete-time transforms called *impulse invariant* transforms. This means that if $x(t)$ is periodically sampled at a rate $f_s = 1/T_s$ to produce a discrete-time time-series $\{x[k]\}$, the Z-transform will preserve the relationship $x[k] = x(kT_s)$. The Z-transform also satisfies a number of other properties that belong to a branch of mathematics called *conformal mapping* (mapping that preserve angles). The connection between the s-plane (domain of s variables) and z-plane (domain of z variables) is interpreted in Figure 7-1. For k, an integer, some important mappings are listed below:

1 The points $s = \pm j2\pi k/T_s$ is mapped to $z = \exp(j2\pi k) = 1$.

2 The points $s = \pm j\pi(2k+1)/T_s$ is mapped to $z = \exp(\pm j\pi(2k+1)) = -1$.

3 The point $s = j\pi/T_s$, which is mapped to $z = \exp(\pm\pi) = -1$, corresponds to $\omega = \pi/T_s = \pi f_s$, which is also the Nyquist frequency in radians per second.

4 The $j\omega$ axis in the s-plane maps onto the unit circle (i.e., $|z| = 1$). It can be seen that the values of z wrap around the unit circle mod 2π. In order to preserve uniqueness of the mapping, one restricts $\pm j\omega T_s < \pm j\pi$. This corresponds to mapping to frequencies bounded by plus/minus the Nyquist frequency.

5 The left-hand plane in the s-plane, for values of $s = \sigma \pm j\omega$, such that $|\omega| < \pi/T_s$, maps to the interior of the unit circle in the z-plane.

6 The right-hand plane in the s-plane, for values of $s = \sigma \pm j\omega$, such that $|\omega| < \pi/T_s$, maps to the exterior of the unit circle in the z-plane.

7 The s-plane region bounded by $s = \sigma + j\omega$, or $\sigma \in [\sigma_0, \sigma_1]$, $\sigma < 0$, and $\omega \in [-\pi/T_s, \pi/T_s]$, maps into an annular ring in the z-plane, with radii $r_0 = \exp(\sigma_0 T_s)$ and $r_1 = \exp(\sigma_1 T_s)$.

FIGURE 7-1   Conformal mapping of the s-plane to the z-plane.

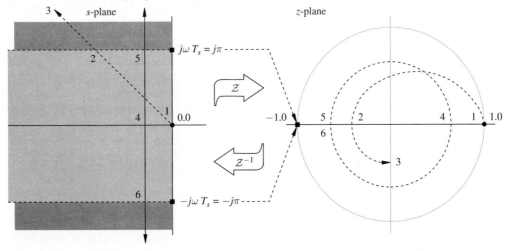

Example 7-1   Conformal mapping

The individual values of $s = \sigma + j\omega$ are mapped into the Z-plane using the rule $z = \exp(sT_s)$. It is also often desirable to interpret mappings in terms of $z^{-1} = \exp(-sT_s)$, where, according to Equation 7-5, z^{-1} represents a delay operator.

Computer Study   S-file `ch7_1.s` and M-file `zplane.m` convert points in the s-plane into their corresponding z-domain locations using the format `zplane (xs,T)`, where `xs` is a set of points in the s-plane and `T` is the sampling period. To convert the six roots of the polynomial $X(s) = 1 + 3s + 7s^2 + 15s^3 + 14s^4 + 12^5 + 8s^6 = (s+1)(s+2)(s+j)(s-j)(s+2j)(s-2j)$ into z-domain locations, for $T = 0.01, 0.1, 1,$ and 10, use the following.

SIGLAB	MATLAB
`>include "ch7_1.s"`	`>> xs = roots([1 3 7 15 14 12 8])`
`>xs=root([1,3,7,15,14,12,8])`	`xs =`
`>xs'`	`   -2.0000`
`  -2+0j`	`    0.0000 + 2.0000i`
`  -1+0j`	`    0.0000 - 2.0000i`
`   0-1j`	`   -1.0000`
`   0+1j`	`   -0.0000 + 1.0000i`
`  ~0-2j`	`   -0.0000 - 1.0000i`
`  ~0+2j`	`>> zplane(xs, 0.01)`
`>zplane(xs,.01)`	`>> zplane(xs, 0.1)`
`>zplane(xs,.1)`	`>> zplane(xs, 1)`
`>zplane(xs,1)`	`>> zplane(xs, 10)`
`>zplane(xs,10)`	

The results are displayed in Figure 7-2. For $T_s > 0$, $z = \exp(sT_s)$ produces trajectories clustered on or inside the unit circle in the z-plane.

To map the complex conjugate pairs into the z-plane for $T_s = \{0.01, 0.1, 1, 10\}$ at $s_0 = \pm j3\pi/4$, $s_1 = -0.25 \pm j\pi/4$, $s_2 = -0.5 \pm j\pi/4$, $s_3 = -1.0 \pm j\pi/8$, and $s_4 = -2 \pm j\pi/16$, use the following.

SIGLAB	MATLAB
`>include "ch7_1.s"`	`>> j = sqrt(-1);`
`>s0=<0,3*pi/4>; s1=<-0.25,pi/2>`	`>> s = [0+j*pi/2, -0.2+j*pi/3,`
`>s2=<-0.5,pi/4>; s3=<-1.0,pi/8>`	`-0.4+j*pi/4, -0.6+j*pi/5,`
`>s4=<-2.0,pi/16>`	`-0.8+j*pi/6];`
`>s=[s0,s1,s2,s3,s4];`	`>> xs = [s conj(s)];`
`xs=[s, conj(s)]`	`>> zplane(xs, 0.01)`
`>zplane(xs,0.01); zplane(xs,0.1)`	`>> zplane(xs, 0.1)`
`>zplane(xs,1); zplane(xs,10)`	`>> zplane(xs, 1)`
	`>> zplane(xs, 10)`

The data is displayed in Figure 7-3 for $T_s = 1$. Observe that complex conjugate pairs in the s-plane remain complex conjugate pairs in the z-plane.

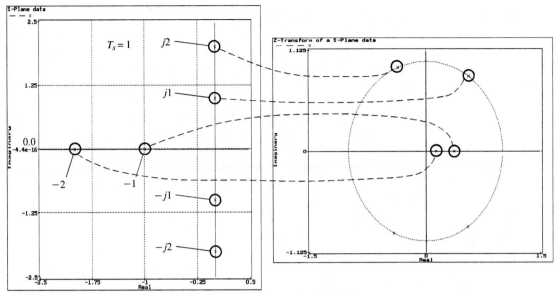

FIGURE 7-2   Mapping of s-plane contours into the z-plane for T=1.

FIGURE 7-3   Mapping of s-domain roots into the z-domain for T=1.

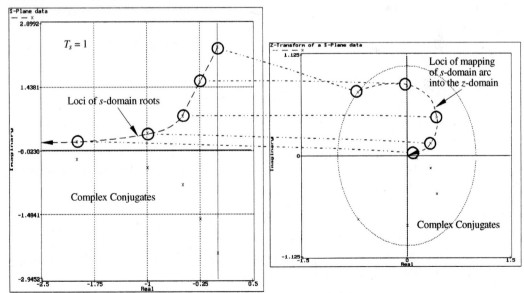

Example 7-2   Conformal mapping

A contour in the s-plane, given by $g(s)$, is mapped into the z-plane using the rule $z = \exp(g(s)T_s)$. The study of $X(s)$ along a predefined contour can be used to predict the performance of a discrete-time system under a specific set of conditions.

Computer Study   To explore how trajectories in the s-plane are mapped into the z-plane, plot the trajectories $g_1(s) = j\pi\omega$, $g_2(s) = -\omega + j\omega$, and $g_3(s) = -\omega + j10\omega$ for $\omega \in [0, 1]$. To display the conformal mappings of these trajectories, use the S-file ch7_2.s, which has a format of circle(g(s)), or M-file zplane.m (see Problem 7-1).

SIGLAB	MATLAB
>include"ch7_2.s" #next, contours	>> j = sqrt(-1);
>w=rmp(500,1/499,500) # spans [0,1]	>> w = (0:511)/512;
>g1=<zeros(500),pi*w>	>> g1 = j*pi*w;
>circle(e^g1) # g1 = j * pi * omega	>> g2 = -w + j*w;
>g2=<-w,pi*w>	>> g3 = -w + j*10*pi*w;
>circle(e^g2)	>> zplane(g1, 1, 1)
>g3=<-w,10*pi*w>	>> zplane(g2, 1, 1)
>circle(e^g3)	>> zplane(g3, 1, 1)

The results are displayed in Figure 7-4.

The mapping rule $z = \exp(sT_s)$ gives rise to the Z-transform for discrete-time signals. The *two-sided Z-transform* of a time-series $\{x[k]\}$ is given by

$$X(z) = \sum_{k=-\infty}^{\infty} x[k]z^{-k} \tag{7-6}$$

and exists for all z, for which $X(z)$ converges. For a causal time-series, the *one-sided Z-transform* applies and is given by

$$X(z) = \sum_{k=0}^{\infty} x[k]z^{-k} \tag{7-7}$$

Using this basic definition, it should also be apparent that Z-transform of a unit step is given by

$$X(z) = \sum_{k=0}^{\infty} z^{-k} = 1 + z^{-1} + z^{-2} + z^{-3} + \cdots \tag{7-8}$$

This infinitely long series can also be represented in closed form using the following useful representation rule

$$\sum_{k=0}^{\infty} x^k = 1 + x^1 + x^2 + x^3 + \ldots = \frac{1}{1+x} \qquad \text{if } |x| < 1 \tag{7-9}$$

If $|x|$ were *not* bounded by unity, then it would be apparent that $x^k \to \infty$ for $k \to \infty$. Therefore, $X(z) = 1 + z^{-1} + z^{-2} + z^{-3} + \cdots$ has a closed form

$$X(z) = \frac{1}{1 - z^{-1}} = \frac{z}{z - 1} \tag{7-10}$$

if $|z^{-1}| < 1$ or $|z| > 1$. A more complicated signal, such as the causal exponential $x[k] = \exp(-\alpha k T_s)u[k] = a^k u[k]$, where $a = \exp(-\alpha T_s)$ and $u[k]$ is the unit-step function, has a Z-transform given by

$$X(z) = \sum_{k=0}^{\infty} a^k z^{-k} = \sum_{k=0}^{\infty} (az^{-1})^k \tag{7-11}$$

which, upon using the representation rule shown in Equation 7-9 for $x^{-1} = az^{-1}$,

FIGURE 7-4   Mapping of s-domain linear contours.

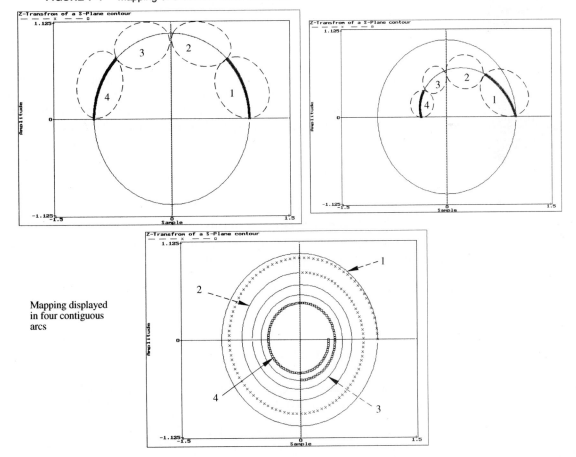

Mapping displayed
in four contiguous
arcs

implies Equation 7-11 can be reduced to

$$X(z) = \frac{1}{1 - az^{-1}} = \frac{z}{z - a} \tag{7-12}$$

if $|az^{-1}| < 1$ or $|z| > |a|$. Therefore, it is seen that there are restrictions on the existence of a Z-transform.

The domain of values of z guaranteeing that a Z-transform of $x[k]$ exists is called the *region of convergence*, or ROC. The ROC consists of an annular ring in the z-plane that is centered about the origin. For example, if $x[k] = [(1/3)^k + (-1/4)^k]u[k]$, then it follows that

$$X(x) = \frac{1}{1 - \frac{1}{3}z^{-1}} + \frac{1}{1 + \frac{1}{4}z^{-1}} = \frac{z\left(2z - \frac{1}{12}\right)}{\left(z - \frac{1}{3}\right)\left(z + \frac{1}{4}\right)} \tag{7-13}$$

For convergence, the individual terms must converge. This means that

$$\frac{1}{1 - z^{-1}/3} = \sum_{k=0}^{\infty}\left(\frac{z^{-1}}{3}\right)^k \Rightarrow \left|\frac{1}{3}z^{-1}\right| < 1$$

$$\frac{1}{1 - z^{-1}/4} = \sum_{k=0}^{\infty}\left(\frac{z^{-1}}{4}\right)^k \Rightarrow \left|\frac{1}{4}z^{-1}\right| < 1 \tag{7-14}$$

From the intersection of the ROCs it can be seen that if $|z| > 1/3$, then $X(z)$ converges. It should be noted, however, that the Z-transform and ROC of an arbitrary time-series $\{x[k]\}$ are rarely computed. Instead, $x[k]$ can normally be expressed as a collection of elementary functions (such as those listed in Table 7-1), whose Z-transforms and ROCs are known and tabled. Along with the basic properties of the Z-transform listed in Table 7-2, the Z-transform of many important time-series can be found as a combination of the Z-transforms listed in Table 7-1.

Example 7-3   Properties of Z-transforms

The Z-transform of $x[k] = k^2u[k]$ can be defined in terms of $x_1[k] = ku[k]$ and $x[k] = kx_1[k]$. From Table 7-1, it follows that

$$X_1(z) = \frac{z}{(z - 1)^2}$$

and from Table 7-2

$$X(z) = -z\frac{d}{dz}X_1(z) = -z\left(\frac{1}{(z - 1)^2} - \frac{2z}{(z - 1)^3}\right) = \frac{z(z + 1)}{(z - 1)^3}$$

An extension to this result is given by $x[k] = ka^k$ for $|a| < 1$. Letting $x_1[k] = k$ and $x[k] = a^kx_1[k]$, and given knowledge of $X_1(z)$, it follows that

$$X(z) = X_1(z/a) = \frac{az}{(z - a)^2}$$

TABLE 7-1 Z-TRANSFORMS OF PRIMITIVE TIME FUNCTIONS

Time-domain	Z-transform	Region of convergence, $|z| > R$
$\delta_K[k]$	1	0
$\delta_K[k - m]$	z^{-m}	0
$u[k]$	$\dfrac{z}{z - 1}$	1
$ku[k]$	$\dfrac{z}{(z - 1)^2}$	1
$k^2 u[k]$	$\dfrac{z(z + 1)}{(z - 1)^3}$	1
$k^3 u[k]$	$\dfrac{z(z^2 + 4z + 1)}{(z - 1)^4}$	1
$\exp(akT_s)u(kT_s)$	$\dfrac{z}{z - \exp(aT_s)}$	$\left| \exp(aT_s) \right|$
$kT_s \exp(akT_s)u(kT_s)$	$\dfrac{zT_s \exp(aT_s)}{(z - \exp(aT_s))^2}$	$\left| \exp(aT_s) \right|$
$(kT_s)^2 \exp(akT_s)u(kT_s)$	$\dfrac{zT_s^2 \exp(aT_s)(z + \exp(aT_s))}{(z - \exp(aT_s))^3}$	$\left| \exp(aT_s) \right|$
$a^k u[k]$	$\dfrac{z}{(z - a)}$	$|a|$
$ka^k u[k]$	$\dfrac{az}{(z - a)^2}$	$|a|$
$k^2 a^k u[k]$	$\dfrac{az(z + a)}{(z - a)^3}$	$|a|$
$\sin(bkT_s)u(kT_s)$	$\dfrac{z \sin(bT_s)}{z^2 - 2z \cos(bT_s) + 1}$	1
$\cos(bkT_s)u(kT_s)$	$\dfrac{z(z - \cos(bT_s))}{z^2 - 2z \cos(bT_s) + 1}$	1
$\exp(akT_s)\sin(bkT_s)u(kT_s)$	$\dfrac{z \exp(aT_s) \sin(bT_s)}{z^2 - 2z \exp(aT_s) \cos(bT_s) + \exp(2aT_s)}$	$\left| \exp(aT_s) \right|$
$\exp(akT_s)\cos(bkT_s)u(kT_s)$	$\dfrac{z(z - \exp(aT_s) \cos(bT_s))}{z^2 - 2z \exp(aT_s) \cos(bT_s) + \exp(2aT_s)}$	$\left| \exp(aT_s) \right|$
$a^k \sin(bkT_s)u(kT_s)$	$\dfrac{az \sin(bT_s)}{z^2 - 2az \cos(bT_s) + a^2}$	$|a|$
$a^k \cos(bkT_s)u(kT_s)$	$\dfrac{z(z - a \cos(bT_s))}{z^2 - 2az \cos(bT_s) + a^2}$	$|a|$
$\dfrac{(\ln(a))^k}{k!} u[k]$	$a^{1/z}$	$a > 0$
$\dfrac{1}{k} \quad (k = 1, 2, 3, \dots)$	$\ln\left(\dfrac{z}{z - 1}\right)$	1
$\dfrac{(k + 1) \cdots (k + m)a^k}{m!}$	$\dfrac{z^{m+1}}{(z - a)^{m+1}}$	$|a|$

TABLE 7-2   PROPERTIES OF Z-TRANSFORMS

Property	Time series	Z-transform
Linearity:	$x_1[k] + x_2[k]$	$X_1(z) + X_2(z)$
Real Scaling:	$ax[k]$	$aX(z)$
Complex Scaling:	$w^k x[k]$	$X(z/w)$
Delay*:	$x[k - L]$	$z^{-L}X(z) + \displaystyle\sum_{n=-L}^{-1} z^{-(L+n)}x[n]$
Time Reversal:	$x[-k]$	$X(1/z)$
Modulation:	$\exp[-ak]x[k]$	$X(\exp(a)z)$
Ramping:	$kx[k]$	$-z\dfrac{d}{dz}X(z)$
Reciprocal:	$\dfrac{x[k]}{k}$	$-\displaystyle\int \dfrac{X(z)}{z}\,dz$
Summation:	$\displaystyle\sum_{n=-\infty}^{k} x[n]$	$\dfrac{zX(z)}{z-1}$
Periodic:	$y[k+N]$	$\dfrac{z^N Y(z)}{z^N - 1}$

Initial Value*:	$x[0] = \lim_{z\to\infty} X(z)$
Final Value**:	$x[\infty] = \lim_{z\to 1}(z-1)X(z)$

* for causal $x[k]$.
** if $X(z)$ has no more than one pole on the unit circle and all other poles are
   interior to the unit circle.

To test the validity of this result, the value of $x[0]$ can be computed using the initial value theorem. Here,

$$x[0] = \lim_{z\to\infty} X(z) = 0$$

as required. The final value is given by

$$x[\infty] = \lim_{z\to 1}(z-1)X(z)$$

which produces $x[\infty] = 0$, as expected.

Signal symmetry properties can sometimes be used to assist in computing the Z-transform of a signal. An *even symmetry* time-series satisfies

$$x[k] = x[-k] \qquad (7\text{-}15)$$

An example of an even symmetry time-series is a sampled, finite-duration, cosine wave. If $x[k]$ is an even symmetry time-series over $k \in [-M, M]$, and zero elsewhere, then

$$X(z) = \sum_{k=-M}^{M} x[k]z^{-k} = \sum_{k=-M}^{-1} x[k]z^{-k} + x(0) + \sum_{n=1}^{M} x[n]z^{-n} \tag{7-16}$$

It follows that, upon substituting $n = -k$ into Equation 7-16, and because $x[k] = x[-k]$

$$X(z) = \sum_{n=1}^{M} x[n]z^{n} + x[0] + \sum_{n=1}^{M} x[n]z^{-n} = x[0] + \sum_{n=1}^{M} x[n](z^{n} + z^{-n}) \tag{7-17}$$

Observe that the Z-transform of an even symmetry time-series satisfies

$$X(z) = X\left(\frac{1}{z}\right) \tag{7-18}$$

Thus, if z_0 is a root of $X(z)$, then so is $1/z_0$.

An odd, or antisymmetric, time-series satisfies the condition that

$$x[k] = -x[-k] \tag{7-19}$$

An example of an antisymmetric time-series is a sampled, finite-duration, sine wave. Following the procedure used to compute the Z-transform for an even symmetry time-series, it follows that

$$X(z) = \sum_{n=1}^{M} x[n](z^{-n} - z^{n}) \tag{7-20}$$

which gives rise to the relationship for an antisymmetric time-series

$$X\left(\frac{1}{z}\right) = \sum_{n=1}^{M} x[n](z^{n} - z^{-n}) = -X(z) \tag{7-21}$$

Thus, if z_0 is a root of $X(z)$, then so is $1/z_0$.

Example 7-4   Symmetry properties of Z-transforms

The roots of even and odd polynomials in z were predicted in Equations 7-18 and 7-21.

Computer Study   An N-sample ($N = 2M + 1$) Hamming window is an even time-series $\{x[k]\}$, having a Z-transform

$$X(z) = \sum_{k=-M}^{M} x[k]z^{-k}$$

Using $N = 7$ samples, the roots of an even and odd symmetry function $X(z)$ are produced using

$$X_{\text{even}}(z) = x[-3]z^3 + x[-2]z^2 + x[-1]z^1 + x[0] + x[1]z^{-1}$$

$$+ x[2]z^{-2} + x[3]z^{-3}$$

where $x[0] = 1.0$, $x[\pm1] = 0.77$, $x[\pm2] = 0.31$, $x[\pm3] = 0.08$. The even Hamming window can be used to create an odd signal by setting $x[0] = 0$ and changing the sign of the data as follows

$$X_{\text{odd}}(z) = x[-3]z^3 + x[-2]z^2 + x[-1]z^1 + x[0] - x[1]z^{-1} - x[2]z^{-2} - x[3]z^{-3}$$

Compute and compare the roots as follows

SIGLAB	MATLAB
>x=ham(7)	>> x = hamming(7);
>xr=rev(x) # x[n] = x[-n]	>> xr = flipud(x); % x(-n)
>max(root(x)-root(xr))	>> max(roots(x) - roots(xr))
0.0	ans =
>#[1,1,1,0,-1,-1,-1]	3.7748e-15 - 4.8850e-15i
>mask=[ones(3),0,-ones(3)]	>> x = [x(1:3); 0; -x(5:7)];
>x=x.*mask	>> xr = [xr(1:3); 0; -xr(5:7)];
>xr=xr.*mask # x[n] = -x[-n]	>> max(roots(x)-roots(flipud(x)))
>max(root(x)-root(-xr))	ans =
~0.0	1.3323e-15 - 1.7764e-15i

This agrees with the predicted results.

7-4   INVERSE Z-TRANSFORM

The process by which a Z-transform of a time-series $\{x[k]\}$, namely $X(z)$, is returned to the time-domain is called the *inverse Z-transform*. Technically, the inverse Z-transform is defined by

$$x[n] = \mathcal{Z}^{-1}[X(z)] = \frac{1}{j2\pi} \oint_C \frac{X(z)z^n}{z} \, dz \qquad (7\text{-}22)$$

where C is a restricted closed path traversed in a counterclockwise direction encircling the origin and residing in the ROC of $X(z)$. Obviously this can be a laborious manual process. There are four practical techniques that can be used to implement an inverse transform. They are

1 long division,

2 partial fraction expansion,

3 residue theorem,

4 difference equation synthesis.

Long division is applicable to the study of the transient response of causal signals. In particular, it can be used to compute the first few sample values of the time-series. Partial fraction expansion is the most popular inversion method, but requires access to Z-transform standard tables. The residue theorem is elegant, but is the most complex. The last technique is a simulation methodology requiring the use of a general purpose computer.

7-5   LONG DIVISION Z-TRANSFORM INVERSION

Suppose $X(z)$ is a causal signal defined as the ratio of two polynomials, say $X(z) = N(z)/D(z)$, where $N(z) = \sum_{i=0}^{N} b_i z^{-i}$ and $D(z) = \sum_{i=0}^{N} a_i z^{-i}$. Thus using long division gives

$$
\begin{array}{r}
b_0/a_0 + (b_1 - a_1 b_0/a_0)/a_0 z^{-1} + \qquad\qquad \cdots \\
\hline
a_0 + a_1 z^{-1} + a_2 z^{-2} + \cdots \overline{)} b_0 \quad + \qquad\qquad b_1 z^{-1} + \qquad\qquad b_2 z^{-2} + \cdots \\
b_0 \quad + \qquad (a_1 b_0/a_0) z^{-1} + \qquad (a_2 b_0/a_0) z^{-2} + \cdots \\
\hline
(b_1 - a_1 b_0/a_0) z^{-1} + (b_2 - a_2 b_0/a_0) z^{-2} + \cdots
\end{array}
$$

$$(7\text{-}23)$$

From the definition of a one-sided Z-transform (Equation 7-7), it is immediately recognized that

$$
X(z) = \frac{b_0}{a_0} + \frac{a_0 b_1 - a_1 b_0}{a_0^2} z^{-1} + \cdots
\tag{7-24}
$$

which implies

$$
x[k] = \left\{ \frac{b_0}{a_0}, \frac{a_0 b_1 - a_1 b_0}{a_0^2}, \cdots \right\}
\tag{7-25}
$$

Long division can be a tedious means of computing an inverse transform if $x[k]$ is needed for $k \gg 1$. The method, therefore, is effective only if the first few initial sample values of $x[k]$ must be known.

Example 7-5   Long division

A discrete causal exponential time-series, given by $x[k] = a^k u[k]$ satisfies

$$
X(z) = 1 + a z^{-1} + a^2 z^{-2} + a^3 z^{-3} + \cdots + a^k z^{-k} + \cdots = \frac{1}{1 - a z^{-1}}
$$

Long division produces

$$1 - az^{-1} \overline{\smash{\big)}\,1} \qquad \begin{array}{l} 1 + az^{-1} + a^2z^{-2} + \cdots \\ \hline 1 - az^{-1} \\ \hline az^{-1} \\ az^{-1} - a^2z^{-2} \\ \hline \cdots \end{array}$$

Then $x[k] = \{1, a, a^2, \ldots\}$, or $x[k] = a^k u[k]$.

Example 7-6   Long division

Consider a third-order example

$$X(z) = \frac{z(z+1)}{(z-1)^3},$$

then

$$z^3 - 3z^2 + 3z - 1 \overline{\smash{\big)}\,z^2 + z} \qquad \begin{array}{l} z^{-1} + 4z^{-2} + 9z^{-3} + \quad \cdots \\ \hline z^2 - 3z^1 + 3z^0 - 1z^{-1} \\ \hline 4z^1 - 3z^0 + 1z^{-1} \\ 4z^1 - 12z^0 + 12z^{-1} - 4z^{-1} \\ \hline 9z^0 - 11z^{-1} + 4z^{-1} \end{array}$$

or

$$X(z) = 0z^0 + 1z^{-1} + 4z^{-2} + 9z^{-3} + \cdots$$

and $x[n] = \{0, 1, 4, 9, \ldots\}$, which implies $x[k] = k^2 u[k]$.

7-6   PARTIAL FRACTION (HEAVISIDE) EXPANSION

An nth-order transfer function $X(z)$ has a general form

$$X(z) = \frac{N(z)}{D(z)} \qquad D(z) = \prod_{i=1}^{L}(z - \lambda_i)^{n(i)} \quad N = \sum_{i=1}^{L} n(i) \qquad (7\text{-}26)$$

where λ_i is a root of $D(z)$ of *multiplicity* $n(i)$. If $n(i) > 1$, then λ_i is said to be a repeated root and if $n(i) = 1$ then λ_i is a distinct root (nonrepeated). The partial fraction, or Heaviside, expansion of $X(z)$ has a form given by

$$X(z) = \sum_{i=1}^{L} \sum_{j=1}^{n(i)} \frac{a_{i,j} z}{(z - \lambda_i)^j} \tag{7-27}$$

which is seen to be defined in terms of having the appearance of $z/(z - \lambda_j)^k$. The rationale for this form is found in Table 7-1. Here, except for $\delta_K[k]$, $\ln(\lambda)^k/k!$, and $1/k$, all causal entries exhibit a Z-transform having the form $z/(z - \lambda)^k$. Therefore, the partial fraction expansion form shown in Equation 7-27 simply mirrors this structure. Observe the partial fraction terms now carry a z in the numerator. Compare this to the partial fraction form used to study continuous signals having a Laplace transform $X(s)$, found in Chapter 2. It should be immediately apparent that the methodology developed to study $X(s)$ can be used to analyze $X(z)$, if $X(z)/z$ is selected for partial fraction expansion. The problem, of course, remains one of computing the *Heaviside coefficients* $a_{i,j}$. Consider creating $X(z)(z - \lambda_i)^{n(i)}/z$ as follows

$$\frac{X(z)(z - \lambda_j)^{n(j)}}{z} = (z - \lambda_j)^{n(j)} \sum_{k=1}^{n(j)} \frac{a_{j,k}}{(z - \lambda_j)^k} + (z - \lambda_j)^{n(j)} \sum_{i=1,i \neq j}^{L} \sum_{k=1}^{n(i)} \frac{a_{i,k}}{(z - \lambda_i)^k}$$

$$= a_{j,1}(z - \lambda_j)^{n(j)-1} + \cdots + a_{j,n(j)-1}(z - \lambda_j)^1 + a_{j,n(j)} + \cdots \tag{7-28}$$

It can be seen that

$$a_{j,n(j)} = \lim_{z \to \lambda_j} \frac{X(z)(z - \lambda_j)^{n(j)}}{z} \tag{7-29}$$

The coefficient $a_{j,n(j)-1}$ is seen to be linear in $(z - \lambda_j)$ and can, therefore, be isolated and evaluated using a simple derivative. In particular

$$\frac{d}{dz} X(z)(z - \lambda_j)^{n(j)} = \frac{d}{dz}(\cdots + a_{j,1}z(z - \lambda_j)^{n(j)-1} + \cdots$$

$$\cdots + a_{j,n(j)-1}z(z - \lambda_j)^1 + za_{j,n(j)} + \cdots) \tag{7-30}$$

where

$$a_{j,n(j)-1} = \lim_{z \to \lambda_j} \frac{d}{dz} \frac{(z - \lambda_j)^{n(j)} X(z)}{z} \tag{7-31}$$

The production rule for $a_{j,n(j)-2}$ given by

$$a_{j,n(j)-2} = \lim_{z \to \lambda_j} \frac{d^2}{dz^2} \frac{1}{2} \frac{(z - \lambda_j)^{n(j)} X(z)}{z} \tag{7-32}$$

Continuing this line of reasoning, the $a_{j,k}$ coefficient is found by

$$a_{j,k} = \lim_{z \to \lambda_j} \frac{d^{n(j)-k}}{dz^{n(j)-k}} \frac{1}{(n(j) - k)!} \frac{(z - \lambda_j)^{n(j)} X(z)}{z} \tag{7-33}$$

Therefore it can be seen that when repeated roots of $X(z)$ are encountered, an elaborate procedure must be followed to produce the Heaviside coefficients.

Example 7-7 Nonrepeated root polynomial

If $X(z) = z^2/[(z - 1)(z - \exp(-T_s)]$, where $x[k]$ is causal and $T_s \neq 0$, then all roots are nonrepeated and

$$X(z) = a_0 + a_1 \frac{z}{(z - 1)} + a_2 \frac{z}{(z - \exp(-T_s))}$$

where a_0 corresponds to a Kronecker delta function, $a_1 z/(z - 1)$ to a step function, and $a_2 z/(z - \exp(-T_s))$ to an exponential. The Heaviside coefficient production rules are, however, specified in terms of factors of $X(z)/z$ where

$$\frac{X(z)}{z} = \frac{z}{(z - 1)(z - \exp(-T_s))} = a_0 \frac{1}{z} + a_1 \frac{1}{(z - 1)} + a_2 \frac{1}{(z - \exp(-T_s))}$$

where

$$a_0 = \lim_{z \to 0} \frac{z X(z)}{z} = 0$$

$$a_1 = \lim_{z \to 1} \frac{(z - 1) X(z)}{z} = \lim_{z \to 1} \frac{z}{z - \exp(-T_s)} = \frac{1}{1 - \exp(-T_s)}$$

$$a_2 = \lim_{z \to \exp(-T_s)} \frac{(z - \exp(-T_s)) X(z)}{z} = \lim_{z \to \exp(-T_s)} \frac{z}{z - 1} = \frac{\exp(-T_s)}{1 - \exp(T_s)}$$

Therefore,

$$X(z) = a_1 \frac{z}{z - 1} + a_2 \frac{z}{z - \exp(-T_s)}$$

and $x(kT_s) = (a_1 + a_2 \exp(-kT_s)) u(kT_s)$.

Computer Study   S-file `pf.s` and M-file `residue.m` perform a partial fraction expansion of $X(z) = N(z)/D(z)$ and were introduced in Example 2-7. They can be used to evaluate $X(z) = z^2/[(z-1)(z - \exp(-T_s)]$ once T_s is numerically specified. For the purposes of numerical analysis, let $\exp(-T_s) = 0.5$. Then the roots of $D(z)$ are $z = 1$ and 0.5. However, what is required is the partial fraction expansion of $X(z)/z = z^2/[z(z-1)(z-0.5)]$, which has roots of $D(z) = z(z-1)(z-0.5) = (z^3 - 1.5z^2 + 0.5z)$ given by $z = 0, 1$ and 0.5.

SIGLAB	MATLAB
`>include "pf.s"`	`>>a=[1,0,0];`
`>a=[1,0,0]; b=[1,-1.5,.5,0]`	`b=[1,-1.5,.5,0];`
`>pfexp(a,b)`	`>>% r= [a1,a2,a3],`
`pole @ 0+0j with multiplicity 1`	`p=[p1,p2,p3], k=a0`
`and coefficients: 0+0j => a_0`	`>>[r,p,k]=residue(a,b)`
`pole @ 0.5+0j with multiplicity 1`	`r=`
`and coefficients: -1-0j => a_2`	`    2`
`pole @ 1+0j with multiplicity 1`	`   -1`
`and coefficients: 2+0j => a_1`	`    0`
	`p=`
	`    1.00000`
	`    0.50000`
	`    0`
	`k=`
	`    []`

The values a_0, a_1, and a_2 are seen to correspond to the computed values. Therefore, the partial fraction expansion of $H(z)$ is

$$H(z) = \frac{0z}{z} + \frac{2z}{z-1} + \frac{-1z}{z-0.5} = \frac{2z}{z-1} - \frac{z}{z-0.5}$$

Example 7-8   Repeated-root polynomial

Suppose that the previous example is modified to read $X(z) = z^2/(z-1)^2$. Then repeated roots arise (i.e., $n(1) = 2$) and

$$X(z) = a_0 + \frac{a_{11}z}{(z-1)} + \frac{a_{12}z}{(z-1)^2}$$

or

$$\frac{X(z)}{z} = \frac{a_0}{z} + \frac{a_{11}}{(z-1)} + \frac{a_{12}}{(z-1)^2}$$

and

$$a_0 = \lim_{z \to 0} \frac{z X(z)}{z} = 0$$

$$a_{12} = \lim_{z \to 1} \frac{(z-1)^2 X(z)}{z} = \lim_{z \to 1} z = 1$$

$$a_{11} = \lim_{z \to 1} \frac{d}{dz} \frac{(z-1)^2 X(z)}{z} = \lim_{z \to 1} \frac{d}{dz} z = 1$$

which results in

$$X(z) = \frac{z}{(z-1)} + \frac{z}{(z-1)^2}$$

which results in the following time-series (see Table 7-1)

$$x[k] = (1+k)u[k]$$

Computer Study   S-file `pf.s` and M-file `residue.m` can be used to provide a partial fraction expansion of $X(z)/z = z^2/[z(z-1)^2]$, which has roots of $D(z) = z(z-1)^2$, given by $z = 0, 1$ and 1.

SIGLAB	MATLAB
`>include "pf.s"`	`>>n=[1,0,0]; d=poly([0,1,1]);`
`>n=[1,0,0]`	`>>% p=[p1,p2,p3],`
`>d=poly([0,1,1])`	`>>% r=[a1,a2,a3],`
`>d # N(z)=z^2; D(z)=z(z-1)(z-1)`	`k=a0`
`1 -2 1 0`	`>>[r,p,k]=residue(n,d)`
`>pfexp(n,d) # evaluate`	`r=`
`partial fraction expansion`	`  1`
`pole @ 0+0j with multiplicity`	`  1`
`1 and coefficients: 0+0j => a_0`	`  0`
`pole @ 1+0j with multiplicity`	`p=`
`2 and coefficients: 1-0j 1+j0`	`  1`
`=> a_2 and a_1`	`  1`
	`  0`
	`k=`
	`  []`

This corresponds to the derived values.

Example 7-9   Higher-order polynomials

A more complicated example is given by

$$X(z) = \frac{3z^3 - 5z^2 + 3z}{(z-1)^2(z-0.5)}$$

which has roots located at $z = 1$ (multiplicity $n(1) = 2$) and $z = 0.5$ (multiplicity $n(2) = 1$). The partial fraction expansion is given by

$$X(z) = a_0 + \frac{a_{11}z}{(z - 0.5)} + \frac{a_{21}z}{(z - 1)} + \frac{a_{22}z}{(z - 1)^2}$$

where a_0 corresponds to a Kronecker delta function, $a_{11}z/(z - 0.5)$ an exponential, $a_{21}z/(z - 1)$ a step function, and $a_{22}z/(z - 1)^2$ a ramp function. What is desired, however, is the partial fraction expansion of $X(z)/z$, where

$$\frac{X(z)}{z} = \frac{a_0}{z} + \frac{a_{11}}{(z - 0.5)} + \frac{a_{21}}{(z - 1)} + \frac{a_{22}}{(z - 1)^2}$$

where

$$a_0 = \lim_{z \to 0} \frac{zX(z)}{z} = 0$$

$$a_{11} = \lim_{z \to 0.5} \frac{(z - 0.5)X(z)}{z} = \lim_{z \to 0.5} \frac{3z^3 - 5z^2 + 3z}{z(z - 1)^2} = 5$$

$$a_{22} = \lim_{z \to 1} \frac{(z - 1)^2 X(z)}{z} = \lim_{z \to 1} \frac{3z^3 - 5z^2 + 3z}{z(z - 1/2)} = 2$$

$$a_{21} = \lim_{z \to 1} \frac{d}{dz} \frac{(z - 1)^2 X(z)}{z}$$

$$= \lim_{z \to 1} \left[\frac{9z^2 - 10z + 3}{z(z - 1/2)} - \frac{2(3z^3 - 5z^2 + 3z)(z - 1/4)}{(z(z - 1/2))^2} \right] = -2$$

which results in

$$x[k] = [5(0.5)^k - 2 + 2k]u[k]$$

This solution can be checked using long division to test the first few samples, say

$$\begin{array}{r} 3 + (5/2)z^{-1} + (13/4)z^{-2} + \cdots \\ \hline z^3 - (5/2)z^2 + 2z - (1/2) \overline{)\,3z^3 - 5z^2 + 3z} \end{array}$$

The initial value theorem also states that

$$x[0] = \lim_{z \to \infty} X(z) = 3$$

which agrees with the known time-series values.

Computer Study   S-file `pf.s` and M-file `residue.m` can be used to provide a partial fraction expansion of $X(z)/z = (3z^3 - 5z^2 + 3z)/[z(z - 0.5)(z - 1)^2]$, which has roots of $z(z - 0.5)(z - 1)^2$, given by $z = 0, 0.5, 1$ and 1.

SIGLAB	MATLAB
`>include "pf.s"`	`>>n=[3,-5,3,0]; d=[1,`
`>n=[3,-5,3,0]`	`-2.5,2,-0.5,0];`
`>d=poly([0,1,1,.5])`	`>>% r= [a1,a2,a3,a4],`
`>d`	`p=[p1,p2,p3,p4], k=a0`
`  1 -2.5 2 -0.5 0`	`>>[r,p,k]=residue(n,d)`
`SIGLAB>pfexp(n,d) # evaluate`	`r=`
`partial fraction expansion`	`   -2.000`
`pole @ 0+0j with multiplicity`	`    2.000`
`1 and coefficients: 0 + 0j`	`  5.000`
`=> a_0`	`    0.000`
`pole @ 0.5+0j with`	`p=`
`multiplicity 1 and coefficients:`	`   1.00000`
`5+0j => a_11`	`   1.00000`
`pole @ 1+0j with`	`   0.50000`
`multiplicity 2 and coefficients:`	`   0`
`2+0j -2+0j =>a_22 and a_21`	`k=`
	`   []`

Example 7-10   Numerator modification

The Z-transform of a causal sequence $x[k]$, given by $X(z) = (z + 1)^3/[(z + 1/2)(z - 1/2)^2]$, may appear closely related to those studied in previous examples. The difference is that the numerator of $X(z)$, in this case, contains no zeros located at $z = 0$. The other examples have, as a minimum, a multiplier of z^1, which defines at least one zero to be located at $z = 0$. The effect of this additional numerator term is apparent once the partial fraction expansion of $X(z)$ is produced

$$X(z) = a_0 + \frac{a_1 z}{(z + 1/2)} + \frac{a_{21} z}{(z - 1/2)} + \frac{a_{22} z}{(z - 1/2)^2}$$

or

$$\frac{X(z)}{z} = \frac{a_0}{z} + \frac{a_1}{(z + 1/2)} + \frac{a_{21}}{(z - 1/2)} + \frac{a_{22}}{(z - 1/2)^2}$$

The Heaviside coefficients are computed as follows

$$a_0 = \lim_{z \to 0} \frac{z X(z)}{z} = 8$$

$$a_1 = \lim_{z \to -1/2} \frac{(z + 1/2) X(z)}{z} = \lim_{z \to -1/2} \frac{(z + 1)^3}{z(z - 1/2)^2} = -\frac{1}{4}$$

$$a_{22} = \lim_{z \to 1/2} \frac{(z - 1/2)^2 X(z)}{z} = \lim_{z \to 1/2} \frac{(z + 1)^3}{z(z + 1/2)} = \frac{27}{4}$$

$$a_{21} = \lim_{z \to 1/2} \frac{d}{dz} \frac{(z-1/2)^2 X(z)}{z} = \lim_{z \to 1/2} \frac{d}{dz} \frac{(z+1)^3}{z(z+1/2)} = -\frac{27}{4}$$

Therefore, the coefficient a_0 now has a nonzero value and

$$X(z) = 8 - \frac{1}{4} \frac{z}{z+1/2} + \frac{-27}{4} \frac{z}{(z-1/2)} + \frac{27}{4} \frac{z}{(z-1/2)^2}$$

It follows that

$$x[k] = \left(8\delta_K[k] - \frac{1}{4} \left(\frac{-1}{2} \right)^k - \frac{27}{4} \left(\frac{1}{2} \right)^k + \frac{27}{4} k \left(\frac{1}{2} \right)^k \right) u[k]$$

Checking, using the initial value theorem, we note

$$x[0] = \lim_{z \to \infty} X(z) = 1$$

and from the derived time-series

$$x[0] = 8 - \frac{1}{4} - \frac{27}{4} = 1$$

Computer Study S-file `pf.s` and M-file `residue` can be used to provide a partial fraction expansion of $X(z)/z = (z+1)^3/[z(z+0.5)(z-0.5)^2]$, which has roots of $z(z+0.5)(z-0.5)^2$, given by $z = 0, -0.5, 0.5$ and 0.5.

SIGLAB	MATLAB

```>include "pf.s"```
```>n=[1,3,3,1]```
```>d=[1,-.5,-.25,.125,0]```
```>pfexp(n,d) # evaluate```
```partial fraction expansion```
```pole @ -0.5+0j with```
```multiplicity 1 and coefficients:```
```-0.2500+0j => a_1```
```pole @ 0+0j with```
```multiplicity 1 and coefficients:```
```8.000+0j => a_0```
```pole @ 0.499985+0j with```
```multiplicity 2 and coefficients: 6.75+0j```
```-6.75+0j =>a_22 and a_21```
	```>>a=[1,3,3,1]; b=[1,```
	```-0.5,-0.25,0.125,0];```
	```>>% r= [a1,a2,a3,a4],```
	```p=[p1,p2.p3,p4], k=a0```
	```>>[r,p,k]=residue(a,b)```
	```r=```
	``` -6.7500```
	``` 6.7500```
	``` -0.2500```
	``` 8.0000```
	```p=```
	``` 0.5000```
	``` 0.5000```
	``` -0.5000```
	``` 0```
	```k=```
	``` []```

This corresponds to the derived values.

Example 7-11   Complex roots

The second-order $X(z) = (3z^2 - 1.5z)/(z^2 - \cos(\pi/6)z + 1/4)$ has nonrepeated complex roots. The partial fraction expansion of $X(z)$ is defined by

$$X(z) = a_0 + a_1 \frac{z}{(z-\alpha)} + a_2 \frac{z}{(z-\alpha^*)}$$

or

$$\frac{X(z)}{z} = \frac{a_0}{z} + \frac{a_1}{(z-\alpha)} + \frac{a_2}{(z-\alpha^*)}$$

where $\alpha = 0.433 \pm j0.25$ and

$$a_0 = \lim_{z \to 0} X(z) = 0$$

$$a_1 = \lim_{z \to \alpha} \frac{(z-\alpha)X(z)}{z} = \lim_{z \to \alpha} \frac{(3z^2 - 1.5z)}{z(z-\alpha^*)}$$

$$a_2 = \lim_{z \to \alpha^*} \frac{(z-\alpha^*)X(z)}{z} = \lim_{z \to \alpha^*} \frac{(3z^2 - 1.5z)}{z(z-\alpha)} = a_1^*$$

Computer Study   S-file `pf.s` and M-file `residue.m` is used to evaluate $X(z) = (3z^2 - 1.5z)/(z^2 - \cos(\pi/6)z + 1/4)$.

SIGLAB	MATLAB
`SIGLAB>include "pf.s"`	`>>n=[3,-1.5,0];`
`SIGLAB>n=[3,-1.5,0]`	`>>d=[1,-cos (pi/6),0.25,0]);`
`SIGLAB>d=[1,-cos (pi/6),1/4,0]`	`>>% r= [a1,a2,a3], p=[p1,`
`SIGLAB>pfexp(n,d) # evaluate`	`p2, p3], k=a0`
`partial fraction expansion`	`>>[r,p,k]=residue(n,d)`
`pole @ 0+0j with`	`r=`
`multiplicity 1 and coefficients:`	`    1.5000+0.4020i`
`0+0j => a_0`	`    1.5000-0.4020i`
`pole @ 0.433013-0.25j`	`         0`
`with multiplicity 1 and`	`p=`
`coefficients: 1.5-0.401924j`	`    0.43301+0.2500i`
`=> a_1`	`    0.43301+0.2500i`
`pole @ 0.433013`	`         0`
`+0.25j with multiplicity 1 and`	`k=`
`coefficients: 1.5+0.401924j=> a_2`	`    []`

Notice that $a_1 = a_2^* = 1.5 - 0.401924j \approx 1.55 \exp(-j\pi/12)$. This can be verified as follows.

SIGLAB	MATLAB
SIGLAB>sqrt(1.5^2+.401924^2)	>>sqrt(1.5^2+.40192^2)
1.55	ans =
SIGLAB>deg #convert to degrees	1.5529
SIGLAB>atan(-.401924/1.5);	>>180*atan(-.401924/1.5)/pi
-15 # degrees ~ pi/12	ans =
	-15.0000

Also note that

$$x_1[k] = \mathcal{Z}^{-1}\left[\frac{1}{z-\alpha}\right] = \alpha^k u[k]$$

and

$$x_2[k] = \mathcal{Z}^{-1}\left[\frac{1}{z-\alpha^*}\right] = (\alpha^*)^k u[k]$$

where $\alpha = 0.433013 - j0.25 = 0.5\exp(-j\pi/6)$. Therefore

$$X(z) = a_1 \frac{z}{z - 0.5\exp(-j\pi/6)} + a_1^* \frac{z}{z - 0.5\exp(j\pi/6)}$$

which corresponds to a time-series, for $k \geq 0$

$$x[k] = 1.55\exp(-j\pi/12)\left(\frac{1}{2}\right)^k \exp(-j\pi k/6)$$

$$+ 1.55\exp(j\pi/12)\left(\frac{1}{2}\right)^k \exp(j\pi k/6)$$

$$= 1.55\left(\frac{1}{2}\right)^k (\exp(-j\pi k/6 - j\pi/12) + \exp(j\pi k/6 - \pi/12))$$

$$= 3.1\left(\frac{1}{2}\right)^k \cos(\pi k/6 + \pi/12)$$

which is seen to be a causal phase-shifted cosine wave with an exponentially decaying envelope. Also observe that $x[0] = 3.1\cos(\pi/12) = 3$, which can be verified using the initial value theorem.

7-7   DIFFERENCE EQUATIONS

Long division can be an intensive and tedious computational process. If a machine-computed inverse transform is desired, the use of difference equations is generally more efficient. Assume that the Z-transform of a time-series $x[k]$ is $X(z)$, where

$$X(z) = \frac{\sum_{i=0}^{M} b_i z^{-i}}{\sum_{i=0}^{N} a_i z^{-i}} \tag{7-34}$$

Recall that the $\mathcal{Z}[\delta_K[k]] = 1$ and $\mathcal{Z}[\delta_K[k-n]] = z^{-n}$. Therefore, it follows that

$$a_0 x[k] + a_1 x[k-1] + \cdots + a_{M-1} x[k-(M-1)] + a_M x[k-M]$$

$$= b_0 \delta_K[k] + b_1 \delta_K[k-1] + \cdots$$

$$+ b_{N-1} \delta_K[k-(N-1)] + b_N \delta_k[k-N] \qquad (7\text{-}35)$$

The response $x[k]$ can be simulated by implementing the difference equation.

Example 7-12   Difference equation

Consider again the causal $X(z)$ investigated in Example 7-9

$$X(z) = \frac{3z^3 - 5z^2 + 3z}{(z-1)^2(z-0.5)} = \frac{3z^3 - 5z^2 + 3z}{z^3 - 2.5z^2 + 2z - 0.5}$$

or

$$X(z) = \frac{3 - 5z^{-1} + 3z^{-2}}{(1 - z^{-1})^2(1 - 0.5z^{-1})} = \frac{3 - 5z^{-1} + 3z^{-2}}{1 - 2.5z^{-1} + 2z^{-2} - 0.5z^{-3}}$$

which produced a time-series

$$x[k] = (5(0.5)^k - 2 + 2k)u[k]$$

Then $x[k]$, for $k \geq 0$, can be simulated using

$$x[k] = 2.5x[k-1] - 2x[k-2] + 0.5x[k-3]$$

$$+ 3\delta_K[k] - 5\delta_K[k-1] + 3\delta_K[k-2]$$

Computer Study   S-file `ch7_12.s` or M-file `ch7_12.m` implements the derived equation for $x[k]$, as well as the simulated value of $x[k]$, using a difference equation implementation of $X(z)$. The first six sample values of each time series are displayed below, along with the maximal error between the two solutions over the first 32 samples.

SIGLAB	MATLAB

```>include "ch7_12.s"```
```x(nT) 3   2.5   3.25   4.625```
```6.312   8.156```
```Difference Equation; 3   2.5```
```   3.25   4.625   6.312   8.156```
```Max Error   0```
	```>> ch7_12```
	```x(nT) =```
	``` 3.0000 2.5000 3.2500```
	``` 4.6250 6.3125 8.1562```
	```Difference eq. =```
	``` 3.0000 2.5000 3.2500```
	``` 4.6250 6.3125 8.1562```
	```Max error =```
	``` 0```

Example 7-13   Population difference equation

The Fibonaccians have sent a scientific team to visit earth to learn about the earthlings' reproduction habits. They find earthlings breeding like, shall we say, rabbits. They collect a vast amount of data and develop the following population model. From their observations they conclude that the female of the species is sexually mature at 18 years of age. Five percent of all adult females die each year and the number of new adult females in year $(N + 18)$ is α times more than those alive in year N. This data was used to produce a model for the female population, given by

$$f[k] = 0.95 f[k - 1] + \alpha f[k - 18]$$

The model is assumed to be initialized in Year 0 of the Fibonaccians' first visit. At that time they also learned the secret of the Z-transform. Having access to Table 7-2 and noting that the delay property states $\mathcal{Z}[f[k - 1]] = z^{-1} F(z) + f[-1]$, $\mathcal{Z}[f[k - 2]] = z^{-2} F(z) + z^{-1} f[-1] + f[-2]$, and so forth, they developed a Z-transform model for the female population (including initial conditions) given by

$$F(z) = 0.95(z^{-1} F(z) + f[-1]) + \alpha \left[z^{-18} F(z) + \sum_{n=-18}^{-1} z^{-(18+n)} f[n] \right]$$

or

$$F(z)(1 - 0.95 z^{-1} - \alpha z^{-18}) = 0.95 f[-1] + \alpha \sum_{n=-18}^{-1} z^{-(18+n)} f[n]$$

or

$$F(z) = \frac{0.95 f[-1] + \alpha \sum_{n=-18}^{-1} z^{-(18+n)} f[n]}{1 - 0.95 z^{-1} - \alpha z^{-18}}$$

The Fibonaccians initialized their model by taking data over 17 consecutive years, but could not agree on the value of α. They observed that if α is small, the female human population appeared to converge toward zero and extinction. For α large, the population was observed to diverge, causing the planet's eventual extinction. In between, some degree of population stability seems to result.
Computer Study   Program S-file `ch7_13.s` or M-file `earth.m` implements the Fibonaccians' population model for 500 years. The program displays the female population for a given α and has the command syntax `earth(alpha)`, for $\alpha = \{1/20, 1/18\}$.

SIGLAB	MATLAB
`SIGLAB>include "ch7-13.s"`	`>> earth(1/20)`
`> earth(1/20)`	`>> earth(1/18)`
`> earth(1/18)`	

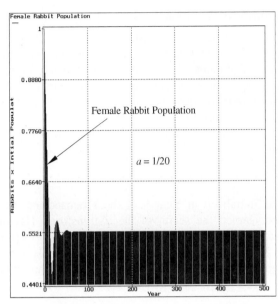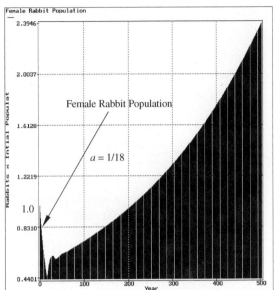

FIGURE 7-5   Female population model.

The graphical results for $\alpha = 1/20$ and $1/18$ are presented in Figure 7-5. Observe that the female population growth rate and relative population size are directly affected by the choice of α, which represents the female repopulation rate.

7-8   RESIDUE CALCULUS (OPTIONAL)

A function $X(z)$ is said to be *analytic* at a point $z = z_0$, if and only if it is single-valued and infinitely differentiable at z_0. If $X(z)$ is analytic at z_0, it can be expressed as a Taylor series, having the form

$$X(z) = \sum_{i=0}^{\infty} a_i (z - z_0)^i \tag{7-36}$$

$$a_0 = X(z_0)$$

$$a_n = \frac{1}{n!} \frac{d^n}{dz^n} X(z) \quad \text{for } z \to z_0$$

If $X(z)$ is not analytic at z_0, then a Laurent series is used, which is given by

$$X(z) = \sum_{i=0}^{\infty} \frac{a_{-i}}{(z - z_0)^i} \tag{7-37}$$

$$a_{-1} = \frac{1}{j2\pi} \oint X(z)\, dz$$

Given knowledge of $X(z)$, the parent time-series is given by

$$x[k] = \sum_{n} \mathrm{res}(X(z)z^{k-1}, p_n) \tag{7-38}$$

where $\mathrm{res}(a, b)$ reads the *residue of a, at point b*, which is restricted to be a simple pole. For our purposes, the Laurent expansion term $1/(z - b)$, is referred to as a simple pole at b.

Example 7-14   Residue theorem

The transform $X(z) = z^2/(z - 1)^2$ was studied, using a partial fraction expansion. To compute the inverse transform using the residue theorem, form

$$X(z)z^{k-1} = \frac{z^{k+1}}{(z - 1)^2} = \frac{a_{-1}}{(z - 1)} + \frac{a_{-2}}{(z - 1)^2}$$

Only one simple pole is found in this equation and corresponds to residue a_{-1}. Computing the residue a_{-1}, using the rule for repeated roots presented in the partial fraction expansion section, produces

$$a_1 = \lim_{z \to 1} \frac{d}{dz}(z - 1)^2 X(z)z^{k-1} = \lim_{z \to 1} \frac{d}{dz} z^{k+1} = k + 1$$

which agrees with the results obtained in Example 7-8.

Example 7-15   Multiple poles

A more complicated $X(z)$, also studied using partial fractions, was given by $X(z) = (z + 1)^3/[(z + 1/2)(z - 1/2)^2]$. Using the residue theorem, form

$$X(z)z^{k-1} = \frac{(z + 1)^3 z^{k-1}}{(z + 1/2)(z - 1/2)^2}$$

A partial fraction expansion must be approached with care. Note that the poles for $k = 0$ are located at $z = 0$ (due to z^{-1} appearing in the numerator), $z = -1/2$ of multiplicity one, and at $z = 1/2$ of multiplicity two. For $k \geq 1$, the poles are located at $z = -1/2$ of multiplicity one and at $z = 1/2$ of multiplicity two. These cases must be individually evaluated.
    For $k = 0$

$$X(z)z^{-1} = \frac{a_0}{z} + \frac{a_{-1}}{(z + 1/2)} + \frac{a_{-21}}{(z - 1/2)} + \frac{a_{-22}}{(z - 1/2)^2}$$

the residues are a_0, a_{-1}, and a_{-21}, and are computed as follows

$$a_0 = \lim_{z \to 0} z(X(z)z^{-1}) = \lim_{z \to 0} \frac{(z+1)^3}{(z+1/2)(z-1/2)^2} = 8$$

$$a_{-1} = \lim_{z \to -1/2} (z+1)(X(z)z^{-1}) = \lim_{z \to -1/2} \frac{(z+1)^3}{z(z-1/2)^2} = -1/4$$

$$a_{-21} = \lim_{z \to 1/2} \frac{d}{dz}(z+1/2)^2(X(z)z^{-1}) = \lim_{z \to 1/2} \frac{d}{dz} \frac{(z+1)^3}{z(z+1/2)} = -6\frac{3}{4}$$

Therefore, $x[0] = \sum$ residues $= a_0 + a_1 + a_{21} = 1$. This can be verified by long division.

For $k \geq 1$, the residue theorem produces

$$X(z)z^{k-1} = \frac{a_1}{(z+1/2)} + \frac{a_{21}}{(z-1/2)} + \frac{a_{22}}{(z-1/2)^2}$$

The residues are now a_{-1}, and a_{-21}. They are computed as follows

$$a_1 = \lim_{z \to -1/2} (z+1/2)(X(z)z^{k-1}) = \lim_{z \to -1/2} \frac{(z+1)^3 z^{k-1}}{z(z-1/2)^2} = -\left(-\frac{1}{2}\right)^k \frac{1}{4}$$

$$a_{21} = \lim_{z \to 1/2} \frac{d}{dz}(z-1/2)^2(X(z)z^{k-1}) = \lim_{z \to 1/2} \frac{d}{dz} \frac{(z+1)^3 z^{k-1}}{(z+1/2)}$$

$$= \frac{27}{2}\left(k - \frac{1}{2}\right)\left(\frac{1}{2}\right)^k$$

Therefore, $x[k] = \sum$ residues $= a_{-1} + a_{-21}$, for $k \geq 1$, or

$$x[k] = -\left(\frac{1}{4}\right)\left(\frac{-1}{2}\right)^k - \frac{27}{4}\left(\frac{1}{2}\right)^k + \left(\frac{27}{2}\right)k\left(\frac{1}{2}\right)^k$$

which agrees with the partial fraction result.

7-9   SUMMARY

In Chapter 7, the principal discrete-time signal analysis tools were introduced. The Z-transform, like the Laplace transform, was shown to reduce a signal to an algebraic expression. The Z-transforms of common signals have, in general, been cataloged and listed in standard tables.

Returning from the transform domain to the discrete-time domain was shown to be accomplished using any number of methods. Long division provided a direct means of analyzing a causal signal locally, around its starting time. The principal transform-inversion method, called partial fraction expansion, can be reduced to a computer program. The residue method can be useful when a signal has roots of high multiplicity. The difference-equation method can be used to simulate the response of a signal having a known Z-transform.

In Chapter 8, it will be shown that the discrete-time signals studied in this chapter can also be represented in the frequency domain. As such, frequency-selective discrete-time systems can be designed to enhance certain signal attributes and suppress others. These discrete-time systems, when implemented with digital hardware or software, become *digital filters*. In Chapter 9 it will be shown that the mechanism by which filtering takes place is discrete-time convolution. Discrete-time convolution will be shown to be more subtle than its continuous-time counterpart. In Chapter 10, the convolution process will be interpreted again in the transform domain. At that time it will become apparent that by transforming both systems and signals, significant analytic efficiencies can be gained.

7-10   COMPUTER FILES

The following computer files were used in Chapter 7.

File Name	Type	Location
Subdirectory CHAP7		
ch7_1.s	C	Example 7-1
zplane.m	C	Examples 7-1, 7-2
ch7_2.s	C	Example 7-2
ch7_12.x	T	Example 7-12
ch7_13.s	C	Example 7-13
earth.m	C	Example 7-13
Subdirectory SFILES and MFILES		
pf.s	C	Examples 7-7, 7-8, 7-9, 7-10, 7-11
Subdirectory SIGNALS		
Subdirectory SYSTEMS		

Here "s" denotes an S-file, "m" an M-file, "x" either/or, "T" a tutorial program, "D" a data file and "C" denotes a computer-aided instruction (CAI) program. CAI programs can be altered and parameterized by the user.

7-11   PROBLEMS

7-1 s-plane to Z-plane mapping   Map the following s-plane trajectories into the Z-plane for $T_s = 1$.

a $\omega \in [-\pi, \pi]$

b $s = a + ja\pi$, (a as above)

c $s = \exp(j\omega)$, $\omega \in [-\pi, \pi]$

7-2 Z-transform   Verify, from Table 7-1, the Z-transform representation of

$$x[k] = k^i u[k]$$

where $i = [0, 1, 2, 3]$.

7-3 Z-transform   Derive the Z-transform for the following signals and compute their ROC.

a $\delta[k - m]$

b $(1/2)^k u[k]$

c $(1/2)^k \cos(2\pi k/10)u[k]$

What conditions must be satisfied to use the initial and final value theorem to numerically check your result?

7-4 Z-transform   Suppose $x[n] = a^n u[n] + b^n u[-n - 1]$. Show that $X(z) = (a - b)z/(z - a)(z - b)$ and the ROC, if $|a| < |z| < |b|$. What happens if $a = b$? Sketch $x[n]$ in both cases.

7-5 Z-transform   Given $x[k] = a^k u[k] - a^{(k-N)}u[k - N]$, $N \geq 0$, $|a| < 1$, what is $X(z)$ and the ROC? Repeat for $x[k] = a^k u[k] - a^k u[k - N]$.

7-6 Z-transform   Given $x[k] = (-a)^{(k-m)}u[k - m]$, what is $X(z)$ for $m \geq 0$ and $|a| < 1$?

7-7 Z-transform   Compute the Z-transform of

$$x_1[k] = a^k \cos(\pi k/8 - \pi/4)u[k] \quad |a| < 1$$

$$x_2[k] = a^k \cos(\pi k/8 + \pi/4)u[k] \quad |a| < 1$$

7-8 Z-transform

a Given, for $k \geq 0$, $y[k]+0.1y[k-1]-0.05y[k-2] = u[k]$, what is $Y(z)$? Simulate $y[k]$ for the first 20 samples assuming the initial conditions are zero.

b How would nonzero initial conditions be represented (suppose $y(-1) = 1$, $y(-2) = 2$).

7-9 Z-transform   A noncausal periodic signal $x[k]$ is converted into a causal signal $y[k]$ using a *window function* $w[k]$, as shown in Figure 7-6. The purpose of the window function is to switch the signal $x[k]$ on or off. The window is assumed to have a duration of 16 samples. What is $Y(z)$ for the situation shown in Figure 7-6? What is the effect of shifting the window in time on $Y(z)$?

7-10 Z-transform   In Chapter 1, zero- and first-order holds were developed. Referring to Equation 1-18, write a Z-transform representation for a first-order hold evaluated at the sample instances $t = (k + 1)T_s$.

7-11 Modulation theorem   Verify the modulation theorem found in Table 7-2.

7-12 Inverse Z-transform   Compute $x[k]$ if $X(z) = (z+1)/z(z-1)$ at $x[0]$, $x[1]$, $x[2]$, $x[3]$, and predict $x[\infty]$. Verify your results using a difference-equation simulation.

7-13 Delay formula   Derive the time-delay formula found in Table 7-2.

7-14 Initial and final value theorem   Derive the initial and final value theorems found in Table 7-2. Why is the condition specified in footnote 2 required?

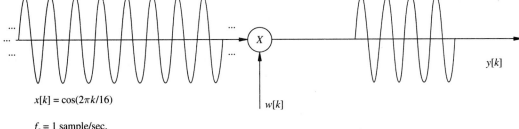

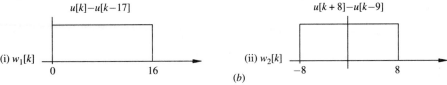

$x[k] = \cos(2\pi k/16)$

$f_s = 1$ sample/sec.

$u[k] - u[k-17]$

$u[k+8] - u[k-9]$

(i) $w_1[k]$ 0   16

(ii) $w_2[k]$ -8 8

(a)

(b)

FIGURE 7-6   Conversion of a noncausal signal into a causal signal.

7-15 Final value theorem   Using the final value theorem, derive a formula for the final, female human population based on the population model presented in Example 7-13.

7-16 Inverse Z-transform   Determine and graph the inverse Z-transform of
a $X(z) = 1 + z^{-1} + 6z^{-5} + 8z^{-10} + 4z^{-12}$
b $X(z) = 1 - z^{-1} + 6z^{-5} - 8z^{-10} + 4z^{-12}$
c $X(z) = 1/(z - 0.5)$
d $X(z) = 1/(z - 0.5)^2$
e $X(z) = 1/(z - 0.5)^3$

7-17 Partial fraction and residue method   Invert and graph, using both the partial fraction and residue-arithmetic methods
a $X_1(z) = (1 + z^{-2})/(1 + .5z^{-1})(1 - .25z^{-1})$
b $X_2(z) = 4(1 - z^{-1} + z^{-2})/(2 + 5z^{-1} + 2z^{-2})$
c $X_3(z) = (1 + z^{-1})^3/(1 - z^{-1})^3$
d $X_4(z) = z + 1 + z^{-1}$

7-18 Inverse Z-transform   The Z-transform of a possibly noncausal signal is given by

$$X(z) = \frac{-1.5z}{(z - 0.5)(z - 2)}$$

Determine the inverse Z-transform of $X(z)$ if the radius of convergence is known to be
a $|z| > 2$
b $|z| < 0.5$
c $0.5 < |z| < 2.0$

7-19 Inverse Z-transform   The Z-transform of a causal signal is given by

$$Y(z) = \frac{z^2(z - \cos(b))}{(z^2 - 2z\cos(b) + 1)(z - 1)}$$

What is $y[k]$?

7-20 Numerical integration   A simple model for a numerical integrator is given by

$$y[k] = x[k] + y[k-1]$$

where $y[-1] = 0$ and $k \geq 0$. What is the Z-transform of a signal that is produced by numerically integrating

a $x[k] = u[k]$
b $x[k] = \cos(\omega_0 k)u[k]$

What is the inverse Z-transform of $Y(z)$ and how do these results agree with your intuition?

7-21 Aliasing   A continuous-time signal $x(t) = \cos(2\pi f_0 t)$ is sampled below the Nyquist rate, using a $f_s = 5f_0/4$ sampler. What is the Z-transform of the resulting time-series?

7-22 Finite wordlength effects   A simple causal exponential signal is given by

$$x[k] = a^k u[k]$$

for $|a| < 1$ and a Z-transform representation $X(z) = z/(z-a)$. The time-series $\{x[k]\} = \{1, 1/a, 1/a^2, \cdots\}$ is sent to an analog, to a 4-bit (1 sign bit, 3-fractional bits) analog, to digital converter. The ADC maps $x[k]$ into $x'[k] = (x[k])_{\text{rounded}}$. What is the resulting time-series $x'[k]$ for $a = 7/8$ and $a = 1/2$? Is $x'[k]$ a valid representation for $x'[k]$? Justify your answer.

7-12   COMPUTER PROJECTS

7-1 Time-series generation   Write an S-file or M-file that modifies `pf.s` or `residue.m`, and that will accept a transform polynomial $X(z) = N(z)/D(z)$ (where $\deg(N(z)) \leq 8$ and $\deg(D(z)) \leq 8$) and return the signal's time-series representation $x[k]$.

7-2 Population model   Using the results from Example 7-13, determine experimentally what the value of α should be for an approximate constant female rabbit population. Suppose further that measurement noise affects the accuracy of the population study. Let the model be given by

$$f[k] = 0.95 f[k-1] + \alpha f[k-18] + \beta n[k]$$

where $n[k]$ is Gaussian noise and β is an adjustable gain. How confident would you be in the accuracy of your estimate of the equipopulation parameter α? For your study, use $\beta = 0.01, 0.1$, and 1.0.

FREQUENCY DOMAIN REPRESENTATION OF DISCRETE-TIME SIGNALS

Yea, from the table of my memory
I'll wipe away all trivial fond records.

—Shakespeare, *Hamlet*

8-1 INTRODUCTION

In Chapter 3, the idea that a continuous-time signal can be represented in the frequency-domain was established. In Chapter 7, the fact that a discrete-time signal can have a transform representation (viz., Z-transform) was introduced. It is logical to assume that these two ideas would provide a gateway to the study of discrete-time signals in the frequency-domain. Once a problem has been cast in a discrete-time framework, it is often assumed that it is a prime candidate for implementation by a general-purpose digital computer. It is interesting to note that the advent of the digital computer had little initial impact on frequency-domain analysis of discrete-time signals. This was due to the high cost of computing in the 50s and early 60s. In 1965, J.W. Cooley and J.W. Tukey published a milestone paper defining what is now called the *fast Fourier transform* or *FFT*. In many respects, it is this work that created the fields we now call digital spectral analysis and digital signal processing (DSP).

The Fourier transform has served, and continues to serve engineering and science well. However, both the Fourier transform and Fourier series can present engineers and scientists with formidable computational barriers when the mathematical description of a signal is complicated, or the signal is obtained experimentally by sampling an arbitrary continuous-time signal. Real-world signals may also contain additive noise or measurement uncertainties that can make the analytic production of a Fourier transform a practical impossibility. However, if the continuous-time signal can be sampled to produce a time-series $x[n]$, an estimate of the Fourier transform can be produced by using a general-purpose digital computer and the procedures developed in this chapter. Obviously, this has enormous importance, since it makes spectral analysis a practical reality.

Example 8-1   Message from Fibonacci

In Chapter 1 a transmission from the planet Fibonacci was received seeking "more bunnies." The message was originally detected because astronomers were monitoring the output of a broadband radio receiver and noted something unusual about the signal. The astronomers were listening to signals arriving from the Jovian satellite Io. The receiver converted a signal from a 640-element dipole antenna array at 26.3 MHz to a signal with 25 kHz bandwidth. Normally they would expect to see a spectrum similar to that shown in Figure 8-1, at the output of the receiver. By closely observing the spectral signature of the received signal with a spectral analyzer, one fateful day it was observed that the spectrum was rich in spectral components normally not expected. The structure of the received spectrum suggested that the signal may have been produced by an intelligent being. A deep-space probe was sent to the source, and signals were monitored along the way. The spectra appearing at the output of the Earth

FIGURE 8-1   Normal signal spectra from Io (data provided by the Department of Astronomy, University of Florida).

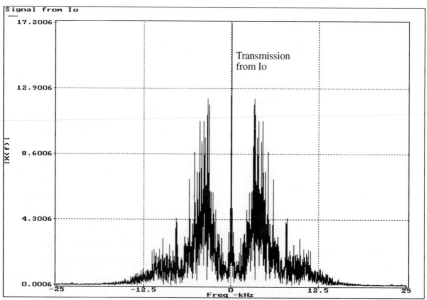

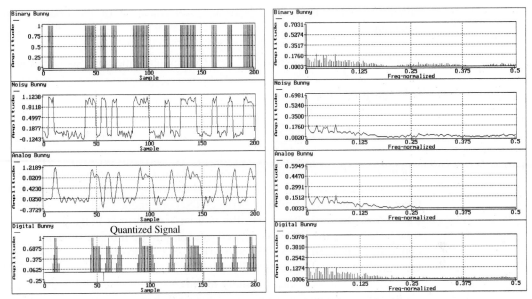

FIGURE 8-2   Fibonacci's message: (left) Fibonacci time-series, (right) normalized frequency response.

digital- and analog-signal processors were observably similar to the spectra measured in route to Fibonacci and at the transmitter. By matching the spectral characteristics, or *signatures*, of the transmitted and received signals, it was concluded that the planet Fibonacci was the signal source.

Computer Study   S-file `ch8_1.s` recreates the time-series studied in Chapter 1 and computes their Fourier representations digitally. The signals and spectra are displayed in Figure 8-2.

SIGLAB	MATLAB
>include "ch8_1.s"	>> not available in Matlab

The broadband signals presented in Figure 8-2 are characteristic of binary messages in that they are aperiodic and rich in high frequencies. It is important that the signatures are closely matched, leading to the conclusion that they were produced by the same source.

8-2 DISCRETE-TIME FOURIER TRANSFORMS

One method of computing a Fourier representation of a signal is to emulate the continuous-time Fourier transform by working with the sample values of $x(t)$

rather than $x(t)$ itself. This leads to a modification of the Fourier transform equation

$$X(j\omega) = \int_{-\infty}^{\infty} x(t) \exp(-j\omega t)\, dt \quad \text{Continuous-time analysis equation} \quad (8\text{-}1)$$

to produce the new equation

$$X(e^{j\Omega}) = \sum_{k=-\infty}^{\infty} x(kT_s) \exp(-j\Omega k) \quad \text{Analysis equation} \quad (8\text{-}2)$$

where $X(e^{j\Omega})$ is called the *discrete-time Fourier transform (DTFT)*. Here T_s is the sampling period and Ω is called the *discrete frequency*. It can be seen that there is considerable similarity between the continuous- and discrete-time Fourier transforms. It can be noted that the Fourier transform is defined over an infinite time history and the DTFT is defined over an infinite number of samples. Another connection between the CTFT and DTFT can be made if one considers letting the sampling period become infinitesimally small (i.e., $T_s \to 0$), or equivalently, the sampling rate becomes infinitely high (i.e., $f_s \to \infty$). Setting aside the mathematical requirements that must be satisfied to have an infinite sum approach the value of an integral (i.e., $\sum \to \int$), consider

$$\lim_{T_s \to 0} X(e^{j\Omega}) = \lim_{T_s \to 0} \sum_{k=-\infty}^{\infty} x(kT_s)(e^{-j\Omega k}) \to \int_{-\infty}^{\infty} x(t)e^{-j\omega t}\, dt = X(j\omega) \quad (8\text{-}3)$$

The DTFT of a time-series $x[k]$ exists if Equation 8-2 is bounded. This condition is guaranteed if $x[k]$ is absolutely summable or is of finite energy. That is, a DTFT exists if

$$\sum_{k=-\infty}^{\infty} |x[k]| < \infty \quad \text{Absolutely summable} \quad (8\text{-}4)$$

or

$$\sum_{k=-\infty}^{\infty} |x[k]|^2 < \infty \quad \text{Finite energy} \quad (8\text{-}5)$$

Another important relationship exists between the DTFT and the Z-transform. In particular, the *two-sided Z-transform* of a time series $x[k]$ was given by

$$X(z) = \sum_{k=-\infty}^{\infty} x[k] z^{-k} \quad (8\text{-}6)$$

Evaluating Equation 8-6 along the contour $z = \exp(j\Omega)$ (i.e., periphery of the unit-circle in the Z-plane), one obtains

$$X(z)|_{z=\exp(j\Omega)} = \sum_{k=-\infty}^{\infty} x[k](e^{j\Omega})^{-k} = \sum_{k=-\infty}^{\infty} x[k] e^{-j\Omega k} = X(e^{j\Omega}) \quad (8\text{-}7)$$

also recognized to be Equation 8-2. As a demonstration, consider the simple time-series $x[n] = a^n u[n]$, which equals $x[n] = a^n$ for $n \geq 0$ and 0 otherwise. From a table of Z-transforms (see Chapter 7), it is known that $X(z) = z/(z - a)$, if $|a| < 1$. Therefore it immediately follows that Equation 8-7 can be reduced to

$$X(e^{j\Omega}) = \left. \frac{z}{z-a} \right|_{z=\exp(j\Omega)} = \frac{\exp(j\Omega)}{\exp(j\Omega) - a} \tag{8-8}$$

Observe that if $\Omega = 2L\pi + \Omega_0$, for L an integer, then $X(e^{j\Omega}) = X(e^{j\Omega_0})$ and the maximum value of $X(e^{j\Omega})$ occurs at frequencies $\Omega = [\cdots, -2\pi, 0, 2\pi, \cdots]$, as suggested in Figure 8-3. Because the DTFT spectrum is intrinsically periodic, with period $T = 1/f_s$, it can be interpreted in the context of the Nyquist sampling theorem. The argument goes that if a continuous-time signal $x(t)$ is properly sampled above the Nyquist rate, then $x[n]$ contains no spectral components at or beyond the Nyquist frequency. The portion of the spectrum residing outside the interval $\Omega \in (-\pi, \pi)$, in Equation 8-8, would therefore be attributable to aliased copies of the spectrum of $x[n]$. Therefore, the replicated copies of the baseband spectrum centered about $\Omega \in (-\pi, \pi)$ normally are considered to be just artifacts.

Example 8-2   Discrete-time Fourier transform

A discrete-time rectangular pulse $x[k] = \text{rect}(k/K)$ is defined to be $x[k] = 1$, if $|k| \leq K/2$ and 0 otherwise. Since $x[k]$ is obviously absolutely summable, the DTFT of $x[k]$ exists and is given by

$$X(e^{j\Omega}) = \sum_{n=-K/2}^{K/2} e^{-j\Omega n}$$

Recall that the finite geometric sum $\sum_{n=0}^{N-1} x^i = N$ if $x = 1$, and $(1-x^N)/(1-x)$ otherwise. It therefore follows that translating the pulse origin to $m = 0$ by replacing n by $m = n + K/2$, the DTFT of a pulse can be written as

$$X(e^{j\Omega}) = \sum_{m=0}^{K} e^{-j\Omega(m-K/2)} = e^{j\Omega K/2} \sum_{m=0}^{K} e^{-j\Omega m}$$

FIGURE 8-3   Discrete-time Fourier transform of $x[n] = a^n u[n]$, showing periodic DTFT behavior in the frequency domain.

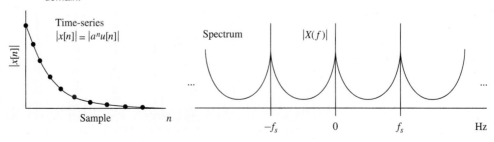

which, using the geometric equality, simplifies to

$$X(e^{j\Omega}) = e^{j\Omega K/2}\frac{(1 - e^{-j\Omega(K+1)})}{(1 - e^{-j\Omega})}$$

$$= e^{j\Omega K/2}\frac{e^{-j\Omega(K/2+1/2)}(e^{j\Omega(K/2+1/2)} - e^{-j\Omega(K/2+1/2)})}{(1 - e^{-j\Omega})}$$

Finally, upon application of Euler's identity, $X(e^{j\Omega})$ becomes

$$X(e^{j\Omega}) = \frac{\sin(\Omega(K/2 + 1/2))}{\sin(\Omega/2)}$$

The shape of the magnitude spectrum is seen to periodically repeat itself with a period of 2π radians (relative to the sampling frequency).

Computer Study   S-file ch8-2.s and M-file dtft.m compute the magnitude DTFT frequency response of the rectangular pulse $x[n]$ using the format dtft(K), where K governs the width of the pulse and is measured in sample instances. For $K = 7$, the magnitude frequency response is computed over the frequency range $f \in (-9f_N, 9f_N)$ ($f_N = f_s/2$) and displayed in Figure 8-4.

SIGLAB	MATLAB
>include "ch8_2.s"	>> dtft(7)
>dtft(7)	

FIGURE 8-4   Discrete-time Fourier transform of a pulse.

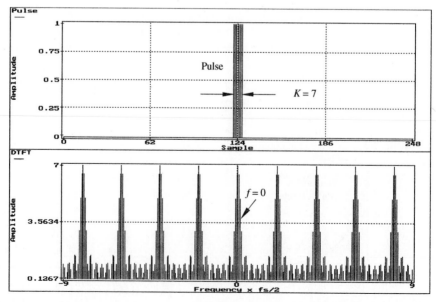

TABLE 8-1   PROPERTIES OF A DISCRETE-TIME FOURIER TRANSFORM

Discrete-time Series	Discrete-time Fourier Transform	Remark
$\displaystyle\sum_{i=0}^{L} \alpha_i x_i[k]$	$\displaystyle\sum_{i=0}^{L} \alpha_i X_i(e^{j\Omega})$	Linearity
$x[k-q]$	$X(e^{j\Omega})\exp(-jq\Omega)$	Time-shift
$x[-k]$	$X^*(e^{j\Omega})$	Time-reversal
$x[k]\exp(j\Omega_0 k)$	$X(e^{j[\Omega-\Omega_0]})$	Modulation, Ω_0 real
$x[k]*h[k]$	$X(e^{j\Omega})H(e^{j\Omega})$	Convolution
$x[k]y[k]$	$X(e^{j\Omega})*Y(e^{j\Omega})$	Multiplication
$\displaystyle\sum_{k=-\infty}^{\infty} x[k]y[k]$	$\dfrac{1}{2\pi}\displaystyle\int_{-\pi}^{\pi} X^*(e^{j\Omega})Y(e^{j\Omega})\,d\Omega$	Parseval
$x_{\text{even}}[k]$	$X(e^{j\Omega}) = x_{\text{even}}[0] + 2\displaystyle\sum_{k=1}^{\infty} x_{\text{even}}[k]\cos(k\Omega)$	x_{even} is an even time-series
$x_{\text{odd}}[k]$	$X(e^{j\Omega}) = x_{\text{odd}}[0] - j2\displaystyle\sum_{k=1}^{\infty} x_{\text{odd}}[k]\sin(k\Omega)$	x_{odd} is an odd time-series

The example emphasizes that the DTFT is intrinsically periodic in the frequency-domain, with a period 2π relative to the sampling frequency. If the signal transformed is sampled above the Nyquist rate, then only those spectral components residing in the interval $\Omega \in (-\pi, \pi)$ will have physical meaning. The *inverse* discrete-time Fourier transform is, therefore, given by the equation

$$x[k] = \frac{1}{2\pi} \int_{-\pi}^{\pi} X(e^{j\Omega})e^{j\Omega k}\, d\Omega \quad \text{Synthesis equation} \qquad (8\text{-}9)$$

where the limits of integration reflect the Nyquist restriction on the sampling rate.

Many of the basic properties applicable to the continuous-time Fourier transform are applicable to the discrete-time Fourier transform, as well. The more important properties are summarized in Table 8-1.

Parseval's theorem for the DTFT is given by

$$\sum_{k=-\infty}^{\infty} x^2[k] = \frac{1}{2\pi} \int_{-\pi}^{\pi} |X(e^{j\Omega})|^2\, d\Omega \qquad (8\text{-}10)$$

which, like its continuous-time counterpart, states that the energy in the time-domain can be computed, given knowledge of the signal's spectrum. The term $|X(e^{j\Omega})|^2$ is referred to as the *energy spectrum* and represents the distribution of energy in the frequency domain on a per Hertz basis.

Again, as in the case of the continuous-time Fourier transform, the discrete-time Fourier transforms of elementary signals are obtained using standard tables.

Fortunately there is a strong link between the discrete-time and continuous-time Fourier transforms, which is supported by extensive tables.

8-3 DISCRETE-TIME FOURIER SERIES (DTFS)

The DTFT, developed for an aperiodic signal $x[n]$, cannot be directly applied to periodic signals without modification. For example, the periodic signal $x[n] = \sin(\Psi n)$ is not, in general, absolutely or magnitude-squared summable (see Equations 8-4 and 8-5). The DTFT of a periodic time-series, having a period N samples such that $x[k] = x[k + mN]$, m an integer, can be expressed by a modification of Equation 8-2, which exploits the known signal periodicity. For discrete-time periodic signals, the *discrete-time Fourier series (DTFS)* is developed in a manner analogous to the derivation of the Fourier series. Recall that the Fourier series was based on the use of periodic *basis functions* $\exp(-jn\omega_0 t)$, where $\omega_0 = 2\pi/T$ and T was the signal's period. The basis of the DTFS is often represented in W, a notation defined by

$$W_N^r = e^{-j2\pi r/N} \tag{8-11}$$

The complex exponential W_N^{-nk} can be seen to satisfy

$$W_N^{-nk} = e^{(j2\pi kn/N)} = e^{j(2\pi(nk \bmod(N))/N)} \tag{8-12}$$

and is seen to be periodic in n and k with period N. As with the Fourier series, assume that the periodic time-series $\{x[n]\}$ can be expressed as a linear combination of periodic discrete-time basis functions to form

$$x[n] = \sum_k c_k e^{j2\pi kn/N} = \sum_k c_k W_N^{-nk} \tag{8-13}$$

The limits of summation of Equation 8-13 can be further refined since only N contiguous values from the set $\{W_N^k\}$ are distinct. Therefore, a baseband signal should be expanded in terms of W_N^k for $k \in [0, 1, \cdots, N-1] =< N >$[1]. Again, turn to the Nyquist sampling theorem to justify this action. Assume that a baseband signal has its highest frequency component, denoted f_m, bounded by the Nyquist frequency $-f_s/2 < f_m < f_s/2$. A sinusoidal signal at a frequency $f = f_s/2$ is seen to correspond to a time series given by $x[n] = W_N^{(N/2)n} = e^{j\pi n}$. Therefore, the first $N/2$ harmonics correspond to the positive frequencies $W_N^k = e^{-j2\pi k/N}$ for $0 \le k < N/2$. The next block of $N/2$ harmonics corresponds to the negative frequencies. In particular, the kth positive harmonic is equivalent to the mth negative harmonic, where $m = N - k$. This can be verified by observing

$$W_N^k = W_N^{N-m} = W_N^N W_N^{-m} = W_N^{-m} \tag{8-14}$$

since $W_N^N = e^{-j2\pi N/N} = e^{-j2\pi} = 1$. Other collections of N contiguous harmonics would correspond to an aliased copy of a baseband signal.

---

[1]The values $k \in [0, 1, \cdots, N - 1] =< N >$ are sometimes referred to as the *residue class of integers modulo(N)*.

Upon summing Equation 8-13 as recommended, the following equation results

$$x[n] = \sum_{k \in <N>} c_k W_N^{-kn} \qquad \text{Synthesis equation} \qquad (8\text{-}15)$$

Finally, the production rule for the DTFS coefficients is given by Equation 8-19 and can be derived by multipling both sides of Equation 8-15 by W_N^{rn} to form

$$x[n]W_N^{rn} = \sum_{k \in <N>} c_k W_N^{(r-k)n} \qquad (8\text{-}16)$$

Summing both sides of Equation 8-16 over n, one obtains

$$\sum_{n \in <N>} x[n]W_N^{rn} = \sum_{n \in <N>} \sum_{k \in <N>} c_k W_N^{(r-k)n} = \sum_{k \in <N>} c_k \sum_{n \in <N>} W_N^{(r-k)n} \qquad (8\text{-}17)$$

The value of the rightmost term in Equation 8-17 can be computed to be

$$\sum_{n \in <N>} W_N^m = \begin{cases} N & \text{if } m = \{0, \pm N, \pm 2N, \cdots\} \\ 0 & \text{otherwise} \end{cases} \qquad (8\text{-}18)$$

Therefore, the rightmost term has a nonzero value only if $m = (r-k) \bmod (N) = 0$. Evaluating Equation 8-17 for $k = r$, one obtains

$$c_k = \frac{1}{N} \sum_{n \in <N>} x[n]e^{j2\pi kn/N} \qquad \text{Analysis equation} \qquad (8\text{-}19)$$

Example 8-3   Discrete-time Fourier series

The periodic sinusoidal signal given by

$$x[n] = \sin(2\pi Kn/N) = \frac{1}{2j}e^{j2\pi Kn/N} - \frac{1}{2j}e^{-j2\pi Kn/N}$$

where $0 < K/N < 1/2$ is a rational number, has a DTFT given by $c_K = 1/2j$, and $c_{-K} = c_{N-K} = -1/2j$. Furthermore, $c_{LN+K} = c_K$ and $c_{-(LN+K)} = c_{-K}$.

The properties of the DTFS follow from those developed for the DTFT and are summarized in Table 8-2.

The analytic study of discrete-time aperiodic and periodic signals from a frequency-domain viewpoint is, in itself, important. However, it is the nature of the modern technologist to closely associate discrete-time formulas with digital (computer) signal processing. Since a computer can efficiently implement equations of finite complexity, such as those found in Equation 8-15 or Equation 8-19, we are not necessarily constrained to work with only deterministic, primitive discrete-time signals. The ability to accept an arbitrary N-sample time-series and map it into the frequency domain has major consequences in how we approach real-

TABLE 8-2   PROPERTIES OF A DISCRETE-TIME FOURIER SERIES

Discrete-time Series	Discrete-time Fourier Series	Remark
$\displaystyle\sum_{i=0}^{L} \alpha_i x_i[n]$	$\displaystyle\sum_{i=0}^{L} \alpha_i c_{i,k}$	Linearity
$x[n-q]$	$c_k W_N^{kq}$	Time-shift
$x[-n]$	c_{-k}	Time-reversal
$x[n]W_N^{-rn}$	c_{k-r}	Modulation
$x[n]$	$c_k = c_k^*$	$x[n]$ Real
$x_{\text{even}}[n]$	$Re(c_k)$	x_{even} is an even time-series
$x_{\text{odd}}[n]$	$jIm(c_k)$	x_{odd} is an odd time-series

world problems. In the next section, the important Discrete Fourier Transform (DFT) will be developed as an extension of the DTFS. Because the DFT is easily computed by a general-purpose digital computer, it has become an important spectral-analysis tool.

8-4   DISCRETE FOURIER TRANSFORM

Summing a time-series $x[n]$ over all $n \in (-\infty, \infty)$ can be a fundamental problem when implementing a DTFT of an arbitrary signal. What is more commonly found is that an analog signal, once presented to analog-to-digital converter (ADC), will fill a buffer of finite size with N words of data. Therefore, only a finite number of samples may be available to represent an arbitrarily long signal that cannot, in general, be assumed to be periodic. The produced DTFT spectrum, nevertheless, remains continuous and requires integration to restore it to the time domain. Unless $X(\exp(j\Omega))$ is analytically known, the required integration is virtually impossible. These realities lead to the important discrete Fourier transform or DFT. The DFT is the principal spectral-analysis computing formula in use today and appears in many valid manifestations.

The DFT is a mapping of an N-sample time-series (generally complex) into the frequency domain consisting of N-harmonics. It is defined by the transform pair

$$X[k] = \sum_{n=0}^{N-1} x[n] \exp\left(\frac{-j2\pi nk}{N}\right)$$

$$= \sum_{n=0}^{N-1} x[n] W_N^{nk} \quad \text{Analysis equation} \tag{8-20}$$

$$x[n] = \frac{1}{N}\sum_{k=0}^{N-1} X[k]\exp\left(\frac{j2\pi nk}{N}\right)$$

$$= \frac{1}{N}\sum_{k=0}^{N-1} X[k]W_N^{-nk} \quad \text{Synthesis equation} \qquad (8\text{-}21)$$

The value of k is the harmonic index. Because the DFT is defined by a finite sum of weighted signal-sample values, it can be directly implemented with a digital computer for an arbitrary signal. In addition, since the spectrum is also discrete and finite, it can be inverted using a digital computer. To produce the N harmonics from an N-sample time-series, at most N^2 complex multiply-accumulations would be required. This translates to a finite, but possibly long, computation time.

Refer to the analysis equation 8-20 and observe that

$$X[k+mN] = \sum_{n=0}^{N-1} x[n]W_N^{n(k+mN)} = \sum_{n=0}^{N-1} x[n]W_N^{nk}W_N^{nmN}$$

$$= \sum_{n=0}^{N-1} x[n]W_N^{nk} = X[k] \qquad (8\text{-}22)$$

Therefore, the $(k+mN)$th harmonic has the same computed value as the kth harmonic. However, since it is assumed that $x[n]$ is sampled above the Nyquist rate, there can be no signal energy actually appearing at these higher harmonic locations. Therefore, only the first N harmonics are retained as physically meaningful.

Along similar lines, consider the synthesis equation 8-21 and note

$$x[n+mN] = \frac{1}{N}\sum_{k=0}^{N-1} X[k]W_N^{-k(n+mN)} = \frac{1}{N}\sum_{k=0}^{N-1} X[k]W_N^{-kn}W_N^{-kmN}$$

$$= \frac{1}{N}\sum_{k=0}^{N-1} X[k]W_N^{-kn} = x[n] \qquad (8\text{-}23)$$

Implicit in the definition of the DFT is the assumption that $x[n]$ is periodic with period N. If $x[n]$ is periodically extended in time (i.e., $-\infty < n < \infty$), then a DTFS will result. The DFT can also be related to the DTFT. If $x[n]$ is an N-sample time-series, then the discrete-time Fourier transform becomes

$$X(e^{j\Omega}) = \sum_{n=-\infty}^{\infty} x[n]e^{-j\Omega n} = \sum_{n=0}^{N-1} x[n]e^{-j\Omega n} \qquad (8\text{-}24)$$

which simplifies to (after Equation 8-2 and 8-20)

$$X[k] = X(e^{j\Omega})\big|_{\Omega=2\pi k/N} = X\left(\frac{2\pi k}{N}\right) \qquad (8\text{-}25)$$

This formula states that when the signal is of finite duration, the DTFT and DFT agree up to a constant at a set of specific frequencies, namely $\Omega = 2\pi k/N$ for $k = \{0, 1, \ldots, N - 1\}$.

Example 8-4   Discrete-time Fourier transform and DFT

A 32-sample, discrete-time, random bandlimited burst of noise of duration T is periodically extended. The DTFT produces a continuous spectrum while the DFT spectrum is a frequency-discrete or line spectrum. The long-record DTFT is implemented using a DFT that approximates the actual DTFT of an arbitrarily long signal burst, having periodicity of 32-samples. The two spectra, however, should also agree at the discrete frequencies of the DFT.

Computer Study   S-file ch8_4.s and M-file ch8_4.m produce a noise burst with a periodic extension. The programs also produce an approximate DTFT using a high-resolution DFT. The spectra are display in Figure 8-5.

SIGLAB	MATLAB
>include "ch8_4.s"	>> ch8_4

The approximate DTFT spectrum appears to be continuous, while the DFT spectrum is discrete.

FIGURE 8-5   DTFT and DFT spectra: (left) time and frequency behavior, (right) spectra comparison (results may vary due to random test signal).

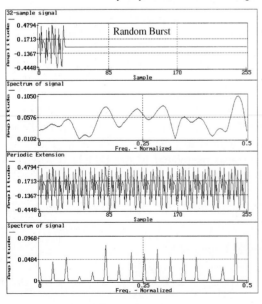
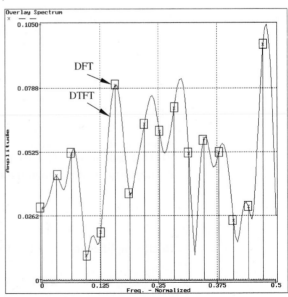

TABLE 8-3   DFT PROPERTIES

Discrete-time Series	Discrete Fourier Transform	Remark
$\sum_{i=0}^{L} \alpha_i x_i[n]$	$\sum_{i=0}^{L} \alpha_i X_i[k]$	Linearity
$x[(n-q) \bmod N]$	$X[k]W_N^{qk}$	Circular time-shift
$x[-n \bmod N]$	$X[-k \bmod N]$	Time-reversal
$x[n]W_N^{-qn}$	$X[(k-q) \bmod N]$	Modulation
$\sum_{n=0}^{N-1} x[n]y[n]$	$\frac{1}{N}\sum_{k=0}^{N-1} X[k]Y^*[k]$	Parseval

A list of DFT properties found in Table 8-3 bears both similarities and differences to other Fourier representations presented. A major difference is that the DFT is based on the analysis of a finite duration N-sample time-series $\{x[k]\}$, which is assumed to be periodically extended in the sample space. Therefore, if $\{x[k]\}$ is loaded into a N-sample circular shift-register network, it can also be assumed that it will be read out in a modulo N manner. This would produce a period N time-series consisting of original N-sample data record replicas. Refer to the two 32-word, shift-register networks shown in Figure 8-6. The shift-register network shown at the top of the Figure 8-6 assumes that data is read from a linear shift-register to produce an output that is the 32 original samples followed by a string of zeros (called zero padding). The network shown at the bottom produces a database with a reoccurring 32-sample pattern. Obviously the frequency-domain representations of these two signals are different, as shown in Figure 8-6. This will be an important observation that will be reinvestigated when the DFT is used to study systems.

FIGURE 8-6   Linear and circular shift register networks.

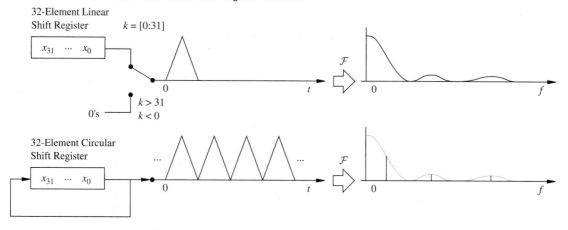

Parseval's theorem, for the DFT, is stated as

$$\frac{1}{N} \sum_{n=0}^{N-1} x^2[n] = \sum_{k=0}^{N-1} |X[k]|^2 \tag{8-26}$$

It is seen to be a weighted sum of squared harmonic values. The term $|X[k]|^2$ is referred to as the *power spectrum* and is a measure of the power in a signal on a per harmonic basis. Other properties of the DFT are listed in Table 8-3.

Example 8-5   Measurement of a 60 Hz component

An EKG signal is also suspected to carry a 60 Hz line-frequency component due to poor signal isolation. It is assumed, due to the biophysics of a human body, that the EKG should contain no appreciable signal components at 60 Hz. Any unusual amount of energy detected at this frequency will, therefore, be classified as 60 Hz noise.

Computer Study   A discrete-time EKG signal record will be used to approximate a continuous-time EKG signal over a 1.89-second interval of time. S-file ch8_5.s computes the complex Fourier component in a signal at frequency $f \in [0, 67.5]$ Hz using the format look(f), displays the approximate magnitude spectrum, and places a marker at frequency f. M-file look.m performs the same operation without inserting a marker.

SIGLAB	MATLAB
> include "ch8_5.s" > look(60) # mark 60Hz activity	>> look(60)

The data in Figure 8-7 clearly demonstrates that an abnormally high 60 Hz signal component is present.

The DFT assumes that the input is an N-sample periodic time series. It is important to remember this when using a DFT, since it explains a phenomenon called *leakage*. Leakage can result when the signal being transformed is not perfectly periodic over an N sample interval. Consider, for example, the two $N = 256$ sample sine waves shown in Figure 8-8. One has a period 12/256 and completes 12 full oscillations in the first 256 samples. The second signal has a period 12.5/256 and completes 12.5 oscillations in the same interval. The DFT assumes that both signals are periodically extended in all directions such that $x[n] = x[n \bmod 256]$. The periodic extension of the period 12/256 signal is smooth into the next 256-sample interval. The period 12.5/256 signal, however, exhibits abrupt changes at every location where $n \bmod 256 = 0$. These discontinuities will introduce new frequencies into the signal's spectral representation as in Figure 8-6. The DFTs for both signals exhibit strong local activity about a frequency corresponding to

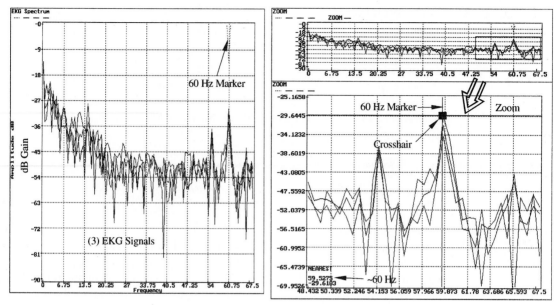

FIGURE 8-7   EKG study of a possible 60 Hz component.

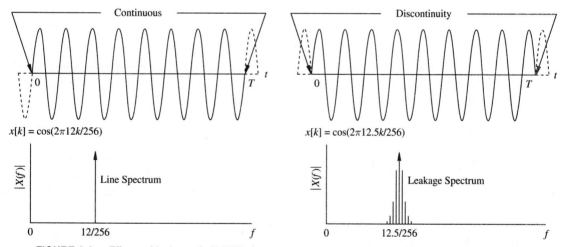

FIGURE 8-8   Effects of leakage: (left) DFT of a period 12/256 cosine; (right) DFT of a period 12.5/256 cosine.

256/12 (i.e, $1/T$). However, the signal with the discontinuous periodic extension exhibits tonal activity in adjacent sidebands, as well. This effect is called leakage. The leakage can be reduced by increasing the value of N, with an attendant increase in computational requirements, or by using *data windows* (covered in Chapter 12).

TABLE 8-4   DFT PARAMETERS

Parameter	Notation
Sample Size	N samples
Sample Period	T_s seconds
Record Length	$T = NT_s$ seconds
Number of Harmonics	N harmonics
Number of Positive (Negative) Harmonics	$N/2$ harmonics
Frequency Spacing between Harmonics	$\Delta f = \dfrac{1}{T} = \dfrac{1}{NT_s} = \dfrac{f_s}{N}$ Hz
DFT Frequency One-Sided Baseband Range	$f \in [0, f_s/2]$ Hz
DFT Frequency Two-Sided Baseband Range	$f \in [-f_s/2, f_s/2]$ Hz
Frequency of the kth Harmonic	kf_s/N Hz

The frequency resolution of an N-sample DFT is established by the values of N and the sample period T_s (equivalently $f_s = 1/T_s$). An N-sample time-series, sampled at a rate $f_s = 1/T_s$, consists of $T = NT_s$ seconds of sampled data. The fundamental frequency of a harmonic oscillation with a period T is $f_0 = 1/T$. From this, the DFT resolution parameters can be derived and are summarized in Table 8-4 and Figure 8-9. The *fundamental harmonic* (i.e., $X[1]$) of a DFT is located at a frequency $f_0 = 1/T = 1/NT_s = f_s/N$, and the spacing between harmonics is $\Delta f = f_0$ Hz.

The DFT spectrum, having a resolution $\Delta f = f_s/N$, consists of $N/2$ positive harmonics that span the discrete baseband frequency range $f \in [0, f_s/2)$, where $f_s/2$ is the Nyquist frequency. The other $N/2$ harmonics span the discrete negative baseband frequency range $f \in [-f_s/2, 0)$. Since $e^{-j2\pi/N} = W_N$, it follows that

$$W_N^{k(N-n)} = W_N^{-kn} = (W_N^{kn})^* \tag{8-27}$$

Therefore, if $x[n]$ is a real time-series, $X[k] = X^*[-k]$. Since W_N is a complex exponential of period N (i.e., $W_N^N = 1$), $X[-k]$ can also be expressed as $X[(N - k) \bmod N]$. If, for example, $N = 50$, then $X[k]$ consists of 25 positive harmonics and 25 negative harmonics. The positive harmonic indices are $[0, 24]$ and those

FIGURE 8-9   DFT resolution parameters.

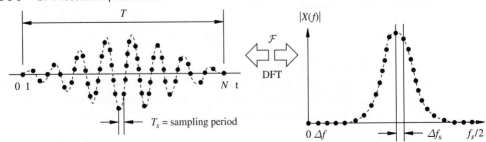

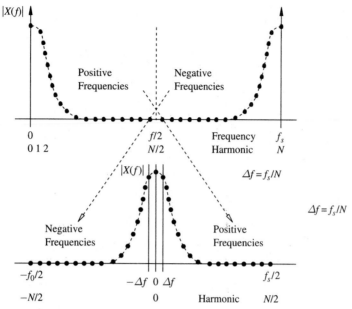

FIGURE 8-10   DFT spectrum.

for the negative harmonics are $[25, 49] = [-25, -1] \bmod 50$. More specifically, if $k = 35$, then $X(35) = X(-15)$. These relationships are summarized in Figure 8-10.

This analysis assumes that the signal being transformed is sampled above the Nyquist rate. If this is not the case, then aliasing errors can occur. Suppose, for purposes of discussion, that an N-sample DFT is taken for two signals. The first is given by $x_1(t) = \cos(2\pi f_1 t) = \cos(2\pi \lambda f_s t)$, where $\lambda \in [0, 0.5)$, if aliasing is not to take place. The harmonic index representing $f_1 = \lambda f_s$ is given by $k = \lambda N$, which has a possible range of $k \in [0, N/2]$. This value also is the location of the discrete frequency, which would reside at $k\Delta f = k f_s / N = \lambda f_s$. The second signal is given by $x_2(t) = \cos(2\pi f_2 t) = \cos(2\pi(n + \lambda) f_s t)$, where n is an integer. If $n > 0$, then aliasing can occur. The discrete frequency, in this case, is located $k'\Delta f = k' f_s / N = (n + \lambda) f_s$ or $k' = N + \lambda N$. The DFT's harmonic index, however, would be computed to be $k = k' \bmod N = (N + \lambda N) \bmod N = \lambda N \bmod N \in [0, N/2)$. Therefore, the spectral signature of $x_2(t)$ impersonates that of $x_1(t)$ (i.e, aliasing). Thus, the effects of aliasing can be studied in a frequency-domain context. Consider the spectral information displayed in Figure 8-11, which consists of a baseband information spectrum along with several extraneous tones at frequencies $\pm f_1$ and $\pm f_2$. The components at $\pm f_2$ Hz are exterior to the alias-free range $f \in (-f_s/2, f_s/2)$ found between the lines labeled *folding frequency* and located at $\pm f_s/2$. The information at $\pm f_2$ "folds" about these critical frequencies into the baseband extending over $f \in (-f_s/2, f_s/2)$.

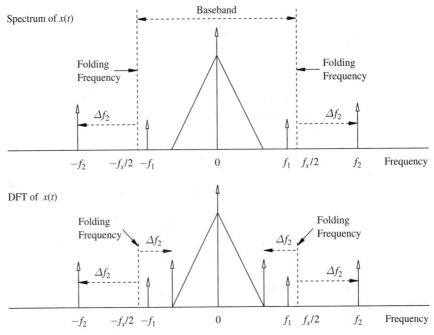

FIGURE 8-11   Aliasing effects in the frequency-domain.

Example 8-6   DFT precision and periodic properties

Three EKG signals studied in Example 3-19, called ekg1, ekg2, and ekg3, are sampled at a rate of 135 samples per second. A 256-sample DFT would resolve the frequency axis into 128 positive and 128 negative harmonics with a resolution of $\Delta f = 135/256 = 0.527$ Hz per harmonic.

If a 60 Hz component is present, then it should appear in the [60/0.527] = 114th harmonic location. The average spectrum of ekg1, ekg2, and ekg3 is computed and displayed in Figure 8-12. The EKG should produce only a narrow baseband spectrum. As previously noted in Example 8-5, the line located at the 114th harmonic is attributable to 60 Hz line-frequency noise. Another spectral line, indicated by the "?" resides at the 28th harmonic, which translates to an unaliased frequency of $28\Delta f = 17.75$ Hz, and has no physical significance. It is suspected, therefore, that this line may be due to aliasing.

Computer Study   S-file `ch8_6.s` computes and displays the average EKG spectrum and superimposes markers representing 60, 120, 180, and 240 Hz. The results are displayed in Figure 8-12.

SIGLAB	MATLAB
>include "ch8_6.s"	>> not available in Matlab

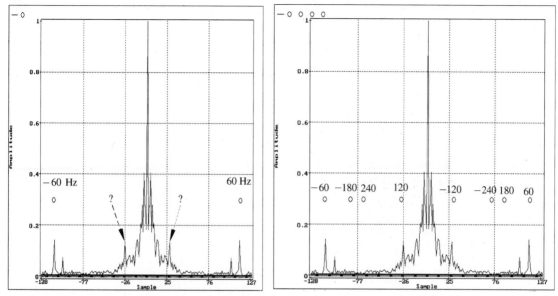

FIGURE 8-12 Average EKG spectrum showing the effects of aliasing: (left) average spectrum, and (right) average spectrum showing markers at 60, 120, 180, and 240 Hz.

From this experiment it is concluded that the second extraneous signal component is probably due to 120 Hz noise. Also, note that the folding operation can map a positive frequency line into an apparent negative aliased frequency line (i.e, 120 and 240 Hz lines).

There are times when the resolution of an N-sample time-series is required to be greater than $\Delta f = f_s/N$ Hz. Since f_s is fixed, the only control parameter is N. The length of a time-series $x[n]$ can be artificially increased to a value $N' = kN$ by adding an array of $N' - N$ zeros to the end of $x[n]$. Such an operation is called *zero padding* and results in a DFT $X[k]$ that has a resolution $\Delta f' = f_s/kN = \Delta f/k$ Hz, or a k-fold improvement. Unfortunately, the DFT computation time increases by a factor usually in excess of k.

Example 8-7 Zero padding

The first 3.5 periods of a 55-sample pulse train $x[n]$ has a DFT $X[k]$. If the sample frequency is $f_s = 256$ samples/second, producing a DFT with a 1 Hz per harmonic resolution would require a 256-sample time-series. To achieve this resolution, the time-series $x[n]$ must be padded by $256 - 55 = 201$ zeros. *Computer Study* S-file ch8_7.s and M-file ch8_7.m produce a pulse train and a zero-padded pulse train. The program generates the spectrum of the padded signal and produces the data shown in Figure 8-13.

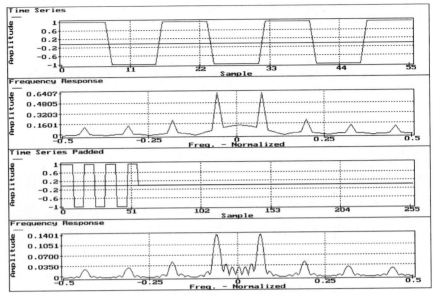

FIGURE 8-13   DFT harmonic relationships $x[n]$, DFT($x[n]$), $x_{pad}[n]$, DFT($x_{pad}[n]$).

SIGLAB	MATLAB
>include "ch8_7.s"	>> ch8_7

The original signal and its spectrum, and the zero-padded original signal and its spectrum, are displayed. It is apparent that the zero-padded spectrum is of higher resolution.

The production of a DFT can be directly implemented using Equation 8-20 and a general-purpose digital computer. In practice, when the DFT is produced using a digital computer, the scaling by $1/N$ found in Equation 8-20 is often eliminated. The rationale is that division is a time-consuming digital operation and that scaling, if required, can be always performed at a later time. Nevertheless, high-resolution N-sample DFTs can be computationally intensive. Each of the N-harmonics would require N complex multiply-accumulates. Such algorithms are classified as $O(N^2)$ (order N squared) in complexity. A faster $O(N \log_2(N))$ algorithm (order $N \log N$), known as the *fast Fourier transform (FFT)*, is usually used to compute a DFT. Appearing in many forms with a number of variants within each type, they achieve higher speeds than the direct DFT. The speedup is, therefore, on the order of $\log_2 N/N$ where N is the time-series sample length.

Example 8-8   Speed measures

A 256-sample DFT of a random signal $x[n]$ can be computed using Equation 8-9 (called the direct form) or by using an FFT. For long signals the FFT is claimed to be faster.

Computer Study   The following program will measure the speed of the direct DFT and FFT (results will vary depending on the CPU).

SIGLAB	MATLAB
`>x=rand(256)`	`>> x = rand(256,1);`
`>time xd=dft(x) # direct DFT`	`>> t0 = clock; fft(x);`
`00:20:88`	`etime(clock, t0)`
`>time xf=fft(x) # FFT`	`ans =`
`00:00:49`	`    0.0161`
	`>> t0 = clock; fft(x(1:255));`
	`etime(clock, t0)`
	`ans =`
	`    0.0499`

This test indicates that at this signal length, the FFT is about 40 times faster.

The DFT and FFT can be used to analyze a two-dimensional signal. A two-dimensional signal has two sample indices $x[n_1, n_2]$. A typical example of a 2-D signal is a graphic image where n_1 and n_2 are X-axis and Y-axis coordinates. The two-dimensional DFT of a two-dimensional time-series $x[n_1, n_2]$ can be computed by representing $x[n_1, n_2]$ as a matrix of data samples. First the rows (columns) of this matrix are transformed and the spectrum returned to its row (column) of origin. Once computed, the columns (rows) of the matrix are then transformed again and the spectrum returned to its column (row) of origin. The complex elements of the matrix are then $X[k_1, k_2]$, the two-dimensional DFT of $x[n_1, n_2]$.

As the name implies, a 2-D DFT is a display of a periodically sampled 2-D signal spectrum. The physical meaning of the spectrum may be better understood with an example. A satellite image of cornfields in the Midwest, for example, would be seen as a collection of interlaced rectangles. Suppose further that the camera resolution is about 0.7 yards. The digitized image is stored in a 10,000 by 10,000 array that corresponds to a 10 by 10 acre area. The fundamental spatial period along either the x or y axis is thus on the order of 7,000 yards, or about 4.0 miles. The first harmonic of the 2-D spectrum is, therefore, one cycle per 4 miles. The fourth harmonic has a period of 1 mile, and so on. If we wish to determine what percent of the view area is occupied by cornfields that are at least 1/4 mile by 1/4 mile, then we would measure all harmonics $X[k_1, k_2]$, for $k_1, k_2 \leq 16$ since they have the required spatial period.

Example 8-9 Two-dimensional DFTs

A symmetric 2-D database can be created by spinning a symmetric 1-D signal about the origin in a 2-D plane. Once the 2-D database is created it can be transformed into the frequency domain, using a DFT.

Computer Study S-file `spin.s` and M-file `spin.m` will convert a symmetric 1-D signal $x[k]$ into a symmetric 2-D $x[k, j]$ signal-matrix using the format `spin(x)`. The 2-D DFT of a 2-D symmetric signal generated by a 1-D symmetric signal `han(16)`, can be computed in the manner explained above. Here `han(16)` represents a 16-sample Hann window.

SIGLAB	MATLAB
`>include "spin.s"`	`>> x = hanning(16);`
`>graph3d(spin(han(16)))`	`>> x = x(8:16);`
`>graph3d(mag(fft(fft`	`>> z = spin(x);`
`(spin(han(16)))')))`	`>> mesh(z);`
	`>> mesh(abs(fftshift(fft2(z))))`

The results are displayed in Figure 8-14. In the left panel, the 2-D time-series is displayed. In the right, the 2-D magnitude DFT is shown. The spectrum is seen to be low-frequency dominated.

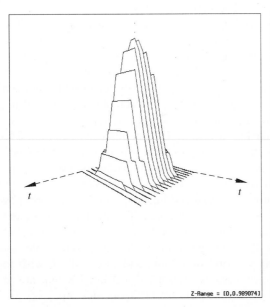
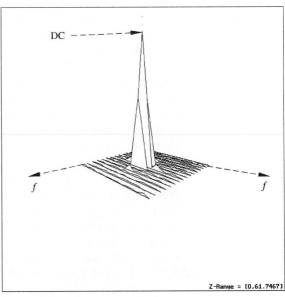

FIGURE 8-14 2-D DFT of a symmetric 2-D symmetric window.

It is important to remember that a DFT is computed on the basis of an N-sample data record. Over this interval, the DFT will compute a harmonic-series approximation of the N-sample time-series. The signal under study may be highly dynamic and change drastically from one N-sample record to another. This means that repeated DFTs may be needed in order to fully understand the frequency-domain properties of a signal. In addition, the frequency content of a signal may be undergoing change during an N-sample record. This is referred to as a *time-varying spectrum*. FM radio and doppler-shifted signals are examples of time-varying spectra. To use a DFT in such cases, the length of the data record N must be kept sufficiently short so that the signal remains harmonically stationary. The faster the frequency content of a signal is changing, the smaller N must be, with an attendant reduction in spectral resolution.

Example 8-10   Time-varying spectral analysis

In Example 8-1, a transmission from the planet Jupiter was received. The typical signal received was a rather broadband signal (spectrum is displayed in Figure 8-1.) There are occasions, however, when a group of electrons moves along Jupiter's magnetic field and produces a unique electromagnetic signature. The frequency of the signal, as received on Earth, is inversely related to the time it takes for the electron group to make a complete trip around Jupiter. Some mechanism is forcing the electrons into coherent orbits which, when changed, take on a new frequency. Such is the case shown in Figure 8-15,

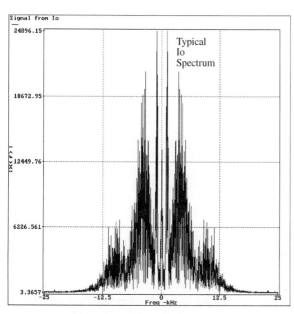
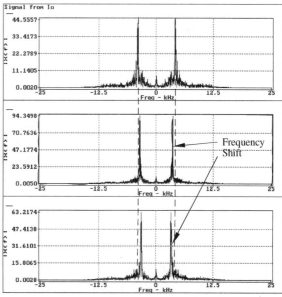

FIGURE 8-15   Jovian spectra.

where a 0.64 msec record is presented to an FFT of length $N = 2048$. The leftmost panel in Figure 8-15 is a typical spectrum. The three on the right are from segmented data records that cover a 2.56 msec interval. Observe that over this brief interval strong tonal activity is present. Also note that the apparent center frequency is changing from record to record. This, therefore, is a signal with a time-varying spectrum.

Computer Study   The change in frequency shown in Figure 8-15 is modest and could, therefore, be analyzed using an FFT. To test the sensitivity of the FFT/DFT to the rate-of-change of frequency, a linearly swept FM signal, called a *chirp*, is used. S-file `chip.s` and M-file `chirp.m` are used to create a chirp time-series. The FM signal has a format given by `chirp(N,f\_0,f\_low,f\_high)` where N is the time-series length. The chirp produces a signal that is linearly swept in frequency over the range $f \in [f_0 + f_{low}, f_0 + f_{high}] f_s$. The time-varying frequency range is seen to span $\Delta f = (f_{high} - f_{low}) f_s$. To test the DFT's sensitivity to time-varying change, set $f_0 = 0.1 f_s$ and $\Delta f = 0.02 f_s$ (narrowband) and $\Delta f = 0.1 f_s$ (broadband). S-file `ch8_10.s` and M-file `timevar.m` accept an $N = 256$ sample time-series and computes and displays DFTs of length 256, 128, 64 and 128, 64 and 32 samples. In this case, the spectral resolution decreases between FFTs by a factor of two.

SIGLAB	MATLAB

```>include "ch8_10.s"```
```>x = chirp(256, 0.1, 0, 0.02)```
```>graph(x, mag(pfft(x)))```
```>timevar(x)```
```>x = chirp(256, 0.1, 0, 0.1)```
```>graph(x, mag(pfft(x)))```
```>timevar(x)```
	```>> x = chirp(256, 0.1, 0, 0.02);```
	```>> timevar(x)```
	```>> x = chirp(256, 0.1, 0, 0.1);```
	```>> timevar(x)```

Observe that in Figure 8-16 there is sufficient change in frequency content of the 256 sample signal to produce notable spread in the FFT. For shorter records the signal appears to be harmonically stationary. For the broadband case, the signal appears to be nonstationary for all but the shortest data records.

8-5   COMPARISON OF FOURIER TRANSFORMS

In this chapter, four types of Fourier transforms were developed. They are summarized in Table 8-5 and in Figure 8-17.

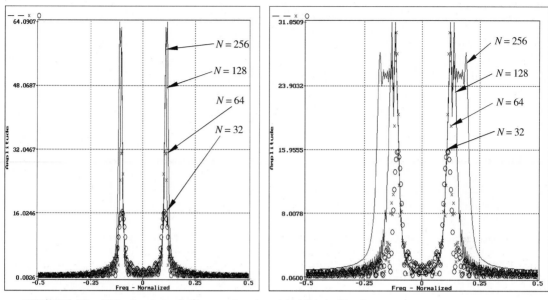

FIGURE 8-16   Chirp spectrum: (left) narrowband, and (right) broadband.

TABLE 8-5   SUMMARY OF IMPORTANT FOURIER RELATIONSHIPS

Fourier Representation	Time-domain Characteristics	Frequency-domain Characteristics
Continuous-time Fourier Transform (CTFT)	Continuous - aperiodic $t \in (-\infty, \infty)$	Continuous - aperiodic $f \in (-\infty, \infty)$
Continuous-time Fourier Series (CTFS)	Continuous - periodic $t \in (-\infty, \infty)$ period - T	Discrete - aperiodic $f \in \{\ldots, -2\Delta f, -\Delta f, 0, \Delta f, 2\Delta f, \ldots\}$ resolution - $\Delta f = 1/T$
Discrete-time Fourier Transform (DTFT)	Discrete - aperiodic $t \in \{\ldots, -2T_s, -T_s, 0, T_s, 2T_s, \ldots\}$	Continuous - periodic $f \in (-\infty, \infty)$ period - 2π
Discrete Fourier Transform (DFT)	Discrete - periodic $t \in \{\ldots, -2T_s, -T_s, 0, T_s, 2T_s, \ldots\}$ period - $T = NT_s$	Discrete - periodic $f \in \{\ldots, -2\Delta f, -\Delta f, 0, \Delta f, 2\Delta f, \ldots\}$ period - $Nf_s = 1/T_s$

8-6   SUMMARY

In Chapter 8 the concept of a discrete-time frequency-domain representation of a signal was introduced. The study introduced the DTFT for arbitrary discrete-

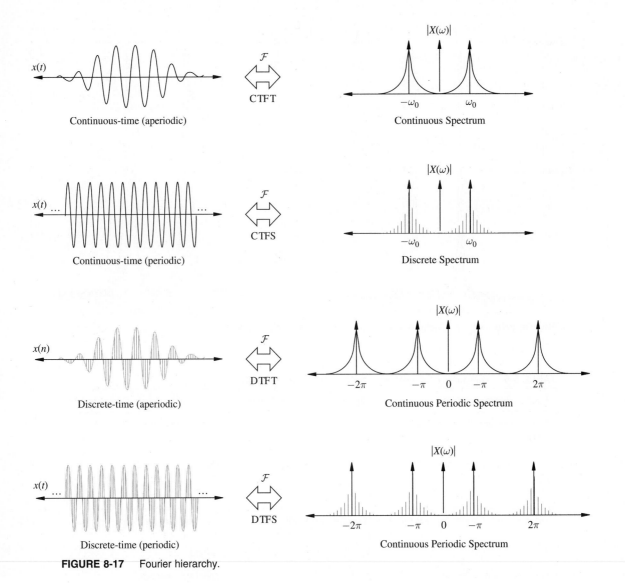

FIGURE 8-17   Fourier hierarchy.

time signals. If the discrete-time signal under study is periodic, then the simpler DTFS can be used. When only a finite duration discrete-time signal is known, the DFT is used. The DFT, however, assumes that the finite duration time-series was periodically extended in all time. The DFT has an important manifestation called the fast Fourier transform, or FFT. The FFT can be implemented on a digital computer with great efficiency. As a result, it has become an invaluable experimental and signal analysis tool.

8-7 COMPUTER FILES

These computer files were used in Chapter 8.

File Name	Type	Location
Subdirectory CHAP8		
ch8_1.s	T	Example 8-1
ch8_2.s	C	Example 8-2
dtft.m	C	Example 8-2
ch8_4.x	T	Example 8-4
ch8_5.s	C	Example 8-5
look.m	C	Example 8-5
ch8_6.s	T	Example 8-6
ch8_7.x	C	Example 8-7
spin.m	T	Example 8-9
ch8-10.s	C	Example 8-10
timevar.m	C	Example 8-10
chirp.m	C	Example 8-10
Subdirectory SFILES or MATLAB		
chirp.s	C	Examples 8-10
spin.s	C	Examples 8-9
packed.s	C	Problem 8-15
doubled.s	C	Problem 8-15
loglin.s	C	Project 8-2
Subdirectory SIGNALS		
bunny	D	Example 8-1
EKG1.imp	D	Examples 8-5, 8-6
EKG2.imp	D	Example 8-6
EKG3.imp	D	Example 8-6
Subdirectory SYSTEMS		
bunny	D	Example 8-1
bunny.arc	D	Example 8-1

Here "s" denotes S-file, "m" an M-file, "x" either a S-file or M-file, "T" denotes a tutorial program, "D" a data file, and "C" denotes a computer-based-instruction (CBI) program. CBI programs can be altered and parameterized by the user.

8-8 PROBLEMS

8-1 DTFT property   Prove that if $x[n]$ is real, $X(e^{j\Omega}) = X^*(e^{-j\Omega})$.

8-2 DTFT existence   The envelopes of a discrete-time signal $x[n]$ is known to be bounded by $1/|a^n|$ for $|a| < 1$. Does the DTFT of $x[n]$ exist? (Test both absolute and magnitude squared summability.)

8-3 DTFT computation   Compute the DTFT of
$$x[n] = \delta[n]$$
$$x[n] = (0.9)^n u[n]$$
$$x[n] = (-0.9)^n u[n]$$

8-4 DTFT computation   Consider the signals shown in Figure 8-18. The signals are assumed to be sampled at a rate of 10 samples per second. Does their DTFTs exist and, if so, what are they?

8-5 Inverse DTFT   Compute the inverse DTFT of

$$X(e^{j\Omega}) = \delta(\Omega - \pi/4) + \delta(\Omega + \pi/4)$$

$$X(e^{j\Omega}) = \delta(\Omega - \pi/4) - \delta(\Omega + \pi/4)$$

if $|\Omega| \leq \pi$.

8-6 Modulation property   Derive the modulation property for the DTFT.

8-7 DTFT properties   Given that $x[n]$ is a periodic signal of period N (samples) having a DTFS given by $\{c_k\}$, what is the DTFS of y[n], where
$$y[n] = x[n - m]$$
$$y[n] = x[n] + x[-n]^*$$
$$y[n] = (-1)^n x[n] \text{ for } N \text{ even}$$
$$y[n] = (-1)^n x[n] \text{ for } N \text{ odd}$$

8-8 Orthogonality   Two periodic time-series $\phi_i[n]$ and $\phi_k[n]$ of period N are said to be *orthogonal* if

$$\sum_{n \in <N>} \phi_i[n]\phi_k^*[n] = K\delta[i,k]$$

Show that $\phi_i[n] = W_N^{in}$ form an orthogonal set.

8-9 DTFS computation   Compute the DTFS of $x[n] = \cos(2\pi n/M + \phi)$ for $n = 0, 1, 2, \ldots$ for $n/M < 1/2$, a rational number.

8-10 DTFT property   A signal $x[k]$ is used to create a new signal $y[n] = \{\ldots, 0, x[-2], 0, x[0], 0, x[2], 0, \ldots\}$ (i.e., $y[n] = x[n]$ if n even, 0 otherwise). Derive a relationship between the DTFT of $x[n]$ and $y[n]$, assuming they exist. What would be the necessary Nyquist sampling rate assumption?

8-11 EKG signal   A 256-sample EKG data record, as found in Example 8-6, was sampled at a rate of 135 Hz. A 256-sample DFT is computed. Specify what the frequency resolution is and which harmonics are closest to 33.2 Hz and −45.7 Hz. If a frequency resolution greater than or equal to 2.2 Hz per harmonic is required, what would be the minimum time-series length N'?

8-12 Leakage   Derive a formula to compute the maximum amount (percent) of leakage that can occur in the DFT of the N-sample time-series $x[n] = \cos(\Omega_0 n)$, where Ω_0 is a normalized frequency such that $0 < \Omega_0 < \pi$.

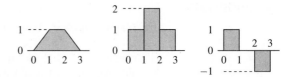

FIGURE 8-18   Example pulse processes.

8-13 Leakage   Leakage was observed when a signal did not complete an integer number of oscillations in an N-sample interval. Create three cosine waves of period 3.25/64, with lengths of 64, 128, and 256. Measure the ratio of the power in the peak spectral line to the leakage power as a function of N. (Hint: use Parseval's equation.) State conclusions about how leakage is related to signal length.

8-14 DFT properties   Verify that the DFT of even and odd signals produce cosine and sine series respectively. Demonstrate.

8-15 Reverse engineering   Often only real signals are transformed. When this is the case, a *real DFT* can be used. The S-files called `packed.s` and `doubled.s` can transform a time-series consisting of $2N$ samples with an N-point DFT. Reverse engineer these algorithms and determine how they work mathematically.

8-16 DFT system   Design a DFT system that can be used to analyze the spectrum of an 8-sample signal record using repetitions of a 2-sample DFT engine. Experimentally verify your design. Represent your solution in block diagram form.

8-17 Power spectrum   Examine the power spectrum of an EKG and locate the contiguous harmonic indices over which are contained 90 percent of the total signal power.

8-18 FFT timing   The FFT is claimed to be an $N \log (N)$ algorithm. Conduct timing tests for $N = 2^n$, $n \in [4,5,6,7,8]$, and plot execution time versus n. Does the $N \log (N)$ claim make sense?

8-19 FFT leakage   A signal $x[n] = \cos(2\pi n/100)$ is known to be product modulated by a carrier $c[n] = \cos(2\pi n/K)$ to form $y[n] = x[n]c[n]$. A 256-point FFT is performed on $y[n]$. What is/are the values of K, if they exist, that will produce a nonaliased leakage-free spectrum?

8-20 FFT bandwidth   A discrete-time signal is sampled at a rate $f_s = 10^3$ samples per second. The 8-bit digitized data is stored in a 1024×8 serial-in parallel-out buffer. An FFT is used that can process 100 1024-point FFTs per second. If a contiguous string of 1024-samples is defined as a data block, what is the percent overlap between data blocks which are processed by the FFT? How could 4-FFT engines be used to increase the FFT bandwidth (FFTs/sec)?

8-21 2-D FFT   A two-dimensional cross is inscribed in a 16×16 matrix X where $X[i, 7] = 1$ for $i \in [0 : 15]$, $X[7, j] = 1$ for $j \in [0 : 15]$, and $X[i, j] = 0$ otherwise. What is the 2-D FFT of X? How is the 2-D FFT image affected by rotations about the center?

8-22 Signal compression   The 256-sample EKG data record, found in Example 8.6, was sampled at a 135 Hz rate. The signal is to be transmitted to a remote diagnostic center along a low bandwidth channel. Rather then passing the signal through a lowpass filter, you decide to compute the FFT of the EKG signal, and transmit only the values of the important harmonics to the diagnostic center. Initially assume that you wish to reconstruct a good facsimile of the EKG across the entire baseband. You also determine that you can transmit thirteen complex harmonics of each 256-sample FFT, quantized to 8-bits of real and complex precision, across the low bandwidth channel. What would the reconstructed EKG signal look like at the remote station? Is it a good facsimile?

8-23 Signal compression and spectral truncation   Again consider the 256-sample EKG data record found in the previous example. Now assume that you wish to reconstruct a good facsimile of the EKG to frequencies out to 12 Hz. You again determine that you can reconstruct the EKG from a few complex harmonics quantized to 8-bits of real and complex precision, sent across a low bandwidth channel.

What would the reconstructed EKG signal look like at the remote station? Is it a good facsimile?

8-24 Signal coding   Repeat the previous problem except now code the data on the basis of a harmonic's relative importance. Transmit five *equivalent* complex harmonics of each 256-sample FFT, quantized to 8-bits of precision, across the low bandwidth channel. This corresponds to 40-bits of real and complex information. You decide to budget the 80-bits on the basis of their relative magnitude. The largest real or complex harmonic value is coded as a 12-bit number. Devise a strategy to use the remaining bits as a percentage of this value. How would the reconstructed EKG signal now look at the remote station? Is it a better facsimile?

8-9   COMPUTER PROJECTS

8-1 ZOOM (band-selectable) FFT   The so-called "zoom DFT" or "band-selectable DFT" can be used to expand the precision of a basic baseband DFT. An N-sample DFT of a time-series being sampled at a rate f_s extends out to $f_s/2$. The frequency resolution of the DFT is, therefore, $\Delta f = 1/T$ Hz per harmonic, where $T = N/f_s$. To improve the spectra resolution (i.e., reduce Δf), T must be increased or, equivalently, f_s decreased. Decreasing the sample rate is called decimation.

The zoom system is diagramed in Figure 8-19. Explain how it functions based on the properties of the DTFT or DFT. Referring to Figure 8-19, assume that one is interested in examining (with high resolution) the spectrum of $x[n]$ over a range of frequencies of width $f_s/8$ Hz. Since the DFT is a baseband analysis tool, a baseband (plus a high-frequency sideband) signal $y[n]$ is created and presented to a lowpass anti-aliasing filter. The filtered output is decimated by 8 and sent to a DFT for processing.

Use a sinusoidal test signal, given by

$$x[n] = \sum_i \cos(2\pi \alpha_i n) \quad \text{for } a_i = \{.21, .215, .22, .225\}$$

The modulating frequency is chosen to be $\cos(2\pi 0.2n)$, which product modulates $x[n]$ to a process having baseband frequencies located at $\{0.01, 0.015, 0.02, 0.025\} f_s$. Choose a decimation rate of 8 and verify that the spectral resolution is changed by a like factor.

Repeat by processing an EKG signal and perform a baseband study of the first $1/8$ of its frequency spectrum. Determine whether or not aliasing errors can, or have, occurred.

FIGURE 8-19   Zoom FFT.

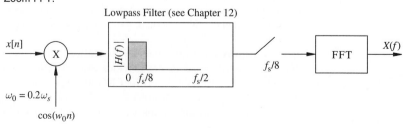

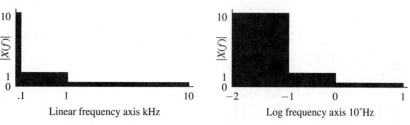

FIGURE 8-20   Linear and logarithmic spectral display.

8-2 **Log interpolation**   An FFT linearly resolves the frequency axis into a line spectrum located on Δf Hz centers. Sometimes a logarithmic cover of the frequency axis, as shown in Figure 8-20, is desired. In general, the mapping of a linear-frequency axis to logarithmic axis is not a straightforward conversion. The resolution over a low-frequency range is far less than that applied to high frequencies. It is difficult, in general, to achieve a valid compromise. S-file loglin.s will map a linear-frequency axis, ranging from $[0, f_s/2]$ Hz, into m-logarithm distributed bins over the same interval.

Examine S-file loglin and explain how it works. Use a low-frequency signal modeled as a ramp function and a broadband-process modeled as a gaussian noise time-series. Both are to be mapped into a 50-bin logarithmic axis. Analyze your results.

9

TIME DOMAIN REPRESENTATION OF DISCRETE-TIME SYSTEMS

In science we resemble children collecting a few pebbles at the beach of knowledge, while the wide ocean of the unknown unfolds itself in front of us.

—Sir Isaac Newton

9-1 INTRODUCTION

In Chapter 4 continuous-time systems were introduced. Continuous-time systems were seen to modify continuous-time signals. Earlier it had been established that discrete-time signals can be derived from continuous-time signals, using sampling. Therefore, it is logical that discrete-time systems can be developed to process discrete-time signals. Whereas the building blocks of continuous-time electronic systems are resistors, capacitors, inductors, and amplifiers, the basic elements of a discrete-time system would be analog delay lines, adders, and multipliers. If these systems were to be designed as true discrete-time systems (e.g., sample data systems), there would be little practical reason to study them, since analog delay lines are, in general, very difficult to design and maintain. Examples of analog delay lines are wires cut to precise lengths, waveguides of known lengths, charge coupled devices (CCDs), and acoustical pipes. They all rely on using the propagation velocity of a signal in a known media, or channel, and therefore are geometric devices. However, if one converts an analog discrete-time system into a digital system, then delay lines become shift registers, which are trivial to implement.

It should be pointed out that discrete-time systems predate the digital- and sample-data era. In fact, they can be traced back to the origins of humanity.

Discrete algorithms were used by our ancestors for a variety of purposes. These procedures were often recursive, like Fibonacci's rabbit population model, and if implemented in hardware today would be called discrete-time systems.

Example 9-1   Population algorithm

The computing algorithm for estimating the rabbit population on the planet Fibonacci was given by

$$\begin{cases} y[0] = y[1] = 1 \\ y[k] = y[k-1] + y[k-2] \end{cases}$$

where $y[k]$ is the rabbit population, in pairs, at the end of the kth sample interval. This formula represents a biological model of the rabbit reproduction system and is also a discrete-time system. It is evident that this equation can be directly implemented with a general-purpose digital computer. The input to this discrete-time model is zero, with all other information supplied in the form of initial conditions (i.e., Adam and Eve Bunny). It can be seen that based on this model, the planet will be knee-deep in rabbits in a very short time. However, on Earth, the model has been modified due to the presence of predators, such as humans. On Earth, only a fraction r of the paired rabbit population survive to the next mating period, and r^2 to the mating season beyond that. The terrestrial rabbit population model is given by

$$\begin{cases} y[k] = ry[k-1] + r^2 y[k-2] \\ y[0] = 1 \\ y[1] = r \end{cases}$$

which yields a completely different solution from the Fibonacci model for $r \in [0, 1)$. In fact, if $r = 0$, the rabbit population would become extinct after the first mating period.

Computer Study   The rabbit reproduction formula for the planet Fibonacci ($r = 1$) and Earth $r \in [0, 1]$ has been implemented by S-file ch9_1.s and M-file fibb.m. The simulation of the rabbit population is given by specifying n, the number of mating periods, and r, the survival factor using the format fibb (n,r). The results are displayed in Figure 9-1 for $r = 1.0, 0.99, 0.95,$ 0.75 and an arbitrary value $0 < r < 1$. The model is initialized and calibrated in rabbit units, where 1 rabbit unit $= 1000$ rabbits.

SIGLAB	MATLAB
>include "ch9_1.s" >fibb(25, 0.5) # n=25, r=0.5	>> fibb(25, 0.5);

For $r = 1.0, r = 0.99, r = 0.95,$ and $r = 0.75$, the rabbit populations are seen to monotonically increase without bound and therefore are said to

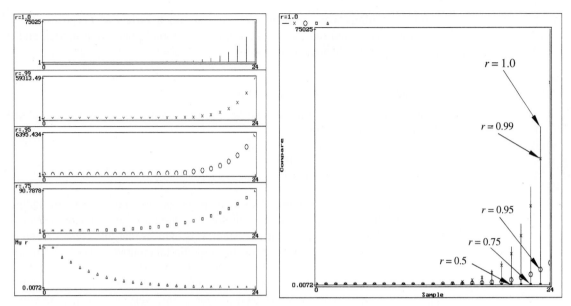

FIGURE 9-1　Rabbit population for $r = 1.0$, 0.99, 0.95, 0.75, and 0.5.

represent unstable biological systems. For $r = 1.0$, over 75,000 rabbit units are present after mating period 25. For $r = 0.75$, the number is reduced to 90 units. For $r = 0.5$, the rabbit population is seen to have an entirely different behavior and, in fact, is headed for extinction.

9-2　SYSTEM CLASSIFICATION

Discrete-time systems can be classified in the manner used to classify continuous-time systems. The primary classification schemes are developed in this section.

9-2-1　Linear and Nonlinear Systems

Linear discrete-time systems obey the *superposition principle* for discrete-time systems. The differential equations used to model continuous-time systems are simply replaced by their discrete-time counterparts. As a result, the discrete-time equation $y[k] = ay[k-1] + x[k]$ is *linear* (with respect to zero initial conditions), whereas $y[k] = ay^2[k-1] + x[k]$ is not. Later in the chapter, methods will be developed to compute the response of a linear system modeled by $y[k] = ay[k-1] + x[k]$, $y[-1] = y_{-1}$. It will be shown that the system response, for all $k \geq 0$, is given by

$$y[k] = a^k y_{-1} + \sum_{i=0}^{k} a^{(k-i)} x[i] \tag{9-1}$$

Two arbitrary input time-series, say $x_1[k]$ and $x_2[k]$, would produce outputs

$$y_1[k] = a^k y_{-1} + \sum_{i=0}^{k} a^{(k-i)} x_1[i] \tag{9-2}$$

$$y_2[k] = a^k y_{-1} + \sum_{i=0}^{k} a^{(k-i)} x_2[i] \tag{9-3}$$

The output $y_3(t)$, to an input $x_3[k] = x_1[k] + x_2[k]$, is computed to be

$$y_3[k] = a^k y_{-1} + \sum_{i=0}^{k} a^{(k-i)} (x_1[i] + x_2[i]) = a^k y_{-1} + \sum_{i=0}^{k} a^{(k-i)} x_1[i] + \sum_{i=0}^{k} a^{(k-i)} x_2[i] \tag{9-4}$$

However, upon comparing the results, it is apparent that

$$y_1[k] + y_2[k] = 2a^k y_{-1} + \sum_{i=0}^{k} a^{(k-i)} x_1[i] + \sum_{i=0}^{k} a^{(k-i)} x_2[i] \neq y_3[k] \tag{9-5}$$

if $y_{-1} \neq 0$. Therefore, as in the continuous-time case, the superposition principle fails for the discrete-time system with nonzero initial conditions.[1] It can be noted, however, that the discrete-time system is linear with respect to zero initial conditions, or equivalently, when the system is at-rest.

9-2-2 Time-Invariant and Time-Varying Systems

For a discrete-time system, time invariance is also called *shift invariance*. A discrete-time system is said to be shift-invariant if a shift in the input produces an identical shift in the output. That is, if $x[k] \rightarrow y[k]$, then $x[k + j] \rightarrow y[k + j]$. These are, in fact, the tests for both time- or shift-invariance.

Example 9-2   Shift-invariance

Suppose a discrete-time system is given by $y[n] = (\alpha n + \beta) x[n]$. To test for shift invariance, substitute for $x[n]$ a shifted time-series $x[n + j]$. The new output is then given by $w[n] = (\alpha n + \beta) x[n + j]$. If the system is shift-invariant, then $w[n] = y[n + j]$. However, $y[n + j] = (\alpha(n + j) + \beta) x[n + j] = \alpha j + w[n]$. Therefore the system is not shift-invariant unless $\alpha = 0$. If $\alpha = 0$ the equation $y[n] = \beta x[n]$ is said to be a constant-coefficient (or a shift-invariant) system.

Computer Study   S-file `ch9_2.s` and M-file `shiftinv.m` set the value of $\beta = 1$ and computes $y[n]$, $w[n - 1]$, and $y[n - 1]$ for a given α using the

---

[1]We will often refer to any system defined by a linear-difference equation as being linear, and thereby make an implicit assumption that the initial conditions are zero.

format `shiftinv(a,k)` where α = a and a one-second delay corresponds to k = 45 samples.

SIGLAB	MATLAB
>include "ch9_2.s"	>> shiftinv(1, 45);
>shiftinv(0, 45); # a=0	>> shiftinv(0, 45);
>shiftinv(1, 45); # a=1	

The results are displayed in Figure 9-2. Observe that if α = 0, the system is shift-invariant. If α = 1, the system is not shift-invariant.

9-2-3 Memory and Memoryless Systems

The simplest of all systems is a device that scales the input by some prespecified constant, such as $y[k] = \alpha x[k]$. Such systems are called *memoryless*, since they have no knowledge of the past. Systems may, however, contain memory. For digital systems, the storage elements are generally shift registers, ROM, and RAM.

9-2-4 Causal and Noncausal Systems

A system is said to be *causal*, or *nonanticipatory*, if it cannot produce an output until an input is present and the system is turned on. If a system is not causal, it is said to be *noncausal* or *anticipatory*.

FIGURE 9-2   Shift-Invariance for α = 0 (left) and α = 1 (right).

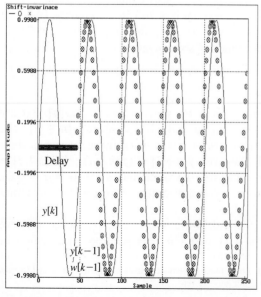
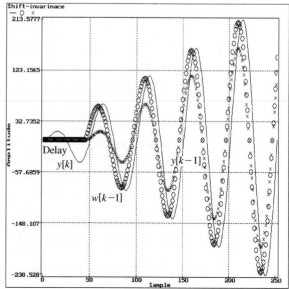

9-3   DISCRETE-TIME SYSTEMS

It was established earlier that differential equations can be used to model a linear continuous-time system and difference equations can model a discrete-time system. A constant coefficient causal discrete-time system, also called *linear shift-invariant (LSI)* system, has the form[2]

$$\sum_{k=0}^{N} a_k y[n-k] = \sum_{k=0}^{M} b_k x[n-k] \qquad \text{for } n \geq 0 \qquad (9\text{-}6)$$

The system output at sample n, namely $y[n]$, is given by

$$y[n] = \frac{1}{a_0} \left[\sum_{k=0}^{M} b_k x[n-k] - \sum_{i=1}^{N} a_k y[n-k] \right] \qquad (9\text{-}7)$$

where $\{y[-1], y[-2], \ldots, y[-N]\}$ are the initial conditions. The homogeneous solution to Equation 9-6 is a time-series $\{y[n]\}$, which satisfies the equation

$$\sum_{k=0}^{N} a_k y[n-k] = 0 \qquad \text{for } n \geq 0 \qquad (9\text{-}8)$$

Based on the study of continuous-time systems, assume that the solution to Equation 9-8 is expressible as a linear combination of discrete-time eigenfunctions of the form $\phi_j[n] = n^{n[j]} \beta_j^n$, $j \in [1, N]$, $n[j] \leq N$, where β_i is called an *eigenvalue* and satisfies the discrete-time system's characteristic difference equation, namely

$$\sum_{k=0}^{N} a_k \beta^{-k} = a_0 \prod_{i=1}^{N} (1 - \beta_i \beta^{-1}) = 0 \qquad (9\text{-}9)$$

if $a_0 \neq 0$. For convenience, assume that the eigenvalues are distinct, or $n[j] = 0$, which simplifies the form of the eigenfunctions to $\phi_j[n] = \beta_j^n$. Later, the case of repeated eigenvalues will be developed. The homogeneous solution can then be written as

$$y[n] = \sum_{i=1}^{N} \gamma_i \beta_i^n \qquad (9\text{-}10)$$

The computation rule for the γ_i will be developed later in this section. The homogeneous solution can then be expressed as

$$\sum_{k=0}^{N} a_k \left(\sum_{i=1}^{N} \gamma_i \beta_i^{[n-k]} \right) = 0 \qquad \text{for } n \geq 0 \qquad (9\text{-}11)$$

---

[2]For the purpose of clarity, systems of the form given by Equation 9-6 are assumed to be proper, which means $N \geq M$. This condition will be relaxed in later sections.

Factoring the constant β^n from Equation 9-11 results in the following simplification

$$\sum_{k=0}^{N} a_k \left(\sum_{i=1}^{N} \gamma_i \beta_i^{-k} \right) = 0 \qquad \text{for } n \geq 0 \qquad (9\text{-}12)$$

Equation 9-12 is seen to be a function of the unknown γ's and the known β's that satisfy the characteristic difference equation given by Equation 9-9. That is

$$\sum_{k=0}^{N} a_k \beta_i^{-k} = 0 \qquad (9\text{-}13)$$

For an Nth-order difference equation, there are N such equations corresponding to each of the β's. Once each equation is solved for a β_i, its value is then substituted into Equation 9-10 and used to solve for the γ's as the next example demonstrates.

Example 9-3   Homogeneous solution to a difference equation

The homogeneous solution to the difference equation $y[n] - (1/4)y[n-1] - (1/8)y[n-2] = 0$, $y[-1] = y[-2] = 1$, can be derived beginning with the characteristic equation

$$1 - \frac{1}{4}\beta^{-1} - \frac{1}{8}\beta^{-2} = 0$$

or

$$\beta^2 - \frac{1}{4}\beta^1 - \frac{1}{8} = 0$$

which factors into

$$\left(\beta - \frac{1}{2} \right) \left(\beta + \frac{1}{4} \right) = 0$$

This specifies the distinct eigenvalues to be $\beta_1 = 1/2$ and $\beta_2 = -1/4$. Furthermore, since $n[1] = n[2] = 0$, it follows that $\phi_i[n] = \beta_i^n$, which can be shown to individually satisfy the system's characteristic equation as follows

$$\beta_1 = \frac{1}{2} \Rightarrow \left(\frac{1}{2} \right)^2 - \frac{1}{4} \left(\frac{1}{2} \right) - \frac{1}{8} = 0$$

$$\beta_2 = -\frac{1}{4} \Rightarrow \left(\frac{-1}{4} \right)^2 - \frac{1}{4} \left(\frac{-1}{4} \right) - \frac{1}{8} = 0$$

The solution $y[n]$ will then be assumed to have the form

$$y[n] = \gamma_1 \left(\frac{1}{2} \right)^n + \gamma_2 \left(\frac{-1}{4} \right)^n$$

Evaluating the candidate solution at the known initial conditions yields the following system of equations

$$y[-1] = \gamma_1 \left(\frac{1}{2}\right)^{-1} + \gamma_2 \left(\frac{-1}{4}\right)^{-1} = 2\gamma_1 - 4\gamma_2 = 1$$

$$y[-2] = \gamma_1 \left(\frac{1}{2}\right)^{-2} + \gamma_2 \left(\frac{-1}{4}\right)^{-2} = 4\gamma_1 + 16\gamma_2 = 1$$

which has a matrix-vector description given by

$$\begin{pmatrix} 2 & -4 \\ 4 & 16 \end{pmatrix} \begin{pmatrix} \gamma_1 \\ \gamma_2 \end{pmatrix} = \begin{pmatrix} 1 \\ 1 \end{pmatrix}$$

and a solution $\gamma_1 = 10/24$ and $\gamma_1 = -1/24$. Therefore, for $n \geq -2$

$$y[n] = \gamma_1 \left(\frac{1}{2}\right)^n + \gamma_2 \left(\frac{-1}{4}\right)^n = \frac{10}{24} \left(\frac{1}{2}\right)^n - \frac{1}{24} \left(\frac{-1}{4}\right)^n$$

Computer Study   The solutions γ_1 and γ_2 are first verified. The S-file ch9_3.s and M-file ch9_3.m compute the homogeneous solution to the difference equation $y[n] = 0.25y[n-1] + 0.125y[n-2]$, $y[-2] = y[-1] = 1$, using simulation. This result is compared to the derived response.

SIGLAB	MATLAB
># matrix A and initial conditions	>> A = [2 -4; 4 16];
>A={[2,-4],[4,16]}; IC={1,1}	>> IC = [1; 1];
>GAMA=inv(A)*IC	>> gamma = A \IC;
>#print solution x 24	>> gamma * 24
>GAMA*24	ans =
10	10
-1	-1
>include "ch9_3.s"	>> ch9_3

The result is displayed in Figure 9-3. The simulated and derived solutions are seen to be in agreement.

The response of a continuous-time LTI system is often predicated on knowledge of the system's impulse response. A similar situation is found in the study of discrete-time LSI systems. In particular, the impulse response, say $h[n]$, of an at-rest Nth order LSI system (i.e., $y[-i] = 0$ for $i \in [1, N]$) is given by

$$\sum_{k=0}^{N} a_k y[n-k] = \sum_{k=0}^{M} b_k x[n-k] \qquad \text{for } n \geq 0 \qquad (9\text{-}14)$$

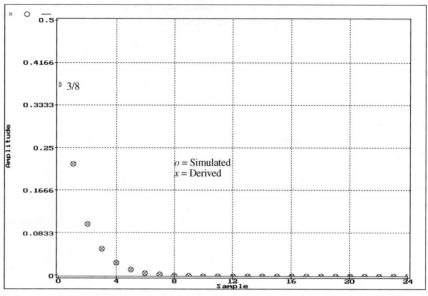

FIGURE 9-3   Homogeneous discrete-time response.

is defined to be the solution to

$$\sum_{k=0}^{N} a_k h[n-k] = \sum_{k=0}^{M} b_k \delta[n-k] \qquad \text{for } n \geq 0 \qquad (9\text{-}15)$$

The solution to Equation 9-15 is obtained in a manner similar to that used to compute the *homogeneous solution* to a continuous-time LTI system. For the *inhomogeneous solution* case, however, the problem will be slightly modified to first produce a set of *pseudo-initial conditions*, namely $\{h[0], h[1], \ldots, h[M]\}$. This set of values can then be used to compute the inhomogeneous impulse response solution for $N \geq M$. The pseudo-initial conditions satisfy the matrix-vector equation

$$\begin{pmatrix} a_0 & 0 & \cdots & 0 \\ a_1 & a_0 & \cdots & 0 \\ \vdots & \vdots & \ddots & \vdots \\ a_M & a_{M-1} & \cdots & a_0 \end{pmatrix} \begin{pmatrix} h[0] \\ h[1] \\ \vdots \\ h[M] \end{pmatrix} = \begin{pmatrix} b_0 \\ b_1 \\ \vdots \\ b_M \end{pmatrix} \qquad (9\text{-}16)$$

Therefore it is seen that Equation 9-16 can be expressed in terms of $\{h[0], h[1], \ldots, h[M-1]\}$, a_i and b_i. If the matrix equation has a solution, then the solutions are the pseudo-initial conditions $\{h[0], h[1], \ldots, h[M-1]\}$. The elements of this array serve as the initial conditions for the impulse-response equation which, for $n \geq M$, satisfies

$$\sum_{k=0}^{N} a_k h[n-k] = 0 \qquad (9\text{-}17)$$

Example 9-4   Inhomogeneous solution to a difference equation

The at-rest difference equation $y[n] - (1/4)y[n-1] - (1/8)y[n-2] = 2x[n] - x[n-1] + x[n-2]$ has an impulse response $h[n]$ that can be derived beginning with the matrix equation for pseudo-initial conditions

$$\begin{pmatrix} 1 & 0 & 0 \\ -1/4 & 1 & 0 \\ -1/8 & -1/4 & 1 \end{pmatrix} \begin{pmatrix} h[0] \\ h[1] \\ h[2] \end{pmatrix} = \begin{pmatrix} 2 \\ -1 \\ 1 \end{pmatrix}$$

which yields a solution $h[0] = 2$, $h[1] = -1/2$, and $h[2] = 9/8$. Since $M = N$ the system is proper (but not strictly proper). There is the possibility of an impulse in the impulse response (see Example 7-10). The assumed solution therefore has the form

$$h[n] = \gamma_0 \delta_K[n] + \gamma_1 \beta_1^n + \gamma_2 \beta_2^n \qquad \text{for } n \geq 0$$

where the values of β_1 and β_1 have been previously established to be $1/2$ and $-1/4$ respectively. Direct substitution yields

$$\begin{pmatrix} 1 & 1 & 1 \\ 0 & 1/2 & -1/4 \\ 0 & (1/2)^2 & (-1/4)^2 \end{pmatrix} \begin{pmatrix} \gamma_0 \\ \gamma_1 \\ \gamma_2 \end{pmatrix} = \begin{pmatrix} 2 \\ -1/2 \\ 9/8 \end{pmatrix}$$

which produces a solution $\gamma_0 = -8.0$, $\gamma_1 = 8/3$, and $\gamma_2 = 22/3$. Therefore the impulse response is given by

$$h[n] = -8.0\delta_K[n] + 8/3(1/2)^n + 22/3(-1/4)^n \text{ for } n \geq 0$$

It can be seen that using this technique to compute the impulse response of a high-order system can rapidly become computationally intensive. In the next chapter, transform methods will be introduced that greatly simplify the production of both the homogeneous and impulse responses.

9-4   DISCRETE-TIME CONVOLUTION

The inhomogeneous response of the discrete-time LSI system shown in Figure 9-4 is defined in terms of a discrete-time convolution sum. The input-output relationship is given by

$$y[n] = \frac{1}{a_0} \left[\sum_{k=0}^{M} b_k x[n-k] - \sum_{i=1}^{N} a_k y[n-k] \right] \qquad (9\text{-}18)$$

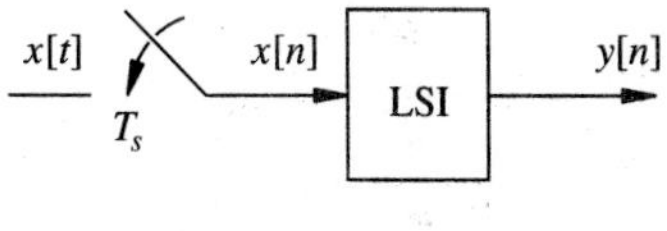

FIGURE 9-4   Discrete LSI system.

It will be assumed that the LSI system has a causal impulse response $h[n]$. The response to an arbitrary causal input $x[n]$ is given by the discrete-time convolution sum (also referred to as *linear convolution*)

$$y[n] = \sum_{i=0}^{\infty} h[n-i]x[i] = \sum_{i=0}^{\infty} h[i]x[n-i] \qquad (9\text{-}19)$$

where $h[n]$ is the system's impulse response. The convolution operation can also be represented with the shorthand notation $y[n] = h[n] * x[n]$. Directly computing the first few terms of Equation 9-19 produces

$$y[0] = h[0]x[0]$$

$$y[1] = h[1]x[0] + h[0]x[1]$$

$$y[2] = h[2]x[0] + h[1]x[1] + h[0]x[2]$$

$$\vdots = \vdots$$

$$y[n] = h[n]x[0] + h[n-1]x[1] + \cdots + h[1]x[n-1] + h[0]x[n] \quad (9\text{-}20)$$

Assume for expository reasons that $h[n] = 0.9^n$ and that $x[n] = u[n]$. The discrete-time step response is then

$$y[0] = h[0]1 = 1$$

$$y[1] = h[0]1 + h[1]1 = 1 + 0.9 = 1.9$$

$$y[2] = h[0]1 + h[1]1 + h[2]1 = 1 + 0.9 + 0.81 = 2.71$$

$$\vdots = \vdots \qquad (9\text{-}21)$$

The solution can be seen to be monotonically increasing in time with the output, at sample index n, given by

$$y[n] = \sum_{i=0}^{n} (0.9)^i \qquad (9\text{-}22)$$

Based on the fact that $\sum_0^{\infty} x^i = 1/(1-x)$ if $|x| < 1$, the final value of $y[n]$, can be computed to be

$$y[\infty] = \sum_{i=0}^{\infty} (0.9)^i = \frac{1}{1-0.9} = 10 \qquad (9\text{-}23)$$

Equation 9-23 can be suitably modified to represent $y[n]$ in closed form as well. In particular, recall that the finite sum

$$\sum_{i=0}^{n} x^i = \sum_{i=0}^{\infty} x^i - \sum_{i=n+1}^{\infty} x^i \tag{9-24}$$

$$= \sum_{i=0}^{\infty} x^i - x^{n+1} \sum_{i=0}^{\infty} x^i = (1 - x^{n+1}) \sum_{i=0}^{\infty} x^i \tag{9-25}$$

$$= \frac{1 - x^{n+1}}{1 - x} \tag{9-26}$$

From this it follows that

$$y[n] = \sum_{i=0}^{n} (0.9)^i = \frac{1 - (0.9)^{n+1}}{1 - 0.9} = 10(1 - (0.9)^{n+1}) \tag{9-27}$$

which produces a time-series given by $\{y[0] = 10(1 - 0.9) = 1, y[1] = 10(1 - 0.81) = 1.9, y[2] = 10(1 - 0.729) = 2.71, \cdots\}$. This agrees with results given in Equation 9-21.

Example 9-5   Discrete convolution

The at-rest proper difference equation $y[n] - (1/4)y[n-1] - (1/8)y[n-2] = 2x[n] - x[n-1] + x[n-2]$, $y[-1] = y[-2] = 0$, has an impulse response computed in Example 9-4 to be

$$h[n] = -8.0\delta_K[n] + 8/3(1/2)^n + 22/3(-1/4)^n \qquad \text{for } n \geq 0$$

The step response, for $x[n] = u[n] = 1$, $n \geq 0$, is defined by the convolution sum

$$y[n] = \sum_{i=0}^{\infty} h[i]u[n-i] = \sum_{i=0}^{n} h[i]$$

or

$$y[n] = -8.0 + \sum_{i=0}^{n} \frac{8}{3}\left(\frac{1}{2}\right)^i + \sum_{i=0}^{n} \frac{22}{3}\left(\frac{-1}{4}\right)^i$$

From Equation 9-26, one obtains

$$\sum_{i=0}^{n} \left(\frac{1}{2}\right)^i = \frac{1 - (1/2)^{n+1}}{1/2} = 2(1 - (1/2)^{n+1})$$

$$\sum_{i=0}^{n} \left(\frac{-1}{4}\right)^i = \frac{1 - (-1/4)^{n+1}}{5/4} = \frac{4(1 - (-1/4)^{n+1})}{5}$$

The output time-series $y[n]$, defined by the convolution sum, is then given by

$$y[n] = -8.0 + \frac{16}{3}\left(1 - \left(\frac{1}{2}\right)^{n+1}\right) + \frac{88}{15}\left(1 - \left(\frac{-1}{4}\right)^{n+1}\right)$$

for $n \geq 0$. The initial value is computed to be $y[0] = -8 + (16/3)(1/2) + (88/15)(5/4) = 2.0$. The correctness of this value can be verified by substituting $u[n]$ into the defining difference equation and computing $y[0]$. The final value of $y[n]$ is computed as $n \rightarrow \infty$ and is seen to be $y(\infty) \rightarrow -8.0 + 16/3 + 88/15 = 48/15 = 3.2$.

Computer Study   S-file ch9_5 and M-file `discrete.m` display the derived system step response, as well as $y[n]$, produced by using numerical methods. The format used is `discrete(n)`, where n is the length of the step $u[n]$, $n \leq 128$. A 21-sample step function is used as a test signal.

SIGLAB	MATLAB
>include "ch9_5.s"	>> u = ones(21, 1);
># 20-sample step function	>> discrete(u);
>discrete(21)	

The results are displayed in Figure 9-5. Both solutions agree and are seen to go through a transient period prior to reaching a steady-state value. The step response is also seen to converge rapidly to its final value of 3.2.

FIGURE 9-5   Comparison of simulated and derived impulse responses.

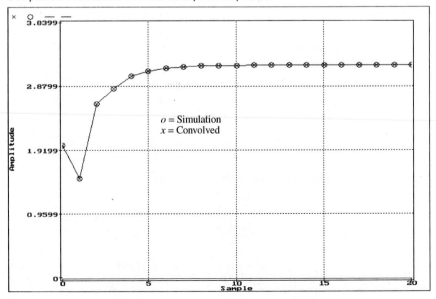

The discrete-time difference equation shown in Equation 9-6 contains coefficients a_i and b_i, which correspond to the weights applied to the signal variables $y[i]$ and $x[i]$. A discrete-time difference equation consisting of feedforward terms only is a special case. Such a system would be characterized by the difference equation

$$y[n] = \sum_{i=0}^{L-1} \alpha_i x[n-i] \qquad (9\text{-}28)$$

The impulse response $h[n]$, in this case, has the simple form

$$h[n] = \alpha_0 \delta_K[n] + \alpha_1 \delta_K[n-1] + \cdots + \alpha_{L-1}\delta_K[n-L-1] \qquad (9\text{-}29)$$

which is seen to be a weighted collection of L Kronecker delta functions. Such a discrete-time model is said to represent an Lth-order *finite impulse-response filter (FIR)* since the impulse response has at most only L nonzero values. Equation 9-29, demonstrates that the linear convolution of an Lth-order finite impulse-response, with an arbitrary N-sample input $x[n]$ will produce $N + L - 1$ output samples, at most. That is, $y[n] = 0$ for all $n \geq L + N$.

Example 9-6   Finite impulse response system

The conjectured length of the output time-series produced by convolving an N-sample input with a Lth-order finite impulse-response system can be experimentally verified to equal its predicted value of $N + L - 1$.

Computer Study   A 32-sample triangle wave and a 256-sample triangle train with period 32 are produced by S-file ch9_6.s and M-file fir.m. The signals are convolved with an impulse response $h[n]$ using the format fir(h). The results are displayed in Figure 9-6. The finite impulse response system is chosen to be length $N = 9$ and is given by $h[n] = [1, 1, 1, 1, 1, 1, 1, 1, 1]/9$. This system is also called a *moving-average filter*, since it calculates the average value of the nine most current input-sample values. Based on the computing formula for the length of a convolved output time-series, namely $N + L - 1$, the output time-series must be of length $9 + 32 - 1 = 40$ and $9 + 256 - 1 = 264$ samples respectively.

SIGLAB	MATLAB
>include "ch9_6.s"	>> h = ones(9, 1) / 9;
>h=[1,1,1,1,1,1,1,1,1]/9; fir(h)	>> fir(h)

Observe that the effect of convoluting $x[n]$ with the moving average filter $h[n]$ is one of data smoothing or averaging.

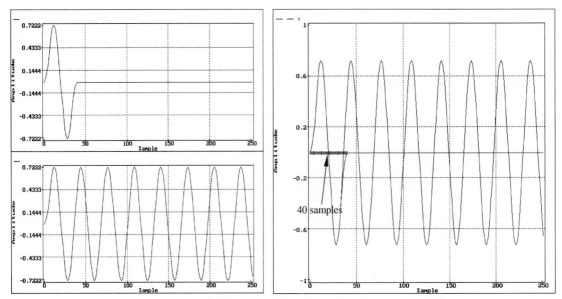

FIGURE 9-6   Linear convolution of a triangular wave and two-finite impulse-responses showing (left) the convolution results, and (right) the convolution results shown with a $N + L - 1 = 40$ sample marker.

Another special case occurs when two periodic signals are convolved. The periodic signals and systems can occur naturally or be generated. For example, sampled data is often read into an N-word shift-register array or buffer. Data can be read from these devices, if desired, in a circular or $\mathrm{mod}(N)$ manner (i.e., circular shift register). That is, $x[n + rN] = x[n]$ where r is an integer. This produces an infinitely long periodic time-series with period N. In particular, if $h[n]$ and $x[n]$ satisfy

$$h[n] = h[n + rN] \qquad (9\text{-}30)$$

$$x[n] = x[n + rN] \qquad (9\text{-}31)$$

for $r > 0$, an integer, and $n \in (-\infty, \infty)$, then the signals are periodic with a period N samples. Observe that the causal linear-convolution formula, given by Equation 9-19, does not apply in this case since $h[n]$ and $x[n]$ are now infinitely long time-series reaching back to $n = -\infty$. Modifying the limits of summation of Equation 9-19 from $i = 0$ to n to $i = -\infty$ to n technically would resolve this problem, but would not be practical. Consider, for purposes of demonstration, that $x[n] = h[n] = 1$ for all $n \in (-\infty, \infty)$. Both time-series are periodic with period 1. The value of $y[n]$, for n finite, would be infinite based on the modified Equation 9-19. Obviously, an alternative method is needed to compute the convolution of periodic signals and systems.

The convolution of two periodic time-series can be defined by the *circular convolution* sum, denoted $y[n] = h[n] \otimes x[n]$, given by

$$y[n] = h[n] \otimes x[n] = \sum_{i=0}^{N-1} h[i \bmod N] x[(n - i) \bmod N] \qquad (9\text{-}32)$$

A circularly convolved output time-series is itself periodic with period N (i.e., $y[n + rN] = y[(n + rN) \bmod N] = y[n]$). The circular convolution process is graphically interpreted in Figure 9-7. Here two length eight time-series $x[n] = h[n] = \{1, 1, 1, 1, 0, 0, 0, 0\}$ are distributed about the periphery of a cylinder and the periphery of a cylinder within that cylinder. The interior cylinder is stationary while the other is free to turn, completing one revolution in $N = 8$ samples. A "1" marks the location of a transparent slit in the cylinder, while a "0" denotes the presence of opaque material. A light source is placed at the center of the assembly. Light is detected by photo detectors placed outside the apparatus, provided that a slit in the inner cylinder overlaps another located on the outer cylinder. The outer wheel is set in motion and the output of the photo detectors is summed and displayed. The summed photo-detector output represents circular convolution and will be itself a periodic process of period 8.

A linear convolver would not see two periodic period-8 time-series, but instead two aperiodic length-8 signals $x[n]$ and $h[n]$. The linear convolution of these two signals produces a time-series of length $8 + 8 - 1 = 15$. The circular and linear results are also displayed in Figure 9-7. It can be seen that the two convolution results are two different waveforms.

The importance of circular convolution will become apparent in Chapter 10 when the discrete Fourier transform (DFT) is used to implement a linear convolution. In Chapter 10 it will be shown that the DFT assumes all signals to be

FIGURE 9-7   Circular convolution.

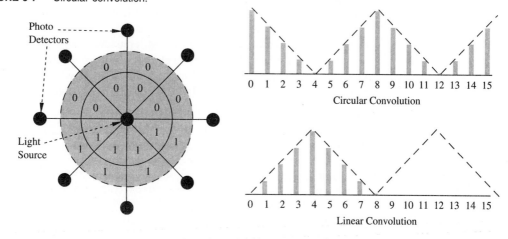

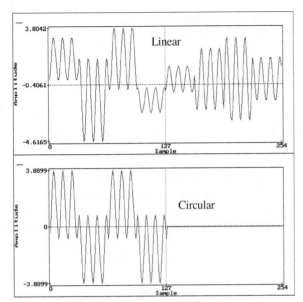

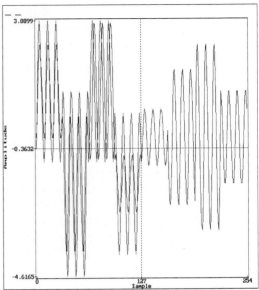

FIGURE 9-8   Circular and linear convolution with 255-sample linear (top), the 128-sample section circular (bottom), and their overlay (left).

periodic and, as such, can be used to perform circular convolution. It can also be used to perform (linear) convolution by being "tricked" into using circular convolution to produce a linear-convolution outcome.

Example 9-7   Circular convolution

Circular and linear convolution are mathematically distinct, serve different purposes, and, in general, produce dissimilar results.

Computer Study   S-file ch9_7.s and M-file circle.m circularly and linearly convolve two time-series $x[n]$ and $h[n]$, which are truncated to a length $N = 2^n$ where $n \leq 8$. The format used is circle(x,h,n). Signals $x[n]$ and $h[n]$ are of length $M \geq N$. For a test, consider $x[n] = \sin(2\pi n/10)$; $h[n]$ is a pulse-train of period 64, duty-cycle 1/2, and $M = 200$; and $n = 7$ (i.e., $200 > 2^7 = 128$). The results are shown in Figure 9-8, where the linear convolution is length $128 + 128 - 1 = 255$ and the circular convolution is length 128 with periodic extension in time implied.

SIGLAB	MATLAB
`>include "ch9_7.s"`	`>> x = sin(2*pi/10*(0:199));`
`>x=mksin(1,1/10,0,200)`	`>> h = square(2*pi/64*(0:199));`
`>h=sq(64,1/2,200)`	`>> circle(x, h, 7)`
`>circle(x,h,7)`	

The linear and circular convolution results can be seen to be distinctly different. Both, however, are correct within the context of their mathematical definitions.

9-5   PROPERTIES OF DISCRETE-TIME CONVOLUTION

The discrete-time convolution sum possesses the properties of the continuous-time convolution integral. They are

Commutative Property: If $x[n]$ is a signal and $h[n]$ an impulse response, then $x[n] * h[n] = h[n] * x[n]$.

Associative Property: If $x[n]$ is a signal and $h_1[n]$ and $h_2[n]$ are impulse responses, then $(x[n] * h_1[n]) * h_2[n] = x[n] * (h_1[n] * h_2[n])$.

Distributive Property: If $x[n]$ is a signal and $h_1[n]$ and $h_2[n]$ are impulse responses, then $x[n] * (h_1[n] + h_2[n]) = x[n] * h_1[n] + x[n] * h_2[n]$.   For an interpretation of these rules, refer to the discussion of linear convolution properties in Section 9-4.

9-6   STABILITY OF DISCRETE-TIME SYSTEMS

A discrete-time system can be stable or unstable. One of the more fundamental definitions of stability is based on the *bounded-input bounded-output (BIBO) principle*. A system is said to be BIBO stable if the output for every possible bounded input is likewise bounded.

Bounded-input bounded-output (BIBO) stability   A discrete-time system with an impulse response $h[n]$ is bounded-input bounded-output stable if and only if

$$\sum_{n=-\infty}^{\infty} |h[n]| < \infty \tag{9-33}$$

That is, the impulse response must be absolutely summable. This, in fact, corresponds to the *worst-case* system response if the input time-series is assumed to have an amplitude bounded by unity (i.e., $|x[n]| \leq 1.0$). The *worst-case input*, denoted $x_w[n]$, is one that produces a maximal output of the convolution sum $y[n] = h[n] * x[n] = \sum h[n - i]x[i]$. It is given by

$$x_w[n] = \frac{h[L - n]}{|h[L - n]|} \tag{9-34}$$

where L is an integer. Note that $|x_w[n]| = 1.0$ (i.e., $x_w[n] = \pm 1$). For some value of N and L, the sample values of $x_w[n]$ will be aligned with those of $h[-i]$ so that every product $h[-i]x[n]$ has *positive* value. This can be seen by substituting Equation 9-34 into the convolution sum to produce

$$y[n] = \sum_{i=-\infty}^{\infty} h(n - i)x[i] = \sum_{i=-\infty}^{\infty} h[n - i](h[L - i]/|h[L - i]|) \tag{9-35}$$

For $n = L$, where $L > 0$, Equation 9-35 simplifies to

$$y[n] = \sum_{i=-\infty}^{\infty} h[n-i]^2/|h[n-i]| = \sum_{i=-\infty}^{\infty} |h[i]| \qquad (9\text{-}36)$$

which is the bound established by Equation 9-33.

The impulse response of a discrete-time LSI was earlier expressed as a linear combination of eigenfunctions from $\phi_j[n] = n^{n[j]}(\beta_i)^n$, $\beta_i = \sigma_i + j\omega_i = |\beta_i|e^{j\phi_i}$. It then follows that the impulse response is of the form $h[n] = \sum \lambda_j \phi_j[n]$. Therefore, in order to be absolutely summable, it must follow that $|\beta_i| < 1.0$. Therefore, if the roots of a discrete-time LSI system's characteristic equation are bounded in absolute value below unity, then the corresponding system is BIBO stable.

Example 9-8   Stability of discrete-time systems

The at-rest discrete-time LSI system $y[n] - 0.875y[n-1] + 0.219y[n-2] - 0.016y[n-3] = x[n]$, has a characteristic equation given by

$$\beta^3 - 0.875\beta^2 - 0.219\beta - 0.016 = 0$$

which factors into

$$\left(\beta - \frac{1}{2}\right)\left(\beta - \frac{1}{4}\right)\left(\beta - \frac{1}{8}\right) = 0$$

The roots of the characteristic equation are $\{1/2, 1/4, 1/8\}$, which are bounded in absolute value below unity. Therefore the system is BIBO stable. To explore this claim further, note that the impulse response of the discrete system can be shown to be given by

$$h[n] = \frac{8}{3}\left(\frac{1}{2}\right)^n - 2\left(\frac{-1}{4}\right)^n + \frac{1}{3}\left(\frac{1}{8}\right)^n \qquad \text{for } n \geq 0$$

Consider using a unit step $x[n] = u[n]$ as a test signal. The step response, given by the linear convolution sum, yields

$$y[n] = \frac{8}{3}\sum_{i=0}^{n}\left(\frac{1}{2}\right)^i - 2\sum_{i=0}^{n}\left(\frac{-1}{4}\right)^i + \frac{1}{3}\sum_{i=0}^{n}\left(\frac{1}{8}\right)^i \qquad \text{for } n \geq 0$$

If plotted against n, $y[n]$ will be seen to monotonically increase from its initial value of $y[0] = 0$ to a final value of $y(\infty)$. Using the identity $\sum_{i=0}^{\infty} a^i = 1/(1-a)$, if $|a| < 1$, the final value can be computed to be

$$y[\infty] = \frac{8}{3}\frac{1}{1-(1/2)} - 2\frac{1}{(1+1/4)} + \frac{1}{3}\frac{1}{(1-1/8)} = 3.04762 < \infty$$

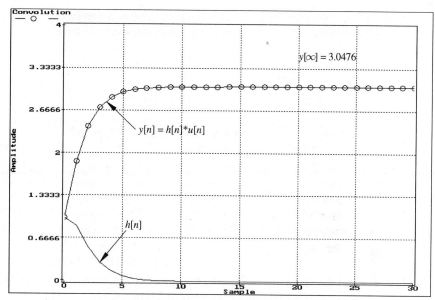

FIGURE 9-9   Impulse-response $h[k]$ and magnitude sum.

The step response is obviously bounded for this individual test input. Unfortunately, one test does not necessarily mean that the system is stable for all arbitrary inputs. Since the impulse-response $h[n]$ is nonnegative, however, the worst-case input (Equation 9-34) is $x_w[n] = u[n]$. Therefore, the bound on the step-response is also the worst-case bound in this case.

Computer Study   S-file `ch9_8.s` and M-file `stable.m` compute the impulse response $h[n]$, as well as the step response for $n \leq N$, using the format `stable(N)`.

SIGLAB	MATLAB

`>include "ch9_8.s"`
`>#evaluate time-series for 31 samples`
`>stable(31)`
	`>> stable(32)`

These results are displayed in Figure 9-9. The step-response $y[n]$ is seen to converge to the predicted worst case bound.

9-7   DISCRETE-TIME TRAJECTORIES

Analyzing high-order discrete-time systems presents the same difficulties that are encountered in the study of high-order continuous-time systems. Therefore, they are often reduced to a collection of low-order subsystems with known behavior and responses. The two basic building blocks for arbitrary discrete-time LSI systems are the first- and second-order subsystems.

9-7-1 First-Order Discrete-Time LSI's

A linear discrete-time first-order system is described by

$$y[n + 1] = a[n]y[n] + b[n]x[n] \tag{9-37}$$

with input $x[n]$ and output $y[n]$. This system is sometimes represented in what is called two-tuple form as $T_2 = \{a[n], b[n]\}$ and is graphically interpreted in Figure 9-10. The time-varying coefficients, while not affecting the linearity of the system, often make computing a closed-form solution a challenging problem. If the coefficients are constant, or can be assumed to be constant over an interval of time, an LSI system results (i.e., a constant coefficient); it is represented by $T_2 = \{a, b\}$

$$y[n + 1] = ay[n] + bx[n] \tag{9-38}$$

The homogeneous response of the LSI system, given by Equation 9-37, can be recursively computed to be:

$$y[1] = a[0]y_0$$

$$y[2] = a[1]y[1] = a[1]a[0]y_0$$

$$y[3] = a[2]y[2] = a[2]a[1]a[0]y_0$$

$$\vdots = \vdots$$

$$y[k] = a[k - 1]y[k - 1] = a[k - 1] \cdots a[2]a[1]a[0]y_0 \tag{9-39}$$

which can be more compactly expressed as

$$y[n] = \left(\prod_{i=0}^{n-1} a[i] \right) y_0 \tag{9-40}$$

For a constant coefficient system, the homogeneous response reduces to

$$y[n] = a^n y_0 \tag{9-41}$$

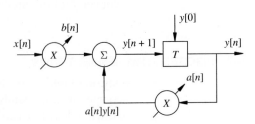

FIGURE 9-10   First-order discrete system where $T = 1/f_s$.

The impulse and inhomogeneous solution, for the system given by the discrete equation $y[n + 1] = ay[n] + bx[n]$, can be shown to be given by

Impulse

$$y[1] = a(y[0] = 0) + b(x[0] = 1) = b$$
$$y[2] = ay[1] + b(x[1] = 0)$$
$$= a(b) = ab$$
$$y[3] = ay[2] + b(x[2] = 0)$$
$$= a(ab) = a^2b$$
$$\vdots$$
$$y[n + 1] = ay[n] + b(x[n] = 0) = a^nb$$

Inhomogeneous

$$y[1] = a(y[0] = 0) + bx[0] = bx[0]$$
$$y[2] = ay[1] + bx[1]$$
$$= abx[0] + bx[1]$$
$$y[3] = ay[2] + bx[2]$$
$$= a^2bx[0] + abx[1] + bx[2]$$
$$\vdots$$
$$y[n + 1] = ay[n] + bx[n] = a^nbx[0]$$
$$+ \cdots + abx[n - 1] + bx[n]$$

$$(9\text{-}42)$$

The inhomogeneous solution is seen to be the convolution sum of the input with an impulse response given by $h[n] = a^nb$. Observe that knowledge of the homogeneous solution is intrinsic to the production of the system's impulse response. Finally, the system is seen to be BIBO stable if $|a| < 1.0$.

Example 9-9   First-order response

A first-order at-rest discrete-time LSI system is modeled as $y[n+1] = ay[n] + x[n]$. It would have an equivalent two-tuple representation given by $T_2 = \{a, 1\}$. If $|a| < 1$, then the system is stable. The impulse response is given by $h[n] = a^n$, which if $0 \le a < 1$, is an exponential solution that monotonically decays to zero. However, if $-1 < a < 0$, then the impulse response decays to zero in an oscillatory manner.

Computer Study   S-file `ch9_9.s` and M-file `ffirst.m` use the format `ffirst(x,a,n)` where x is the input time-series, a is the system-damping coefficient, and $n < 250$ is the observation interval in samples. The programs produce impulse responses for $h_1[n] = a^n$ and $h_2[n] = (-a)^n$, plus their convolution with $x[n]$. Finally, for $a = 0.95$, the stability of the system is predicted by computing $\sum |h_1[n]|$ and $\sum |h_2[n]|$ both of which have computed worst-case values of $1/(1 - |a|) = 1/0.05 = 20$. Begin with a step function of duration 100-samples and a 100-sample random time-series.

SIGLAB	MATLAB
```>include "ch9_9.s"```	```>> x = ones(100, 1)```
```>x=ones(100)```	```>> ffirst(x, 0.95, 151)```
```>ffirst(x,.95,151); # a=0.95```	```sum |h1(k)| = 19.99```
```sum |h1[n]|=```	```sum |h2(k)| = 19.99```
```19.9901 ~ 20```	
```sum |h1[n]|=```	
```19.9901 ~ 20```	

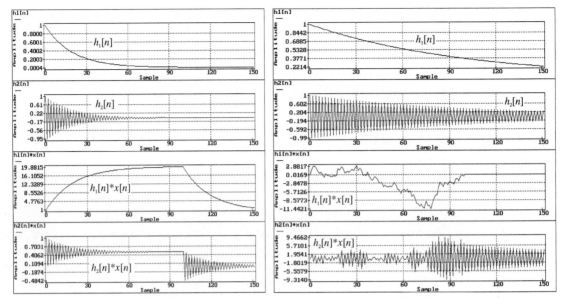

FIGURE 9-11   First-order system response showing: (left) from top to bottom $h_1[k]$, $h_2[k]$, $h_1[k] * x[k]$, $h_2[k] * x[k]$ for $x[k]$, a step, and (right) from top to bottom $h_1[k]$, $h_2[k]$, $h_1[k] * x[k]$, $h_2[k] * x[k]$ for $x[k]$, a random time-series (results may vary due to random data).

The results are displayed in Figure 9-11. It can be seen that $h_1[n]$ is exponential and $h_2[n]$ oscillates. The response to a step 100-samples in length slowly builds in the case of convolution with $h_1[n]$ and more rapidly in the case of $h_2[n]$. The second uses a random input (results may vary due to random data).

SIGLAB	MATLAB
>x=gn(100);	>> rand('normal');
>ffirst(x,.99,151)	>> x = rand(100, 1);
sum |h1[n]|= 77.8544	>> ffirst(x, 0.99, 151)
sum |h2[n]|= 77.8544	sum |h1(k)| = 78.08
	sum |h2(k)| = 78.08

It can be now seen that $h_1[k]$ has a smoothing effect while the oscillatory behavior of $h_2[k]$ gives rise to additional rapid movements at the output, as shown in Figure 9-11.

9-7-2   Second-Order Discrete-Time LSI

The next level of sophistication is a linear second-order system, as shown in Figure 9-12, having the form

$$y[n] + a[n]y[n - 1] + b[n]y[n - 2] = c[n]v[n] \qquad (9\text{-}43)$$

which has a three-tuple representation $T_3 = \{a[n], b[n], c[n]\}$. Again, solving

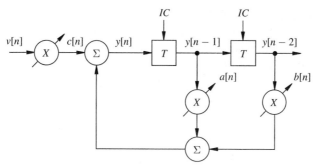

FIGURE 9-12   Second-order discrete-time system.

a time-varying difference equation can be difficult. If the coefficients are constant, then a solution can be easily computed. A typical BIBO stable second-order discrete-time LSI system has the form

$$y[n] - 2r\cos(\phi)y[n-1] + r^2 y[n-2] = x[n] \qquad (9\text{-}44)$$

where ϕ is called the normalized *natural resonant frequency*, $(0 \le \phi \le \pi)$, and r is the *damping factor*, $(0 \le r < 1)$. The characteristic equation is given by

$$\beta^2 - 2r\cos(\phi)\beta + r^2 = 0 \qquad (9\text{-}45)$$

and has roots given by

$$\beta_1 = r(\cos(\phi) - \sqrt{(\cos(\phi)^2 - 1)})$$

$$\beta_2 = r(\cos(\phi) + \sqrt{(\cos(\phi)^2 - 1)}) \qquad (9\text{-}46)$$

The values of β_i are real if $\cos(\phi) = \pm 1$ and imaginary otherwise. The homogeneous solution will converge to its equilibrium value in an exponential manner if $|\beta_i| < 1$. If $\cos(\phi) = 1$, the solution is said to be *critically damped*; otherwise, the homogeneous solution is called *underdamped*.

Example 9-10   Second-order system

A second-order system is modeled as

$$y[n] - 0.2215y[n-1] + 0.00118y[n-2]$$

$$= 0.5(x[n] - 0.4425x[n-1] + x[n-2])$$

The discrete-time system is clocked at a rate of $f_s = 200$ Hz. This establishes a delay value of 5 ms. The system is reported to suppress frequencies over the range $f \in [40, 50]$ Hz. Such systems are called bandstop filters and will be studied in Chapter 12. The system will be tested using a multitone signal, then a linearly-swept FM signal, called a *chirp*.

Computer Study S-file 9_10.s convolves a signal $x[n]$ with the impulse response of the given second-order LSI system using the format second(x). To test the hypothesis that the system is frequency selective, a multitone signal is convolved with the system.

Time interval	$x[n] = \cos(2\pi f n/100)$
[0.0, 0.5) sec.	$f = 0$ Hz
[0.5, 1.0) sec.	$f = 25$ Hz
[1.0, 1.5) sec.	$f = 50$ Hz
[1.5, 2.0) sec.	$f = 75$ Hz

SIGLAB	MATLAB
| ```
>include "ch9_10.s"
>x1=mkcos(1,0,0,101) # DC 101 samples
>x2=mkcos(1,1/8,0,100) # 25Hz
>x3=mkcos(1,1/4,0,100) # 50 Hz
>x4=mkcos(1,3/8,0,100) # 75Hz
>x=x1 & x2 & x3 & x4;
>#concatenate 0.5 second records
>second(x)
``` | >> not available in Matlab |

The results are presented in Figure 9-13. The solution can be seen to move from one 0.5-second epoch to the next with a brief transient build-up to a steady-state regime. The filter's output is highest for a DC input and lowest at 50 Hz.

A test is also conceived to measure the system with one signal that is swept over the frequency range of interest, say $f \in [0.0, 100]$ Hz. A signal that can produce this range of frequencies is a linearly swept FM signal, a chirp (i.e., $x[n] = \cos(k\omega_0 n^2)$).

SIGLAB	MATLAB
| ```
>include "chirp.s"
>second(chirp(401,0,0,.5))
># chirp covers range 0 to 0.5fs=100Hz
``` | >> not available in Matlab |

The results are also presented in Figure 9-13. The response can be seen to decrease in amplitude until the input signal achieves an instantaneous frequency in the neighborhood of 40–50 Hz, and then increases to 100 Hz. From

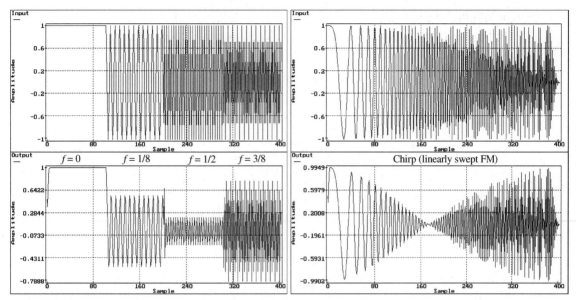

FIGURE 9-13   Response of a frequency selective second-order system to: (left) a multitone sinusoid, and (right) a chirp.

this test it can be seen that the system has a frequency dependent gain (or loss).

9-8   SUMMARY

In Chapter 9 discrete-time systems were introduced. The study focused on linear shift-invariant or LSI models. LSI at-rest systems were shown to interact with signals through a process called convolution. Convolution was defined in terms of a given input signal and the system's impulse response. Two types of convolution were considered, linear and circular. Linear convolution can be used whenever the signal and system are causal. Circular convolution can be used with periodic (noncausal) signals and/or systems. In the next chapter, transform-domain methods will be used to convert the complicated time-domain implementation of a convolution sum into a set of simple algebraic operations.

Stability was also studied in the context of a bounded-input bounded-output (BIBO) criteria. The stability of an LSI system was seen to be established by the factors of the system's characteristic equation. In the next chapter, this concept will be extended using transform-domain methods.

As was found with continuous-time systems, the fundamental building blocks of many discrete-time LSI systems are first- and second-order subsystems. Wherever the eigenvalues β_i of a second-order system appear as complex conjugate pairs, it is normally assumed that the second-order system's coefficients will be collected as $(\beta - \beta_1)(\beta - \beta_1^*) = \beta^2 - 2\operatorname{Re}(\beta_1) + |\beta_1|^2 = \beta^2 - a\beta + b$ to form a second-order section having only real coefficients a and b.

9-9 COMPUTER FILES

These computer files were used in Chapter 9.

File Name	Type	Location
Subdirectory CHAP9		
ch9_1.s	C	Example 9-1
fibb.m	C	Example 9-1
ch9_2.s	C	Example 9-2
shiftinv.m	C	Example 9-2
ch9_3.x	C	Example 9-3
ch9_5.s	C	Example 9-5
discrete.m	C	Example 9-5
ch9_6.s	C	Example 9-6
fir.m	C	Example 9-6
ch9_7.s	C	Example 9-7
circle.m	C	Example 9-7
ch9_8.s	C	Example 9-8
stable.m	C	Example 9-8
ch9_9.s	C	Example 9-9
ffirst.m	C	Example 9-9
ch9_10.s	C	Example 9-10
Subdirectory SFILES or MATLAB		
chirp.s	C	Example 9-10
Subdirectory SIGNALS		
Subdirectory SYSTEMS		
ch9_10.arc	D	Example 9-10

Here "s" denotes an S-file, "m" an M-file, "x" either/or, 'T" denotes a tutorial program, "D" a data file, and "C" denotes a computer-based-instruction (CBI) program. CBI programs can be altered and parameterized by the user.

9-10 PROBLEMS

9-1 Linearity   Show that a first-order, at-rest time-varying system, given by $y[n+1] = a[n]y[n] + b[n]x[n]$, is linear.

9-2 Discrete-time time-varying model   Consider the Fibonacci rabbit population model given by

$$y[n] = y[n-1] + y[n-2]$$

for $y[-1] = y[-2] = 1$ and $n \geq 0$. Assume that the model has been modified to become a time-varying biological system given by

$$y[n] = a[n]y[n-1] + y[n-2].$$

What is $y[n]$ if $a[n] = (-1)^n$ and $n \in [0, 4]$. Optionally use simulation to produce $x[n]$ for all $n \geq 0$.

9-3 Nonlinear discrete-time system   An at-rest nonlinear system is assumed to be modeled by

$$y[n] = \text{sign}(y[n]) + x[n]$$

For $n \in [0, 4]$, what is $y[n]$ if $x[n] = u[n]$? If $x[n] = (-1)^n u[n]$? Optionally, use simulation to produce $x[n]$ for all $n \geq 0$.

9-4 Homogeneous solution   Consider again the Fibonacci rabbit population model given by

$$y[n] = y[n-1] + y[n-2]$$

for $y[-1] = y[-2] = 1$ and $n \geq 0$. Derive the closed-form homogeneous solution for $y[n]$.

9-5 Convolution   Linearly convolve the following impulse response and signal pairs.

$$h[n] = (0.9)^n u[n]$$

$$h[n] = \delta_K[n] - \delta_K[n-1]$$

$$h[n] = u[n] - u[n-1]$$

$$x[n] = \delta_K[n-1]$$

$$x[n] = u[n-1]$$

9-6 Convolution   Suppose the signals and system in Problem 9-5 are of finite length. Suppose they are defined for $0 \geq n \geq 15$ and 0 elsewhere. Repeat the previous analysis. How do the outputs compare over $0 \geq n \geq 15$? How do the outputs compare for $n \geq 16$?

9-7 Numerical differentiator   A numerical differentiator is sometimes modeled as

$$y[n] = \frac{2}{T_s} x[n] - \frac{2}{T_s} x[n-1] - y[n-1]$$

where T_s is the sample period and $y[-1] = 0$. What is the system's impulse response? Determine the value of $y[\infty]$ if $x[n] = u[n]$. Experimentally determine how well the algorithm differentiates $x[n] = \cos(\omega_0 n)$ for various values of ω_0.

9-8 Square root algorithm   A discrete-time system defined by

$$y[n] = \left(\frac{1}{2}\right) \left(y[n-1] + \frac{x}{y[n-1]}\right)$$

is claimed to produce an estimate $y[\infty] \to \sqrt{x}$. For $x = 2$ and using initial estimates $y[-1] > 0$, compute $y[10]$. How well does $y[10]$ approximate $\sqrt{2}$?

9-9 Properties   Prove the properties list found in Section 9-5.

9-10 Stability   Determine the stability of the following system:

$$y[n] - 3.5y[n-1] + 4.85y[n-2] - 3.325y[n-3] + 1.1274y[n-4]$$

$$- 0.1512y[n-5] = x[n-2] - x[n-3] + x[n-4] - x[n-5]$$

What are the system's eigenvalues? Simulate the system's response to a step input.

9-11 Stability   An at-rest FIR filter (see Chapter 12) is assumed to have a discrete-time model given by

$$y[n] = \sum_{i=0}^{9} \left(\frac{1}{i+1} \right) x[n-i]$$

Prove that the system is stable. What is the worst-case input drawn from all possible inputs $|x[n]| \leq 1$, and what is the worst-case value of $y[n]$? Simulate the system's response to the worst-case input and compare to the computed value.

9-12 Stability   Given

$$y[n] - 1.9\cos(\pi/4)y[n-1] + (0.95)^2 y[n-2] = x[n] + x[n-2]$$

where $|x[n]| \leq 1$, determine the stability of the system and its impulse response. Assume that the impulse response can be approximated by its first 100 samples. What is the approximate worst-case input and output? Simulate the worst-case response and estimate the value of the worst-case output.

9-13 Circular convolution   Circularly convolve the period 20 signals $h[n] = \sin(\pi n/10)$ and $x[n] = \cos(\pi n/10)$. Sketch the result.

9-14 Convolution   An at-rest system has an impulse response $h[n] = \sin(\pi n/10)$ for $n \in [0, 19]$ and zero elsewhere. The input of the system is $x[n] = \cos(\pi n/10)$ for $n \in [0, 19]$ and zero elsewhere. What is the output of the system? Sketch the output. Compare to the previous problem's result.

9-15 Convolution   An at-rest system has an impulse response $h[n] = \sin(\pi n/10)$ for $n \in [0, 19]$ and zero elsewhere. The input signal $x[n]$ is sent to a modulo (20) circular shift register network that creates a new periodic input signal of period 20 (i.e., $x[n \bmod 20] = \{\dots, x[0], \dots, x[19], x[0], \dots, x[19], \dots\}$). The input of the system is $x[n] = \cos(\pi n/10)$ for $n \in [0, 19]$ and zero elsewhere. What is the output of the system? Sketch the output. Compare to the previous problem's result.

9-16 Second-order LSI system   Design a stable second-order at-rest discrete-time LSI system satisfying $y[n] - ay[n-1] + 0.25y[n-2] = x[n] + 2x[n-1] + x[n-2]$, which is sampled at a rate of 1000 samples per second and whose output is $z[n] = y[n] - 1$. The rise time is to be less than 10 ms. What is the minimum overshoot of your solution?

9-17 Nth-order LSI system   Derive a discrete-time LSI model for the system shown in Figure 9-14.

FIGURE 9-14   *N*th-order LSI system.

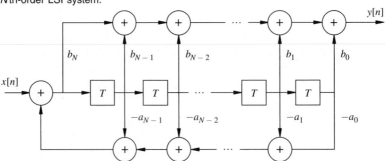

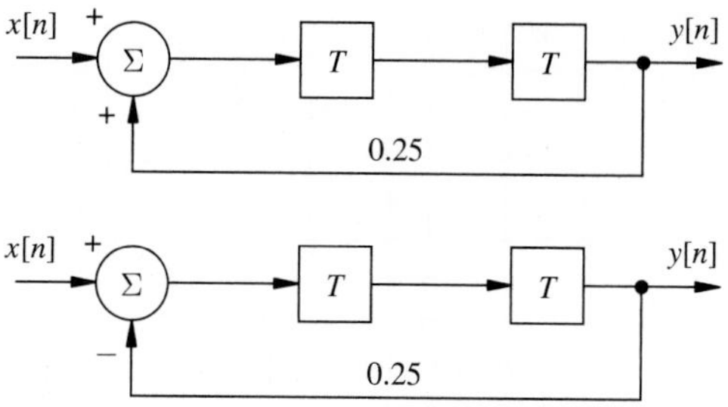

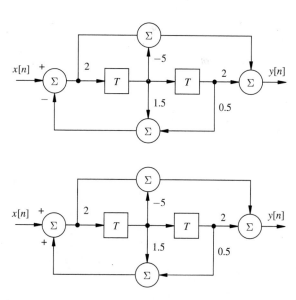

FIGURE 9-15   Second-order LSI system.

9-18 Second-order LSI system   Derive a difference equation model for the systems shown in Figure 9-15. Are the systems stable? What are the system's homogeneous responses? If the systems are at-rest, what are their step responses?

9-19 Second-order LSI system   Derive a discrete-time LSI model for the systems shown in Figure 9-16. Are the systems stable? What are the system's homogeneous responses? If the systems are at-rest, what are their step responses?

9-20 2-D discrete-time system   A 2-D filter is defined below and in Figure 9-17.

$$y[n, m] = \sum_{i=-1}^{1} \sum_{j=-1}^{1} a[i : j] x[n - i : m - j]$$

$$a[1 : 1] = a[-1 : 1] = a[1 : -1] = a[-1 : -1] = 1/16$$

$$a[0 : 1] = a[0 : -1] = a[1 : 0] = a[-1 : 0] = 2/16$$

$$a[0 : 0] = 4/16$$

FIGURE 9-16   Second-order LSI system.

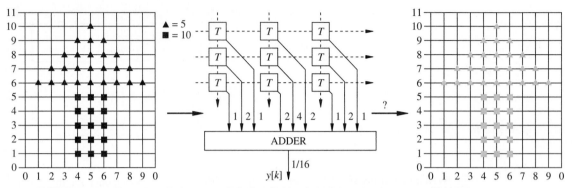

FIGURE 9-17   Example 2-D data-base X (left), LSI filter (middle), and output space Y (right).

It is claimed that the 2-D filter can map the sample values $x[i : j]$ from the 2-D data space X into an output space Y, which contains the locally weighted average value of the samples. Perform the mapping using the defined 2-D filter. Assume that the signal-sample values are all ones over $n \in [-3, 3]$ and $m \in [-3, 3]$ and zero elsewhere. What will be the output mapping? What purpose does the filter serve?

9-11   COMPUTER PROJECTS

9-1 Filter design and analysis   The EKG signal previously studied indicated that most of the biosignal's energy was concentrated in the first few hertz. An FIR was designed (see Chapter 12) using an FIR CAD tool and is given by

$$y[n] = \sum_{i=0}^{10} b[i]x[n - i]$$

$$b(0) = b(10) = 0.05086$$

$$b(1) = b(9) = 0.07634$$

$$b(2) = b(8) = 0.10016$$

$$b(3) = b(7) = 0.11959$$

$$b(4) = b(6) = 0.13230$$

$$b(5) = 0.13571$$

Verify mathematically and experimentally that the system is stable. Compute the worst-case gain and verify using simulation.

Conduct an experiment that has an EKG input and compare the output spectrum with that of the input. Discuss the filter's effect in the frequency-domain.

Experimentally measure the group delay at 10 distinct frequency locations and analyze.

Return to the worst-case analysis and assume that your filter is to be built using n-bit *fixed-point* arithmetic units that have a data format $(x\lfloor 0 \rfloor, x\lfloor 1 \rfloor, \ldots, x\lfloor n-1 \rfloor)$, where $x\lfloor i \rfloor$ is the ith bit, $x\lfloor i \rfloor = \{0, 1\}$ and $x\lfloor 0 \rfloor$ is the sign bit. A number x would be constructed from its binary representation as follows

$$x = \left(-x\lfloor 0 \rfloor + \sum_{i=1}^{n-1} 2^{-i} x\lfloor i \rfloor \right) 2^{K}$$

where $K = n - b - 1$, and b denotes the location of the binary point. For example, if $n = 8$ and $b = 4$, then $K = 3$, and

$$x = -2^{3} x\lfloor 0 \rfloor + 2^{3} \left(\sum_{i=1}^{7} 2^{-i} x\lfloor i \rfloor \right)$$

Here, $x = 15/16 \rightarrow [0 : 000 \diamond 1111]$ where ":" denotes the location of the sign bit and $\diamond$ indicates the location of the binary point. Implement your filter from $n = 8$ and $b = \{7, 4, 1\}$. Use as the input $x_1[n] = $ EKG and $x_2[n] = 4x_1[n]$. Explain your results.

9-2 Filter design and analysis   Repeat Project 9-1 using the fifth-order Butterworth filter (see Chapter 12) that satisfies

$$y[n] - 4.967y[n-1] + 6.788y[n-2] - 5.66y[n-3]$$

$$+ 2.381y[n-4] - 0.0403y[n-5]$$

$$= K(x[n] + 5x[n-1] + 10x[n-2] + 10x[n-3] + 5x[n-4] + x[n-5])$$

where $K = 3.490 \times 10^{-5}$. Estimate the worst-case condition based upon truncating the infinite impulse response to some finite number of samples. Beyond that sample range, it is assumed that the impulse response has essentially decayed to zero.

10

TRANSFORM DOMAIN REPRESENTATION OF DISCRETE-TIME SYSTEMS

We shall not cease from exploration,
and the end of all our exploring,
will be to arrive where we started,
and know the place for the first time.

—T. S. Elliot

10-1 INTRODUCTION

In the previous chapter, discrete-time LSI systems were studied in the time domain. The response of a discrete-time LSI system was defined in terms of a homogeneous solution, impulse response, and convolution sum. BIBO stability was defined on the basis of time-domain computations that were predicated on knowledge of the system impulse response. Unfortunately, this type of analysis can become complicated whenever the system under study is of high order. It was earlier shown that a continuous-time LTI system can be analyzed efficiently in a transform domain based on the Laplace and Fourier transforms. In this chapter, discrete-time systems will also be studied in the context of transforms and, in particular, the Z-transform and the discrete Fourier transform.

10-2 DISCRETE-TIME SYSTEMS

The behavior of an at-rest discrete-time LSI system was shown to be defined in terms of a system's impulse response in Chapter 9. The impulse response contains sufficient information to compute the at-rest system response to an arbitrary input

$x[n]$ using the convolution sum given by Equation 9-19. Suppose, for the sake of discussion, assume that both $h[n]$ and $x[n]$ possess Z-transforms given by $H(z)$ and $X(z)$, respectively. Then, the Z-transform of the causal output time-series $y[n] = h[n] * x[n]$ is given by

$$\mathcal{Z}[y[n]] = \mathcal{Z}[h[n] * x[n]] = \mathcal{Z}\left[\sum_{k=0}^{\infty} h[k]x[n-k]\right] = \sum_{n=0}^{\infty}\left[\sum_{k=0}^{\infty} h[k]x[n-k]\right]z^{-n}$$

$$= \sum_{k=0}^{\infty} h[k]\left[\sum_{n=0}^{\infty} x[n-k]z^{-n}\right] \tag{10-1}$$

Upon making a substitution of variables $m = n - k$, it follows that

$$\sum_{k=0}^{\infty} h[k]z^{-k}\left[\sum_{m=0}^{\infty} x[m]z^{-m}\right] = H(z)X(z) = Y(z) \tag{10-2}$$

It can be seen that there exists a strong similarity between the discrete-time result presented in Equation 10-2 and the continuous-time formula found in Equation 5-2. In particular, note that the Z-transform of the convolution of $h[n]$ and $x[n]$ is mathematically equivalent to the product of their Z-transforms.

Equation 10-2 is also known by its popular name, the *convolution theorem* for Z-transforms. This theorem provides a bridge between time-domain convolution and transform operations.

Convolution Theorem   The output of a discrete-time LSI system having a Z-transformable impulse response $h[n]$ and an input signal $x[n]$ having a Z-transform $X(z)$, is $y[n] = h[n] * x[n]$ whose Z-transform is given by

$$Y(z) = H(z)X(z) \tag{10-3}$$

If the regions of convergence for $X(z)$ and $H(z)$ are $\mathcal{R}_x$ and $\mathcal{R}_h$ respectively, then the region of convergence of $Y(z)$ is given by $\mathcal{R}_y$ where

$$\mathcal{R}_y \supset \mathcal{R}_x \cap \mathcal{R}_H \tag{10-4}$$

Finally, the desired output time-series $y[n]$ is given by

$$y[n] = \mathcal{Z}^{-1}[Y(z)] \tag{10-5}$$

This process is graphically interpreted in Figure 10-1 and is seen to consist of two paths. For the same reason that transforms are used to implement convolution in continuous-time systems, the discrete-time convolution theorem converts the often cumbersome derivation of a system's impulse response, followed by an infinite convolution summation, into a more efficient set of algebraic operations. The efficacy of the convolution theorem presumes that the Z-transforms of $x[n]$ and $h[n]$, as well as the inverse Z-transform of $Y(z)$, can be computed with

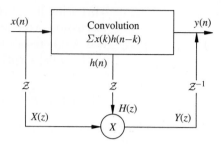

FIGURE 10-1   Discrete-time convolution theorem.

relative ease. If the input time-series can be modeled as a primitive signal $x[n]$, then the Z-transform $X(z)$ can be directly inferred from the information found in Tables 7-1 and 7-2. If, in addition, the discrete-time at-rest LSI system is modeled by the Lth-order linear-difference equation

$$a_0 y[n] + a_1 y[n-1] + \cdots + a_{L-1} y[n-L+1] + a_L y[n-L]$$

$$= b_0 x[n] + b_1 x[n-1] + \cdots + b_{L-1} x[n-L+1] + b_L x[n-L] \quad (10\text{-}6)$$

then Equation 10-6 can be directly mapped into the Z-domain as

$$a_0 Y(z) + a_1 z^{-1} Y(z) + \cdots + a_{L-1} z^{-L+1} Y(z) + a_L z^{-L} Y(z)$$

$$= b_0 X(z) + b_1 z^{-1} X(z) + \cdots + b_L z^{-L} X(z) \quad (10\text{-}7)$$

Therefore, the Z-transform of the impulse response $h[n]$ is given by the solution to Equation 10-7 where it is assumed that $x[k] = \delta_K[k]$, or equivalently $X(z) = 1$, $Y(z) = H(z)$. Upon making this substitution, the Z-transform of the system's impulse response $h[n]$ can be immediately written as

$$a_0 H(z) + a_1 z^{-1} H(z) + \cdots + a_{L-1} z^{-L+1} H(z) + a_L z^{-L} H(z)$$

$$= b_0 + b_1 z^{-1} + \cdots + b_{L-1} z^{-L+1} + b_L z^{-L} \quad (10\text{-}8)$$

or

$$H(z) = \frac{\sum_{i=0}^{L} b_i z^{-i}}{\sum_{i=0}^{L} a_i z^{-i}} \quad (10\text{-}9)$$

where $h[n] = \mathcal{Z}^{-1}[H(z)]$. Therefore, it can be seen that in a typical situation, the Z-transform of a system's impulse response can be produced directly without the need for complicated formulas or calculations. Finally, the simulation diagram representation of a system characterized by Equation 10-7 is shown in Figure 10-2, in what will be later called a Direct II form (see Chapter 12). It is important to note

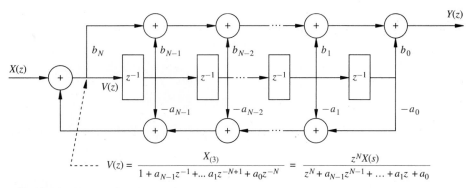

FIGURE 10-2 Simulation diagram of a discrete-time system.

that the operator z^{-1} represents the signal being delayed by one sample period. In hardware, this would be implemented with a digital-shift register.

Example 10-1 RLC simulation

In Chapter 6, second-order causal continuous-time LTI systems were studied. The second-order ODE equation $d^2y(t)/dt^2 + 3dy(t)/dt + 2y(t) = x(t)$ has an impulse response given by $h(t) = \exp(-t) - \exp(-2t)$. The discrete-time version, based on a sample period of T_s seconds, would satisfy $h(nT_s) = \exp(-nT_s) - \exp(-2nT_s)$, which can be simplified to read $h[n] = a^n u[n] - b^n u[n]$, where $a = \exp(-T_s)$ and $b = \exp(-2T_s)$. Furthermore, the Z-transform of $h[n]$ is $H(z) = z/(z-a) - z/(z-b) = (a-b)z/(z-a)(z-b)$. Assume also that the input is a unit step $x[n] = u[n] = 1$, such that $U(z) = z/(z-1)$. It immediately follows that $Y(z) = H(z)Y(z) = ((a-b)z^2)/(z-1)(z-a)(z-b)$. Using methods presented in Chapter 7, it can be shown that the inverse Z-transform of $Y(z)$ will produce a time-series

$$n[n] = \left[\frac{(a-b)}{(1-a)(1-b)} + \frac{a^{n+1}}{a-1} - \frac{b^{n+1}}{b-1} \right] u[n]$$

Computer Study S-file `ch10_1.s` and M-file `zrtrans.m` compute $y[n]$ using linear convolution and transform methods. The results are displayed using the format `zrtrans(Ts)` where T_s is the sample period in seconds.

SIGLAB	MATLAB
>include "ch10_1.s"	>>zrtrans(1/50)
>zrtrans(1/50)	

The results are displayed in Figure 10-3. It can be observed that the results produced by both methods are identical.

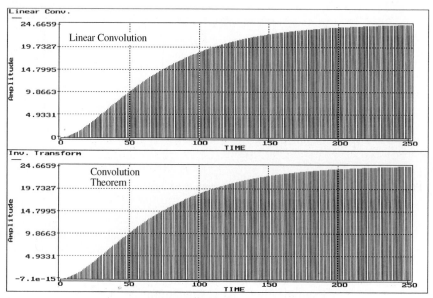

FIGURE 10-3   Example of the convolution theorem for $T_s = 1/50$.

Referring again to the convolution theorem, namely $Y(z) = H(z)X(z)$, observe that if $x[k] = \delta_K[k]$ (i.e., $X(z) = 1$), it follows that $Y(z) = H(z)$. Therefore $H(z)$ is the Z-transform of the system's impulse response. $H(z)$ is also known by its popular name, *transfer function*, since it defines how a Z-transformed input signal is mapped (transferred) into a Z-transformed output signal. The transfer function is specified in terms of the system's coefficients and is independent of the input signal. In particular, if the system's impulse response is given by the discrete-time equation

$$\sum_{i=0}^{N} a_i y(n-i) = \sum_{i=0}^{M} b_i \delta_K(n-i) \tag{10-10}$$

the transfer function satisfies the equation

$$H(z) = \frac{N(z)}{D(z)} = \frac{\sum_{i=0}^{M} b_i z^{-i}}{\sum_{i=0}^{N} a_i z^{-i}} = \frac{z^{N-M} \sum_{i=0}^{M} b_i z^{M-i}}{\sum_{i=0}^{N} a_i z^{N-i}} \tag{10-11}$$

The roots of the polynomial $N(z)$ are called the system's *zeros* and the roots of the polynomial $D(z)$ are called the system's *poles*. Both play an important role in characterizing and analyzing a system. It can also be noted that the poles are the eigenvalues of the system's characteristic equation.

Equation 10-11 is more general than Equation 10-9 in that it allows the numerator's order to be independently specified. If the order of the numerator is less than that of the denominator, the system is said to be *strictly proper*. If the orders are equal, then the system is said to be *proper*. If the order of the numerator equals or exceeds the order of the denominator by an amount $P = M - N \geq 0$, then the transfer function will be assumed to be factored into the form

$$H(z) = \frac{N(z)}{D(z)} = Q(z) + \frac{R(z)}{D(z)} \qquad (10\text{-}12)$$

The quotient polynomial $Q(z)$ is of order P, the remainder polynomial $R(z)$ is of order less than N, and $D(z)$ is order N. It therefore follows that $R(z)/D(z)$ must be strictly proper. The quotient polynomial can be computed using long division. In particular, if $P > 0$, then

$$h[n]=h[n-P]\delta_K[n+P]+\cdots+h[-2]\delta_K[n+2]+h[-1]\delta_K[n+1]+\sum_{i=0}^{\infty}h[i]\delta_K[n-i]$$

$$(10\text{-}13)$$

Observe that such a system would be noncausal. Such systems do not, in general, occur in practice. If, however, $P = M - N \leq 0$, then the system is causal, which is the common case.

Discrete-time LSI system models, such as a transfer function, can also be experimentally determined. A number of techniques can be used to produce an approximate transfer function model of an LSI system having measured input/output spectral behavior. The modeling procedures produce approximate transfer functions that attempt to reproduce the observed input/output measurements to varying degrees of accuracy. The quality of the approximation is dependent on a number of factors too numerous to mention here. Some of the more popular modeling techniques, however, belong to a class of estimators called *auto-regressive moving-average (ARMA)* algorithms. This subject is treated in-depth in advanced system studies but, for now, will be thought of as simply a tool.

Example 10-2 Impulse response

In Example 8-5, the spectrum of a typical EKG signal was presented. The EKG signal, for the purpose of discussion, will be assumed to be produced by a custom signal generator. The experimental apparatus used to develop a

FIGURE 10-4 ARMA modeling process.

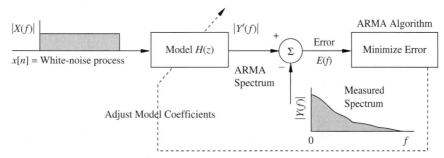

model of the signal generator is shown in Figure 10-4. It consists of an ARMA algorithm that produces a discrete-time LSI model of a system. The model coefficients are adjusted to produce an output spectrum, denoted $Y'(f)$, which is similar (in a least-squares sense) to the measured spectrum $Y(f)$. The input presented to the ARMA model is assumed to be a white-noise process (see Example 3-10).

Computer Study   Using an ARMA routine[1], a fifth-order model of an EKG signal generator was computed. The model is given by

$$H(z) = \frac{N(z)}{D(z)}$$

where it was determined that[2]

$$N(z) = 1.595z^5 - 0.321z^4 + 0.938z^3 - 0.053z^2 + 0.126z - 0.193$$

$$D(z) = 1.0z^5 - 1.395z^4 + 1.291z^3 - 0.79z^2 + 0.048z + 0.025$$

The fifth-order ARMA model and target EKG magnitude frequency-responses are shown in Figure 10-5. It can be seen that the ARMA model has the general shape of the parent EKG signal over the baseband frequency range $f \in [0, 67.5]$ Hz. In addition, the system model is seen to be causal, with a magnitude frequency response that reasonably agrees with the measured spectrum (see Figure 10-5). An alternative display format for $H(z)$ is in log-log, or Bode plot. To produce a Bode plot representation of $H(z) = N(z)/D(z)$, for $z = \exp(j2\pi f)$, the following procedure is used.

SIGLAB	MATLAB
```>include "ch10_2.s"```	not available in Matlab
```>include "bode.s"```	
```>h=rf("arma.coe")```	
```>n=h[0,]; d=h[1,]```	
```>graph(bode(n/128,d,67.5,100))```	
```># see ex: 5.9```	

The result is displayed in Figure 10-5 in a log-log format. Since the Bode plot is flat for frequencies above $f = 10^{0.2} = 1.6$ Hz, it can be argued that the ARMA model interprets the original EKG signal spectrum as having a $1/f$ decay.

Generally, due to the existence of the convolution theorem, the actual production of an impulse response $h[n]$ is not required to compute the output of a

---

[1] The ARMA routine used is a shareware S-file supplied by The Athena Group Inc., Gainesville, FL.

[2] For users of SIGLAB, the ARMA model polynomial coefficients are saved under the filename `arma.coe`.

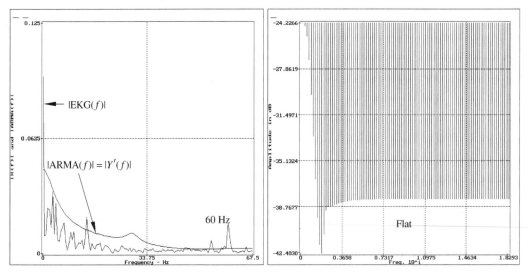

FIGURE 10-5   An ARMA model of an EKG signal generator and its Bode plot.

discrete-time LSI system having a Z-transformable input $x[n]$. Once a transfer function $H(z)$ is known, and the input transform $X(z)$ is specified, the output transform immediately follows. If $Y(z)$ is then known, then so is $y[n]$. The techniques of inverting $Y(z)$ into a time-series developed in Chapter 7 were based on a partial fraction expansion. If $Y(z)$ is the ratio of polynomials, where the order of the numerator (i.e., M) exceeds the order of the denominator (i.e., N) by an amount $P = M - N \geq 0$, then $Y(z)$ will be assumed to be factored into two polynomials, namely

$$Y(z) = H(s)X(s) = Q(z) + \frac{R(z)}{D(z)} \tag{10-14}$$

where the quotient polynomial $Q(z)$ satisfies

$$Q(z) = q_P z^P + \cdots + q_1 z + q_0 \qquad P = N - M \tag{10-15}$$

which has an inverse Z-transform

$$q[n] = q_P \delta_K(n + P) + \cdots + q_1 \delta_K(n + 1) + q_0 \delta_K[n] \tag{10-16}$$

Including $q[n]$ in the final system response, one obtains

$$y[n] = q_P \delta_K[n + P] + \cdots + q_1 \delta_K[n + 1] + q_0 \delta_K[n] + \mathcal{Z}^{-1}\left[\frac{R(z)}{D(z)}\right] \tag{10-17}$$

In general, it shall be assumed that $Y(z) = H(z)X(z)$ is proper and has the general form

$$Y(z) = H(z)X(z) = \frac{R(z)}{D(z)}$$

$$D(z) = \prod_{i=1}^{L}(z - p_i)^{n(i)} \tag{10-18}$$

$$N = \sum_{i=1}^{L} n(i)$$

where $D(z)$ has roots p_i of multiplicity $n(i)$. If $n(i) > 1$, then p_i is said to be a repeated root and if $n(i) = 1$, p_i is called distinct. Based on the partial fraction expansion techniques developed in Chapter 7, the *Heaviside expansion* or partial fraction expansion of $Y(z)$ would have the form

$$Y(z) = \sum_{j=1}^{L}\sum_{i=1}^{n(j)} \frac{\alpha_{j,i}z}{(z - p_j)^i} = \cdots + \frac{\alpha_{j,1}z}{(z - p_j)} + \frac{\alpha_{j,2}z}{(z - p_j)^2} + \cdots + \frac{\alpha_{j,n(j)}z}{(z - p_j)^{n(j)}} + \cdots \tag{10-19}$$

The individual terms, explicitly displayed in Equation 10-19, are seen to be of the form $z/(z - \lambda_j)^k$, and are found in Table 7-1. The mechanics of computing the Heaviside coefficients α_{jk} were developed in detail in the z-domain in Chapter 7 and are given by

$$\alpha_{jn(j)} = \lim_{z \to p_j} \frac{(z - p_j)^{n(j)}Y(z)}{z}$$

$$\alpha_{j,n(j)-1} = \lim_{z \to p_j} \frac{d}{dz}\frac{(z - p_j)^{n(j)}Y(z)}{z}$$

$$\alpha_{j,n(j)-2} = \lim_{z \to p_j} \frac{d^2}{dz^2}\frac{1}{2}\frac{(z - p_j)^{n(j)}Y(z)}{z}$$

$$\vdots = \vdots$$

$$\alpha_{j,k} = \lim_{z \to p_j} \frac{d^{n(j)-k}}{dz^{n(j)-k}}\frac{1}{(n(j) - k)!}\frac{(z - p_j)^{n(j)}Y(z)}{z}$$

These calculations can be implemented using a general-purpose digital computer as demonstrated in the next example.

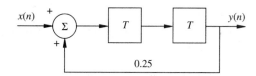

FIGURE 10-6   Second-order discrete-time system.

Example 10-3   Nonrepeated roots

Suppose $H(z)$ represents a simple second-order system given by $H(z) = z/((z+0.5)(z-0.5))$. The system can be equivalently expressed as $H(z^{-1}) = 1/((1+0.5z^{-1})(1-0.5z^{-1})) = 1/(1-0.25z^{-2})$ and is interpreted as the simulation diagram presented in Figure 10-6. The step response to $x[n] = u[n]$, or equivalently $X(z) = z/(z-1)$, is given by the convolution theorem to be $Y(z) = z^2/[(z-1)(z+0.5)(z-0.5)]$ in the z-domain. Observe that all roots of $Y(z)$ are nonrepeated, and

$$Y(z) = \alpha_0 + \alpha_1 \frac{z}{(z-1)} + \alpha_2 \frac{z}{(z-0.5)} + \alpha_3 \frac{z}{(z+0.5)}$$

or

$$\frac{Y(z)}{z} = \alpha_0 \frac{1}{z} + \alpha_1 \frac{1}{(z-1)} + \alpha_2 \frac{1}{(z-0.5)} + \alpha_3 \frac{1}{(z+0.5)}$$

which represents the polynomials studied in Chapter 2. Here α_0 corresponds to a Kronecker delta function, $\alpha_1 z/(z-1)$ to a step function, and $\alpha_2 z/(z+0.5)$ and $\alpha_3 z/(z-0.5)$ to exponentials. Since $Y(z)$ is strictly proper, α_0 will have a computed value of zero.

The Heaviside coefficients are given by

$$\alpha_0 = \lim_{z \to 0} \frac{zY(z)}{z} = 0$$

$$\alpha_1 = \lim_{z \to 1} \frac{(z-1)Y(s)}{z} = \lim_{z \to 1} \frac{z}{(z+0.5)(z-0.5)} = \frac{4}{3}$$

$$\alpha_2 = \lim_{s \to .5} \frac{(z-0.5)Y(s)}{z} = \lim_{z \to .5} \frac{z}{(z-1)(z+0.5)} = -1$$

$$\alpha_3 = \lim_{s \to -.5} \frac{(z+0.5)Y(s)}{z} = \lim_{z \to .5} \frac{z}{(z-1)(z-0.5)} = -\frac{1}{3}$$

From Table 7-1, the inverse Z-transform $Y(z)$ can be expressed as $y[n] = 0\delta_K[n] + 4/3u[n] - (0.5)^n u[n] - 1/3(-0.5)^n u[n]$. Note that $y[0] = 0$ and $y[\infty] \to 4/3$.

Computer Study   S-file `pf.s` and M-file `residue.m` perform a partial fraction expansion of $Y(z) = N(z)/D(z) = z^2/[(z-1)(z-0.5)(z+0.5)]$. The roots of $D(z)$ are $z = 1$, 0.5 and -0.5. However, what is needed is the partial fraction expansion of $Y(z)/z = z^2/[z(z-1)(z-0.5)(z+0.5)]$, which has as roots of $D(z) = z(z-1)(z-0.5)(z+0.5)$, located at $z = 0$, 1, 0.5 and -0.5.

SIGLAB	MATLAB
`>include "pf.s"`	`>> n = [1, 0, 0];`
`>n=[1,0,0] # N(s)=z^2`	`>> d = poly([0,1,0.5,-0.5])`
`>d=poly([0,1,.5,-.5])`	`d = 1.0000 -1.0000 -0.2500`
`>#d(z)=z(z-1)(z-0.5)(z+0.5)`	`0.25000 >> % r = [a1a2a3a4]`
`># 1 -1 -0.25 0.25 0`	`>> % p = [p1 p2 p3 p4], k = a0`
`>pfexp(n,d) # Heaviside`	`>> [r p k] = residue(n, d)`
`pole @ -0.5+0j`	`r =`
`with multiplicity 1 and`	`1.3333`
`coefficients: -0.333333-0j`	`-1.0000`
`=> a3`	`-0.3333`
`pole @ 0+0j with`	`0`
`multiplicity 1 and coefficients:`	`p =`
`0+0j => a0`	`1.0000`
`pole @ 0.5+0j with`	`0.5000`
`multiplicity 1 and coefficients:`	`-0.5000`
`-1-0j => a2`	`0`
`pole @ 1+0j with`	`k =`
`multiplicity 1 and coefficients:`	`[]`
`1.33333+0j => a1`	

This corresponds to the derived values.

Example 10-4   Repeated roots

Suppose a second-order system is given by

$$H(z) = \frac{3z^2 - 5z^1 + 3}{(z-1)(z-0.5)}$$

Then $H(z^{-1})$ is given by

$$H(z^{-1}) = \frac{3 - 5z^{-1} + 3z^{-2}}{(1 - z^{-1})(1 - 0.5z^{-1})} = \frac{3 - 5z^{-1} + 3z^{-2}}{1 - 1.5z^{-1} + 0.5z^{-2}}$$

which has a simulation diagram shown in Figure 10-7.

The Z-transform of the system's unit-step response is given by $Y(z) = H(z)(z/(z-1))$ or

$$Y(z) = \frac{3z^3 - 5z^2 + 3z}{(z-1)^2(z-0.5)}$$

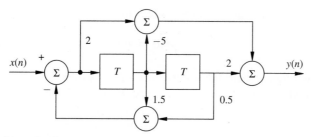

FIGURE 10-7 Second-order system.

which has a partial fraction expansion given by

$$Y(z) = \alpha_0 + \frac{\alpha_{11}z}{(z - 0.5)} + \frac{\alpha_{22}z}{(z - 1)} + \frac{\alpha_{21}z}{(z - 1)^2}$$

or

$$\frac{Y(z)}{z} = \frac{\alpha_0}{z} + \frac{\alpha_{11}}{(z - 0.5)} + \frac{\alpha_{21}}{(z - 1)} + \frac{\alpha_{22}}{(z - 1)^2}$$

where the coefficients are given by

$$\alpha_0 = \lim_{z \to 0} \frac{zY(z)}{z} = 0$$

$$\alpha_{11} = \lim_{z \to 0.5} \frac{(z - 0.5)Y(z)}{z} = \lim_{z \to 0.5} \frac{(3z^3 - 5z^2 + 3z)}{z(z - 1)^2} = 5$$

$$\alpha_{22} = \lim_{z \to 1} \frac{(z - 1)^2 Y(z)}{z} = \lim_{z \to 1} \frac{(3z^3 - 5z^2 + 3z)}{z(z - 1/2)} = 2$$

$$\alpha_{21} = \lim_{z \to 1} \frac{d}{dz} \frac{(z - 1)^2 Y(z)}{z}$$

$$= \lim_{z \to 1} \left[\frac{9z^2 - 10z + 3}{z(z - 1/2)} - \frac{(3z^3 - 5z^2 + 3z)(2z - 1/2)}{[z(z - 1/2)]^2} \right] = -2$$

This results in

$$y[n] = [5(0.5)^n - 2.0 + 2n]u[n]$$

for $n \geq 0$. The first few sample values of the solution can be checked using long-division as follows:

$$\begin{array}{r} 3 + (5/2)z^{-1} + (13/4)z^{-2} + \cdots \\ z^3 - (5/2)z^2 + 2z - 1/2 \overline{)3z^3 + 5z^2 + 3z} \end{array}$$

The initial value theorem states that

$$y(0) = \lim_{z\to\infty} Y(z) = 3$$

which agrees with the known time-series values.

Computer Study   S-file pf.s and M-file residue.m can again be used to produce the partial fraction expansion of $Y(z)/z = (3z^3 - 5z^2 + 3z)/[z(z - 0.5)(z - 1)^2]$, which has as roots of $D(z) = z(z - 0.5)(z - 1)^2$, $z = 0, 0.5, 1$ and 1.

SIGLAB	MATLAB

>include "pf.s"
>n=[3,-5,3,0];
d=poly([0,1,1,.5])
>#d; 1 -2.5 2 -0.5 0
>pfexp(n,d) # Heaviside
pole @ 0+0j
with multiplicity 1 and
coefficients: 0+0j
=> a0
pole @ 0.5+0j
with multiplicity 1 and
coefficients: 5+0j
=> a11
pole @ 1+0j
with multiplicity 2 and
coefficients: 2+0j
-2+0j => a22 and a21
	>> n = [3, -5, 3, 0];
	>> d = poly([0, 1, 1, 0.5])
	d =
	   1.0000 -2.5000 2.0000
	-0.5000 0
	0
	>> % r=[a1 a2 a3 a4]
	>> % p = [p1 p2 p3 p4], k = a0
	>> [r p k] = residue(n, d)
	r =
	   -2.0000
	   2.0000
	   2.0000
	   0
	p =
	   1.0000
	   1.0000
	   0.5000
	   0
	k =
	   []

---

which corresponds to the derived values.

Example 10-5   Impulse response

A fifth-order EKG signal generator was derived in Example 10-2. The response of the signal generator to an impulse $x[n] = \delta_K[n]$ can be computed to be $y[n] = h[n] = \mathcal{Z}^{-1}[H(z)]$. The fifth-order transfer function can be shown to have distinct poles and would therefore have a partial fraction expansion given by

$$H(z) = \alpha_0 + \alpha_1 \frac{z}{(z - p_1)} + \cdots + \alpha_5 \frac{z}{(z - p_5)}$$

where the production rules for computing α_i have been previously developed.

Computer Study   The modified transfer function $H(z)/z$ can be analyzed in transfer-function form as follows

SIGLAB	MATLAB
| ```
>p=rf("arma.coe") # ARMA polynomial
>n=p[0,];d=p[1,];dd=
[d,0], pfexp(n,dd)
pole @ -0.1342821+0j with multiplicity
1 and coefficients:
4.56223+1.07359e-09j => a1~4.56223
pole @ 0+0j with multiplicity
1 and coefficients:
-7.86828-2.34486e-09 => a0~-7.86828
pole @ 0.208882-0.873852j with
multiplicity 1 and coefficients:
0.0626029+0.126792j => a2
pole @ 0.208882+0.873852j with
multiplicity 1 and coefficients:
0.0626029-0.126792j => a3 = a2^*
pole @ 0.266432 to with
multiplicity 1 and coefficients:
in 3.05123=>0.4
pole @ ~0.845691 + 0j with
multiplicity 1 and coefficients:
~1.72443 => a5
``` | not available in Matlab |

From this analysis, it follows that

$$H(z) = -7.86828 + \frac{4.56223z}{(z+0.134821)} + \frac{3.0512z}{(z-0.266432)}$$

$$+ \frac{(0.0626029 + j0.126792)z}{(z - 0.208882 - j0.873852)}$$

$$+ \frac{(0.0626029 - j0.126792)z}{(z - 0.208882 + j0.873852)} + \frac{1.72443z}{(z - 0.845691)}$$

which has an inverse Z-transform defined by the elements in Table 7-1. The solution is seen to consist of a weighted impulse, two exponentials of the form $(-0.134812)^n$ and $(0.845691)^n$, and a damped-complex exponential.

10-3   STABILITY OF DISCRETE-TIME SYSTEMS

The concept of BIBO stability for a discrete-time system was introduced in Chapter 9. There were additional stability definitions introduced for continuous-time systems in Chapters 4 and 5. These concepts also apply to discrete-time systems, as well. If, for all possible bounded initial conditions, the homogeneous solution
```

$y[n] \to 0$ as $n \to \infty$, then the system is said to be asymptotically stable. If a discrete-time system is asymptotically stable, it is also BIBO stable. If the homogeneous response of a discrete-time system can only be guaranteed to be bounded, namely $|y[n]| < M$, for all $n \geq 0$, then the system is said to be marginally stable. A system is obviously unstable if, for at least one possible initial condition, $|y[n]| \to \infty$ as $n \to \infty$.

The stability of a discrete-time LSI system can also be determined in the transform domain. It was previously noted that there is a correspondence between the roots of a characteristic difference equation of a discrete-time LSI system and its time-domain behavior. The roots of the characteristic equation, called eigenvalues, were also shown to be the system poles. Therefore, the stability of an LTI system can be established by analyzing the pole locations of the system's transfer function.

Referring to Equation 10-19, it can be noted that the N poles of $H(z)$ are given by $z = p_i$, where p_i may be real or complex, distinct or repeated. In general, the partial fraction expansion of a strictly proper $H(z)$ is given by

$$H(z) = \sum_{k=1}^{L} \sum_{i=1}^{n[k]} \frac{\alpha_{k,i} z}{(z - p_k)^i} = \cdots + \frac{\alpha_{k,1} z}{(z - p_k)} + \frac{\alpha_{k,2} z}{(z - p_k)^2} + \cdots + \frac{\alpha_{k,n[k]} z}{(z - p_k)^{n[k]}} + \cdots$$

$$(10\text{-}20)$$

which has an inverse Z-transform (see Table 7-1) given by

$$h[n] = \cdots + \alpha_{k1}(p_k)^n + \alpha_{k2} n (p_k)^n + \cdots + \beta_{ki} n^{q(i)} (p_k)^n + \cdots \qquad (10\text{-}21)$$

where $q(i)$ is a nonnegative integer and β is defined by an algebraic collection of the α's. If p_k is complex, then $p_k = r_k \exp(j\phi_k[k])$ and $(p_k)^n = r_k^n \exp(jn\phi_k[k])$ where $|p_k| = r_k$. If p_k is real, then $\phi[k] = 0$. A discrete-time LSI system is asymptotically stable if and only if the magnitude of all the poles of $H(z)$, denoted r_k, are bounded by unity. This will ensure that $|p_k|^n = |r^n \exp(jn\phi[k])| = |r^n| |\exp(jn\phi[k])| \leq |r^n| \to 0$ as $n \to \infty$. The LSI system is marginally stable if and only if $|p_k| \leq 1$ for all nonrepeated roots and $|p_i| < 1$ if p_e is repeated. The LSI system is unstable if there exists at least one pole such that $|p_k| > 1$, or if p_k is repeated, $|p_k| = 1$. One must be careful to insure that the location of potentially troublesome poles are not masked by a zero (called masking zero) located at the same point. If, for example, a discrete-time LSI system is specified by the difference equation $y[n+1] - y[n] = x[n+1] - x[n]$, $y[0] = y_0$, then the transfer function (which assumes the system is at rest) is given by

$$H(z) = \frac{Y(z)}{X(z)} = \frac{(z-1)}{(z-1)} = 1 \qquad (10\text{-}22)$$

The system would, therefore, appear to be stable to all inputs. However, for a nonzero initial condition the homogeneous solution is $y[n] = n y_0$, which diverges to infinity as $n \to \infty$. In general, an LSI system having a transfer function $H(z)$,

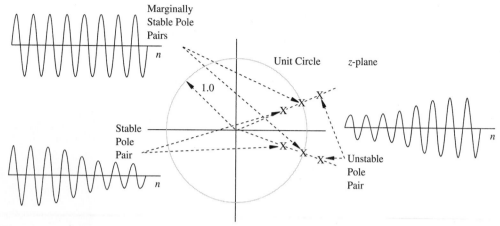

FIGURE 10-8   Z-domain stability requirements.

with no marginally stable or unstable poles uncovered or canceled by masking zeros, is asymptotically stable.

The transform-domain stability tests are seen to be based on the location of the system's poles in the z-plane. The stability of an LSI system can, therefore, be determined by partitioning the z-plane as shown in Figure 10-8. Stability is guaranteed if all the poles reside interior to the unit circle. Instability will occur if any pole is exterior to the unit circle. If poles are located along the periphery of the unit circle, they are marginally stable or unstable depending on their multiplicity.

Example 10-6   Stability

The stability of a discrete-time LSI system can be established by the location of the system poles. Three strictly proper filters were created with a transfer function $H(z)$, where

$$H(z) = K \frac{\prod_{i=1}^{4}(z - z_i)}{\prod_{i=1}^{5}(z - p_i)}$$

Their pole-zero distributions are listed here and displayed in Figure 10-9, along with the $|H(e^{j\Omega})|$, $\Omega\varepsilon[0, \pi]$, and impulse responses. Observe that the impulse-response record for the stable system decays, the marginal system achieves a constant amplitude, and divergence occurs for the unstable system. In each case the system's magnitude frequency response has a local peak in the vicinity of the pole located at $\sqrt{2}+j\sqrt{2}$, which is located at a phase angle of $2\pi/8$ radians in the z-plane. This corresponds to a frequency of $f_{sample}/8 = 0.125 f_{sample}$.

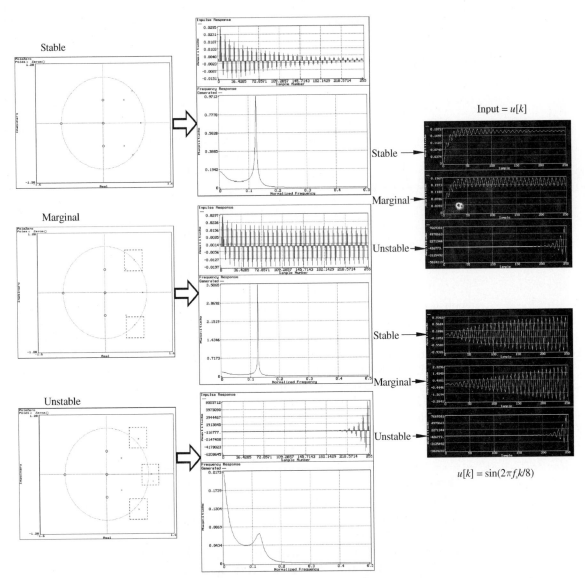

FIGURE 10-9   Stability study of stable, marginally stable, and unstable systems showing the pole-zero distributions, impulse, and magnitude-frequency response. In addition the step and sinusoidal responses are displayed showing stable, marginally stable, and unstable behavior.

Filter Name	K	z_1	z_2	z_3	z_4	Stability
stable.arc	0.002	0	−1	$j0.5$	$−j0.5$	Stable
marg.arc	0.002	0	−1	$j0.5$	$−j0.5$	Marginal
unstab.arc	0.002	0	−1	$j0.5$	$−j0.5$	Unstable

Filter Name	p_1	p_2	p_3	p_4	p_5	Stability
stable.arc	$5.+j.5$	$.5−j.5$	$.7+j.7$	$.7−j.7$	.9	Stable
marg.arc	$.5+j.5$	$.5−j.5$	$.707+j.707$	$.707−j.709$	.9	Marginal
unstab.arc	$.35+j.53$	$.35−j.53$	$.767+j.766$	$.767−j.776$	1.07	Unstable

The equivalent transfer functions are given by[3]

$$H_{\text{stable}} = \frac{0.002(z^4 + z^3 + 0.25z^2 + 0.25z)}{z^5 − 3.3z^4 + 5.04z^3 − 4.272z^2 + +2.002z − 0.441}$$

$$H_{\text{marginal}} = \frac{0.002(z^4 + z^3 + 0.25z^2 + 0.25z)}{z^5 − 3.314z^4 + 5.086z^3 − 4.329z^2 + 2.036z − 0.45}$$

$$H_{\text{unstable}} = \frac{0.002(z^4 + z^3 + 0.25z^2 + 0.25z)}{z^5 − 3.304z^4 + 4.886z^3 − 3.869z^2 + 1.572z − 0.308}$$

Computer Study   S-file ch10_6.s computes the forced response of each system for a test signal $x[n]$ of length n. The format is given by stable(x,n). The stability of the three systems can be tested using as inputs, $x[n] = u[n]$ and $x[n] = \sin(2\pi n/8)u[n]$.

SIGLAB	MATLAB
>include "ch10_6.s"	not available in Matlab

The results are displayed in Figure 10-9. The stability, marginal stability, and instability of the filters can be observed. Note the impulse response of the stable system decays, persists for the marginally stable system, and diverges for the unstable system. The unstable system is seen to have a divergent inhomogeneous response, while the marginally stable system becomes unstable when driven at its critical frequency (defined by the pole located at $\pm 2\pi/8$ radians).

10-4   DISCRETE-TIME STEADY STATE RESPONSE

It was previously noted that the forced response of a stable LSI system can be divided into transient and steady-state regimes. If the system input is a one-sided

---

[3]For users of MONARCH, the filters can be implemented using using the TRANSFER option.

discrete-time periodic process, then the steady-state response will likewise be periodic, possibly with an altered amplitude and phase. The verification of this claim follows the same line of reasoning used in the study of continuous-time systems.

The steady-state frequency response of a discrete-time LSI system is given by

$$|H(e^{j\omega})| = |H(z)|_{z=e^{j\omega}} \quad \text{Magnitude response} \qquad (10\text{-}23)$$

$$\phi(e^{j\omega}) = \arg(H(e^{j\omega})) \arctan \frac{\text{Im}[H(e^{j\omega})]}{\text{Re}[H(e^{j\omega})]} \quad \text{Phase response} \quad (10\text{-}24)$$

where $\omega\varepsilon[-\pi, \pi]$. The magnitude frequency response is also called the system gain to a one-sided periodic input of the form $x[n] = \cos(\omega_o n)u[n] = \cos(\omega_0 n T_s)u[n]$, at steady state. The phase angle $\phi(\exp(j\omega_0))$ represents the amount of phase shift imparted by the discrete-time LSI system to $x[n]$ at steady state. If $\phi(\exp(j\omega_0))$ is positive, the system is called a *lead system*. If $\phi(\exp(j\omega_0))$ is negative, the system is called *lag system*.

Assume that a discrete-time LSI system has a transfer function representation given by

$$H(z) = K \frac{\prod_{i=1}^{N} z - z_i}{\prod_{i=1}^{N} z - p_i} \qquad (10\text{-}25)$$

where z_i and p_i are, in general, complex numbers. Evaluating $H(z)$ along the contour $z = \exp(j\omega T_s)$ produces the system's steady-state response to a sinusoidal input of frequency ω. In particular, $H(\exp(j\omega T_s))$ satisfies

$$H(e^{j\omega T_s}) = K \frac{\prod_{i=1}^{N} e^{j\omega T_s} - z_i}{\prod_{i=1}^{N} e^{j\omega T_s} - p_i} = K \frac{\prod_{i=1}^{N} \alpha_i(j\omega)}{\prod_{i=1}^{N} \beta_i(j\omega)} \qquad (10\text{-}26)$$

where

$$\alpha_i = |\alpha_i(j\omega)|e^{j\theta_i}$$

$$(10\text{-}27)$$

$$\beta_i = |\beta_i(j\omega)|e^{j\phi_i}$$

Then

$$|H(e^{j\omega T_s})| = K \frac{\prod_{i=1}^{N} |\alpha_i(j\omega)|}{\prod_{i=1}^{N} |\beta_i(j\omega)|} \qquad (10\text{-}28)$$

$$\arg(H(e^{j\omega T_s})) = \sum_{i=1}^{N} \theta_i - \sum_{i=1}^{N} \phi_i + (0 \text{ if } K > 0 \text{ and } \pi \text{ if } K < 0) \quad (10\text{-}29)$$

The values of $|H(\exp(j\omega T_s))|$ and $\arg(H(\exp(j\omega T_s)))$ can be graphically determined, as suggested in Figure 10-10. The parameters $|H(\exp(j\omega T_s))|$ and $\arg(H(\exp(j\omega T_s)))$ can be determined by first locating the point on the periphery of the unit circle corresponding to $z_0 = \exp(j\omega T_s)$. By connecting all the

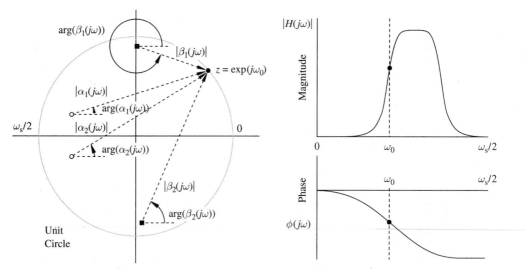

FIGURE 10-10   Graphical interpretation of an LSI system's steady-state response.

poles and zeros to the point z_0, and then measuring the length of the rays and their angles, the data needed to evaluate Equation 10-29 can be graphically produced. Equation 10-29 also establishes a qualitative relationship between the value of $|H(\exp(j\omega T_s))|$ and the proximity of z_i or p_i to a point z_0 residing on the periphery of the unit circle. It is apparent that as z_0 approaches z_i, the value of $|H(\exp(j\omega T_s))|$ will rapidly approach zero. As z_0 approaches p_i, the value of $|H(\exp(j\omega T_s))|$ will correspondingly move toward infinity. Therefore, if a system has a given target-frequency response, then its poles should be placed in regions where the gain is to have a high value and zeros placed where high attenuation is desired. In between, poles and zeros will interact to produce a gain residing somewhere between these two extremes. This technique is used by experienced designers to manipulate the frequency response of a discrete-time LSI system in order to achieve a desired effect.

Example 10-7   Manual design

A classic sixth-order lowpass Chebyshev I filter (see Chapter 12) is designed to have a 250 Hz passband at a sample frequency of 1000 Hz and has a transfer function given by

$$H(z) = K \frac{\prod_{i=1}^{6}(z - z_i)}{\prod_{i=1}^{6}(z - p_i)}$$

where

$$K = 0.009631124$$

$$z_1 = z_2 = \cdots = z_6 = -1.0$$

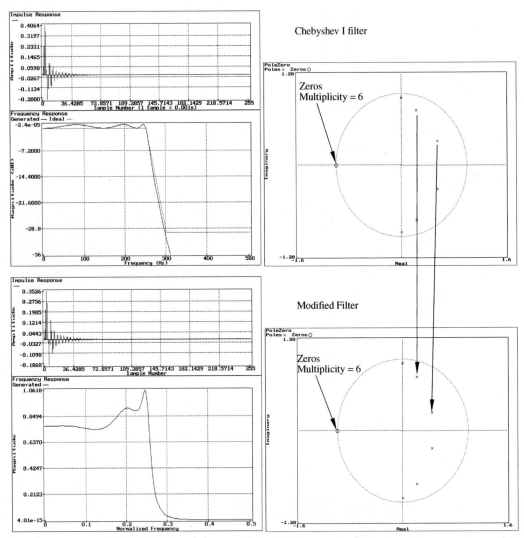

FIGURE 10-11 Sixth-order Chebyshev I filter (top) showing impulse- and magnitude-frequency responses and the pole-zero distribution, the modified Chebyshev I Lowpass Filter response (bottom), and pole-zero distribution.

$$p_1 = 0.004381697 + j0.939354 = p_2^*$$

$$p_3 = 0.2330879 + j0.766518 = p_4^*$$

$$p_5 = 0.5509034 + j0.335068 = p_6^*$$

The pole-zero distribution is shown in Figure 10-11 along with the filter's response in a linear format. The passband is seen to be relatively flat.

Assume that the designer knew that the system was intended to filter signals with a slight 20% amplitude "roll-off" near the high end of the passband. Desiring a flat, or equalized, output response, the designer chooses to "pre-emphasize" the high frequency side of the passband relative to the filter's low frequency response. Two options are available. The poles controlling the high frequency support, located near $\pm\pi/2$ radians in Figure 10-11, could be moved closer to the periphery of the unit circle. Alternatively, the other poles that govern the low-frequency behavior of the filter could be moved away from the unit circle (i.e., towards the origin). If the latter approach is taken and an iterative approach is used, a filter with the slightly modified pole locations will be accepted for the final design. The modified design is given by

$$K = 0.009631124$$

$$z_1 = z_2 = \cdots = z_6 = -1.0$$

$$p_1 = 0.004381697 + j0.939354 = p_2^* \quad \text{(unchanged)}$$

$$p_3 = 0.22 + j0.75 = p_4^* \quad \text{(versus } 0.2330879 + j0.76651\text{)}$$

$$p_5 = 0.45 + j0.25 = p_6^* \quad \text{(versus } 0.5509034 + j0.33506\text{)}$$

The original Chebyshev filter is summarized in Figure 10-11. The pole-zero distribution of this filter has been modified and is shown in Figure 10-11[4]. The pre-emphasis of the passband can immediately be seen. By continuing this processes, the shape to the passband can be manipulated to achieved a desired effect.

10-5   SUMMARY

In Chapter 10 discrete-time systems were studied using transform-domain methods. As was discovered for continuous-time systems, complicated time-domain operations, such as convolution, can be reduced to simple algebraic tasks. It was shown the inhomogeneous solution can be characterized in terms of a system's transfer function. In general, the qualitative behavior of a linear system was shown to be established by a linear system's pole distribution. In addition, a connection was made between poles and eigenvalues. The stability of linear systems was shown to be definable in terms of the system's pole locations. System stability, based on pole location, was then interpreted in terms of the unit circle in the z-plane (for discrete-time systems). It was shown that if continuous- and discrete-time stable systems are forced by a one-sided periodic process, it was shown that the total solution would consist of a transient and a periodic steady-state solution. Graphical analysis and display techniques were presented to support steady-state system analysis.

---

[4]For users of MONARCH, the nominal filter is saved under filename "manual" and the modified under filename "modify."

Chapter 11 will use this information to discuss the problem of realizing a system transfer function in hardware. This study will introduce the concepts of architecture and state variables and will provide a foundation upon which systems, having prespecified attributes, can be synthesized and implemented. In Chapter 12 digital-filter synthesis procedures will be presented that put to practical use the material found in Chapters 7 through 11.

10-6 COMPUTER FILES

These computer files were used in Chapter 10.

File Name	Type	Location
Subdirectory CHAP10		
ch10_1.s	C	Example 10-1
zrtrans.m	C	Example 10-1
ch10_2.s	T	Example 10-2
ch10_6.s	C	Example 10-6
Subdirectory SFILES or MATLAB		
bode.s	C	Example 10-2
pf.s	C	Examples 10-3, 10-4, 10-5
Subdirectory SIGNALS		
EKG1.imp	D	Example 10-2
Subdirectory SYSTEMS		
aram.coe	D	Example 10-2
stable.arc	D	Example 10-6
marginal.arc	D	Example 10-6
unstable.arc	D	Example 10-6
manual.arc	D	Example 10-7
modify.arc	D	Example 10-7

Here "s" denotes an S-file, "m" an M-file, "x" either an S-file or M-file, "T" denotes a tutorial program, "D" a data file, and "C" denotes a computer-based-instruction (CBI) program. CBI programs can be altered and parameterized by the user.

10-7 PROBLEMS

10-1 Transfer function If a discrete-time at-rest system has an impulse response $h[n] = ((1/2)^n + (-1/4)^n)u[n]$, what is the system's transfer function? What is its domain of convergence? What are the initial and final values of the system's impulse response?

10-2 Domain of convergence A discrete-time LSI system $h[k] = a^k u[k]$, where $|a| < 1$, is convolved with $x[k] = u[k]$ to produce $y[k] = h[k] * x[k]$. What is $Y(z)$ and what is its region of convergence?

10-3 Impulse response Find the impulse response of the following discrete-time systems

$$H_1(z) = \frac{(z+1)^2}{z(z+0.5)(z-0.5)}$$

$$H_2(z) = \frac{(z+1)^4}{z(z+0.5)(z-0.5)}$$

$$H_3(z) = \frac{(z+1)^2}{(z-.5)(z^2+0.5z+0.25)}$$

$$H_4(z) = \frac{z^2+2z+1}{z^2-z+0.25}$$

$$H_5(z) = 1 + z^{-1} + z^{-2} + z^{-3} + z^{-4}$$

and sketch their frequency responses in formats 1 through 5 found in Section 5-6.

10-4 Stability Determine the stability of each system defined in Problem 10-3 and sketch the pole-zero distribution.

10-5 Inhomogeneous response For each of the systems defined in Problem 10-3 to determine the forced (inhomogeneous) response

$$x_1(t) = u[n]$$

$$x_2[n] = a^n u[n]; |a| < 1$$

$$x_3[n] = \cos(\pi n/16)u[n]$$

10-6 Simulation diagram Produce the simulation diagrams for the systems shown in Problem 10-3.

10-7 Bode plot Compute and sketch a straight-line approximate Bode plot for each system defined in Problem 10-3. Verify with a computer.

10-8 Stability What is the transfer function of $y[n+1]-1.1y[n] = x[n+1]-1.1x[n]$? Is this an unstable discrete-time LSI system?

10-9 Marginal stability An LSI system was said to be marginally stable if and only if $|p_k| \leq 1$ for all nonrepeated roots, and $|p_i| < 1$ if p_i is repeated. Why are there two cases?

10-10 All-pass filter Show that the following is an all-pass filter.

$$H(z) = \pm z^{-N} \frac{P(z^{-1})}{P(z)}$$

where

$$P(z) = \sum_{i=0}^{N} a_i z^{-i}$$

and a_i is real. What is the system response if $P(z) = 1.0-0.2z^{-1}-0.3z^{-2}+0.4z^{-3}$?

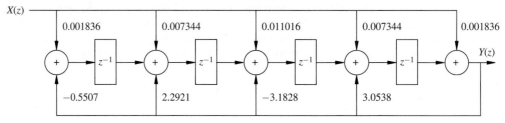

FIGURE 10-12   Fourth-order LSI system.

10-11 LSI system analysis   Consider the two discrete-time LSI systems given by

$$H_1(z) = (z+1)^2/(z - r\exp(j\pi/4))(z - r\exp(-j\pi/4)))$$

and

$$H_2(z) = (z+1)^2/(z - (1/r)\exp(j\pi/4))(z - (1/r)\exp(-j\pi/4)))$$

where $r \in (0, 1)$. Sketch the pole-zero distribution and magnitude-frequency response of each system.

10-12 LSI system analysis   Consider the discrete-time LSI system

$$y[n] = y[n-1] - 0.9y[n-2] + x[n] - 0.9x[n-1]$$

Compute the system's transfer function. Classify the system stability. What is the system's step response?

10-13 LSI system analysis   For

$$H(z) = \frac{1 - \sqrt{2}z^{-1} + z^2}{1 + k1.5z^{-1} + 0.75z^{-2}}$$

plot the pole locations that are functions of k. How is stability related to k? For $k = 1$, what is the frequency that maximizes the steady-state magnitude-frequency response?

10-14 LSI system analysis   Refer to the system shown in Figure 10-12. What is its stability classification? What is its steady-state magnitude-frequency response?

10-15 Continuous-time to discrete time conversion   A linear at-rest continuous-time system is represented by the ODE

$$\sum_{i=0}^{N} a_i \frac{d^i y(t)}{dt^i} = \sum_{i=0}^{N} b_i \frac{d^i x(t)}{dt^i}$$

The ith derivative is to be approximated by the difference equation

$$\nabla^i(q[n]) = \nabla^{i-1}(\nabla^1(q[n])$$

$$\nabla^1(q[n]) = \frac{q[n+1] - q[n]}{T_s}$$

where $q[n] = q(nT_s)$. Therefore, the ODE can be approximated by

$$\sum_{i=1}^{N} a_i \frac{\nabla^i (y[n])}{dt^i} = \sum_{i=1}^{N} b_i \frac{\nabla^i (x([n])}{dt^i}$$

If $dy(t)/dt = -y(t) + x(t)$, what are the step responses of the continuous- and discrete-time systems for $T_s = \{1, 10^{-1}, 10^{-2}\}$. How do the step responses compare?

10-16 Graphical response Graphically compute and sketch the response $H(e^{j\omega})$ for

$$H(z) = \frac{z^2}{z^2 - 1.0z + 0.8}$$

10-17 Discrete-time oscillator A discrete-time LSI given by $y[n] = 2\cos(\omega_0)y[n-1] - y[n-2] + x[n] - \cos(\omega_0)x[n-1]$ is offered as a possible sinusoidal signal generator. What is the system's transfer function? Discuss how the initial conditions can be used to create a homogeneous signal $y[n] = \cos(\omega_0 n + \phi)$. Classify the stability of the system.

10-18 Discrete-time oscillator Consider again Problem 10-17. Discuss the hardware implementation of the homogeneous signal generator. Sketch its simulation diagram. Simulate the fixed-point system performance if an eight fractional bit technology is used. How does fixed-point arithmetic affect the signal generator?

10-19 ARMA model Consider the ARMA model of an EKG signal generator developed in Example 10-5. What is the impulse response of the generator if the system is initially at rest? Compare the resulting time-series with the original EKG signal. How well do they correlate?

10-20 ARMA model Manually manipulate the pole-zero distribution of the ARMA model developed in Example 10-2 to make the low-frequency behavior of the model more accurately correspond to the measured-frequency response.

10-21 Manual design Redistribute the poles and zeros of discrete-time LSI studied in Example 10-7 to produce a filter having a nearly linear magnitude-frequency response over the frequency range $f \in [0.0.25]f_s$ with the passband gains restricted to $|H(e^{j0})| = 0.707$ and $|H(e^{j0.5\pi f_s})| = 1.1$.

10-22 Manual design Construct a seventh-order stable discrete-time LSI system, having a sample rate of $f_s = 10^4$ Hz, which approximates the magnitude-frequency response shown in Figure 10-13.

FIGURE 10-13 Desired magnitude-frequency response.

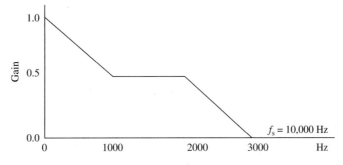

10-23 Modified design Suppose the poles and zeros of the stable filter are located at $r_i \exp(j\theta_i)$. What would the magnitude-frequency response of the system be if the poles and zeros were replaced by ones located at $(1/r_i) \exp(j\theta_i)$?

10-8 PROJECTS

10-1 ARMA modeling The ARMA model of an EKG signal was developed in Example 10-2. The system was represented in transfer-function form. Conduct a set of experiments to validate the model. Use white noise to produce an output $y'[n]$, as shown in Figure 10-4. Compare the experimentally produced $y'[n]$ and its DFT $Y'[k]$, with the original EKG time-series and spectrum.

The system is to be implemented using a 16-bit DSP microprocessor. Conduct experiments to determine how well the system will process the original EKG signal if it uses 15, 12, 8, and 4-bits of fractional precision.

Add up to two additional poles to improve the frequency response of the ARMA model about the EKG spectral peaks located at 0 Hz and 0.22×135 Hz in Figure 10-5. Display your results. Repeat the first part of the problem using your modified model.

10-2 Digital filter In Chapter 12 digital filters will be introduced. Refer to Example 12-16 in which a sixth-order Butterworth filter is designed. Using its pole-zero distribution, graphically produce the filter's magnitude-frequency response. Compare to the computer-generated response.

Using a Heaviside expansion, derive the filter's response to $x[n] = \cos(\omega_0 n)$, where $0 \leq \omega_0 < \pi$, for selected values of ω_0. Determine the transient and steady-state responses. How does the steady-state response compare to the graphically predicted response?

Devise a test that can be applied to the system to directly measure the system's frequency response. Conduct the experiment and compare the result to that achieved by analytical predictions.

11

DISCRETE-TIME SYSTEM ARCHITECTURE

I don't want to insist on it, Dave, but I am incapable of making an error.

—Dialogue from *2001*, Arthur C. Clarke

11-1 SYSTEM ARCHITECTURE

In the previous chapters, discrete-time systems were described in terms of difference equations and transfer functions. These systems, once designed, can be implemented in hardware, software, or both. For digital hardware systems, the information is stored in shift-registers and memory. In software, the information is stored in arrays or as named variables.

Continuous-time system architecture was studied in Chapter 6. The study of system architecture becomes especially important when discrete-time or digital systems are to be developed. The performance of a digital system, in particular, is often highly predicated on the choice of architecture. This can be attributed to the fact that computer (digital) arithmetic is approximate, therefore introduces errors into the system. Furthermore, latency and complexity are functions of the number of shift registers and arithmetic operations associated with a specific architecture.

The three most important architecture representation forms for discrete-time systems are block diagrams, signal-flow graphs, and state variables. These tools and methods will now be developed.

11-2 BLOCK DIAGRAMS AND SIGNAL FLOW GRAPHS

Discrete-time systems can be represented in *block diagram* or *signal-flow graph* form. Signal-flow and block diagram representations of discrete-time systems are based on the methodology developed for continuous-time systems (see Chapter 6). The fundamental difference is that the Laplace transform operator "*s*" is replaced by the Z-transform operator "*z*." As before, Mason's Gain Formula can be used to reduce the signal-flow diagram of a discrete-time LSI system into a transfer function.

11-3 MASON'S GAIN FORMULA

Mason's Gain Formula   Mason's Gain Formula for discrete-time systems is given by

$$H(z) = \sum_k \frac{M_k(z)\Delta_k(z)}{\Delta(z)} \tag{11-1}$$

where the path gain of the $k = k$th feedforward path is $M_k(z)$, $\Delta(z)$ is called the *characteristic equation* of $H(z)$, and is given by

$$\Delta(z) = 1 - \left[\sum \text{ gains from all individual loops } (r = 1)\right]$$

$$+ \left[\sum \text{gains of all possible combinations of two nontouching loops}\right.$$

$$\left. (r = 2)\right] \pm \dots$$

or formally

$$\Delta(z) = 1 - \sum_m P_{m1}(z) + \sum_m P_{m2}(z) - \sum_m P_{m3}(z) + \dots \tag{11-2}$$

where $P_{mr}(z) = $ gain of the m possible combinations of r nontouching loops (i.e., no common nodes), and $\Delta_k(z) = $ value of Equation 11-2 for the part of the graph not touching the kth forward path.

These relationships are identical to those developed for the continuous-time case. Example 11-1 will demonstrate this.

Example 11-1   Computer-generated discrete-time transfer function

A digital filter possesses a transfer function $H(z)$ that can be expressed in signal-flow diagram form only after it has been reduced to an architecture. Once in signal-flow graph form, Mason's Gain Formula can be used to convert the signal-flow graph back into transfer-function form.

Computer Study   A third-order lowpass filter is given by[1]

$$H(z) = \frac{1 + 3z^{-1} + 3z^{-2} + z^{-3}}{1 - 1.75914404524z^{-1} + 1.18199399636z^{-2} - 0.277785502803z^{-3}}$$

[1] Users of MONARCH will find $H(z)$ saved under filename "Cascade."

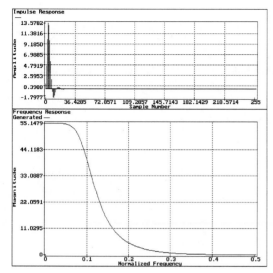

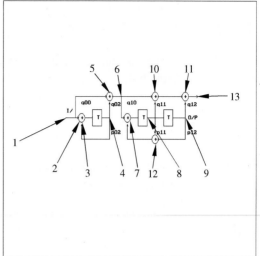

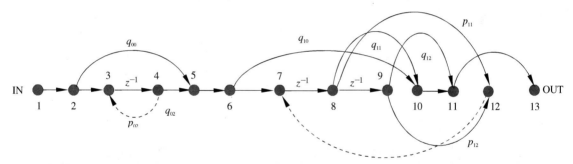

FIGURE 11-1   Third-order example showing system schematic, indexed node locations, and signal flow graph.

The filter was then implemented using what is called a cascade architecture. The filter schematic and node indexing are shown in Figure 11-1. The coefficients are $q_{00} = q_{10} = 1.0$, $q_{11} = 3.249912$, $q_{02} = 1.509232$, $q_{12} = 0.45450154$, $p_{11} = 1.249912$, $p_{02} = 0.5092324$, and $p_{12} = -0.5454985$. The system, having a signal-flow graph representation shown in Figure 11-1, can be analyzed using Mason's Gain Formula.

Unity Gain Paths
$1 = [1, 2]$, $1 = [2, 3]$, $1 = [2, 5]$, $1 = [5, 6]$, $1 = [6, 7]$, $1 = [6, 10]$, $1 = [10, 11]$, $1 = [11, 13]$, $1 = [12, 7]$

Delay Paths
$z^{-1} = [3, 4]$, $z^{-1} = [7, 8]$, $z^{-1} = [8, 9]$

Nonunity Paths
$1.509232 = [4, 5]$, $0.509232 = [4, 3]$, $3.249912 = [8, 10]$, $1.249912 = [8, 12]$,
$-0.5454985 = [9, 12]$, $0.454501 = [9, 11]$

The transfer function is produced using program `sfg` as introduced in Example 6-4.

SIGLAB	MATLAB
>path\sfg path\cas.sfg	Not available in Matlab

The resulting transfer function is computed and displayed as follows, and agrees with the known result.

SIGLAB	MATLAB
echo print of branch and node data messages generated by program execution loops and paths in matrix form	Not applicable

```
Signal flow graph transfer function
------------------------------------
Numerator Polynomial Coefficients
1; 3; 3; 1
Denominator Polynomial Coefficients
-0.277786; 1.18199; -1.75914; 1
------------------------------------
```

It can be seen that block and signal-flow diagrams can be used to show how subsystems exchange and internally manage information. Later in this chapter this concept will be directly related to the concept of architecture. Another efficient means of representing information is based on the *state-variable method*.

11-4 DISCRETE-TIME STATE DETERMINED SYSTEMS

A discrete-time system can also be modeled in state-variable form. In general, a multiple-input multiple-output discrete-time system, consisting of p-inputs, r-outputs, and n-states, has a state-variable representation given by

$$\mathbf{x}[k + 1] = \mathbf{A}[k]\mathbf{x}[k] + \mathbf{B}[k]\mathbf{u}[k] \quad \mathbf{x}[0] = \mathbf{x}_0 \quad \text{State equation} \quad (11\text{-}3)$$

$$\mathbf{y}[k] = \mathbf{C}^T[k]\mathbf{x}[k] + \mathbf{D}[k]\mathbf{u}[k] \quad \text{Observation (output) equation} \quad (11\text{-}4)$$

where $\mathbf{A}[k]$ is an $n \times n$ matrix of coefficients, $\mathbf{B}[k]$ is an $n \times p$ matrix, $\mathbf{C}[k]$ is an $n \times r$ matrix, and $\mathbf{D}[k]$ is a $r \times p$ matrix, $\mathbf{u}[k]$ is an arbitrary $p \times 1$ input vector, $\mathbf{x}[k]$ is a $n \times 1$ state vector, and $\mathbf{y}[k]$ is an $r \times 1$ output vector. Such a system can also be represented in four-tuple form as $S = [\mathbf{A}[k], \mathbf{B}[k], \mathbf{C}[k], \mathbf{D}[k]]$. If the

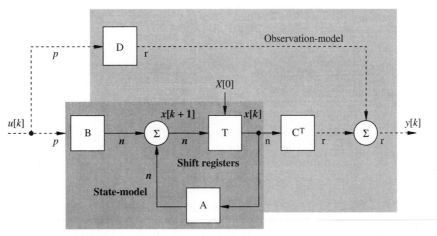

FIGURE 11-2   Discrete state-variable system model.

discrete-time system is also a discrete-time LSI system, then the state four-tuple entries are constant coefficients and would have the form $S = [\mathbf{A}, \mathbf{B}, \mathbf{C}, \mathbf{D}]$.

The discrete-time LSI state-determined system, based on Equations 11-3 and 11-4, is graphically interpreted in Figure 11-2. In discrete-time systems, state variables serve the same function as they did in continuous-time systems, namely that of information management. Discrete-time systems store information, thus state variables, in the memory and/or shift registers. Therefore, the system shown in Figure 11-2 contains a block of n shift registers.

Many important discrete-time systems are also single-input single-output systems that can be modeled using a monic nth-order difference equation of the form

$$y[k]+a_1 y[k-1]+\cdots+a_n y[k-n] = b_0 u[k]+b_1 u[k-1]+\ldots+b_n u[k-n] \quad (11\text{-}5)$$

In Chapter 10, the discrete-time LSI system, defined by Equation 11-5, was expressed in terms of a transfer function $H(z)$, where

$$H(z) = \frac{b_0 + b_1 z^{-1} + \cdots + b_n z^{-n}}{1 + a_1 z^{-1} + a_2 z^{-2} + \cdots + a_n z^{-n}}$$

$$= b_0 + \frac{(b_1 - b_0 a_1)z^{-1} + \cdots + (b_n - b_0 a_n)z^{-n}}{1 + a_1 z^{-1} + a_2 z^{-2} + \cdots + a_n z^{-n}} \quad (11\text{-}6)$$

$$= b_0 + \frac{c_1 z^{-1} + \cdots + c_n z^{-n}}{1 + a_1 z^{-1} + a_2 z^{-2} + \cdots + a_n z^{-n}}$$

Procedures exist by which the system defined by Equation 11-5 or 11-6 can be placed in state-variable form. One method results in what is called a *direct II* system model (called extended-companion form for continuous-time systems) with a state model given by

$$x_1[k + 1] = x_2[k]$$

$$x_2[k + 1] = x_3[k]$$

$$\vdots \quad = \quad \vdots \qquad\qquad (11\text{-}7)$$

$$x_{n-1}[k + 1] = x_n[k]$$

$$x_n[k + 1] = -a_n x_1(k) - a_{n-1} x_2(k) - \cdots - a_1 x_n[k] + u[k]$$

The observation (output) equation is given by

$$y[k] = (b_n - b_0 a_n) x_1[k] + (b_{n-1} - b_0 a_{n-1}) x_2[k] + \cdots + (b_1 - b_0 a_1) x_n[k] + b_0 u[k]$$
$$(11\text{-}8)$$

The state-variable representation of these equations is

$$\mathbf{x}[k] = \begin{pmatrix} x_1[k] \\ x_2[k] \\ \cdots \\ x_n[k] \end{pmatrix} \qquad \text{State vector}$$

$$\mathbf{A} = \begin{pmatrix} 0 & 1 & 0 & \cdots & 0 \\ 0 & 0 & 1 & \cdots & 0 \\ 0 & 0 & 0 & \cdots & 0 \\ \vdots & \vdots & \vdots & \ddots & \vdots \\ 0 & 0 & 0 & \cdots & 1 \\ -a_n & -a_{n-1} & -a_{n-2} & & -a_1 \end{pmatrix}$$

$$\mathbf{b}^T = (0 \quad 0 \quad 0 \quad \cdots \quad 0 \quad 1) \qquad\qquad (11\text{-}9)$$

$$\mathbf{c}^T = (b_n - b_0 a_n \quad b_{n-1} - b_0 a_{n-1} \quad \cdots \quad b_1 - b_0 a_1)$$

$$d = (b_0)$$

This is just one of many architectural manifestations a discrete-time LSI system may take. However, now it is only necessary to know that the discrete-time LSI system, defined by Equation 11-5 or 11-6, can be directly mapped into the discrete-time state-variable model given in Equation 11-9.

Example 11-2   High-order discrete-time system

Many digital filters are specified in terms of high-order transfer functions. An eighth-order digital filter, having the frequency response shown in

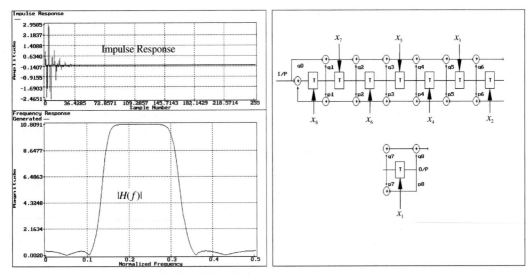

FIGURE 11-3   High-order digital filter reduced to state-variable form: (left) filter response, and (right) direct II architecture showing state assignments.

Figure 11-3, has a transfer function given by[2]

$$H(z) = \frac{1 - 0.44z^{-1} - 1.36z^{-2} + 0.04z^{-3} + 2.04z^{-4} + 0.04z^{-5}}{1 - 0.88z^{-1} + 1.35z^{-2} - 0.91z^{-3} + 1.12z^{-4} - 0.46z^{-5}}$$

$$\frac{-1.36z^{-6} - 0.446z^{-7} + z^{-8}}{+0.30z^{-6} - 0.078z^{-7} + 0.049z^{-8}}$$

The system can be directly mapped into a state-variable model using Equation 11-9, thereby placing it in a direct II form that possesses the architecture shown in Figure 11-3. The filter is seen to consist of eight shift registers and a set of feedforward and feedback paths. The state variables reside in the shift registers that are indexed as shown in Figure 11-3. The state variable model $\mathcal{S} = [\mathbf{A}, \mathbf{b}, \mathbf{c}, d]$, in this case, is given by

$$\mathbf{A} = \begin{pmatrix} 0 & 1 & 0 & 0 & 0 & 0 & 0 & 0 \\ 0 & 0 & 1 & 0 & 0 & 0 & 0 & 0 \\ 0 & 0 & 0 & 1 & 0 & 0 & 0 & 0 \\ 0 & 0 & 0 & 0 & 1 & 0 & 0 & 0 \\ 0 & 0 & 0 & 0 & 0 & 1 & 0 & 0 \\ 0 & 0 & 0 & 0 & 0 & 0 & 1 & 0 \\ 0 & 0 & 0 & 0 & 0 & 0 & 0 & 1 \\ -0.049 & 0.078 & -0.30 & 0.46 & -1.12 & 0.91 & -1.35 & 0.88 \end{pmatrix}$$

[2]Users of MONARCH will find $H(z)$ saved under filename "HighOrd."

$$\mathbf{b}^T = (0 \quad 0 \quad 0 \quad 0 \quad 0 \quad 0 \quad 0 \quad 1)$$

$$\mathbf{c}^T = (0.95 \quad -0.36 \quad -1.67 \quad 0.50 \quad 0.92 \quad 0.95 \quad -2.71 \quad 0.44)$$

$$d = 1$$

11-5   STATE-TRANSITION MATRIX

The forms taken by the state-determined homogeneous solution (unforced) and inhomogeneous solution (forced) bear a strong similarity to their continuous-time counterparts. The solution to the homogeneous state equation is given by $\mathbf{x}[k + 1] = \mathbf{A}[k]\mathbf{x}[k]$, $\mathbf{x}[0] = \mathbf{x}_0$, where $\mathbf{A}[k]$ is, in general, a matrix of time-varying coefficients. The homogeneous solution is given by $\mathbf{x}[k] = \Phi(k, 0)\mathbf{x}_0$, where $\Phi(k + 1, 0) = \mathbf{A}[k]\Phi(k, 0)$. The matrix $\Phi(k, m)$ is called the discrete-time state-transition matrix. The state-transition matrix satisfies a number of important properties, including

$$\begin{array}{lll} \Phi(k, m) = \Phi(k, j)\Phi(j, m) & \text{Transition property} \\ \Phi(k, m) = \Phi^{-1}(m, k) & \text{Inversion property} \end{array} \qquad (11\text{-}10)$$

which are graphically interpreted in Figure 11-4.

If $\mathbf{A}$ is a constant coefficient matrix, then a discrete-time LSI system results and the state-transition matrix reduces to

$$\mathbf{x}[1] = \mathbf{A}\mathbf{x}_0$$

$$\mathbf{x}[2] = \mathbf{A}\mathbf{x}[1] = \mathbf{A}^2\mathbf{x}_0$$

$$\mathbf{x}[3] = \mathbf{A}\mathbf{x}[2] = \mathbf{A}^2\mathbf{x}[1] = \mathbf{A}^3\mathbf{x}_0 \qquad (11\text{-}11)$$

$$\cdots = \cdots$$

$$\mathbf{x}[k] = \mathbf{A}\mathbf{x}[k - 1] = \mathbf{A}^2\mathbf{x}[k - 2] = \ldots = \mathbf{A}^k\mathbf{x}_0$$

The discrete-time transition matrix, in this case, can be expressed as $\Phi(k, 0) = \Phi(k) = \mathbf{A}^k$. Equation 11-11 is useful as a computational tool only if used over

FIGURE 11-4   Properties of the state-transition matrix.

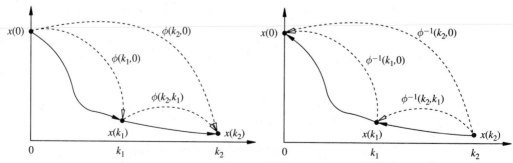

a limited sampling range. If knowledge of $\Phi[k]$ is required over a wide range of sample indices, then other computational methods will be required.

Example 11-3   Homogeneous response

The rabbits of Fibonacci have developed a passion for Earth technology. They have been having trouble keeping track of their population back on their home planet and hope that state variables will provide the answer. The Fibonacci sequence, given in Equation 1-9, is modeled as a homogeneous state-determined system as follows. Let $x_1[k] = F[k-1]$, $x_2[k] = F[k]$, and $y[k] = x_1[k] + x_2[k] = F[k] + F[k-1]$. The initial conditions are given by $\mathbf{x}_0 = [0, 1]^T$. The state-variable four tuple is therefore given by

$$\mathbf{A} = \begin{pmatrix} 0 & 1 \\ 1 & 1 \end{pmatrix}$$

$$\mathbf{b}^T = (0 \quad 0)$$

$$\mathbf{c}^T = (1 \quad 1)$$

$$d = (0)$$

The homogeneous state response is given by

$$\mathbf{x}[k] = \mathbf{A}^k \mathbf{x}_0$$

and the observed output, which is the rabbit population, satisfies

$$y[k] = (1, 1)\mathbf{x}[k]$$

Computer Study   The homogeneous state and output responses, over the first six mating periods, can be computed in the following manner, which agrees with the known result.

SIGLAB	MATLAB
| ```
>a={[0,1],[1,1]}; c=[1,1]
>x0=[0,1]'; xnext=x0
>for (i=0:5)
>>y=c*a*xnext
>>xnext=a*xnext; y
>>end
2; 3; 5; 8; 13; 21
``` | ```
>>a=[0,1;1,1]; c=[1,1];
>>x0=[0,1]'; xnext=x0;
>>for (i=0:5); y=c*a*xnext;
  xnext=a*xnext; y
  end
y=
      2
      3
      5
      8
     13
     21
``` |

The computation of the discrete-time state-transition matrix can be modeled after the techniques developed to compute the continuous-time state transition matrix. As was the case for a continuous-time system, the Cayley-Hamilton Method can be used to produce a discrete-time state transition matrix. The study of discrete-time state-determined systems requires that $\mathbf{A}^k$ be computed. To achieve this, refer to Equation 6-27 and replace $f(\lambda) = \exp(\lambda t)$ with $f(\lambda) = \lambda^k$. Upon making this substitution, and assuming for the time being that the $n \times n$ matrix $\mathbf{A}$ has distinct eigenvalues, it immediately follows that

$$\Phi[k] = \mathbf{A}^k = \alpha_0 \mathbf{I} + \alpha_1 \mathbf{A} + \alpha_2 \mathbf{A}^2 + \cdots + \alpha_{n-1} \mathbf{A}^{n-1} \tag{11-12}$$

Since the eigenvalues of $\mathbf{A}$ are assumed distinct, that is $\lambda = \lambda_1, \ldots, \lambda_n$ and $\lambda_i \neq \lambda_j$ for $i \neq j$, it then follows that

$$\lambda_1^k = \alpha_0 + \alpha_1 \lambda_1 + \alpha_2 \lambda_1^2 + \cdots + \alpha_{n-1} \lambda_1^{n-1}$$

$$\lambda_2^k = \alpha_0 + \alpha_1 \lambda_2 + \alpha_2 \lambda_2^2 + \cdots + \alpha_{n-1} \lambda_2^{n-1}$$

$$\vdots \quad = \quad \vdots \tag{11-13}$$

$$\lambda_n^k = \alpha_0 + \alpha_1 \lambda_n + \alpha_2 \lambda_n^2 + \cdots + \alpha_{n-1} \lambda_n^{n-1}$$

After solving this system of n-equations for the n unknowns $\{\alpha_i\}$, and upon substituting this result into Equation 11-12, the discrete-time state-transition matrix can be explicitly computed.

Example 11-4   Discrete-time state transition matrix

A discrete-time system where $\mathbf{A}$ is given by

$$\mathbf{A} = \begin{pmatrix} 0 & 1 \\ 1/8 & -1/4 \end{pmatrix}$$

produces a characteristic equation

$$\det(\lambda \mathbf{I} - \mathbf{A}) = (\lambda + 1/2)(\lambda - 1/4)$$

which is seen to have eigenvalues of $\lambda = -1/2$ and $1/4$. The Cayley-Hamilton Method states that

$$\mathbf{A}^k = \alpha_0 \mathbf{I} + \alpha_1 \mathbf{A}$$

where the coefficients α_0 and α_1 are computed as follows

$$(-1/2)^k = \alpha_0 + (-1/2)\alpha_1$$
$$(1/4)^k = \alpha_0 + (1/4)\alpha_1$$

which translates to

$$\alpha_0 = (2/3)(1/4)^k + (1/3)(-1/2)^k$$

$$\alpha_1 = (4/3)(1/4)^k - (4/3)(-1/2)^k$$

Finally, upon substituting this result in the generating equation for $\mathbf{A}^k$, the state-transition matrix can be explicitly represented as

$$\Phi[k] = \mathbf{A}^k = \begin{pmatrix} (2/3)(1/4)^k + (1/3)(-1/2)^k & (4/3)(1/4)^k - (4/3)(-1/2)^k \\ (1/6)(1/4)^k - (1/6)(-1/2)^k & (1/3)(1/4)^k + (2/3)(-1/2)^k \end{pmatrix}$$

As a check, note that the value of $\Phi[0] = \mathbf{I}$.

If the eigenvalues of $\mathbf{A}$ are repeated, then the previous method must be modified. Assume that the eigenvalue λ_i occurs with a multiplicity $n(i)$. Then Equation 11-13 will contain $n(i)$ linearly dependent equations. As before, these equations can be successively differentiated to produce a linearly independent system of equations. In particular, one would form

$$\lambda_i^k = \alpha_0 + \alpha_1\lambda_i + \alpha_2\lambda_i^2 + \cdots + \alpha_{n-1}\lambda_i^{n-1}$$

$$\frac{d}{d\lambda}\lambda_i^k = k\lambda_i^{k-1} = \alpha_1 + 2\alpha_2\lambda_i + \cdots + (n-1)\alpha_{n-1}\lambda_i^{n-2}$$

$$\vdots \quad = \quad \vdots \qquad\qquad\qquad\qquad (11\text{-}14)$$

$$\frac{d^{n(i)-1}}{d\lambda^{n(i)-1}}\lambda_i^k = \cdots + (n-1)(n-2)\cdots(n-n(i)+1)\alpha_{n-1}\lambda_i^{n-n(i)}$$

from which the coefficients $\{\alpha_i\}$ can be computed.

Example 11-5   Repeated eigenvalues

A discrete-time system where $\mathbf{A}$ is given by

$$\mathbf{A} = \begin{pmatrix} 1/2 & 1 & 0 \\ 0 & 1/2 & 1 \\ 0 & 0 & 1/2 \end{pmatrix}$$

$$\det(\lambda\mathbf{I} - \mathbf{A}) = (\lambda - 1/2)^3$$

is seen to have an eigenvalue of $\lambda = 1/2$ of multiplicity $n[1] = 3$. The Cayley-Hamilton Method states that

$$\mathbf{A}^k = \alpha_0\mathbf{I} + \alpha_1\mathbf{A} + \alpha_2\mathbf{A}^2$$

where the coefficients α_0, α_1, and α_2 are computed as follows

$$\alpha_0 + (1/2)\alpha_1 + (1/2)^2\alpha_2 = (1/2)^k$$
$$\alpha_1 + 2(1/2)\alpha_2 = k(1/2)^{k-1}$$
$$2\alpha_2 = k(k-1)(1/2)^{k-2}$$

which translates to

$$\alpha_0 = (1/2)^k \left(1 - k + \frac{k(k-1)}{2}\right)$$

$$\alpha_1 = (1/2)^{k-1}(2k - k^2)$$

$$\alpha_2 = (1/2)^{k-1}(-k + k^2)$$

As a simple check, consider computing the values of $\mathbf{A}^k$ for $k = \{0, 1, 2\}$. The results are summarized here and are seen to be correct.

k	α_0	α_1	α_2	$\mathbf{A}^k$
0	1	0	0	$\mathbf{I}$
1	0	1	0	$\mathbf{A}$
2	0	0	1	$\mathbf{A}^2$

Finally, the state-transition matrix is computed to be $\Phi[k] = \mathbf{A}^k = \alpha_0\mathbf{I} + \alpha_1\mathbf{A} + \alpha_2\mathbf{A}^2$, or

$$\Phi[k] = \mathbf{A}^k = \begin{pmatrix} \phi_{11} & \phi_{12} & \phi_{13} \\ \phi_{21} & \phi_{22} & \phi_{23} \\ \phi_{31} & \phi_{32} & \phi_{33} \end{pmatrix}$$

$$= \begin{pmatrix} \alpha_0 + \alpha_1/2 + \alpha_2/4 & \alpha_1 + \alpha_2 & \alpha_2 \\ 0 & \alpha_0 + \alpha_1/2 + \alpha_2/4 & \alpha_1 + \alpha_2 \\ 0 & 0 & \alpha_0 + \alpha_1/2 + \alpha_2/4 \end{pmatrix}$$

Observe that

$$\phi_{11} = \phi_{22} = \phi_{33} = 1 \quad \text{if } k = 0 \text{ and } (1/2)^k \text{ otherwise}$$

$$\phi_{12} = \phi_{23} = 0 \quad \text{if } k = 0 \text{ and } k(1/2)^{k-1} \text{ otherwise}$$

$$\phi_{13} = 0 \quad \text{if } k = 0 \text{ and } (k^2 - k)(1/2)^{k-1} \text{ otherwise}$$

As a check, note that the value of $\Phi[0] = \mathbf{I}$.

Transform methods can also be used to compute the state-transition matrix. In particular, the Z-transform of the *homogeneous-state equation* $\mathbf{x}[k + 1] = \mathbf{A}\mathbf{x}[k]$; $\mathbf{x}[0] = \mathbf{x}_0$ is given by

$$z(\mathbf{X}(z) - \mathbf{x}_0) = \mathbf{A}\mathbf{X}(z) \tag{11-15}$$

$$z\mathbf{X}(z) = \mathbf{A}\mathbf{X}(z) + z\mathbf{x}_0 \Rightarrow \mathbf{X}(z) = z(z\mathbf{I} - \mathbf{A})^{-1}\mathbf{x}_0 \tag{11-16}$$

Therefore, it immediately follows that for $k \geq 0$

$$\Phi[k] = \mathbf{A}^k = Z^{-1}(z(z\mathbf{I} - \mathbf{A})^{-1}) \tag{11-17}$$

Example 11-6   Homogeneous state-transition solution using a Z-transform

Consider again the discrete-time system with a constant coefficient (plant) matrix $\mathbf{A}$ given by

$$\mathbf{A} = \begin{pmatrix} 0 & 1 \\ 1/8 & -1/4 \end{pmatrix}$$

$$\det(\lambda \mathbf{I} - \mathbf{A}) = (\lambda + 1/2)(\lambda - 1/4)$$

Then

$$z(z\mathbf{I} - \mathbf{A})^{-1} = \frac{z \begin{pmatrix} z + 1/4 & 1 \\ 1/8 & z \end{pmatrix}}{(z + 1/4)(z - 1/2)}$$

Upon performing a partial fraction expansion on the four individual terms of $z(z\mathbf{I} - \mathbf{A})^{-1}$, $\Phi[k]$ can be computed. Each element of the matrix $z(z\mathbf{I} - \mathbf{A})^{-1}$, namely $\Phi_{ij}(z)$, will have the form $\Phi_{ij}(z) = az/(z - 1/2) + bz/(z + 1/4)$.
Computer Study   The partial fraction expansion of $z(z\mathbf{I} - \mathbf{A})^{-1}$ can be computed using S-file `pf.s` or M-file `residue.m` as follows.

SIGLAB	MATLAB
`>d=poly([-1/2,1/4]);d`	`>>d=poly([-1/2,1/4])`
`1  .25  -.125`	`1  .25  -.125`
`>n11=[1,1/4]; n12=[1]`	`>>n11=[1,1/4]; n12=[1];`
`>n21=[1/8]; n22=[1,0]`	`>>n21=[1/8]; n22=[1,0];`

This produces the following data.

Item	a, pole @ $[-0.5 + 0j]$	b, pole @ $[0.25 + 0j]$
z11(n11,d)	1/3	2/3
z12(n12,d)	$-$4/3	4/3
z21(n21,d)	$-$1/6	1/6
z22(n22,d)	2/3	1/3

It results in

$$\Phi[k] = \mathbf{A}^k = \begin{pmatrix} (2/3)(1/4)^k + (1/3)(-1/2)^k & (4/3)(1/4)^k - (4/3)(-1/2)^k \\ (1/6)(1/4)^k - (1/6)(-1/2)^k & (1/3)(1/4)^k + (2/3)(-1/2)^k \end{pmatrix}$$

which agrees with the results presented in Example 11-4.

The previous example demonstrates that once the matrix $z(z\mathbf{I} - \mathbf{A})^{-1}$ is known, a computer can perform the partial fraction expansion of the elements of $\Phi(z)$ to produce the matrix $\Phi[k]$. It would be desirable to automate the production of $(z\mathbf{I} - \mathbf{A})^{-1}$. Unfortunately, standard matrix-inversion computer programs are designed to invert a constant coefficient matrix, and $(z\mathbf{I} - \mathbf{A})^{-1}$ is defined in terms of the symbolic operator z. There is, however, a technique that can be used to invert a matrix of this form; it is called *Leverrier's formula*.

Leverrier's Formula   Given an $n \times n$ matrix $\mathbf{A}$

$$(z\mathbf{I} - \mathbf{A})^{-1} = \frac{\text{adj}(z\mathbf{I} - \mathbf{A})}{\Delta(\mathbf{z})} \tag{11-18}$$

where $\text{adj}(z\mathbf{I} - \mathbf{A})$ denotes an adjoint matrix (see Appendix A) and $\Delta(z) = \det(z\mathbf{I} - \mathbf{A})$, then

$$\text{adj}(z\mathbf{I} - \mathbf{A}) = \mathbf{H}_1 z^{n-1} + \mathbf{H}_2 z^{n-2} + \cdots + \mathbf{H}_n \tag{11-19}$$

where

$$\mathbf{H}_1 = \mathbf{I}$$

$$d_1 = -\text{trace}(\mathbf{A}) \tag{11-20}$$

and for $k \in [2, n]$, the matrix $\mathbf{H}_k$ can be recursively computed as follows

$$\mathbf{H}_k = \mathbf{A}\mathbf{H}_{k-1} + d_{k-1}\mathbf{I}$$

$$d_k = -\frac{1}{k}\text{trace}(\mathbf{A}\mathbf{H}_k) \tag{11-21}$$

where $\text{trace}(\mathbf{A}\mathbf{H}_k)$ is equal to the sum of the on-diagonal elements of the matrix product $(\mathbf{A}\mathbf{H}_k)$.

Notice that the production of the adjoint matrix is completely specified by the elements of the matrix $\mathbf{A}$, which are known. Leverrier's formula, therefore, provides a mechanism by which the adj $(z\mathbf{I} - \mathbf{A})$ can be computed using a general-purpose digital computer. It should be apparent that the same method can be used to compute $\text{adj}(s\mathbf{I} - \mathbf{A})$, as well. This tool can, therefore, be used to automate the computation of both continuous- and discrete-time state-transition matrices. Regardless, once $\text{adj}(z\mathbf{I} - \mathbf{A})$ is computed, $\Phi(z) = z\, \text{adj}(z\mathbf{I} - \mathbf{A})/\Delta(\mathbf{z})$ follows.

Example 11-7   Leverrier's formula

Leverrier's formula is given in Equation 11-19. It can be programmed as a simple recursive routine and used in conjunction with a partial fraction expansion tool to automate the computation of a state-transition matrix.

Computer Study   Consider a discrete-time system where $\mathbf{A}$ is given by

$$\mathbf{A} = \begin{pmatrix} 14/8 & -7/8 & 1/8 \\ 1 & 0 & 0 \\ 0 & 1 & 0 \end{pmatrix}$$

Given $\mathbf{A}$, an $n \times n$ matrix, S-file `lever.s` and M.file `lever.m` can be used to return an $n \times n^2$ packed matrix containing the values of $[\mathbf{H}_1, \mathbf{H}_2, \cdots, \mathbf{H}_n]$.

SIGLAB	MATLAB
| ```
># matrix A
>A={[14/8,-7/8,1/8],[1,0,0],
[0,1,0]}
>include "lever.s"
>lev=lever(A); lev #packed data
``` | ```
>>A=[14/8,-7/8,1/8;1,0,0;
0,1,0]
>>lever(A)
``` |

This results in the coefficient display

```
1.000   0.000   0.000   0.000   -0.875   0.125   0.000   0.125   0.000
0.000   1.000   0.000   1.000   -1.75    0.000   0.000   0.000   0.125
0.000   0.000   1.000   0.000   1.000    -1.75   1.000   -1.75   0.875
```

At this point the adjoint matrix can be written as

$$\text{adj}(z\mathbf{I} - \mathbf{A}) = \begin{pmatrix} z^2 & -0.875z + 0.125 & 0.125z \\ z & z^2 - 1.75z & 0.125 \\ 1 & z - 1.75 & z^2 - 1.75z + 0.875 \end{pmatrix}$$

11-6   STATE-DETERMINED SOLUTIONS

Regardless of the method used to compute the state-transition matrix, once defined it can be used to compute the homogeneous-state and observation responses defined in Equations 11-22 and 11-23.

$$\mathbf{x}[k] = \Phi(k, 0)\mathbf{x}_0 \qquad \text{State equation} \tag{11-22}$$

$$\mathbf{y}[k] = \mathbf{c}^T \Phi(k, 0)\mathbf{x}_0 \qquad \text{Observation equation} \tag{11-23}$$

It is, however, the inhomogeneous solution that is most often sought. The inhomogeneous-state solution to the discrete-time state-determined system is given by the convolution sum

$$\mathbf{x}[k] = \sum_{i=0}^{k-1} \Phi(k - 1 - i, 0)\mathbf{bu}[i] \qquad \text{for } k > 0 \tag{11-24}$$

and it follows that the output is defined as

$$\mathbf{y}[k] = \mathbf{c}^T \sum_{i=0}^{k-1} \Phi(k-1-i, 0)\mathbf{b}u[k] + \mathbf{d}u[k] \qquad \text{for } k > 0 \qquad (11\text{-}25)$$

The total solution is the linear combination of the homogeneous and inhomogeneous solutions. In particular, the total solution is given by

$$\mathbf{y}[k] = \mathbf{c}^T \left(\Phi(k, 0)\mathbf{x}_0 + \sum_{i=0}^{k-1} \Phi(k-1-i, 0)\mathbf{b}u(i) \right) + \mathbf{d}_0\mathbf{u}[k] \qquad \text{for } k > 0$$

$$(11\text{-}26)$$

For a discrete-time LSI system, the state-transition matrix $\Phi(j, 0)$, can be replaced by $\mathbf{A}^j$. Here, for an at-rest system, Equations 11-24 and 11-25 reduce to

$$\mathbf{x}[k] = \sum_{i=0}^{k-1} \mathbf{A}^{k-1-i}\mathbf{b}u[i] \qquad \text{for } k > 0 \qquad (11\text{-}27)$$

$$\mathbf{y}[k] = \mathbf{c}^T \sum_{i=0}^{k-1} \mathbf{A}^{k-1-i}\mathbf{b}u[k] + \mathbf{d}u[i] \qquad \text{for } k > 0 \qquad (11\text{-}28)$$

Equation 11-27 can be easily verified by computing the states of an at-rest LSI system defined by $S = [\mathbf{A}, \mathbf{b}, \mathbf{c}, d]$

$$\mathbf{x}[1] = \mathbf{A}\mathbf{x}[0] + \mathbf{b}u[0] = \mathbf{b}u[0]$$

$$\mathbf{x}[2] = \mathbf{A}\mathbf{x}[1] + \mathbf{b}u[1] = \mathbf{A}\mathbf{b}u[0] + \mathbf{b}u[1]$$

$$\mathbf{x}[3] = \mathbf{A}\mathbf{x}[2] + \mathbf{b}u[2] = \mathbf{A}^2\mathbf{b}u[0] + \mathbf{A}\mathbf{b}u[1] + \mathbf{b}u[2] \qquad (11\text{-}29)$$

$$\cdots = \cdots$$

$$\mathbf{x}[k] = \mathbf{A}^{k-1}\mathbf{b}u[0] + \cdots + \mathbf{b}u[k]$$

which agrees with Equation 11-27. Assuming that Equation 11-29 can be computed with high precision, it can be used to simulate the state response of a system to an arbitrary input over a small range of k. However, as the sample index k becomes large, the accumulated numerical errors will begin to affect the accuracy of the simulation.

The analysis of a discrete-time LSI system generally focuses on producing the inhomogeneous solution to the state and observation equations. Earlier in this section, homogeneous solutions were efficiently computed using Z-transforms and the convolution theorem. The Z-transform of the *inhomogeneous-state solution* can also be directly computed by taking the Z-transform of Equation 11-3. In particular

$$\mathbf{x}[k+1] = \mathbf{A}\mathbf{x}[k] + \mathbf{b}u[k] \Rightarrow z\mathbf{X}(z) = \mathbf{A}\mathbf{X}(z) + \mathbf{b}U(z) \tag{11-30}$$

or

$$\mathbf{X}(z) = (z\mathbf{I} - \mathbf{A})^{-1}\mathbf{b} \tag{11-31}$$

When Equation 11-30 and Equation 11-17 are compared, a slight difference between them can be noted. If $\Phi(k, 0) = \mathcal{Z}^{-1}[z(z\mathbf{I} - \mathbf{A})]$, then $\Phi(k-1, 0) = \mathcal{Z}^{-1}(z\mathbf{I}-\mathbf{A})$. Whereas $\Phi(k, 0)$ is used to study homogeneous systems, $\Phi(k-1, 0)$ is used in the convolution sum that characterizes inhomogeneous solutions. Knowledge of $\Phi(k, 0)$ is obviously sufficient to specify $\Phi(k-1, 0)$. Therefore, the state-transition matrix needs to be computed only once. Finally, it follows that the transform of the observation equation satisfies

$$\mathbf{Y}(z) = [\mathbf{c}^T(z\mathbf{I} - \mathbf{A})^{-1}\mathbf{b} + \mathbf{d}]U(z) \tag{11-32}$$

From this, the transfer function $H(z)$ can be computed and it is

$$H(z) = \frac{Y(z)}{X(z)} = \mathbf{c}^T(z\mathbf{I} - \mathbf{A})^{-1}\mathbf{b} + \mathbf{d} \tag{11-33}$$

From Equations 11-17 and 11-33 it follows that the system's impulse response is given by $y[k] = \mathbf{d}$ for $k = 0$ and $y[k] = \mathbf{c}^T\mathbf{A}^{k-1}\mathbf{b}$ of $k \geq 1$. The matrix $(z\mathbf{I} - \mathbf{A})^{-1}$ can be also computed in adjoint form as

$$(z\mathbf{I}-\mathbf{A})^{-1} = \frac{1}{\Delta(z)}\mathbf{H}(z) = \frac{1}{\Delta(z)}[z^{n-1}\mathbf{H}_1 + z^{n-2}\mathbf{H}_2 + \cdots + z\mathbf{H}_{n-1} + \mathbf{H}_n] \tag{11-34}$$

where $\Delta(z)$ is the determinate of $(z\mathbf{I} - \mathbf{A})$, $\mathbf{H}(z)$ is the adjoint of the matrix $(z\mathbf{I} - \mathbf{A})$ (see Appendix A), and $\mathbf{H}_i$ is an $n \times n$ matrix. The roots of $\det(z\mathbf{I} - \mathbf{A})$ are also the poles of the system. Substituting this result into the transfer-function equation, the following results

$$H(z) = \mathbf{c}^T\left(\frac{1}{\Delta(z)}[z^n\mathbf{H}_1 + z^{n-1}\mathbf{H}_2 + \cdots + z^2\mathbf{H}_{n-1} + \mathbf{H}_n]\mathbf{b} + \mathbf{d}\right)$$

$$= \frac{\mathbf{c}^T([z^n\mathbf{H}_1 + z^{n-1}\mathbf{H}_2 + \cdots + z^2\mathbf{H}_{n-1} + z\mathbf{H}_n]\mathbf{b} + \Delta(z)\mathbf{d})}{\Delta(z)} \tag{11-35}$$

The stability of a discrete-time LSI system was defined in Chapter 10 in terms of the system's pole locations. The stability of the system is, therefore, seen to be established by the roots of $\Delta(z) = \det(z\mathbf{I}-\mathbf{A})$, which is completely characterized by the matrix $\mathbf{A}$. That is, stability is a function only of the feedback structure of the filter.

11-7   DISCRETE-TIME SYSTEM ARCHITECTURE

Architecture has been defined as the study of how a system is constructed by interconnecting a collection of primitive elements or building blocks. One of the major advantages of a state-variable model is its usefulness in describing an architecture. The formal study of architecture becomes increasingly important for discrete-time and digital systems, since their basic building blocks can be either software or hardware entities (i.e., shift registers, multiplies, and adders). A large number of commercially available programmable hardware products have been developed to implement a discrete-time system in a variety of architectures (e.g., DSP microprocessors). These hardware systems are often fixed-point (limited precision) digital devices. Fixed-point digital systems possess finite precision and can introduce rounding, overflow, and a number of other arithmetic-related errors that are not modeled by the transfer function. These errors, called *finite wordlength effects*, are studied in advanced courses on DSP. Because of the precision limitations associated with fixed-point DSP microprocessors, the study of architecture is more sharply focused for digital systems.

In previous sections much attention was given to developing a transfer-function representation of a discrete-time system. The transfer-function specification is not necessarily the final stage in the system-design paradigm. If the system is to be physically implemented in hardware or software, then it must be given a structure (i.e., architecture). Many distinctly different structures can be used to implement the monic Nth-order transfer function of the form

$$H(z) = K \frac{\sum_{i=0}^{M} b_i z^{-i}}{1 + \sum_{i=1}^{N} a_i z^{-i}} \tag{11-36}$$

where $M \leq N$. Each unique factorization of $H(z)$ represents a different architecture. An architecture that requires no more than N shift-registers to implement an Nth-order LSI system is called a *canonic* architecture. Since they all represent the same transfer function, architectures differ only in the style in which they implement $H(z)$. The general structure of an Nth transfer function can be uniquely specified in terms of its state-variable representation

$$w[k] = Ku[k] \qquad \text{Input scaling} \tag{11-37}$$

$$\mathbf{x}[k+1] = \mathbf{A}\mathbf{x}[k] + \mathbf{b}w[k] \qquad \text{State equations} \tag{11-38}$$

$$y[k] = \mathbf{c}^T\mathbf{x}[k] + \mathbf{d}w[k] \qquad \text{Observation equation} \tag{11-39}$$

which is shown schematically in Figure 11-5. It is similar to the diagram shown in Figure 11-2, except for the added input-gain term, denoted K. The gain K is usually chosen so that the maximum passband gain of a frequency-selective filter is unity or 0 dB. Regardless, each distinct four tuple $S = [\mathbf{A}, \mathbf{b}, \mathbf{c}, d]$ gives rise to special architectures, with the three most common being the direct II, cascade, and parallel.

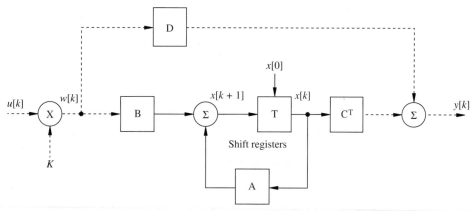

FIGURE 11-5   General state-variable model for a discrete-time LSI system.

All discrete LSI systems can be modeled in state-variable form. A state-variable model is, in itself, an architecture specification. The indexing of the elements of the state-variable four-tuple [**A**, **b**, **c**, d] for an Nth order system is given by

$$\mathbf{A} = \begin{pmatrix} a_{1,1} & a_{1,2} & \cdots & a_{1,n} \\ a_{2,1} & a_{2,2} & \cdots & a_{2,n} \\ \vdots & \vdots & \vdots & \vdots \\ a_{n,1} & a_{n,2} & \cdots & a_{n,n} \end{pmatrix}$$

$$\mathbf{b} = \begin{pmatrix} b_1 \\ b_2 \\ \vdots \\ b_n \end{pmatrix}$$

$$\mathbf{c}^T = (c_1 \quad c_2 \quad \cdots \quad c_n)$$

$$d = (d_0)$$

(11-40)

The coefficients of the model specify precisely how information is passed from one register to another. Assume that states $x_i[k]$ and $x_j[k]$ reside in the ith and jth shift registers, respectively. Then a nonzero coefficient a_{ij} establishes that there is a path between the ith and jth shift register and specifies the path gain, as well. If $a_{ij} = 1$, then the connection is direct and $x_j[k] = x_i[k]$. If $a_{ij} = -1$, then the connection passes through an inverter and $x_j[k] = -x_i[k]$. Otherwise, a nonzero a_{ij} would represent a need for a digital multiplier required to form $x_j[k] = a_{ij}x_i[k]$. For example, refer to Figure 11-3 and note that $a_{81} = 0.490$, which corresponds to the path gain connecting shift register SR_1 to shift register SR_8.

Direct II Architecture   The simplest implementation of an Nth-order transfer function $H(z)$ is the direct II architecture, which requires virtually no factoring of $H(z)$. The direct II architecture is similar to the analog-extended companion form and is given by

$$H(z) = K \frac{\sum_{i=0}^{M} b_i z^{-i}}{1 + \sum_{i=1}^{N} a_i z^{-i}} = K \left(r_0 + \frac{\sum_{i=1}^{N} r_i z^{-i}}{1 + \sum_{i=1}^{N} a_i z^{-i}} \right) \qquad (11\text{-}41)$$

where the quotient term $r_0 = b_0$ and the remainder terms equal (see Equation 11-9)

$$r_1 = b_1 - a_1 b_0$$

$$\vdots \qquad \vdots \qquad\qquad (11\text{-}42)$$

$$r_N = b_N - a_N b_0$$

The production rules give rise to an Nth-order direct II state-variable model $S = [\mathbf{A}, \mathbf{b}, \mathbf{c}, d]$, given by

$$\mathbf{A} = \begin{pmatrix} 0 & 1 & 0 & \cdots & 0 & 0 \\ 0 & 0 & 1 & \cdots & 0 & 0 \\ \vdots & \vdots & \vdots & \ddots & \vdots & \vdots \\ 0 & 0 & 0 & \cdots & 0 & 1 \\ -a_N & -a_{N-1} & -a_{N-2} & \cdots & -a_2 & -a_1 \end{pmatrix} \qquad N \times N \text{ matrix}$$

$$\mathbf{b} = \begin{pmatrix} 0 \\ 0 \\ \vdots \\ 0 \\ 1 \end{pmatrix} \qquad N \text{ vector} \qquad\qquad (11\text{-}43)$$

$$\mathbf{c} = \begin{pmatrix} r_N \\ r_{N-1} \\ \vdots \\ r_2 \\ r_1 \end{pmatrix} \qquad N \text{ vector}$$

$$d = r_0 \qquad \text{scalar}$$

which is diagramed in Figure 11-6.

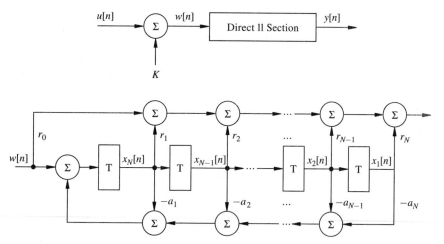

FIGURE 11-6   Direct II architecture.

Example 11-8   Direct II architecture

In Example 11-2, an eighth-order digital filter, having the frequency response shown in Figure 11-3 and a transfer function given by[3]

$$H(z) = \frac{1 - 0.44z^{-1} - 1.36z^{-2} + 0.04z^{-3} + 2.04z^{-4} + 0.04z^{-5} - 1.36z^{-6}}{1 - 0.88z^{-1} + 1.35z^{-2} - 0.91z^{-3} + 1.12z^{-4} - 0.46z^{-5} + 0.30z^{-6}}$$

$$\frac{-0.446z^{-7} + z^{-8}}{-0.078z^{-7} + 0.049z^{-8}}$$

was shown in direct II form in Figure 11-3.

Subsystem Factorization   Other architecture forms are also available. They all rely on factoring $H(z)$ into lower order subsystems having various attributes. Many of these methods presume that $H(z)$ can be factored into a collection of first- and second-order subsystems. Assume that the denominator of the transfer function $H(z)$ is given by

$$H(z) = \frac{N(z)}{D(z)} = K \frac{\sum_{i=0}^{M} b_i z^{-i}}{1 + \sum_{i=1}^{N} a_i z^{-i}} \qquad (11\text{-}44)$$

where K, b_i, and a_i are real. Further assume that $D(z)$ can be factored as

$$D(z) = \prod_{i=0}^{N-1} (z - \lambda_i) \qquad (11\text{-}45)$$

---

[3]Users of MONARCH will find the state-variable system description saved under filename "HighOrd.RPT."

where the factors of $D(z)$, namely $z = \lambda_i$, are either real or complex. In addition, assume that $H(z)$ can be expressed as a product of a $Q \leq N$ low-order subsystem, which satisfies

$$H(z) = K \prod_{i=0}^{Q} H_i(z) \tag{11-46}$$

where $H_i(z)$ is either a first- or second-order system.

Suppose that λ_i is real. Then $H_i(z)$ can be expressed as $H(z) = (q_{i0}z + q_{i1})/(z - \lambda_i)$, where q_{i0} and q_i are real. This corresponds to a system having a difference equation of the form $y_i[n+1] - \lambda_i y_i[n] = q_{i0}x_i[n+1] + q_{i1}x_i[n]$. Since λ_i is itself real, all the system coefficients can be implemented using only real multiplications and accumulators. As such, whenever real poles are encountered, they will be used to create first-order subsystems of the form

$$H_i(z) = \frac{q_{i0} + q_{i1}z^{-1}}{1 + \lambda_i z^{-1}} = q_{i0} + \frac{r_i z^{-1}}{1 + \lambda_i z^{-1}} \tag{11-47}$$

where $r_{i1} = q_{i1} - p_{i1}q_{i0}$.

If λ_i is complex, then complex multiplication will be required to implement a first-order complex subsystem. This is highly undesirable, since a digital computer would implement a complex multiplier using four real products and two additions (i.e., $(a + jb)(c + jd) = (ac - bd) + j(ad + bc)$).

If λ_i is complex, then there generally also exists its complex conjugate λ_i^*. Individually, they would define two first-order systems, each requiring complex multipliers and adders. However, if they are combined as follows

$$(z - \lambda_i)(z - \lambda_i^*) = z^2 + 2\operatorname{Re}(\lambda_i)z + |\lambda_i|^2 = z^2 + p_{i1}z + p_{i2} \tag{11-48}$$

then a polynomial is formed, having only real coefficients. A second-order subfilter, say $H_i(z)$, can therefore be expressed as

$$H_i(z) = \frac{w_{i0} + w_{i1}z^{-1} + w_{i2}z^{-2}}{1 + p_{i1}z^{-1} + p_{i2}z^{-2}} = q_{i0} + \frac{r_{i1}z^{-1} + r_{i2}z^{-2}}{1 + p_{i1}z^{-1} + p_{i2}z^{-2}} \tag{11-49}$$

where all the coefficients, including $q_{i0} = w_{i0}$, $r_{i1} = w_{i1} - p_{i1}w_{i0}$, and $r_{i2} = w_{i2} - p_{i2}w_{i0}$, are now real. Because the transfer function is the ratio of two second-order polynomials, this form is called *biquadratic*, or simply *biquad*. Both subfilters are graphically interpreted in Figure 11-7.

The first- and second-order subsystems also possess compact state-variable models.

First-order state model

$$A_i = (-p_{i1})$$

$$b_i = (1)$$

$$c_i = (r_{i1})$$

$$d_i = (q_{i0}) \tag{11-50}$$

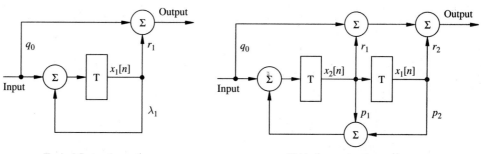

Typical first-order section   Typical second-order section

FIGURE 11-7   First- and second-order sections.

Second-order state model

$$\mathbf{A}_i = \begin{pmatrix} 0 & 1 \\ -p_{i2} & -p_{i1} \end{pmatrix}$$

$$\mathbf{b}_i = \begin{pmatrix} 0 \\ 1 \end{pmatrix}$$

$$\mathbf{c}_i = \begin{pmatrix} r_{i2} \\ r_{i1} \end{pmatrix}$$

$$d_i = (q_{i0}) \tag{11-51}$$

where, again, all coefficients are real.

Example 11-9   Second-order sections

The filter studied in Example 11-2, as a direct II system, has a transfer function given by

$$H(z) = \frac{1 - 0.44z^{-1} - 1.36z^{-2} + 0.04z^{-3} + 2.04z^{-4} + 0.04z^{-5} - 1.36z^{-6}}{1 - 0.88z^{-1} + 1.35z^{-2} - 0.91z^{-3} + 1.12z^{-4} - 0.46z^{-5} + 0.30z^{-6}}$$

$$\frac{-0.446z^{-7} + z^{-8}}{-0.078z^{-7} + 0.049z^{-8}}$$

The poles and zeros of this transfer function are graphically displayed in Figure 11-8. It can be seen that the poles and zeros in this case are all complex. The system can be decomposed into a collection of second-order systems with real coefficients by simply combining a pair of complex conjugate poles with a nearby pair of complex conjugate zeros.

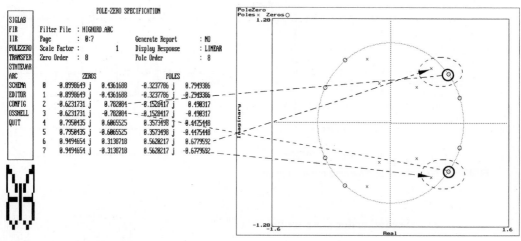

FIGURE 11-8   Pole-zero distribution of $H(z)$.

Computer Study   The partial fraction expansion of $H(z)$ can be produced using S-file `pf.s` with this result.

$$H(z) = k_0 + \frac{k_1 z}{(z + 0.1528 + j0.49031)} + \frac{k_1^* z}{(z + 0.1528 - j0.49031)}$$

$$+ \frac{k_2 z}{(z - 0.35734 + j0.44754)} + \frac{k_2^* z}{(z - 0.35734 - j0.44754)}$$

$$+ \frac{k_3 z}{(z + 0.32377 + j0.79493)} + \frac{k_3^* z}{(z + 0.32377 - j0.79493)}$$

$$+ \frac{k_4 z}{(z - 0.56202 + j0.67795)} + \frac{k_4^* z}{(z - 0.56202 + j0.67795)}$$

Here the Heaviside coefficients (note complex conjugate relationship) are computed.

SIGLAB	MATLAB
># transfer function	not available in Matlab
>h=rarc("highord.arc",0)	
>n=h[0,]; d=h[1,]	
>include "pf.s"	
>pfexp(n,d)	

The resulting data is summarized below.

k_0	k_1	k_2	k_3	k_4
0.0925	$-0.5299 + 0.0383j$	$-0.4711 + 0.1150j$	$-0.0174 - 0.1592j$	$-0.0209 + 0.1350j$

Based on this complex-conjugate symmetry relationship, a second-order section with only real coefficients can also be derived as

$$H_i(z) = \frac{k_i}{(z - \lambda_i)} + \frac{k_i^*}{(z - \lambda_i^*)} = \frac{(k_i + k_i^*)z - (k_i\lambda_i^* + k_i^*\lambda_i)}{(z^2 - (\lambda_i^* + \lambda_i)z + |\lambda_i|)}$$

$$= \frac{w_{i1}z + w_{i1}}{z^2 + p_{i1}z + p_{i2}}$$

where the coefficients w_{i1}, w_{i1}, p_{i1}, and p_{i2} are all real and are given in Equation 11-49.

Second-order sections can also be formed, as suggested, by combining neighboring complex-conjugate pole and zero pairs. Combining poles and zeros on the basis of their proximity will generally result in a filter with a smaller coefficient dynamic range requirement than other pairing policies. Choosing, for example, to pair the pole and zero pairs indicated in Figure 11-8 would result in a second-order filter given by

$$H_i(z) = \frac{(z - 0.79504 - j0.60655)(z - 0.79504 + j0.60655)}{(z - 0.56202 - j0.67795)(z - 0.56202 + j0.67795)}$$

which simplifies to

$$H_i(z) = \frac{z^2 - 1.59z + 1}{z^2 - 1.124z + 0.7755}$$

and a second-order state-variable model given by

$$\mathbf{A}_i = \begin{pmatrix} 0 & 1 \\ -0.7755 & 1.124 \end{pmatrix}$$

$$\mathbf{b}_i = \begin{pmatrix} 0 \\ 1 \end{pmatrix}$$

$$\mathbf{c}_i = \begin{pmatrix} 0.2245 \\ -0.466 \end{pmatrix}$$

$$d_i = (1) \tag{11-52}$$

Cascade Architecture   For a cascade architecture, an Nth-order $H(z)$ is factored into a multiplicative collection of first- and second-order subsystems of real coefficients, such as

$$H(z) = K \prod_{i=1}^{Q} H_i(z) \tag{11-53}$$

where K is the overall scale factor and $\sum \text{order}(H_i(z)) = N$. The Q subsystems $\mathcal{S} = [A_i, b_i, c_i, d_i]$, when configured as a cascade architecture, have a state-variable system model $\mathcal{S} = [\mathbf{A}, \mathbf{b}, \mathbf{c}, d]$, which is given by

$$\mathbf{A} = \begin{pmatrix} A_1 & 0 & 0 & \cdots & 0 \\ b_2 c_1^T & A_2 & 0 & \cdots & 0 \\ b_3 d_2 c_1^T & b_3 c_2^T & A_3 & \cdots & 0 \\ \vdots & \vdots & \vdots & \ddots & \vdots \\ b_Q(d_{Q-1}d_{Q-2}\ldots d_2)c_1^T & b_Q(d_{Q-1}d_{Q-2}\ldots d_3)c_2^T & b_Q(d_{Q-1}d_{Q-2}\ldots d_4)c_3^T & \cdots & A_Q \end{pmatrix}$$

where $\mathbf{A}$ is an $N \times N$ matrix and

$$\mathbf{b} = \begin{pmatrix} b_1 \\ d_1 b_2 \\ \vdots \\ (d_{Q-1}\cdots d_1)b_Q \end{pmatrix} \qquad N \text{ vector}$$

$$\mathbf{c} = \begin{pmatrix} (d_Q d_{Q-1}\cdots d_2)c_1 \\ (d_Q d_{Q-1}\cdots d_3)c_2 \\ \vdots \\ c_Q \end{pmatrix} \qquad N \text{ vector}$$

$$d = d_Q d_{Q-1}\cdots d_2 d_1 \qquad \text{scalar} \tag{11-54}$$

The resulting system is graphically interpreted in Figure 11-9.

Example 11-10   Cascade architecture

The eighth-order digital filter studied in Example 11-2, as a direct II, has a transfer function given by

$$H(z) = \frac{1 - 0.44z^{-1} - 1.36z^{-2} + 0.04z^{-3} + 2.04z^{-4} + 0.04z^{-5} - 1.36z^{-6}}{1 - 0.88z^{-1} + 1.35z^{-2} - 0.91z^{-3} + 1.12z^{-4} - 0.46z^{-5} + 0.30z^{-6}}$$

$$\frac{0.446z^{-7} + z^{-8}}{-0.078z^{-7} + 0.049z^{-8}}$$

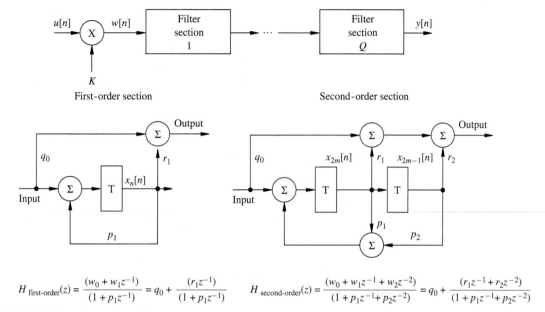

FIGURE 11-9   Cascade architecture.

This filter can be factored into a cascade filter and implemented in state-variable form.

Computer Study   The eighth-order filter[4] was implemented as a cascade filter and is graphically displayed in Figure 11-10 (see Problem 11-19 for additional details).

$$\mathbf{A} = \begin{pmatrix} 0 & 1 & 0 & 0 & 0 & 0 & 0 & 0 \\ -0.26 & -0.30 & 0 & 0 & 0 & 0 & 0 & 0 \\ 0 & 0 & 0 & 1 & 0 & 0 & 0 & 0 \\ 0.73 & 0.94 & -0.32 & 0.71 & 0 & 0 & 0 & 0 \\ 0 & 0 & 0 & 0 & 0 & 1 & 0 & 0 \\ 0.73 & 0.94 & 0.67 & -0.87 & -0.73 & -0.64 & 0 & 0 \\ 0 & 0 & 0 & 0 & 0 & 0 & 0 & 1 \\ 0.73 & 0.94 & 0.67 & -0.87 & 0.26 & 1.15 & -0.77 & 1.12 \end{pmatrix}$$

$$\mathbf{b}^T = (0 \quad 1 \quad 0 \quad 1 \quad 0 \quad 1 \quad 0 \quad 1)$$

$$\mathbf{c}^T = (0.73 \quad 0.94 \quad 0.67 \quad -0.87 \quad 0.26 \quad 1.152 \quad 0.22 \quad -0.77)$$

$$d = 1$$

[4]Users of MONARCH will find the cascade filter saved under filename "highcas."

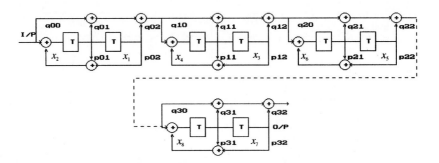

Cascade collection of second-order sections

FIGURE 11-10   High-order digital filter reduced to state variable form: cascade architecture.

Parallel Architecture   A parallel architecture consists of an additive combination of first- and second-order sections of real coefficients. The result of the factorization is a parallel architecture of the form

$$H(z) = K\left(d_0 + \sum_{i=1}^{Q} H_i(z)\right) \tag{11-55}$$

where K is the scale factor and $H_i(z)$ is either a first- or second-order subsystem. The sum of the individual subsystem orders must equal the order of the transfer function $H(z)$. The parallel architecture, based on the model in Example 11-10, is shown in Figure 11-11. The direct input-output path, denoted d_0 in Figure 11-11, is given by

$$d_0 = \sum_{i=1}^{Q} d_i = \sum_{i=1}^{Q} q_{i0} \tag{11-56}$$

which can be used to replace the individual paths denoted q_{i0} in each elemental first- and second-order filter. The Q subsystems, each given by $\mathcal{S}_i = [\mathbf{A}_i, \mathbf{b}_i, \mathbf{c}_i, d_i]$, can be configured as a parallel architecture, having a complete state-variable system model $\mathcal{S} = [\mathbf{A}, \mathbf{b}, \mathbf{c}, d]$, given by

$$\mathbf{A} = \begin{pmatrix} A_1 & 0 & 0 & \cdots & 0 \\ 0 & A_2 & 0 & \cdots & 0 \\ 0 & 0 & A_3 & \cdots & 0 \\ \vdots & \vdots & \vdots & \ddots & \vdots \\ 0 & 0 & 0 & \cdots & A_Q \end{pmatrix} \qquad N \times N \text{ matrix}$$

$$\mathbf{b} = \begin{pmatrix} b_1 \\ b_2 \\ \vdots \\ b_{Q-1} \\ b_Q \end{pmatrix} \qquad N \text{ vector}$$

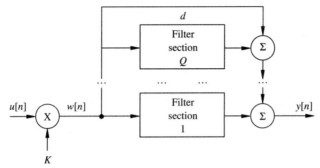

First-order section   Second-order section

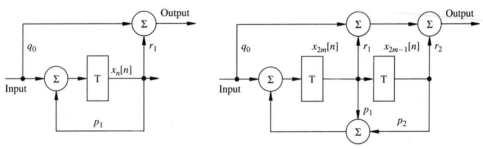

$$H_{\text{first-order}}(z) = \frac{(w_0 + w_1 z^{-1})}{(1 + p_1 z^{-1})} = q_0 + \frac{(r_1 z^{-1})}{(1 + p_1 z^{-1})} \qquad H_{\text{second-order}}(z) = \frac{(w_0 + w_1 z^{-1} + w_2 z^{-2})}{(1 + p_1 z^{-1} + p_2 z^{-2})} = q_0 + \frac{(r_1 z^{-1} + r_2 z^{-2})}{(1 + p_1 z^{-1} + p_2 z^{-2})}$$

FIGURE 11-11   Parallel architecture.

$$\mathbf{c} = \begin{pmatrix} c_1 \\ c_2 \\ \vdots \\ c_{Q-1} \\ c_Q \end{pmatrix} \qquad N \text{ vector}$$

$$d = \sum d_i \qquad \text{scalar} \tag{11-57}$$

Example 11-11   Parallel architecture

The eighth-order digital filter studied in Example 11-2 as a direct II filter and in Example 11-10 as a cascade filter, has a transfer function given by

$$H(z) = \frac{1 - 0.44z^{-1} - 1.36z^{-2} + 0.04z^{-3} + 2.04z^{-4} + 0.04z^{-5} - 1.36z^{-6}}{1 - 0.88z^{-1} + 1.35z^{-2} - 0.91z^{-3} + 1.12z^{-4} - 0.46z^{-5} + 0.30z^{-6}}$$

$$\frac{-0.446z^{-7} + z^{-8}}{-0.078z^{-7} + 0.049z^{-8}}$$

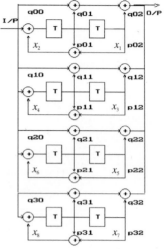

FIGURE 11-12 High-order digital filter reduced to state-variable form: Parallel architecture.

Parallel collection of second-order sections

The filter can be factored into a parallel filter and implemented in state-variable form.

Computer Study The eighth-order filter[5] was implemented as a parallel filter and is graphically displayed in Figure 11-12 (see Problem 11-20 for additional details).

$$
\mathbf{A} = \begin{pmatrix}
0 & 1 & 0 & 0 & 0 & 0 & 0 & 0 \\
-0.26 & -0.30 & 0 & 0 & 0 & 0 & 0 & 0 \\
0 & 0 & 0 & 1 & 0 & 0 & 0 & 0 \\
0 & 0 & -0.32 & 0.71 & 0 & 0 & 0 & 0 \\
0 & 0 & 0 & 0 & 0 & 1 & 0 & 0 \\
0 & 0 & 0 & 0 & -0.73 & -0.64 & 0 & 0 \\
0 & 0 & 0 & 0 & 0 & 0 & 0 & 1 \\
0 & 0 & 0 & 0 & 0 & 0 & -0.77 & 1.12
\end{pmatrix}
$$

$$\mathbf{b}^T = (0 \quad 1 \quad 0 \quad 1 \quad 0 \quad 1 \quad 0 \quad 1)$$

$$\mathbf{c}^T = (1.34 \quad 11.45 \quad 2.52 \quad -10.18 \quad 2.61 \quad -0.37 \quad 2.23 \quad -0.45)$$

$$d = 1$$

The three architectures presented here are only representative of many possible architectures that are covered in advanced studies of digital filter architectures. The three architectures presented generally offer the following general trade-offs.

---

[5] For users of MONARCH, the parallel filter is saved under filename "`highpar`."

Type	Advantages	Disadvantages
Cascade	Good fixed-point performance	Multiply-accumulate intensive
Direct II	Few multiply-accumulates	Poor fixed-point precision
Parallel	Fault tolerant	Often poor fixed-point precision

A filter with few multiply and accumulates per filter cycle, versus another with many, would have a higher potential bandwidth, since arithmetic delays represent the most time-consuming operation in a digital filter. Therefore, a direct II would normally be chosen if maximum bandwidth was the primary design objective. For example, the multiplier count for the eighth-order system previously studied are 17 (direct II), 20 (cascade), and 20 (parallel). A parallel filter has the advantage that if one of n parallel subsystems fails, the filter would continue to function, albeit degraded. Other architectures would produce poor or no results if a component were to fail. The cascade filter is generally a good compromise if maintaining minimum precision in a fixed-point implementation is required.

The effects of numerical errors, such as roundoff and overflow, are mathematically studied in an advanced course in DSP.

11-8 SUMMARY

Discrete-time architecture was shown to be expressed as a block diagram, a signal-flow graph, or in state-variable form. Specific architectures were introduced in Chapter 11. They included the direct II, cascade, and parallel forms. All these can be used to implement a given transfer function. In Chapter 12 methods of deriving transfer functions of systems that exhibit prespecified frequency selectivity will be presented. Once such a filter has been derived, the methods presented in this chapter will be used to implement it in hardware or software.

11-9 COMPUTER FILES

These computer files were used in Chapter 11.

File Name	Type	Location	File Name	Type	Location
		Subdirectory CHAP11			
		Subdirectory SFILES or MATLAB			
pf.s	C	Example 11-9	lever.x	C	Example 11-7
		Subdirectory SIGNALS			
		Subdirectory SYSTEMS			
sfg.exe	C	Example 11-1	highord.rpt	D	Examples 11-2, 11-8, 11-9
cas.sfg	D	Example 11-1	highord.arc	D	Examples 11-7, 11-9
cascade.arc	D	Example 11-1	highcas.arc	D	Example 11-10
cascade.iir	D	Example 11-1	highpar.arc	D	Example 11-11
highord.irr	D	Example 11-2	ladder.arc	D	Project 11-2

Where "s" denotes an S-file, "m" an M-file, "x" either an S-file or M-file, "T" denotes a tutorial program, "D" a data file, and "C" a computer-based instruction (CBI) program. CBI programs can be altered and parameterized by the user.

11-10   PROBLEMS

11-1 Signal-flow graph   Represent the following discrete-time transfer function

$$H(z) = \frac{1 - 1.414z^{-1} + z^{-2}}{1 + 1.5z^{-1} + 0.75z^{-2}}$$

in signal-flow graph form. Use Mason's Gain Formula to verify $H(z)$.

11-2 Block diagram reduction   Reduce the system shown in Figure 11-13 to a signal-flow graph and then a transfer function, using Mason's Gain Formula. Determine $H(z)$.

11-3 Direct II architecture   Place the system described in Figure 11-13 in direct II form.

11-4 Nonproper system   A nonproper discrete-time system has a transfer function

$$H(z) = \frac{\sum_{i=0}^{4} b_i z^{-i}}{\sum_{i=0}^{3} a_i z^{-i}}$$

where $a_0 = 1$. Display $H(z)$ in signal-flow graph form. Use Mason's Gain Formula to verify your signal-flow diagram.

11-5 Signal-flow and block diagram   Given the discrete-time system

$$H(z) = 0.001836 \frac{(1 + z^{-1})^4}{(1 - 1.4996z^{-1} + 0.8482z^{-2})(1 - 1.5548z^{-1} + 0.6493z^{-2})}$$

represent $H(z)$ in block-diagram form and in signal-flow graph form.

11-6 Mason's Gain Formula   Using the signal-flow graph developed in Problem 11-5 and Mason's Gain Formula, derive and verify $H(z)$.

11-7 Direct II architecture   Place $H(z)$ from Problem 11-5 in a direct II form. Discuss the stability of $H(z)$ based on the eigenvalues of $(z\mathbf{I} - \mathbf{A})$.

11-8 Impulse response   An at-rest discrete-time LSI system has a state four-tuple given by $\mathcal{S} = [\mathbf{A}, \mathbf{b}, \mathbf{c}, d]$. Derive a formula defining the system's impulse response.

FIGURE 11-13   Block-diagram representation of a discrete-time system used in Problem 11-2.

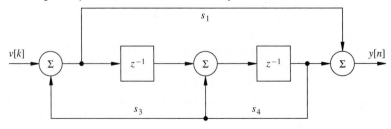

11-9 Similarity transform   The state-determined system defined in Problem 11-8 is to undergo a change in variables. The state vector of the new system, $w[n]$, is related to the original state vector $x[n]$ through $w[k] = \mathbf{T}x[k]$, where $\mathbf{T}$ is a nonsingular square matrix. Derive the state-transition matrix for the new system in terms of the original state-transition matrix. Derive a formula for the inhomogeneous solution.

11-10 State transition matrix   The state model for a discrete-time system is shown by

$$\mathbf{A} = \begin{pmatrix} 0 & 1 & 0 & 0 \\ -0.26 & -0.30 & 0 & 0 \\ 0 & 0 & 0 & 1 \\ 0.73 & 0.94 & -0.32 & 0.71 \end{pmatrix}$$

$$\mathbf{b} = \begin{pmatrix} 0 \\ 1 \\ 0 \\ 1 \end{pmatrix}$$

$$\mathbf{c}^T = (0.73 \quad 0.94 \quad 0.67 \quad -0.87)$$

$$d = (1)$$

What is the state-transition matrix? What is the homogeneous response to $\mathbf{x}_0^T = (1, 1, 1, 1)$? Is the system stable? Why?

11-11 Inhomogeneous response   For the at-rest system defined in Problem 11-10, what is the step response?

11-12 Architecture   Sketch the architecture described by the state variable equations shown below.

$$\mathbf{A} = \begin{pmatrix} r\cos(\theta) & r\sin(\theta) \\ -r\sin(\theta) & r\cos(\theta) \end{pmatrix}$$

$$\mathbf{b} = \begin{pmatrix} 0 \\ 1 \end{pmatrix}$$

$$\mathbf{c}^T = (c_1 \quad c_2)$$

$$d = (1)$$

11-13 Transfer function   Sketch the signal-flow diagram for the system studied in Problem 11-12. Compute the system's transfer function.

11-14 Direct II architecture   Implement the system defined in Problem 11-12 with a direct II architecture. Verify that it represents the known transfer function.

11-15 Direct II architecture   Implement $H(z) = 0.002(1 + z^{-1})^4/((1 - 1.5z^{-1} + 0.85z^{-2})(1 - 1.55z^{-1} + 0.65z^{-2}))$ as a direct II system.

11-16 Cascade architecture   Implement $H(z) = 0.002(1+z^{-1})^4/((1-1.5z^{-1}+0.85z^{-2})(1 - 1.55z^{-1} + 0.65z^{-2}))$ as a cascade system.

11-17 Parallel architecture   Implement $H(z) = 0.002(1+z^{-1})^4/((1-1.5z^{-1}+0.85z^{-2})(1 - 1.55z^{-1} + 0.65z^{-2}))$ as a parallel system.

11-18 Cascade architecture   Verify that Equation 11-54 is a valid model for a cascaded system of Q-sections. Refer to Example 11-10 and observe that not all the coefficients found in the state four-tuple $\mathcal{S} = [\mathbf{A}, \mathbf{b}, \mathbf{c}, d]$ are actually found in the system simulation diagram shown in Figure 11-10. The coefficients found in the simulation diagram are sometimes called the *digital filter coefficients*. Locate the digital filter coefficients in $\mathcal{S} = [\mathbf{A}, \mathbf{b}, \mathbf{c}, d]$. What is the purpose of the other coefficients $S = [\mathbf{A}, \mathbf{b}, \mathbf{c}, d]$?

11-19 Parallel architecture   Repeat Problem 11-18 for the parallel architecture defined by Equation 11-57.

11-20 Stability   Given a causal second-order discrete-time system, having a plant model given by

$$\mathbf{A} = \begin{pmatrix} 0 & 1 \\ -0.375 & -1.25 \end{pmatrix}$$

$$\mathbf{b}^T = [0, 1]$$

is the system stable? What is the state transition matrix? If the initial conditions are $x_1[0] = 0$ and $x_2[0] = 1$, find inputs $v[1]$ and $v[2]$, which will take the state vector $\mathbf{x}[k]$ to the origin at sampling index $k = 2$.

11-21 Finite wordlength effects   The fourth-order discrete time system shown here is to be implemented digitally, using an arithmetic unit with only eight and four-bits of fractional precision.

$$H(z) = \frac{(1 + z^{-1})^4}{(1 - 3.0536z^{-1} + 3.8281z^{-2} - 2.2921z^{-3} + 0.5507z^{-4})}$$

Implement $H(z)$ as a direct II system. Compare the floating-point precision transfer function of $H(z)$ with that obtained using coefficients with only eight and four-bits of fractional precision. Compare the two systems' magnitude frequency responses and discuss their differences.

11-11   COMPUTER PROJECTS

11-1 Internal structure and behavior of a state-determined system   In Example 8-6 it was shown that the EKG signal displayed in Figure 8-12 contained 60 Hz harmonic contamination. A filter can be used to remove the unwanted signal components leaving the signal energy clustered about 0 Hz unaffected. Assume that the signal bandwidth of the EKG signal shown in Figure 8-12 (sampled at 135 Hz) is, at most, 12 Hz wide. Design an IIR digital filter (see Chapter 12) that will pass the information spectrum and attenuate everything beyond 15 Hz by at least 50 dB, with no more than a 0.25 dB passband ripple deviation. Verify the design using an FFT of the filter's impulse response.

Implement this filter as cascade, direct II, and parallel systems. Create a program to isolate the signal found in the ith shift register (HINT: consider changing $\mathbf{c}^T$ to read $\mathbf{c}^T = (0, 0, \cdots, 1, \cdots, 0, 0)$, where the 1 is in the ith location of $\mathbf{c}^T$ and $d = 0$). Run a frequency response measured at the output of each shift-register location (i.e., state location) and compare it to overall system frequency response.

Based on this analysis, what is the worst-case unit amplitude, steady-state sinusoidal input for each system (i.e, the input that produces the highest internal gain)?

Present each system with a worst-case unit amplitude, steady-state sinusoidal input. Assume also that the filter is to be implemented using 16-bit hardware. Determine the maximal output precision for each architecture. Produce a noise-gain graph for each architecture that displays the log of the square roundoff error $\varepsilon = (x_i - x_i[f])^2$, where x_i is the floating-point output of the filter measured at the ith shift-register location; $x_i[f]$ is the same variable produced by a 16-bit machine using f fractional bits of precision, versus fractional wordlength f. For your study, use $f \in [8, 15]$.

11-2 Ladder architecture Many architectures exist for implementing discrete-time systems. One of the classic architectures is called the Gray-Markel ladder form. A simple fourth-order bandpass filter design specification is saved in file "`ladder. arc`." The filter's magnitude frequency response is also shown in Figure 11-14. The filter implemented is a fourth-order transfer function given by

$$H(z) = 0.11989 \frac{z^4 - 0.36656z^3 - 0.75270z^2 - 0.36656z + 1}{z^4 - 0.98467z^3 + 1.28447z^2 - 0.57842z + 0.37484}$$

The filter is implemented as a Gray-Markel ladder as shown in Figure 11-15 (see page 406).

FIGURE 11-14 Magnitude frequency response of H(z).

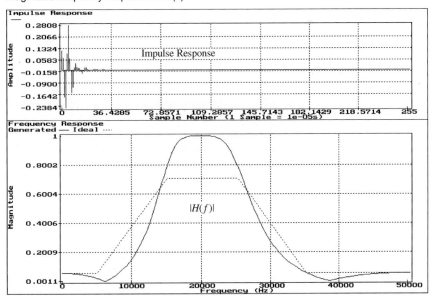

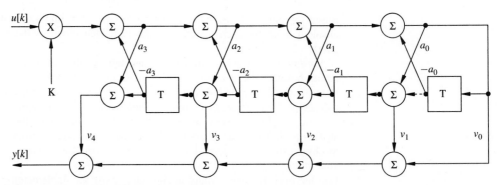

FIGURE 11-15   Fourth-order Gray-Markel ladder filter.

The coefficients are given by[6]

$$a_0 = 0.40175$$

$$a_1 = -0.76187$$

$$a_2 = 0.24354$$

$$a_3 = -0.37484$$

$$v_4 = 1.33788$$

$$v_3 = -1.41673$$

$$v_2 = -1.48496$$

$$v_1 = 0.61810$$

$$v_0 = 1.0000$$

Experimentally determine the effect of coefficient rounding if the coefficients are rounded to 8-bits of significance, using 4, 5, and 6 bits of fractional precision. Compare these results to a direct II implementation of the same filter. Explain any major differences that may exist. Given these results, would there be any advantage in using the more complex ladder filter?

---

[6]For users of MONARCH, the actual filter coefficients have been saved in file "ladder.rpt."

FOUR

APPLICATIONS

INTRODUCTION
TO DIGITAL SIGNAL
PROCESSING

Take me to the magic of the moment, on a glory night, where the children of tomorrow dream away in the wind of change.

—*The Wind of Change*, Klaus Meine

12-1 INTRODUCTION

Digital signal processing (DSP) is a relatively young technology that has brought the winds of change to many fields. Beginning in the 1970s, DSP began to replace traditional analog signal-processing systems and subsystems. The popularity and importance of DSP technology has continued to grow. It is now considered to be a discipline unto itself, replete with its own mathematics, analysis and synthesis methodology, and technology. The future of DSP is, in some sense, limited only by human creativity and our ability to successfully apply DSP to original problems. Some contemporary applications of DSP are:

General Purpose
Filtering (Convolution)
Detection (Correlation)
Spectral analysis (Fourier transforms)
Adaptive filtering
Instrumentation
Waveform generation
Transient analysis

Graphics
Rotation
Image transmission and compression
Image recognition
Image enhancement
Control
Servo control
Disk control

409

Instrumentation (*Continued*)
Steady-state analysis
Biomedical instrumentation
Information systems
Speech processing
Audio processing
Voice mail
Facsimile [FAX]
Modems
Cellular telephones
Modulators, demodulators
Line equalizers
Data encryption
Spread-spectrum
Digital and LAN communications

Control (*Continued*)
Printer control
Engine control
Guidance and navigation
Vibration (modal) control
Power-system monitors
Robots
Others
Radar and sonar
Radio and television
Music and speech synthesis
Entertainment

A classic DSP study consists of two major topics, called spectral analysis and filtering. Discrete-time spectral analysis was introduced in Chapter 8 when the DFT and FFT were introduced. The DFT and FFT are but two of many basic digital spectral analysis tools. The study of spectral analysis and its application is pervasive in most scientific and engineering disciplines. Each discipline brings to this field its own terminology, methodology, and standards. Time and space limitations, however, will prohibit the development of this branch of DSP beyond that found in Chapters 8 and 10.

The second major DSP field of study is digital filters. Early examples of digital filters are the numerical analysis routines used to perform numerical integration and linear regression. These algorithms, in one sense or another, filter data. Initially, digital filters were implemented as software programs or with discrete-digital components. If DSP had remained in this state, digital filtering would rarely be found in commercial applications today. However, a revolution began in the late 70s due to the introduction of a device called the *DSP microprocessor*. The DSP microprocessor is a digital electronic device that places a heavy emphasis on performing the most fundamental DSP operation, namely multiply-accumulate, at high speeds. This technological revolution continues today and has become a primary force in making DSP an important commercial technology. The DSP microprocessor has led to reliable, high-precision, and cost-effective digital replacements for many analog filters. DSP microprocessors have also led to economical implementation of programmable digital signal and image processors, as well as adaptive filters that are difficult to realize as analog systems. While digital filters are aggressively challenging traditional analog filters in the marketplace, in many respects they owe their existence to the theory developed to design analog filters earlier in the twentieth century.

The existence of analog electrical and electronic filters can be traced to the beginning of radio. Here they were essential elements of a technology that radically altered our life and life style. The purpose of these filters was to perform frequency-selective filtering of radio and audio-frequency signals. In this capacity, filters removed unwanted signal energy and amplified desired signal information. Before the advent of the general-purpose digital computer, the design of filters was based on the use of carefully assembled standard tables, charts, and graphs. Because the production of these tables was labor intensive, only a few important filter types were ever reduced to a standardized database. From this short menu of filter types, the radio engineer would use simple scaling rules to convert a standard filter model (called an *analog prototype*) into a final-analog filter model. This filter would then be implemented in hardware.

Continuous-time or analog filters can be classified as being *passive* or *active*. Passive filters use resistor, inductor, and capacitor elements as their primitive building blocks. Active filters add high-gain operational amplifiers to this mix. For example, the transfer function

$$H(s) = \frac{1}{s^3 + 2s^2 + 2s + 1}$$

can be implemented using the passive and active filters shown in Figure 12-1. As the following list of guidelines suggest, a number of factors determine whether a filter should be implemented as a passive or active analog system.

Frequency Limitations: Large valued LC components are required to design filters that can operate at subaudio frequencies. Active filters, however, can be designed to operate in this range. Above 50 Hz most commercial operational amplifiers (*op amps*) have an open-loop gain too small to produce high precision

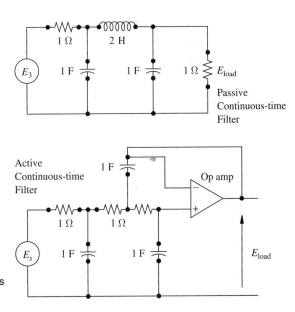

FIGURE 12-1   Passive and active implementations of an identical transfer function.

filters. Specialized amplifiers, however, can be found with a bandwidth extended out to much higher frequencies. LC filters can be designed to operate out to the megahertz range. At higher frequencies, integrated filters are used.

Packaging: Active filters are, in general, smaller than LC filters.

Cost: Since precision capacitors and inductors are expensive, active filters are generally more economical. Active filters usually use off-the-shelf components while LC filters often must be customized.

Adjustment and Calibration: Analog filters must be tuned periodically. Some must be adjusted continually. Inductors and capacitors below a few hundred pico-farads can be made to be adjustable. Designing tunable active filters is difficult, since this often requires that two or more resistors be simultaneously adjusted.

12-2 CLASSIC ANALOG FILTERS

While the importance of analog filters is continuously being reduced by their digital counterparts, they remain an important study, if for no other reason than they provide a gateway to the study of digital filters. The design of a contemporary analog filter, in many cases, remains today as it was during the early days of radio. The design objective of the radio engineers was to shape the frequency spectrum of a received or transmitted signal using modulators, demodulators, and frequency-selective filters. The frequency-selective filters were defined in terms of a mathematical ideal. The ideal models represent lowpass, highpass, bandpass, bandstop, and all-pass filters. These are graphically interpreted in Figure 12-2. Their shape represents the steady-state magnitude-frequency response of a filter with a transfer function of $H(\Omega) = H(s)|_{s=j\Omega}$, where Ω denotes an *analog frequency* measured in radians per second. The mathematical specification of each ideal filter is summarized as

$$\text{Ideal Lowpass} \quad |H(\Omega)| = \begin{cases} 1 & \text{if } \Omega \in [-B, B] \\ 0 & \text{otherwise} \end{cases} \tag{12-1}$$

$$\text{Ideal Highpass} \quad |H(\Omega)| = \begin{cases} 0 & \text{if } \Omega \in [-B, B] \\ 1 & \text{otherwise} \end{cases} \tag{12-2}$$

$$\text{Ideal Bandpass} \quad |H(\Omega)| = \begin{cases} 1 & \text{if } \Omega \in [-B_2, -B_1] \text{ or } \Omega \in [B_1, B_2] \\ 0 & \text{otherwise} \end{cases} \tag{12-3}$$

$$\text{Ideal Bandstop} \quad |H(\Omega)| = \begin{cases} 0 & \text{if } \Omega \in [-B_2, -B_1] \text{ or } \Omega \in [B_1, B_2] \\ 1 & \text{otherwise} \end{cases} \tag{12-4}$$

$$\text{All-pass} \quad |H(\Omega)| = 1 \text{ for all } \Omega \in [-\infty, \infty] \tag{12-5}$$

Analog filter design is often based on the use of several well-known models called Butterworth, Chebyshev, and elliptic (Cauer) filters. To standardize the design procedure, a set of normalized analog prototype filter models was agreed upon and reduced to tables, charts, and graphs. These models, called *prototypes*,

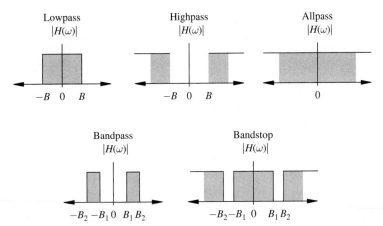

FIGURE 12-2   Basic ideal filter types.

were all developed as lowpass systems having a known gain (typically -1 dB or -3 dB passband attenuation) at a known critical cut-off frequency (typically 1 radian/second). The transfer function of an analog prototype filter, denoted $H_p(s)$, would be encapsulated in a standard table as a function of filter type and order. The prototype filter $H_p(s)$ would then be mapped into a final filter $H(s)$ having critical frequencies specified by the designer. The mapping rules, called *frequency-frequency transforms*, are summarized here and interpreted in Figure 12-3.

Nth order Prototype	Frequency-to-Frequency Transform	Order	
Lowpass to Lowpass	$s \Leftarrow s/\Omega_p$	N	(12-6)
Lowpass to Highpass	$s \Leftarrow \Omega_p/s$	N	(12-7)
Lowpass to Bandpass	$s \Leftarrow (s^2+(\Omega_{p1}\Omega_{p2}))/(s(\Omega_{p2}-\Omega_{p1}))$	$2N$	(12-8)
Lowpass to Bandstop	$s \Leftarrow (s(\Omega_{p2}-\Omega_{p1}))/(s^2+(\Omega_{p2}\Omega_{p1}))$	$2N$	(12-9)

Example 12-1   Frequency-frequency scaling

The analog filter implemented in Figure 12-1 and defined by $H_p(s) = 1/(s^3 + 2s^2 + 2s + 1)$ and has poles located at $s = -1.0$ and $s = 0.5 \pm j0.866$. The magnitude-frequency response of the prototype filter is shown in Figure 12-4(a). Observe that the gain at $\Omega = 1$ rad/sec is $|H_p(\Omega)| = |1/(1 + j1)| = 1/\sqrt{2} \Rightarrow -3$ dB. Assume that the desired filter has a similar shape, except that the desired -3 dB critical frequency is translated out to 1 kHz. The desired filter

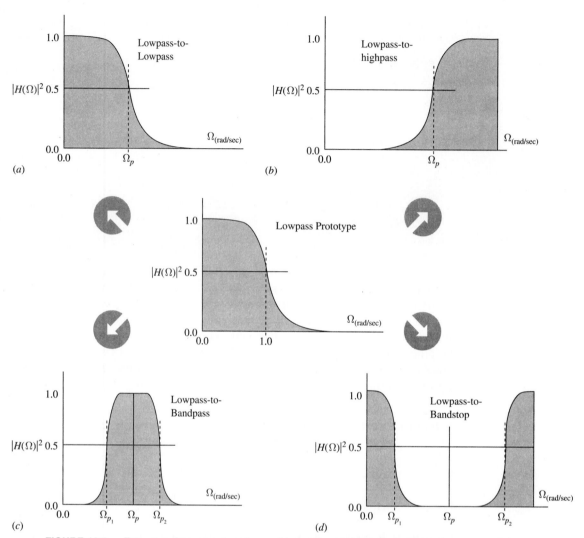

FIGURE 12-3   Frequency-frequency transforms showing the mapping of a prototype lowpass to: (a) lowpass, (b) highpass, (c) bandpass, and (d) bandstop mappings.

can be realized by scaling the prototype model $H_p(\Omega)$, under the appropriate lowpass-to-lowpass frequency-frequency transformation, into its final form. In particular

$$H(s) = \frac{1}{s^3 + 2s^2 + 2s + 1}\bigg|_{s \rightarrow s/\Omega_p}$$

where Ω_p is given by $\Omega_p = 2\pi 1000 (\text{rad/sec})/1(\text{rad/sec}) = 6283$. The final

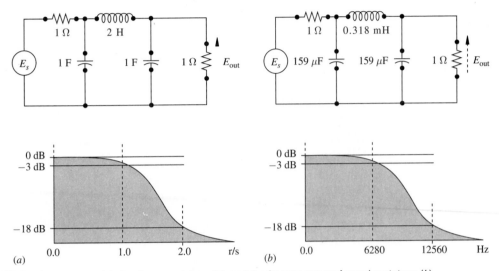

FIGURE 12-4   Lowpass model showing a prototype (*a*) and the frequency transformed prototype (*b*).

filter $H(s)$ then becomes

$$H(s) = \frac{1}{4.03 \times 10^{-12}s^3 + 5.07 \times 10^{-8}s^2 + 3.18 \times 10^{-4}s + 1}$$

The new poles are located at $s = -6283$ and $s = -3142 \pm j5441$. Note also that the poles are scaled by $s = s/\Omega_p$. The frequency response of the final filter, and its attendant circuit elements are shown in Figure 12-4(*b*).

12-3 LOWPASS PROTOTYPE FILTERS

Normally, analog prototype filters are lowpass systems with a critical cutoff frequency of $\Omega = 1$ rad/sec. The magnitude-squared frequency response, at a frequency of $\Omega = 1$ rad/sec, is often specified to be

$$|H(s)|_{s=j1}|^2 = |H(j1)|^2 = \frac{1}{1 - \epsilon^2} \qquad (12\text{-}10)$$

If $\epsilon^2 = 0.5$, the system is said to be a -3 dB filter. The specifications of an arbitrary filter define the filter's passband gain limit, stopband gain limit, and the frequencies as described in Figure 12-5. The frequencies Ω_p, Ω_a, Ω_{p1}, Ω_{p2}, Ω_{a1}, and Ω_{a2} are called *critical* or *cut-off* frequencies. The passband and stopband gains are specified in terms of the parameters ϵ and A. The steepness of the filter skirt is measured in terms of the *transition gain ratio*, which is given by $\eta = \epsilon/(A^2 - 1)^{1/2}$. The *frequency transition ratio*, denoted $k_d < 1$, is a measure

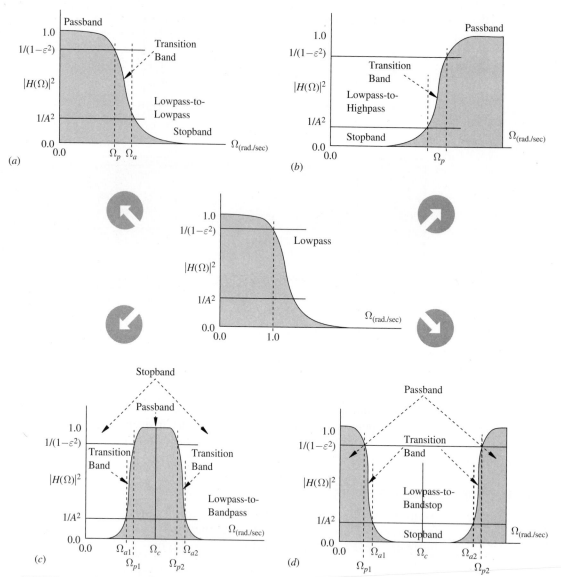

FIGURE 12-5   Frequency-frequency transforms: (a) lowpass-to-lowpass, (b) lowpass-to-highpass, (c) lowpass-to-bandpass, and (d) lowpass-to-bandstop.

of the transition bandwidth and is given by

$$\text{Lowpass}\qquad k_d = \Omega_p/\Omega_a \qquad\qquad (12\text{-}11)$$

$$\text{Highpass}\qquad k_d = \Omega_a/\Omega_p \qquad\qquad (12\text{-}12)$$

$$\text{Bandpass} \qquad k_d = \begin{cases} k_1 & \text{if } \Omega_u^2 \geq \Omega_v^2 \\ k_2 & \text{if } \Omega_u^2 < \Omega_v^2 \end{cases} \tag{12-13}$$

$$\text{Bandstop} \qquad k_d = \begin{cases} 1/k_1 & \text{if } \Omega_u^2 \geq \Omega_v^2 \\ 1/k_2 & \text{if } \Omega_u^2 < \Omega_v^2 \end{cases} \tag{12-14}$$

where

$$\Omega_c = (\Omega_{p1}\Omega_{p2})/(\Omega_{p2} - \Omega_{p1}) \tag{12-15}$$

$$k_1 = (\Omega_{a1}/\Omega_c)(1/(1 - \Omega_{a1}^2/\Omega_v^2)) \tag{12-16}$$

$$k_2 = -(\Omega_{a2}/\Omega_c)(1/(1 - \Omega_{a2}^2/\Omega_v^2)) \tag{12-17}$$

$$\Omega_u^2 = \Omega_{a1}\Omega_{a2} \tag{12-18}$$

$$\Omega_v^2 = \Omega_{p1}\Omega_{p2} \tag{12-19}$$

If $\delta^2 = A^2 - 1$, the *gain ratio* is defined as $d = \delta/\epsilon$. From these parameters the order and transfer function of several classic analog filters can be determined. The most important of these are summarized in this next section as analog prototype lowpass filters with a critical frequency $\Omega = 1$ rad/sec.

Butterworth filters   The *Butterworth filter* is also called a *flat* filter, since its magnitude frequency response leaving 0 Hz and its $(N-1)$st derivatives (N is the order of the transfer function) are flat. An Nth-order lowpass Butterworth filter has a magnitude-squared frequency response given by

$$|H(j\Omega)|^2 = \frac{1}{1 + \epsilon^2 \Omega^{2N}} \tag{12-20}$$

The order of a Butterworth lowpass filter is estimated to be

$$N = \log\left(\frac{A^2 - 1}{\epsilon^2}\right) \frac{1}{2\log(1/k_d)} \tag{12-21}$$

The $2N$ poles of $|H(s)|^2 = |H(s)H(-s)|$ are located on a circle of radius $r = 1/\epsilon^{1/N}$ in the s-plane. In particular, the poles of $|H(s)|^2$ are located at $s_k = re^{-j(2k-1)\pi/2N}$, $k \in [0, 2N - 1]$. N of the poles are located in the stable left-hand plane and belong to the realizable part of $|H(s)|^2$, which is denoted $H(s)$. The other N poles belong to $H(-s)$ in Figure 12-6. The magnitude frequency response of $H(s)$, as a function of order N, is displayed in Figure 12-6.

Chebyshev filters   An Nth-order lowpass *Chebyshev I filter* has a magnitude-squared frequency response of

$$|H(j\Omega)|^2 = \frac{1}{1 + \epsilon^2 C_N^2(\Omega)} \tag{12-22}$$

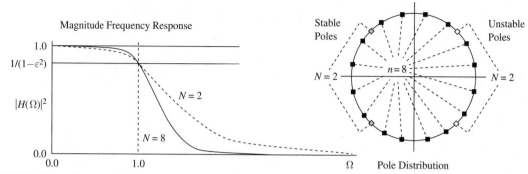

FIGURE 12-6   Butterworth filter magnitude frequency response.

where $C_N(s)$ is a Chebyshev polynomial of order N. The first few Chebyshev polynomials are listed below

$$C_1(\Omega) = \Omega$$

$$C_2(\Omega) = 2\Omega^2 - 1$$

$$C_3(\Omega) = 4\Omega^3 - 3\Omega$$

$$C_4(\Omega) = 8\Omega^4 - 8\Omega^2 + 1$$

$$C_5(\Omega) = 16\Omega^5 - 20\Omega^3 + 5\Omega$$

The order of a Chebyshev I filter is estimated to be

$$N = \frac{\log[(1 + \sqrt{1 - \eta^2})/\eta]}{\log[1/k_d + \sqrt{(1/k_d)^2 - 1}]} \tag{12-23}$$

The $2N$ poles of $|H(s)|^2 = |H(s)H(-s)|$ lie on an ellipse with geometry determined by ϵ and N. Again, N of the poles of $|H(s)|^2$ belong to the stable left-hand plane and are assigned to $H(s)$. In particular, the left-hand plane poles are located at $s_k = -\sin(x[k])\sinh(y[k]) + j\cos(x[k])\cosh(y[k])$ where $x[k] = (2k - 1)\pi/2N$, and $y[k] = \pm\sinh^{-1}(1/\epsilon)/N$. The other N poles belong to $H(-s)$.

The magnitude frequency response of a Chebyshev I filter is displayed in Figure 12-7. The Chebyshev I filter exhibits ripple in the passband and a smooth response through the transition band into the stopband. A variation on the Chebyshev I model, called the Chebyshev II filter, is given by

$$|H(j\Omega)|^2 = \frac{1}{1 + [\epsilon^2 C_N^2(\Omega_a/\Omega)]^{-1}} \tag{12-24}$$

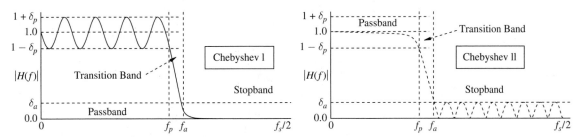

FIGURE 12-7   Magnitude frequency response for Chebyshev filter models showing a Chebyshev I (left) and Chebyshev II (right).

where Ω_a defines the stopband edge. The Chebyshev II filter, shown in Figure 12-7, is seen to have ripple in the stopband and smooth response through the transition band into the passband. The order of a Chebyshev II filter is estimated using the formula that applies to the Chebyshev I filter.

Elliptic filters   The attenuation of an Nth-order *elliptic filter* is given by

$$A_N(\Omega) = \frac{\Omega(a_2^2 - \Omega^2)(a_4^2 - \Omega^2) \cdots (a_m^2 - \Omega^2)}{(1 - a_2^2\Omega^2)(1 - a_4^2\Omega^2) \cdots (1 - a_m^2\Omega^2)} \qquad (12\text{-}25)$$

if N is odd and $m = (N-1)/2$. If N is even and $m = N/2$, then

$$A_N(\Omega) = \frac{(a_2^2 - \Omega^2)(a_4^2 - \Omega^2) \cdots (a_m^2 - \Omega^2)}{(1 - a_2^2\Omega^2)(1 - a_4^2\Omega^2) \cdots (1 - a_m^2\Omega^2)} \qquad (12\text{-}26)$$

The zeros of the elliptic filter are located at $a_2, a_4, \ldots, a_m$ and the poles are found at $1/a_2, 1/a_4, \cdots, 1/a_m$. The values of $a_2, a_4, \ldots, a_m$ are defined in terms of an elliptic integral given by

$$K[k] = \int_0^{\pi/2} \frac{d\theta}{\sqrt{1 - k^2 \sin^2(\theta)}} \qquad (12\text{-}27)$$

The integral is, in general, difficult to evaluate and therefore has been reduced to standard tables.

The order of an elliptic filter is estimated to be

$$N \geq \frac{\log(16D)}{\log(1/q)} \qquad (12\text{-}28)$$

where

$$k' = \sqrt{(1 - k_d^2)}$$

$$q_0 = 0.5(1 - \sqrt{k'})/(1 + \sqrt{k'})$$

$$q = q_0 + 2q_0^5 + 15q_0^9 + 150q_0^{13}$$

$$D = d^2$$

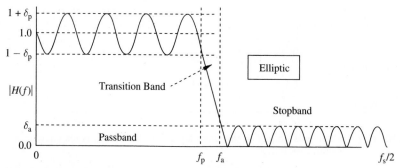

FIGURE 12-8 Magnitude frequency response for an elliptic filter model.

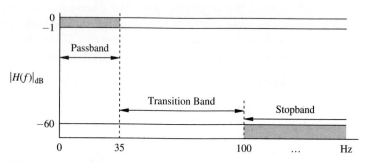

FIGURE 12-9 Desired lowpass filter model.

The magnitude frequency response of a typical elliptic lowpass filter is shown in Figure 12-8. It is characterized by ripple behavior in both the passband and stopband.

Example 12-2 Classic analog filter

The lowpass filter shown in Figure 12-9 has a stopband bounded by $A = 1000$ (-60 dB) and a passband bounded by $\epsilon = 0.508$ (-1 dB). The remaining design parameters are given by

$$\eta = \frac{\epsilon}{\sqrt{A^2 - 1}} = 0.508 \times 10^{-3}$$

$$k_d = 35/100 = 0.35$$

The filter order of classic analog filters can be estimated to be as follows:

Butterworth

$$N = \log\left(\frac{10^6 - 1}{0.258064}\right) \frac{1}{2\log(2.857)} = 7.18 < 8$$

Chebyshev I or Chebyshev II

$$N = \frac{\log(1 + \sqrt{1 - 0.258 \times 10^{-6}})/0.508 \times 10^{-3}}{\log(2.857 + \sqrt{7.163})} = 4.83 < 5$$

Elliptic

$$k' = (1 - 0.35^2)^{1/2} = 0.9367$$

$$q_0 = 0.5(0.0321)/1.967 = 0.01631$$

$$q \approx q_0$$

$$D = 1968^2$$

$$N \approx \log(16D)/\log(1/q) = 3.68 < 4$$

Therefore an eighth-order Butterworth, fifth-order Chebyshev, and fourth-order elliptic would be required to satisfy the same design specifications. Assuming that all filters that meet the given specifications are equally acceptable, the elliptic filter would probably be the filter of choice, due to its lower complexity (order).

As a general rule, for a classic filter specified in terms of the magnitude frequency response parameters as found in Figure 12-3, the orders N required to meet or exceed the design objectives are related as order(Butterworth) $\geq$ order(Chebyshev I) = order(Chebyshev II) $\geq$ order(elliptic).

Example 12-3   Analog filter design

The filters studied in Example 12-2 can be investigated in terms of their proto-type and final designs. The lowpass cutoff frequency is 35 Hz and the passband gain is bounded from below by -1 dB.

Computer Study   The required Butterworth, Chebyshev I and II, and elliptic filters that meet the stated specifications[1] are of order 8, 5, 5, and 4, respectively. Their impulse and magnitude frequency responses are shown in Figure 12-10. The lowpass-to-lowpass frequency-frequency transform is given by $s = s/\Omega_p = 2\pi s/35$.

Prototype Butterworth

$$H_p(s) =$$

$$\frac{1}{s^8 + 5.125s^7 + 13.137s^6 + 21.846s^5 + 25.688s^4 + 21.846s^3 + 13.137s^2 + 5.125s + 1}$$

---

[1]For users of MONARCH, the filters are saved under filenames "butter", "cheby1", "cheby2," and "ellip."

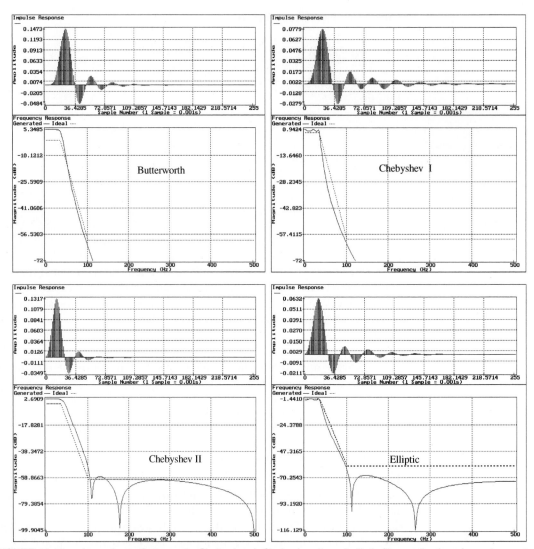

FIGURE 12-10   Approximate Butterworth, Chebyshev I, Chebyshev II, and elliptic filter realizations for a given set of design specifications.

Prototype Chebyshev I

$$H_p(s) = \frac{1}{s^5 + 0.936s^4 + 1.688s^3 + 0.974s^2 + 0.580s + 0.122}$$

Prototype Chebyshev II

$$H_p(s) = 0.0147 \frac{s^4 + 34.653s^2 + 240.174}{s^5 + 4.092s^4 + 8.373s^3 + 10.636s^2 + 8.46s + 3.534}$$

Prototype Elliptic

$$H_p(s) = 0.000461 \frac{s^4 + 22.502s^2 + 65.237}{s^4 + 0.553s^3 + 0.498s^2 + 0.149s + 0.0337}$$

Upon performing a lowpass-to-lowpass transform, the following filters result.

Butterworth

$$H(s) =$$

$$\frac{1.11 \times 10^{19}}{s^8 + 1.231 \times 10^3 s^7 + 7.583 \times 10^5 s^6 + 3.029 \times 10^8 s^5 + 8.559 \times 10^{10} s^4 +}$$

$$1.748 \times 10^{13} s^3 + 2.526 \times 10^{15} s^2 + 2.368 \times 10^{17} s + 1.110 \times 10^{19}$$

Chebyshev I

$$H(s) =$$

$$\frac{6.446 \times 10^{10}}{s^5 + 2.068 \times 10^2 s^4 + 8.233 \times 10^4 s^3 + 1.048 \times 10^7 s^2 + 1.379 \times 10^9 s + 6.446 \times 10^{10}}$$

Chebyshev II

$$H(s) =$$

$$\frac{3.249(s^4 + 1.689 \times 10^6 s^2 + 5.709 \times 10^{11})}{s^5 + 9.036 \times 10^2 s^4 + 4.082 \times 10^5 s^3 + 1.145 \times 10^8 s^2 + 2.011 \times 10^{10} + 1.855 \times 10^{12}}$$

Elliptic

$$H(s) = \frac{0.000461(s^4 + 3.229 \times 10^6 s^2 + 1.342 \times 10^{12})}{s^4 + 2.096 \times 10^2 s^3 + 7.148 \times 10^4 s^4 + 8.149 \times 10^6 s + 6.961 \times 10^8}$$

The pole and zero distribution of each $|H(s)|^2 = H(s)H(-s)$ can be computed and displayed using S-file `ch12_3` and M-file `analog.m` with a format `analog(n,d)`, where n and d are the numerator and denominator polynomials of $H(s)$. The pole and zero distributions of $|H(s)|^2$ are displayed in Figure 12-11 for Chebyshev II and elliptic filter prototypes. The realizable poles of $H(s)$ are shown in the left-hand plane.

SIGLAB	MATLAB
`>include "ch12_3.s"`	`>> n = [1 0 34.653 0 240.174];`
`>n=[1,0,34.653,0,240.174]`	`>> d = [1 4.092 8.373 10.636`
`>d=[1,4.092,8.373,10.636,`	`8.46 3.534];`
`8.46,3.534]`	`>> analog(n, d)`
`>analog(n,d) # Chebyshev-II`	`>> n = [1 0 22.502 0 65.237];`
`>n=[1,0,22.502,0,65.237]`	`>> d = [1 0.553 0.498 0.149`
`>d=[1,0.553,0.498,0.149,.0337]`	`0.0337];`
`>analog(n,d) # Elliptic`	`>> analog(n, d)`

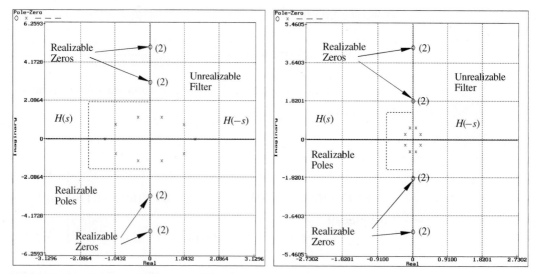

FIGURE 12-11   Pole-zero distributions of Chebyshev II and elliptic prototypes.

12-4   DIGITAL FILTERS

The design of a DSP system often begins with the conversion of an analog signal to a digital word, using an analog to digital converter (ADC). This process is shown in Figure 12-12. Observe that an analog anti-aliasing filter has been placed in front of the ADC. This filter ensures that the highest frequency presented to the ADC is bounded below the Nyquist frequency. Following the ADC is the DSP system or set of subsystems. The output of the subsystems may or may not be converted back into an analog signal, depending on the application.

The typical system occupying the DSP system block shown in Figure 12-12 is assumed to be a digital filter. Classic digital filters have evolved along two major paths called *finite impulse response filter (FIR)* and *infinite impulse response filter (IIR)*. FIRs are purely feed-forward systems, and will be shown to be stable and to have a simple structure. However, high order FIRs are required to meet even the most basic frequency-domain specifications. An IIR filter has both feedforward

FIGURE 12-12   DSP signal train.

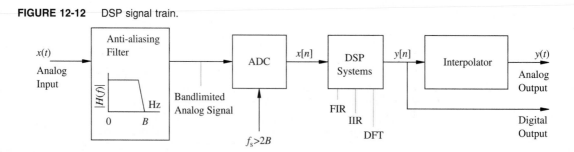

and feedback networks. Unlike a FIR, an IIR meets a set of frequency-selective specifications with a much lower order design, but introduces a host of other potential design problems including instability. Both classes of filters will be developed in this chapter.

12-5   FINITE IMPULSE RESPONSE (FIR) FILTERS

A finite impulse response (FIR) filter, as the name implies, has an impulse response consisting of only a finite number of sample values. The impulse response of an Nth-order FIR is given by

$$\{h[n]\} = \{h_0, h_1, \ldots, h_{N-1}\} \tag{12-29}$$

and is graphically interpreted in Figure 12-13. The FIR's time-series response to an arbitrary input $u[n]$, is given by the convolution sum. In particular, the output time-series is given by

$$y[n] = \sum_{i=0}^{N-1} h_i u[n-i] \tag{12-30}$$

The production of $y[n]$ is shown in the simulation diagram found in Figure 12-13. It can be seen that the FIR consists of nothing more than a shift-register array of length $N - 1$, N multipliers (called *tap-weight multipliers*), and an accumulator.

In Chapter 10 discrete-time systems were also expressed in terms of transfer functions. Formally, the Z-transform of a filter having the impulse response given in Equation 12-29 is given by

$$H(z) = \sum_{i=0}^{N-1} h_i z^{-i} \tag{12-31}$$

Example 12-4 explores several simple FIRs in terms of their transfer functions.

FIGURE 12-13 Nth-order FIR filter showing: (left) impulse response, and (right) simulation diagram.

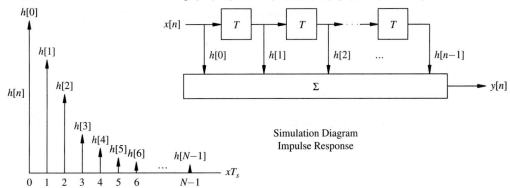

Example 12-4   The moving average and comb FIRs

The simplest FIR to implement is that which contains no tap-weight multipliers (i.e., multiplier-free FIR). One such filter is called the *moving-average filter* (see Figure 12-14) and is given by

$$H_{MA}(z) = \frac{1}{N}\sum_{i=0}^{N-1} z^{-i} = \frac{1}{N}\sum_{i=0}^{\infty} z^{-i} - \frac{1}{N}\sum_{i=N}^{\infty} z^{-i}$$

$$= \frac{1}{N}\frac{1}{1-z^{-1}} - \frac{1}{N}\frac{z^{-N}}{1-z^{-1}} = \frac{1}{N}\frac{1-z^{-N}}{1-z^{-1}}$$

Another multiplier-free FIR is called a *comb filter* (see Figure 12-14) and is given by

$$H_C(z) = 1 \pm z^{-N}$$

The moving-average FIR is seen to produce an output that is the average value of the most current N sample values. The comb filter simply adds or subtracts a delayed sample value from the current sample value. It can be immediately seen that both filters have a very simple structure and require no multipliers; only shift-registers and adders are needed. In addition, observe that both FIRs have similar numerators, namely $N(z) = 1 \pm z^{-N}$. The N roots of $1 \pm z^{-N} = 0$ are given by

$$1 + z^{-N} = 0 \Rightarrow z_i = \exp(j2\pi i/N + j\pi) \qquad i \in [0, N-1] \quad (12\text{-}32)$$

$$1 - z^{-N} = 0 \Rightarrow z_i = \exp(j2\pi i/N) \qquad i \in [0, N-1] \quad (12\text{-}33)$$

which can be verified by direct substitution. The pole-zero distributions of the moving-average and comb filters can be readily computed. They are displayed in Figure 12-15 for the case where $N = 7$.

Observe that the moving-average filter has no zeros at $z = 1$, establishing the fact that this filter would have a finite DC value (i.e., $H(1) \neq 0$). The comb filter given by $H(z) = 1 + z^{-7}$ would also have a finite DC gain by virtue of having no zero at $z = 1$. The comb filter would also have zero gain at the Nyquist frequency (i.e., $z = -1$). The comb filter given by $H(z) = 1 - z^{-7}$ has a zero at $z = 1$, but none at $z = -1$. This filter would, therefore, have a zero gain at DC but a finite gain at the Nyquist frequency that corresponds to $z = -1$.

FIGURE 12-14   Simple FIR filters: (left) moving average and (right) comb.

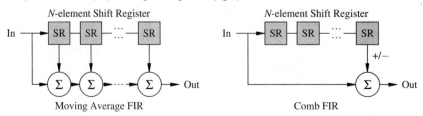

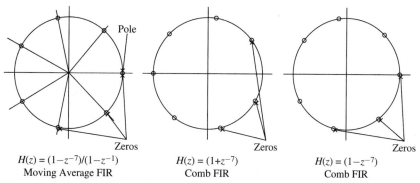

FIGURE 12-15   Pole-zero distributions of simple $N = 7$ moving-average and comb FIR filters.

Computer Study   The frequency-domain responses of a $N = 7$ moving-average and $N = 6$ comb filters are shown in Figure 12-16. The magnitude frequency responses, were created using S-file `ch12_4.s` or M-file `comb.m` with a format given by `comb(M,u)`, where $u[n]$ is a 128-sample input-time series and $M = N - 1$. Both the moving-average and comb filters' impulse responses are convolved with an impulse and an EKG signal.

SIGLAB	MATLAB
`>x=impulse(128)`	`>> x = [1; zeros(255,1)];`
`>y=rf("ekg1.imp")`	`>> load ekg1.imp`
`>y=dec(y,2) # reduce 256-samples to 128`	`>> y = ekg1(1:2:256);`
`>include "ch12_4.s"`	`>> comb(7, x)`
`>comb(7,x); comb(7,y)`	`>> comb(7, y)`

The results are shown in Figure 12-16. The moving-average filter is seen to behave like a lowpass filter, while the comb filter functions as a bank of frequency-selective subfilters. This observation can be reinforced by noting that the filtered EKG signal has little high-frequency activity when passed through a moving-average filter, but continues to exhibit localized activity at high frequencies when processed by a comb filter.

FIRs owe their popularity to a number of factors. The more important of these are listed below

Implementation   The FIR's simple structure makes implementation straightforward in either hardware or software.

Stability   If the input to an FIR is bounded by unity (i.e., $|u[n]| \le 1.0$), then the maximal (or worst-case) output is given by $G_{max} = \max\left(\left|\sum h_i u[n-i]\right|\right) = \sum |h_i|$, where $i \in [0, 1, \ldots, n - 1]$. If all the tap weights $|h_i| < M$, then $G_{max} < \infty$, and the FIR is bounded-input bounded-output (BIBO) stable. If

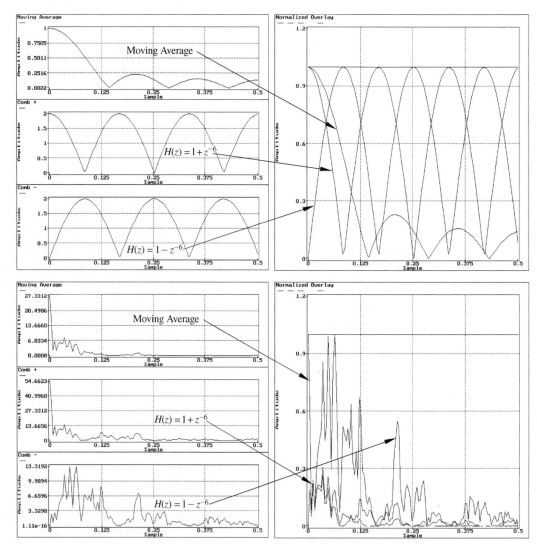

FIGURE 12-16   Magnitude frequency responses for moving-average and comb FIRs for an impulse-forcing function (top) and an EKG forcing function (bottom).

the input time-series is, on a sample-by-sample basis, bounded by unity (i.e., $|u[k]| \leq 1$), then the input that maximizes the FIR output (see Equation 9-34) is given by $u[k] = \text{sgn}(h_{L-k}) = \pm 1.0$, for L an integer, and is called the *worst-case input*. This can be verified by referring to Equation 12-30 and noting, for $n = L$,

$$y[n] = \sum_{k=0}^{N-1} h_{n-k} u[k] \leq \sum_{k=0}^{N-1} h_{n-k} \text{sgn}(h_{n-k}) = \sum_{k=0}^{N-1} |h_{n-k}| \tag{12-34}$$

Linear Phase   The two-sided frequency response of a FIR having a transfer function $H(z)$ is denoted $H(\omega)$, where $z = \exp(j\omega)$ and $\omega \in [-\pi, \pi]$. A system is said to have a *linear phase response* if the measured phase response has the form $\phi(\omega) = \alpha\omega + \beta$ (see Figure 12-17). Linear phase filters are important in a number of applications, such as: (1) a phase lock loop (PLL) system used to synchronize data and decode phase-modulated signals; and (2) linear-phase anti-aliasing filters placed in front of signal-analysis subsystems (e.g., DFT).

Linear phase filtering is easily achieved with an FIR. If the FIR tap-weight coefficients are symmetrically distributed about the filter's midpoint (see Figure 12-18), then the phase response is given by $\phi(\omega) = -\lfloor (N-1)/2 \rfloor \omega + c$, where c is a constant and $\lfloor x \rfloor$ denotes the integer part of x. The group delay is given by $\tau = -d\phi(\omega)/d\omega$ and has, in this case, a numerical value equal to $\tau = \lfloor (N-1)/2 \rfloor$. The value of τ, measured in clock delays, corresponds to the midpoint of the FIR. If the tap weights are distributed antisymmetrically, such that $-h_{N-1-n} = h_n$, then $\phi(\omega) = -\lfloor (N-1)/2 \rfloor \omega \pm \pi/2$. Therefore, it follows that the group delay is again equal to $\tau(\omega) = -d\phi(\omega)/d\omega = \lfloor (N-1)/2 \rfloor$.

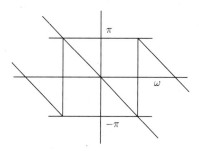

FIGURE 12-17   Linear phase response ($\beta = 0$) plotted as phase versus frequency modulo 2π.

FIGURE 12-18   FIR symmetry conditions.

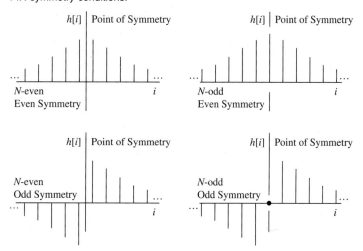

To verify that symmetric filters exhibit a linear phase behavior, again consider Equation 12-31 in a slightly modified form. For N odd, let $h_i = \pm h_{-i}$

$$H(z) = \sum_{i=0}^{N-1} h_i z^{-i} = z^{-L} \left(h_0 + \sum_{i=1}^{L-1} (h_i z^{-i} \pm h_{-i} z^i) \right) \qquad (12\text{-}35)$$

where $L = (N-1)/2$, which is also the point of symmetry. If the filter has even symmetry, then upon substituting $z = \exp(j\omega)$ into Equation 12-35, one obtains

$$H(e^{j\omega}) = e^{-i\omega L} \left(h_0 + \sum_{i=1}^{L-1} h_i(e^{-j\omega i} + e^{j\omega i}) \right) = e^{-j\omega i} \left(h_0 + \sum_{i=1}^{L-1} 2h_i \cos(\omega i) \right)$$

$$= e^{-j\omega L} \left(h_O + \sum_{i=1}^{L-1} 2h_i \cos(\omega i) \right)$$

$$= e^{-j\omega L} C(\omega) \qquad (12\text{-}36)$$

where $C(\omega)$ is a real function of ω. Therefore, the phase of an even-symmetry FIR is given by

$$\arg(H(e^{j\omega})) = -\omega L + \arg(C(\omega)) \qquad (12\text{-}37)$$

where $\arg(C(\omega)) = 0$ if $C(\omega) > 0$ and $\arg(C(\omega)) = \pm\pi$ if $C(\omega) < 0$. If the filter has odd symmetry, then the phase profile has a similar form, except

$$\arg(H(e^{j\omega})) = \frac{\pi}{2} - \omega L + \arg(C(\omega)) \qquad (12\text{-}38)$$

Example 12-5   Linear phase FIR filters

The phase response of a 51st-order symmetric FIR[2] is displayed in Figure 12-19. Note that the filter's impulse response possesses even coefficient symmetry and, therefore, satisfies the conditions for linear phase.

Computer Study   The filter's magnitude frequency, phase, and group-delay responses are also shown in Figure 12-19. The group delay is seen to have a value of 25 delays (the small deviations about 25 are due to numerical errors), which is equal to the theoretical value $(N-1)/2$. The data displayed in Figure 12-19 is produced by S-file ch12_5.s, which has a format given by linphase(h) where h is the FIR impulse response.

SIGLAB	MATLAB
>h=rf("LinPhase.imp") >include "ch12_5.s" >linphase(h)	Not available in Matlab

[2]For users of MONARCH, the FIR is saved under filename "LinPhase."

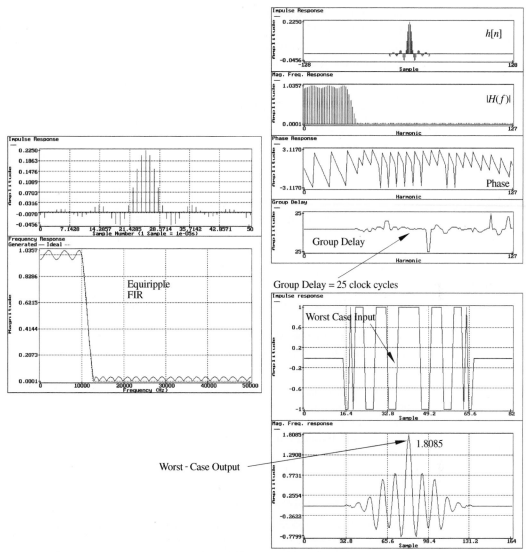

FIGURE 12-19   Symmetric $N = 51$ FIR properties showing a basic symmetric FIR impulse and magnitude frequency responses, frequency, phase, group-delay responses (note that the group delay has a value of 25 samples), and worst-case input and its convolution with FIR showing maximal gain of 1.8085.

Close inspection of the FIR passband in Figure 12-19 shows that the gain fluctuates between 1 ± 0.0357 and achieves these values at a number of discrete frequency locations. However, 1.0357 is not the worst-case maximal gain, which is given by $G_{\max} = \sum |h_i| = 1.80858$. This dynamic range requirement can be directly verified. Observe that the worst-case filter response to a unit-bound input was defined to be sgn(h_{L-k}). The worst-case input can, therefore, be modeled as $u[k] = \text{sgn}(h_{-k}) = \text{rev}(\text{sgn}(h_k))$, where "rev" simply reverses the order of the time-series. The worst-case response, displayed in Figure 12-19, is computed using ch12_5a.s.

SIGLAB	MATLAB
>h=zeros "ch12_5a.s" Not available in Matlab	
>maxima(h)	
worst-case gain	
1.8085	

The response to the worst-case input is seen to have a value of $y[n] = 1.8085$ in Figure 12-19.

12-6   FIR SYNTHESIS

Normally, FIRs are designed to achieve a prespecified frequency-domain response. If $|H_d(\omega)|$ denotes a filter's *desired magnitude frequency response*, then the design objective becomes one of determining the FIR impulse response (i.e., coefficient set $\{h_i\}$) such that $|H(\omega)|$ approximates $|H_d(\omega)|$ in some acceptable sense.

Direct Synthesis   The simplest synthesis technique is called the *direct method* and involves computing an M-sample inverse DFT (IDFT) of $H_d(\omega)$. An Nth-order direct FIR, where $N \leq M$, is defined by the N center-oriented sample values of the M-harmonic IDFT of $H_d(\omega)$. The direct filter design procedure is diagramed in Figure 12-20.

Example 12-6   Direct FIR design

A direct FIR, designed having an ideal magnitude frequency response $|H_d(\omega)| = \cos(\omega)$, $0 \leq \omega \leq \pi/2$, can be approximated in a piecewise constant sense with an 8-band 63rd-order direct FIR, where the band gains are given by

$$|H(\omega)| = \cos(0); 0 \leq \omega < \pi/8$$
$$|H(\omega)| = \cos(\pi/14); \pi/8 \leq \omega < 2\pi/8$$
$$|H(\omega)| = \cos(2\pi/14); 2\pi/8 \leq \omega < 3\pi/8$$
$$|H(\omega)| = \cos(3\pi/14); 3\pi/8 \leq \omega < 4\pi/8$$
$$|H(\omega)| = \cos(4\pi/14); 4\pi/8 \leq \omega < 5\pi/8$$
$$|H(\omega)| = \cos(5\pi/14); 5\pi/8 \leq \omega < 6\pi/8$$
$$|H(\omega)| = \cos(6\pi/14); 6\pi/8 \leq \omega < 7\pi/8$$
$$|H(\omega)| = \cos(7\pi/14); 7\pi/8 \leq \omega \leq 8\pi/8$$

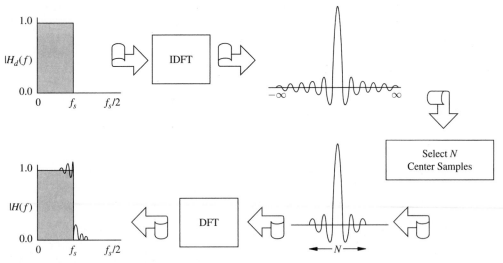

FIGURE 12-20   Direct synthesis method.

The FIR[3] assumes the sample frequency is $f_s = 1$ Hz. The normalized critical frequencies are:

Band	f_{lower}	f_{upper}	Gain
1	0.000	0.063	1.00
2	0.073	0.125	0.97
3	0.135	0.188	0.90
4	0.198	0.250	0.78
5	0.260	0.312	0.62
6	0.322	0.375	0.43
7	0.385	0.438	0.22
8	0.448	0.500	0.00

When generated, the filter produces the magnitude frequency response shown in Figure 12-21. Note that it follows the general shape of the designed filter, but also exhibits some "ringing" (due to Gibb's phenomenon) about the desired response. In this case it can be seen that the direct method produces a filter that may have a low mean-squared error, but also may have a large localized maximum error. Therefore, the direct method is best applied to those designs that are modeled as continuous, rather than piecewise-constant, magnitude frequency response.

Equiripple FIR Design   The principal problem with a Direct FIR filter is that local errors, given by $\epsilon(\omega) = |H_d(\omega) - H(\omega)|$, can become relatively large compared with the average or mean-squared error. The frequencies at which $\epsilon(\omega)$

[3]For MONARCH users, the FIR is saved under the filename "Direct."

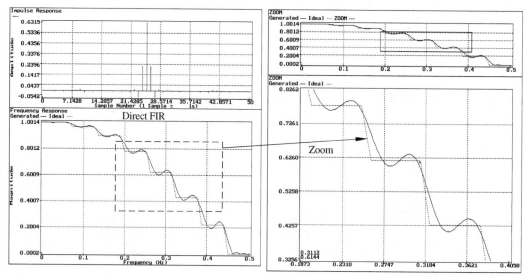

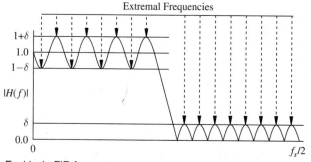

FIGURE 12-21   Direct FIR design.

FIGURE 12-22   Equiripple FIR frequency response.

is maximized are called the *extremal frequencies*. In many applications a superior criterion is one that will minimize the largest error. This is called the *mini-max design criteria*. The objective of an equiripple design criterion is to create a filter with an impulse response of $\{h[n]\}$, such that the maximal errors occuring at **all** extremal frequencies ω_i, namely $\max[\epsilon(\omega_i)] = |H_d(\omega_i) - H(\omega_i)| = \delta$, is minimized. The mini-max criterion is graphically interpreted in Figure 12-22. Since the maximal errors occuring at the extremal frequencies ω_i are equal, the filter is called an *equiripple FIR*.

Example 12-7   Equiripple FIR
A 51st-order bandpass equiripple FIR is required meeting the specifications

$f_{\text{sample}} = 100$ kHz:

Band 1: $f \in [0, 10]$ kHz; Desired gain $= 0$

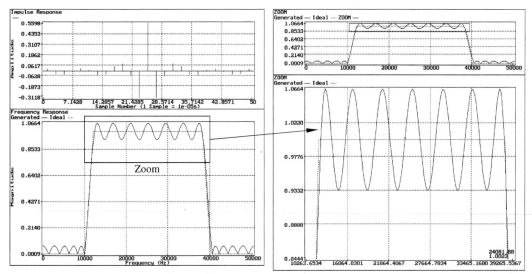

FIGURE 12-23   51st order equiripple FIR filter showing (left) impulse and magnitude frequency responses, and (right) zoom expansion of the passband.

Band 2: $f \in [12, 38]$ kHz; Desired gain $= 1$

Band 3: $f \in [40, 50]$ kHz; Desired gain $= 0$

This FIR must have a passband and stopband deviation from the ideal bounded by $\delta_p = \delta_a < -20$ dB or $\delta_p = \delta_a < 0.1$.

Computer Study   The equiripple FIR[4] that meets or exceeds the established specifications is shown in Figure 12-23. The impulse response is seen to have even symmetry and the magnitude frequency response has an equiripple deviation from the ideal in both the passband and stopband. Referring to the zoom expansion, the extremal error is seen to be about $\delta_p = 0.0664$ $(-23.5$ dB).

12-7 WINDOWS

The presented FIR design method indicates that a FIR filter may exhibit relatively large approximation errors at isolated frequencies. The direct method, for example, was seen to produce filter responses that can have significant overshoot errors at the transition band edges. This is typical of systems with a finite time-series that attempt to represent discrete-time phenonena, defined by an infinitely long time-series. This problem is pervasive in DSP where, due to finite memory limitations, arbitrarily long signals may be characterized by a time-series containing a finite number of samples.

---

[4]For users of MONARCH, the FIR is saved under the filename "Equi."

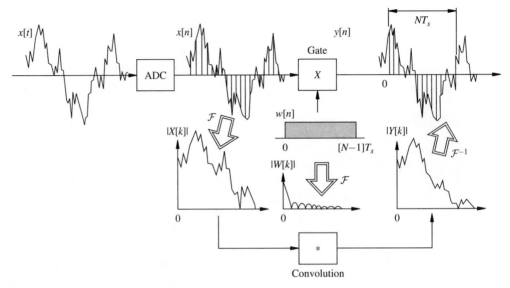

FIGURE 12-24   Windows and their effect on signals.

Consider the system shown in Figure 12-24. In this diagram an arbitrarily long signal $x[n]$ is gated by a finite duration window function $w[n]$. The output is given by $y[n] = x[n]w[n]$. The discrete-time Fourier transform of $x[n]$ shall be denoted $X(f)$ and that of the window $w(n)$ denoted $W(f)$. Using the Duality Theorem, it can be seen that the Fourier transform of the gated signal $y[n]$ is given by $Y(f) = X(f)*W(f)$. Ideally, the spectral image of $X(f)$ and $Y(f)$ should be in maximum agreement. However, the spectrum of the rectangular gating pulse, shown in Figure 12-24, has a $\text{sinc}(f)$ shape. As a result, the spectrum $Y(f) = X(f) * \text{sinc}(f)$ will generally differ from $X(f)$. An ideal window is one where $Y(f) = X(f) * W(f) = X(f)$, which would require that $W(f) = \delta(f)$, or $w[n] = 1$ for all time. Obviously this is not a window of finite duration and therefore is impractical.

A number of window approximation functions have, however, been developed that approximate $W(f) \approx \delta(f)$ to various degrees of accuracy. The more popular of these are the Blackman, Hamming, Hann, and Kaiser windows. Their discrete-time window specifications are listed here.

Rectangular Window

$$w[n] = \begin{cases} 1 & \text{if } |n| \leq \frac{(N-1)}{2} \\ 0 & \text{otherwise} \end{cases} \tag{12-39}$$

Hamming/Hann Window

$$w[n] = \begin{cases} a + (1-a)\cos\left(\frac{2\pi n}{(N-1)}\right) & \text{if } |n| \leq (N-1)/2 \\ 0 & \text{otherwise} \end{cases} \tag{12-40}$$

where $a = 0.54$ for a Hamming window and $a = 0.5$ for a Hann window.

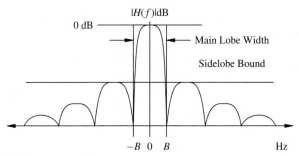

FIGURE 12-25   Windows parameters.

Blackman Window

$$w[n] = \begin{cases} 0.42 + 0.5 \cos\left(\frac{2\pi}{N-1}\right) + 0.08 \cos\left(\frac{4\pi n}{N-1}\right) & \text{if } |n| \leq (N-1)/2 \\ 0 & \text{otherwise} \end{cases}$$

(12-41)

Kaiser Window

$$w[n] = \begin{cases} \dfrac{I_0(\beta(1 - (2n/(N-1))^2)^{0.5})}{I_0(\beta)} & \text{if } |n| \leq (N-1)/2 \\ 0 & \text{otherwise} \end{cases}$$

(12-42)

where $I_0(x)$ is the zeroth-order modified Bessel function.

These basic windows can be compared using a number of measures such as the width of the double-sided main-lode and maximum sidelobe height as shown in Figure 12-25 and summarized here. This comparison presumes that the window duration is T seconds, where $T = N/f_s$ for f_s being the sample frequency.

Window	Transition width f_s/N	Highest sidelobe in dB
Rectangular	1.0	-13
Hann	2.07	-31
Hamming	2.46	-41
Blackman	3.13	-58
Kaiser ($\beta = 2.0$)	1.21	-19

Applying a window often improves the appearance and interpretability of the spectrum derived from a finite length section of an infinitely long signal. However,

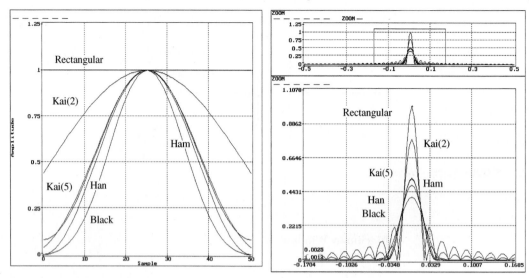

FIGURE 12-26   Basic windows: (left) Rectangular, Hamming, Hann, Blackman, and Kaiser ($\beta = 2.0, 5.0$), and (right) their magnitude spectral profile (shown with zoom expansion).

it should be remembered that the naturally periodic signal $x[n]$ possesses an ideal line spectrum and if a nonrectangular window is applied a distorted spectrum will result.

Example 12-8   Effects of windows

Many windows have been developed that make subtle tradeoffs between the width of the main lobe, the transition bandwidth, and the height of the sidelobes (see Figure 12-25). These parameters can be analytically or experimentally measured.

Computer Study   S-file `ch12_8.s` and M-file `window.m` use the format `window(M)` to create an *M*-sample Rectangular, Hann, and Hamming windows. The S-file also computes Blackman and Kaiser ($\beta = 2, 5$) windows. The data is displayed in Figure 12-26 for $M = 51$.

SIGLAB	MATLAB
>include "ch12_8.s" >> window(51)	
>window(51)	

It can be immediately seen that each window establishes a different trade-off between the main lobe width and the stopband suppression.

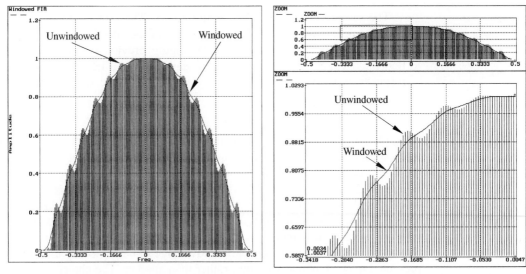

FIGURE 12-27 Effect of a window on an FIR magnitude frequency response showing (left) direct and windowed direct two-sided responses, and (right) zoom expansion.

Applying a symmetric window to a symmetric FIR will preserve the filter's linear phase response. The effect of the window is measurable in the magnitude frequency responses where the window will tend to suppress ringing.

Example 12-9   Windowed FIR

In Example 12-6 the direct synthesis method was used to approximate a FIR filter having an ideal piecewise-constant magnitude frequency response. The resulting impulse response, shown in Figure 12-21, was defined by the $M = 2N + 1$ center samples of IDFT($H(k)$). The displayed direct filter, from Example 12-6, was produced using a rectangular window of length M samples. Windows can, however, be used to make the realized spectral image appear more like that of the ideal model, even though technically it would require an infinitely long time-series representation.

Computer Study   S-file `ch12_9.s` accepts an M-sample FIR impulse response $h[n]$, applies a Blackman window $w[n]$ to $h[n]$, and then displays the two-sided magnitude frequency response of $h[n]$ and $h[n]w[n]$. The format is `winfir(h)`.

SIGLAB	MATLAB
>h=rf("direct.imp") #Direct Ex.12.6 Not available in Matlab >include "ch12_9.s" >winfir(h)	

The results are displayed in Figure 12-27. The original filter response exhibits considerable ringing about the values of the specified target filter model. The window is seen to suppress the ringing previously found at the transition band edges. Compared with the unwindowed FIR, the windowed filter response is seen to be in greater agreement with the desired cosine-shaped response.

Applying a window to a FIR's impulse response has the effect of smoothing the resulting filter's magnitude frequency response. The effect of a window on the steepness of the FIR's skirt and stopband suppression can be readily seen in Figure 12-27. The width of the main lobe of the window controls the amount of smoothing that takes place in the windowed filter's transition band. Which window to use, if any, is an open question. It should be remembered that the precomputed window weights are applied to a signal using sample-by-sample multiplication. Therefore, there is no implementation advantage to be gained by using any of the presented nonrectangular windows rather than another. The choice often is one of personal preference.

12-8   FIR ARCHITECTURE

The design of a FIR begins with the definiton of the filter's impulse response $\{h[n]\}$ or transfer function $H(z)$, and ends with the specification of an architecture. The standard architectural options for an Nth-order FIR are summarized in Figure 12-28(a). The *direct form* FIR consists of $N - 1$ delays, N multipliers, and $N - 1$ adders. If the coefficients are symmetric, the architecture shown in Figure 12-28(b) can be used. In this form the number of multiplications can be nearly halved by preprocessing the shift-delayed data before applying the tap weights, as shown in Figure 12-28. A third type of FIR architecture, called a *lattice* FIR, is shown in Figure 12-28(c). The coefficients k_i are referred to as reflection or partial correlation *(PARCOR)* coefficients. The coefficients of the lattice FIR filter associated with an Nth-order transfer function given by

$$H(z) = [1.0 + \sum_{j=1}^{N-1} a_j z^{-j}] \tag{12-43}$$

can be computed recursively. For $i = 1, 2, \ldots, N-1$, define $H_i(z) = 1 + \sum a_m^{[i]} z^{-m}$ for $m = 1, 2, \ldots, i$ where

$$a_i^{[i]} = k_i \tag{12-44}$$

$$a_m^{[i]} = a_m^{[i-1]} - k_i a_{i-m}^{[i-1]}$$

$$a_m = a_m^{[N-1]} \tag{12-45}$$

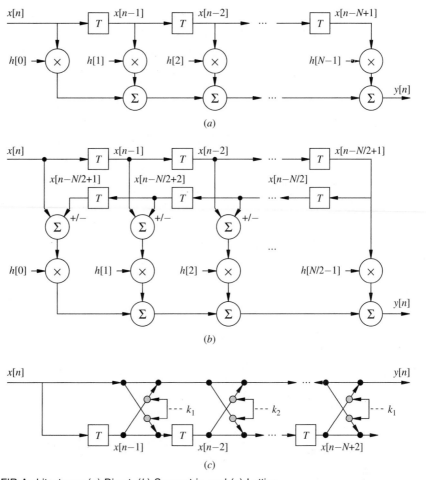

FIGURE 12-28   FIR Architectures: (*a*) Direct, (*b*) Symmetric, and (*c*) Lattice.

Then, for $j = N - 1, N - 2, \ldots, 2, 1$, define

$$k_{N-1} = a_{N-1}^{[N-1]} \tag{12-46}$$

$$k_j = a_j^{[j]}$$

$$a_m^{[j-1]} = a_m^{[j]} + k_j a_{j-m}^{[j]}/[1 - k_j^2] \tag{12-47}$$

It is important to note that the transfer function $H(z) = 1 + \sum a_i z^{-i}$ does not represent a linear phase filter since the coefficients do not possess the required symmetry property. As a result, the lattice filter has other purposes, such as that of an adaptive filter.

Example 12-10 Lattice FIRs

The conversion of a transfer function $H(z) = [1.0 + \sum a_i z^{-i}]$, $i = 1, 2, \ldots,$ $N - 1$ to and from lattice filter coefficients is defined by Equations 12-44 through 12-47.

Computer Study S-file `lattice.s` and M-file `lattice.m` have formats `lattice(x,c)`, where x is a coefficient array, $c = 0$ corresponds to direct architecture-to-lattice conversion, and $c = 1$ corresponds to lattice to direct architecture conversion. For example, to convert $H(z) = 1 - 0.9z^{-1} + 0.64z^{-2} - 0.575z^{-3}$ to a lattice filter and back again, use the following procedure.

SIGLAB	MATLAB
`>include "lattice.s"`	`>a=[-.9,.64,-.575]`
`>a=[1,-.9,.64,-.575]`	`>b=lattice(a,0)`
`>b=lattice(a,0) #0=>direct to lattice`	`-0.672,.183,-.575`
`>b`	`>lattice(b,1)`
`-0.672,.183,-.575`	`-.9,.64,-.575`
`>lattice(b,1) #1=>lattice to direct`	
`1,-.9,.64,-.575`	

12-9 INFINITE IMPULSE RESPONSE (IIR) FILTERS

The second common type of digital filter is an infinite impulse response or IIR filter. As the name suggests, an IIR filter has an impulse response that can persist forever. A causal discrete-time LSI IIR filter, having a discrete-time impulse response $h[k] = \{h_0, h_1, \ldots\}$, can be expressed in a transfer-function form as follows

$$H(z) = \sum_{k=0}^{\infty} h[k]z^{-k} = \frac{\sum_{j=0}^{M} b_j z^{-j}}{\sum_{j=0}^{N} a_j z^{-j}} \tag{12-48}$$

where N is called the filter order. It shall be assumed that $M \leq N$. If $M > N$, then the transfer function found in Equation 12-48 could be alternatively expressed as

$$H(z) = \sum_{k=1}^{M-N} c_k z^{k} + \frac{\sum_{j=0}^{N} b_j' z^{-j}}{\sum_{j=0}^{N} a_j z^{-j}} \tag{12-49}$$

which is the transfer function of a noncausal system. The steady-state frequency response of the system specified in Equation 12-48 can be determined by evaluating $H(z)$ along the periphery of the unit circle given by $z = \exp(j\omega)$, where $\omega \in [-\pi, \pi]$. Knowledge of the frequency-domain behavior of an IIR is fundamentally important since an IIR is normally designed to meet or exceed a prespecified set of frequency-domain specifications. A typical steady-state IIR filter's magnitude frequency response (i.e., $|H(\exp(j\omega)|)$, along with the filter's presumed desired response $H_D(\exp(j\omega))$, are interpreted in Figure 12-29. The desired magnitude frequency response is often specified to be a piecewise constant function having a unity passband and zero stopband gain. An acceptable

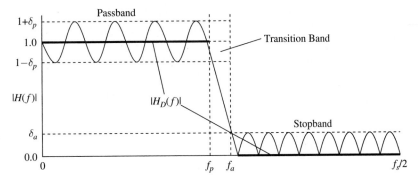

FIGURE 12-29   Typical IIR magnitude frequency response.

filter is one with a magnitude frequency response deviating from the ideal by no more than some prespecified amount. In Figure 12-29, the allowable deviations from the ideal are given by δ_p in the passband and δ_a in the stopband. These parameters also could have been defined in logarithmic (dB) units.

IIR filters achieve high stopband attenuation with a low-order design. IIRs also implement narrow transition band (i.e., steep-skirt) filters. Both high attenuation and steep-skirt filters are, in general, difficult to achieve with a low-order FIR filter. An empirical formula often used to predict the order of an FIR given amplitude and frequency design specifications, is

$$N_{\text{FIR}} \approx \frac{-10\log_{10}(\delta_a\delta_p) - 15}{14\Delta f} + 1 \qquad (12\text{-}50)$$

where δ_p and δ_p have been previously defined and Δf is the normalized transition bandwidth given by $\Delta f = (f_a - f_p)/f_s$, for a given sampling frequency f_s. What an IIR lacks is the ability to produce a linear phase response such as that generated by a symmetric FIR. As a result, IIR filters are generally not used in phase-sensitive applications. However, if a system already possesses a nonlinear phase response $\phi(\omega)$, then cascading this system with an all-pass IIR having a phase response $\theta(\omega) = k\omega - \phi(\omega)$ will result in an overall compensated system with linear phase. Such an all-pass filter is called a *phase compensator* or *compensator*.

Example 12-11   Comparison of FIR and IIR filter order

Given the following design objectives

$$f_s = 8 \text{ kHz}$$
$$f_p = 1.75 \text{ kHz}$$
$$f_a = 2.0 \text{ kHz}$$
$$\delta_a(\text{dB}) \leq -40 \text{ dB}$$
$$\delta_p(\text{dB}) \geq -0.0875 \text{ dB}$$

a 63rd-order equiripple FIR would be needed to satisfy the design criteria based on Equation 12-50. However, a sixth-order elliptic IIR can meet the same specifications.

Computer Study   FIR and an IIR filters[5] have been designed and their magnitude, log magnitude, phase, and group-delay responses are shown in Figure 12-30. The data displayed in Figure 12-30 was produced using S-file `ch12_11.s`, which has a format `firiir(fir,iir)`.

SIGLAB	MATLAB
`>fir=rf("firiir.imp")`	Not available in Matlab
`>iir=rarc("iirfir.arc",1)`	
`>include "ch12_11.s"`	
`>firiir(fir,iir)`	

Both filters are seen to meet or exceed the design specifications. The principal advantage of the FIR design is its linear phase. The principal advantage of the IIR is its low order that translates into fewer computations per filter cycle (i.e., increased bandwidth).

12-10   BILINEAR *Z*-TRANSFORM

IIR filters generally appear in two forms, known as user-defined and classic filters. User-defined filters are specified in terms of transfer functions, pole-zero distributions, or state-variable representations. The design of an arbitrary IIR is, in general, a challenging problem. The design of classic IIR filters, based on the analog filter models developed earlier in Chapter 12, is comparatively straightforward. In particular, the classic analog Butterworth, Chebyshev I and II, and elliptic filters all have digital filter counterparts. To realize a classic digital filter, a continuous-time filter model $H(s)$ must be converted into a discrete-time model $H(z)$. The method by which a continuous-time transfer function is mapped into a discrete-time transfer function is fundamentally important. For example, the standard Z-transform was used to model FIR filters. It is very tempting to use the standard Z-transform to map a classic analog transfer function $H(s)$ into a discrete-time filter $H(z)$. However, this mapping can often produce undesirable results, as Example 12-12 illustrates.

Example 12-12   Standard *Z*-transform mapping

The RC circuit shown in Figure 12-31 has a continuous-time model given by

$$v_0(t) + RC\frac{dv_0(t)}{dt} = v(t)$$

[5]For MONARCH users, the filters are saved under filenames "FIRIIR" and "IIRFIR".

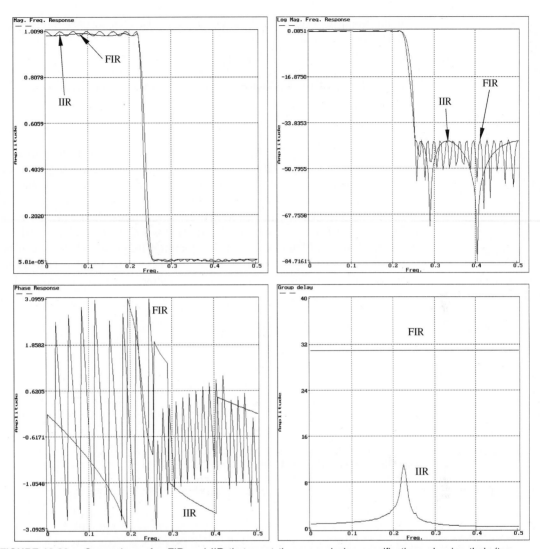

FIGURE 12-30   Comparison of a FIR and IIR that meet the same design specifications showing their (top left) magnitude frequency responses, (top right) log magnitude frequency responses, (bottom left) phase responses, and (bottom right) group-delay responses.

and a transfer function $H(s)$ given by

$$H(s) = \frac{1}{1 + RCs}$$

The system's impulse response is given by

$$h(t) = \frac{1}{RC}e^{-t/RC}$$

If the impulse were sampled every T_s seconds, then the resulting time-series would be

$$h[k] = \frac{1}{RC}e^{-kT_s/RC}$$

Using the standard Z-transform, the discrete-time system transfer function becomes

$$H(z) = \frac{b}{1 - az^{-1}} = \frac{bz}{z - a}$$

where $a = \exp(-T_s/RC)$ and $b = 1/RC$. The first-order discrete-time system is interpreted as a simulation diagram in Figure 12-31. Note that if $0 < a < 1$ for all positive values of R, C, and T_s, the filter $H(z)$ is BIBO stable.

Assume, for numerical purposes, that $RC \le 1$ and $T_s \le 1$. The magnitude frequency responses of the resulting transfer functions $H(s) = 1/(1+RCs)$ and $H(z) = (1/RC)/(1 - \exp(-T_s/RC)z^{-1})$, for $RC = 0.1$ and $T_s = \{0.1, 0.05\}$, are displayed in Figure 12-32 over a common frequency range. It can be seen that the digital lowpass filter produced by using a standard Z-transform lacks the high frequency roll-off of the original analog model. The difference between

FIGURE 12-31   Simple first-order RC system and discrete-time model.

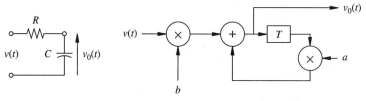

FIGURE 12-32   Magnitude frequency response of a simple first-order IIR system mapped using the standard Z-transform, where $RC = \{0.1\}$ and $T_s = \{0.1, 0.05\}$.

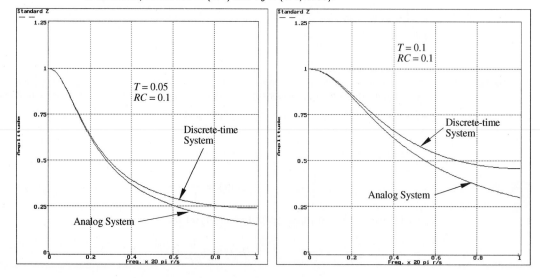

the continuous-time and discrete-time magnitude frequency responses is seen to increase as the bandwidth of the analog lowpass filter increases. As a general rule, the standard Z-transform is used to map $H(s)$ into $H(z)$ if the filter's lowpass cutoff frequency $f_p \ll f_s/2$ ($f_s/2$ is the Nyquist frequency). This implies that many lowpass (wide baseband), most bandpass and bandstop, and virtually all highpass analog filters should not be converted into discrete-time filters using the standard Z-transform.

The previous example demonstrates that the standard Z-transform has limitations in producing a useful continuous-time to discrete-time IIR filter model that preserves the shape of the magnitude frequency response of the analog filter. A Z-transform that more faithfully maps the frequency-response profile of a continuous-time LTI system (defined by a linear ODE) into a discrete-time system (defined by a linear difference equation) exists; it is called the *bilinear Z-transform*. The bilinear Z-transform is based on a discrete-time model of an integrator which makes it well-suited for use in mapping systems defined by linear ODEs into discrete-time equivalent systems. In particular, the bilinear Z-transform is based on a Riemann integral. Formally, the Riemann integral of a time-series $x[n]$ was given by

$$y[n+1] = y[n] + \frac{T}{2}(x[n] + x[n+1]) \qquad (12\text{-}51)$$

where $y[n]$ is a smoothed estimate of $x[n]$ defined at sample index k. The Z-transform of Equation 12-51 is given by

$$Y(z) = z^{-1}Y(z) + \frac{T}{2}[z^{-1}X(z) + X(z)] \qquad (12\text{-}52)$$

or, in terms of a transfer function, the Riemann integrator can be modeled as

$$H(z) = \frac{T}{2}\frac{(z+1)}{(z-1)} \qquad (12\text{-}53)$$

The discrete-time integrator, defined by Equation 12-53, has an ideal continuous-time counterpart given by $H(s) = 1/s$ (i.e., an ideal integrator). Equating $H(s)$ and $H(z)$, one obtains

$$s = \frac{2}{T}\frac{(z-1)}{(z+1)} \qquad (12\text{-}54)$$

Equation 12-54 is called the bilinear Z-transform. Before establishing what the bilinear Z-transform is, it is worthwhile pointing out what it is not. The standard Z-transform was previously defined as an impulse invariant transform (see Chapter 7). The bilinear Z-transform does not have this property as Example 12-13 demonstrates.

Example 12-13   Impulse invariance

The integral of a step function is a ramp function, denoted $r(t) = t$. If the ramp function is sampled at a rate of one sample per second, a time series $r[n] = n$ results. The Laplace transform of $r(t)$ is $R(s) = 1/s^2$ and the standard Z-transform of $r[n]$ is given by $R(z) = z^2/(z^2 - 2z + 1)$. Recall that a time sample of an impulse-invariant transform will agree with that of the original signal. The time series derived from $R(z)$ is given by

$$
\begin{array}{r}
1 \quad +2z^{-1}+3z^{-2} \\
\hline
z^2 - 2z + 1 \overline{\smash{)}z^{-2}} \\
\end{array}
$$

$$
z^{-2}- \ 2z+ \quad 1
$$

$$
2z- \quad 1
$$

$$
2z- \quad 4+2z^{-1}
$$

$$
\cdots \quad \cdots
$$

which can be seen to preserve the value of $r(t)$ at the sample instances.

The bilinear Z-transform of the double integrator for $T_s = 1$ satisfies

$$
\frac{1}{s^2} = \left(0.5\frac{z+1}{z-1}\right)^2 = 0.25\frac{z^2 + 2z + 1}{z^2 - 2z + 1}
$$

The bilinear Z-transform, however, produces a time-series given by

$$
\begin{array}{r}
0.25 \quad +1z^{-1}+2z^{-2}+ \ \cdots \\
\hline
z^2 - 2z + 1 \overline{\smash{)}0.25z^{-2}+0.5z+0.25} \\
\end{array}
$$

$$
0.25z^{-2}-0.5z+0.25
$$

$$
z
$$

$$
z- \quad 2+ \ z^{-1}
$$

$$
2- \ z^{-1}
$$

$$
2-4z^{-1}+2z^{-2}
$$

$$
\cdots \quad \cdots
$$

which does not align itself with the sample values of $r(t)$. Thus, the bilinear Z-transform is not an impulse-invariant transform.

Refer again to Equation 12-54. It immediately follows that the bilinear Z-transform can also be expressed as

$$
z = \frac{(2/T_s) + s}{(2/T_s) - s} \tag{12-55}
$$

Based on Equations 12-54 and 12-55, it can be seen that moving from the s-domain to the Z-domain, or vice versa, is a straightforward algebraic exercise. This is not true for a standard Z-transform based on the complex exponential $z = \exp(sT_s)$. This, however, is an insufficient reason to select the bilinear Z-transform over the standard Z-transform in the design of IIR filters. The true advantage of the bilinear Z-transform is found in its ability to define a digital filter $H(z)$, which has a frequency response close to that of an analog filter $H(s)$ parent. Obviously, this is important in the design of classic IIR filters, which are based on analog models.

12-11 WARPING

Recall that the frequency response of an analog filter can be determined by evaluating $H(s)$ along the $j\Omega$-axis in the s-domain. Similarly, the frequency response of a discrete-time filter can be determined by evaluating $H(z)$ along the contour $z = \exp(j\omega)$, for $\omega \in [0, \pi]$, in the z-domain. Therefore, in order to be useful in designing frequency selective filters, the bilinear Z-transform must be able to map $j\Omega$-axis onto the contour $\exp(j\omega)$, where Ω is called the analog frequency and ω is the digital frequency. Direct substitution into Equation 12-54 yields

$$j\Omega = \frac{2}{T_s}\frac{e^{j\omega} - 1}{e^{j\omega} + 1} = \frac{2}{T_s}\frac{j\sin(\omega/2)}{\cos(\omega/2)} = \frac{2}{T_s}j\tan(\omega/2) \qquad (12\text{-}56)$$

Equation 12-56 is called the *prewarping equation*, which can be simplified to read

$$\Omega = \frac{2}{T_s}\tan\left(\frac{\omega T_s}{2}\right) \qquad (12\text{-}57)$$

Equivalently, the *warping equation* is given by

$$\omega = \frac{2}{T_s}\tan^{-1}\left(\frac{\Omega T_s}{2}\right) \qquad (12\text{-}58)$$

Equations 12-57 and 12-58 establish a nonlinear relationship between the analog and digital frequency axes, which is graphically interpreted in Figure 12-33. The analog to digital frequency axis relationship established by the warping Equation

FIGURE 12-33   Warping of the analog and discrete-time frequency axes under the bilinear Z-transform.

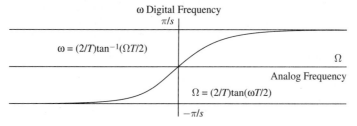

12-57, and that defined by the standard Z-transform, namely $z = \exp(j\Omega T_s)$, is quite different. The bilinear Z-transform, for example, is a single-valued mapping between the frequency axis for $\omega T_s \in (-\pi, \pi)$, where $\omega/T_s = \pi$ corresponds to the Nyquist frequency and $\Omega \in (-\infty, \infty)$. Thus the bilinear Z-transform is capable of resolving even the highest frequency components (albeit warped) of an analog filter.

12-12   CLASSIC DIGITAL IIR FILTER DESIGN

A classic digital IIR is defined in terms of the Z-transform of a classic analog model. In general, the transform of preference is the bilinear Z-transform. Because of warping, the critical frequencies of the analog-filter model will differ from those of the target digital filter. The actual values of the analog system's critical frequencies are referred to as the prewarped critical frequencies and are defined by Equation 12-57, where ω is a digital (filter) frequency and Ω is an analog (filter) frequency. The classic IIR design paradigm is presented in Figure 12-34 and consists of the following steps:

1 Specify the desired digital filter requirements
2 Prewarp digital frequencies into analog frequencies
3 Design an analog prototype filter

FIGURE 12-34   Classic IIR design paradigm using a bilinear Z-transform.

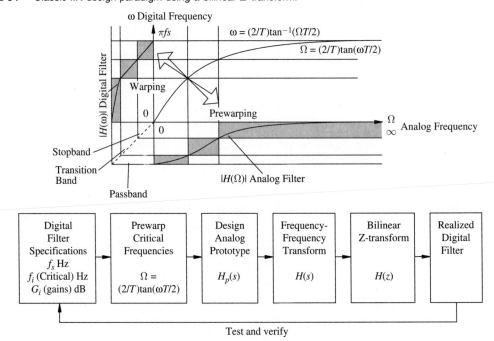

4 Design a final analog filter $H(s)$ using frequency-frequency transforms

5 Design a digital filter $H(z)$ using a bilinear Z-transform of $H(s)$

6 (Optional) Implement digital filter in hardware or software (see architecture)

While this method may seem to be complicated at first glance, it can be easily implemented with a general-purpose digital computer. Several examples follow to demonstrate how classic IIRs are designed using the bilinear Z-transform. It is worth noting that if a classic digital IIR filter was to be designed using a standard Z-transform, then step 2 of the above list (prewarping) would be omitted.

Example 12-14 Step-by-step lowpass IIR design

A Butterworth filter is required which meets or exceeds the following specifications:

Maximum passband attenuation: 3 dB

Passband $f \in [0, 1]$ kHz

Minimum stopband attenuation: 10 dB

Passband $f \in [2, 5]$ kHz

Sample frequency $f_s = 10$ kHz

Given these desired filter specifications, the design steps previously developed were used to derive the digital filter shown in Figure 12-35. First, they consist of determining the prewarped analog frequencies which are given by

$$\Omega_p = \frac{2}{T} \tan\left(\frac{\omega_p T}{2}\right) = 20,000 \tan(0.1\pi) = 6498.4 \text{r/s} \Rightarrow 1.0345 \text{ kHz}$$

$$\Omega_a = \frac{2}{T} \tan\left(\frac{\omega_a T}{2}\right) = 20,000 \tan(0.2\pi) = 14531 \text{r/s} \Rightarrow 2.312 \text{ kHz}$$

From this, the order of the analog prototype can be computed to be

$$N = \frac{\log((A^2 - 1)/\epsilon^2)}{2\log(1/k)} = 1.319 < 2$$

and second-order Butterworth analog prototype model is given by

$$H(s) = \frac{1}{s^2 + 1.414s + 1}$$

Upon applying the lowpass-to-lowpass frequency transform, namely $s \rightarrow s/6498.4$, the second-order analog model results

$$H(s) = \frac{4.229 \times 10^7}{s^2 + 9.2 \times 10^3 s + 4.23 \times 10^7}$$

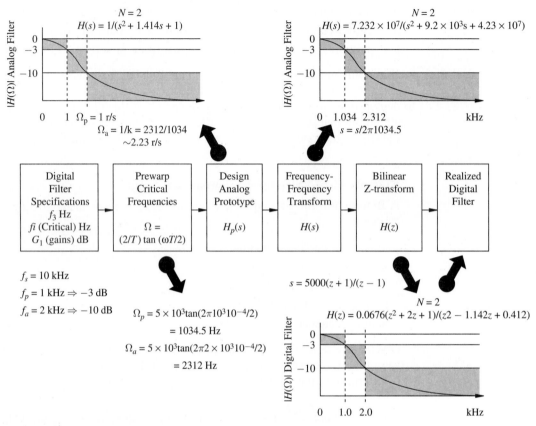

FIGURE 12-35   Filter design procedure.

Finally, applying the bilinear Z-transform results in the digital filter[6]

$$H(z) = 0.0676 \frac{z^2 + 2z + 1}{z^2 - 1.142z + 0.412}$$

The magnitude frequency response of H(z) is shown in Figure 12-36, as is the pole and zero distribution of H(z). Since the poles are seen to reside interior to the unit circle, the filter is BIBO stable.

Example 12-15   Step-by-step Bandpass IIR design

A digital Butterworth filter is required that meets or exceeds the following specifications:

---

[6]For MONARCH users, the Butterworth lowpass filter that meets the design criteria is saved under filename B12-14 with analog prototype, frequency transformed analog, and digital filter report saved under filename B12-14.RPT.

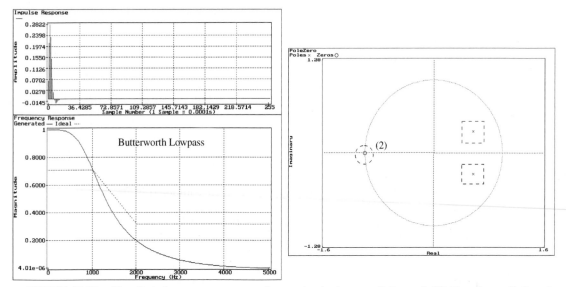

FIGURE 12-36   Magnitude frequency response of second-order lowpass Butterworth IIR filter along with its pole and zero distribution.

Maximum passband attenuation: 1 dB

Passband $f \in [2, 3]$ kHz

Minimum stopband attenuation: 20 dB

Stopband 1 $f \in [0, 1]$ kHz

Stopband 2 $f \in [4, 5]$ kHz

Sample frequency $f_s = 10$ kHz

Again the first step is one of determining the prewarped analog frequencies given by

$$\Omega_{p1} = \frac{2}{T} \tan\left(\frac{\omega_{p1} T_s}{2}\right) = 20,000 \tan(0.2\pi) = 14526 \text{ r/s} \Rightarrow 2.312 \text{ kHz}$$

$$\Omega_{p2} = \frac{2}{T} \tan\left(\frac{\omega_{p2} T_s}{2}\right) = 20,000 \tan(0.3\pi) = 27520 \text{ r/s} \Rightarrow 4.38 \text{ kHz}$$

$$\Omega_{a1} = \frac{2}{T} \tan\left(\frac{\omega_{a1} T_s}{2}\right) = 20,000 \tan(0.1\pi) = 6500 \text{ r/s} \Rightarrow 1.0345 \text{ kHz}$$

$$\Omega_{a2} = \frac{2}{T} \tan\left(\frac{\omega_{a2} T_s}{2}\right) = 20,000 \tan(0.4\pi) = 61554 \text{ r/s} \Rightarrow 9.79 \text{ kHz}$$

Observe that the analog bandwidth is 2.07 kHz (vs. 1 kHz for the digital filter). Since

$$\Omega_u^2 = \Omega_{a1}\Omega_{a2} = 10.08 \times 10^6 < \Omega_{p1}\Omega_{p2} = \Omega_v^2 = 10.12 \times 10^6$$

the transition ratio, in this case, is given by $k = k_2$. For $\Omega_0 = \Omega_{p1}\Omega_{p2}/(\Omega_{p1} - \Omega_{p2}) = 10.12 \times 10^6/(2.07 \times 10^3) = 4.888 \times 10^3$, the transition ratio can be computed to be

$$k_2 = -\frac{\Omega_{a2}}{\Omega_0} \frac{1}{(1 - \Omega_{a2}^2/\Omega_v^2)} = -\frac{9.79}{4.888} \frac{1}{1 - 95.84/10.12} = 0.236$$

The order of the analog prototype is then computed to be

$$N = \frac{\log((A^2 - 1)/\epsilon^2)}{2\log(1/k_2)} = \frac{\log((10 - 1)/0.35^2)}{\log(1/0.236)} = \frac{1.886}{0.627} = 2.975 < 3$$

A third-order Butterworth analog prototype is given by

$$H_p(s) = \frac{1}{s^3 + 2s^2 + 2s + 1}$$

Upon applying the lowpass-to-bandpass frequency transform, namely $s \rightarrow (s^2 + \Omega_{p1}\Omega_{p2})/s(\Omega_{p2} - \Omega_{p1}) = (s^2 + 4 \times 10^8)/s(1.3 \times 10^4)$, the following sixth-order analog model results

$$H(s) = \frac{4.31 \times 10^{12}s^3}{s^6 + 3.25 \times 10^4 s^5 + 1.73 \times 10^9 s^4 + 3.03 \times 10^{13} s^3 +}$$

$$6.92 \times 10^{17} s^2 + 5.20 \times 10^{21} s + 6.39 \times 10^{25}$$

Finally, applying the bilinear Z-transform results in the digital filter[7]

$$H(z) = 0.0304 \frac{z^6 - 3z^4 + 3s^2 - 1}{z^6 + 1.48z^4 + 0.92s^2 + 0.20}$$

The magnitude frequency response of $H(z)$ is shown in Figure 12-37. Also shown in Figure 12-37 is the pole and zero distribution of $H(z)$. Since the poles are seen to reside interior to the unit circle, the filter is BIBO stable.

Example 12-16   IIR Comparisons

Classic IIR lowpass filters are required to meet or exceed the following specifications:

---

[7]For MONARCH users, the Butterworth lowpass filter that meets the design criteria is saved under filename B12-15 with analog prototype, frequency transformed analog, and digital filter report saved under filename B12-15.RPT.

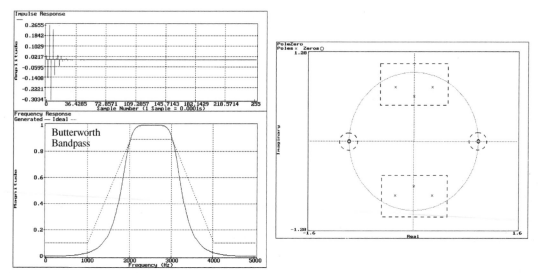

FIGURE 12-37   Magnitude frequency response of sixth-order bandpass Butterworth IIR filter along with its pole and zero distribution.

Maximum passband attenuation: 1 dB

Passband $f \in [0, 1500]$ Hz

Minimum stopband attenuation: 40 dB

Stopband $f \in [2500, 5000]$ Hz

Sample frequency $f_s = 10$ kHz

The methods used in the previous examples are also applied here to derive the appropriate analog prototype. The filter orders and digital filter models are computed as follows

Butterworth ($N = 8$)

$$H(z) = 5.7 \times 10^{-4} \frac{z^8 + 8z^7 + 28z^6 + 56z^5 + 70z^4 + 56z^3}{z^8 - 2.82z^7 + 4.30z^6 - 4.09z^5 + 2.61z^4 - 1.13z^3}$$

$$\frac{+28z^2 + 8z + 1}{+0.32z^2 - 0.05z + 0.004}$$

Chebyshev I ($N = 5$)

$$H(z) = 2.02 \times 10^{-3} \frac{z^5 + 5z^4 + 10z^3 + 10z^2 + 5z + 1}{z^5 - 3.16z^4 + 4.76z^3 - 4.05z^2 + 1.93z - 0.41}$$

Chebyshev II ($N = 5$)

$$H(z) = 5.23 \times 10^{-2} \frac{z^5 + 2.07z^4 + 3.17z^3 + 3.17z^2 + 2.07z + 1}{z^5 - 1.20z^4 + 1.21z^3 - 0.49z^2 + 0.16z^2 - 0.013}$$

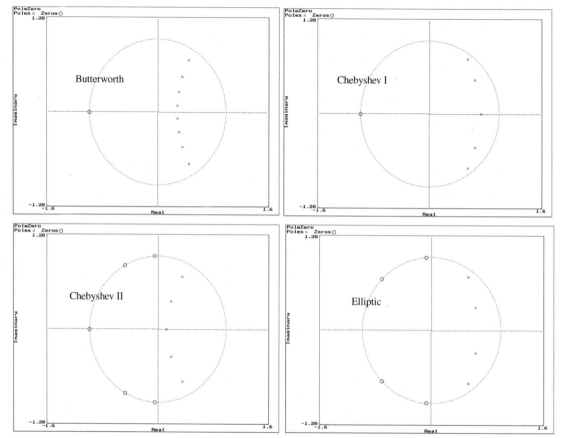

FIGURE 12-38   Pole-zero distribution of eighth-order Butterworth, fifth-order Chebyshev I, fifth-order Chebyshev II, and fourth-order elliptic IIR filters.

Elliptic $(N = 4)$

$$H(z) = 2.09 \times 10^{-2} \frac{z^4 + 1.56z^3 + 2.19z^2 + 1.56z + 1}{z^4 - 2.34z^3 + 2.69z^2 - 1.58z + 0.413}$$

Computer Study   The pole-zero distribution of each classic IIR filter can be displayed using the POLE-ZERO display option as reported in Figure 12-38[8]. The frequency response of the four IIR filters can be compared using S-file 12_16.s. The S-file generates the magnitude and log magnitude frequency responses and displays as an overlay graph.

---

[8]For MONARCH users, the IIR filters are saved under filenames "zbutt.iir," "zcheb1.iir," "zcheb2.iir," and "zellip.iir."

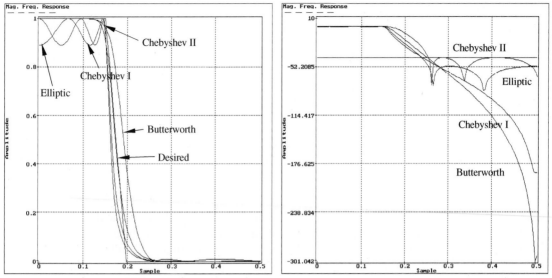

FIGURE 12-39   Comparison of Classic IIRs showing: (left) magnitude frequency responses, and (right) log magnitude frequency responses.

SIGLAB	MATLAB
>include "ch12_16.s"	Not available in Matlab

The results are reported in Figure 12-39. It can be immediately seen that each filter exhibits the classic shape of their analog-filter class. In all cases, the design specifications can been seen to be met.

A classic IIR produces a magnitude frequency response that is an approximation of a piecewise-constant ideal filter (see Figure 12-2). These filters can be linearly combined to form a more complex system with an arbitrary shape. In such cases, a set of IIRs are built where each filter is only responsible for a band of frequencies. Generally, the passband of each filter does not significantly overlap the passband of another. The output of each filter is weighted, as shown in Figure 12-40, to shape the final system response. The quality of the realized filter is a function of the number of sub-band filters used and their overlap.

Example 12-17   Ten-band equalizer

Audio reproduction systems are often provided with a system of N band-selectable filters called an N-band equalizer. If the center frequency of each subsystem differs from its neighbor by a factor of two (i.e., $\times 2$ or $\times 1/2$), then the filters are said to be *octave filters* or *sub-band filters*. The gain of each subsystem can be manually adjusted to suit the listener's preference.

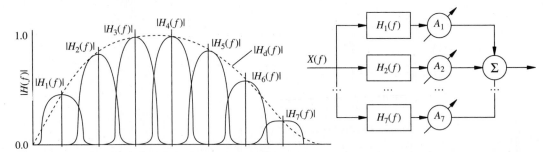

FIGURE 12-40   Arbitrary 7-band filter designed by combining the weighted outputs of a collection of non-overlapping sub-band bandpass IIRs.

Assume that a signal is sampled at a 44.1 kHz rate and the frequencies of interest span the entire audio spectrum. This range, beginning near 20 Hz and extending toward 20 kHz, is to be covered by a 10-band equalizer. The center frequencies of interest are

$$f \in [30, 60, 120, 240, 480, 960, 1920, 3840, 7680, 15360] \text{ Hz}$$

which are seen to be separated by an octave of frequency. The passband of each filter is defined to extend to midpoint between a particular filter and its two nearest neighbors. That is, if the ith filter has a center frequency f_i, then its passband is defined over $f \in [3f_i/4, 3f_i/2]$ Hz. As a result, each filter is separated from its neighbor by an octave with nonoverlapping passbands.

Computer Study   Each of the 10 IIRs is designed as a simple second-order passband filter[9]. For demonstration purposes, four of the sub-band filters are used to study the equalization over the frequency range $f \in [720, 9600]$ Hz. The pole-zero distribution for the four sub-band filters are shown in Figure 12-41. The equalizer's magnitude frequency response is produced using S-file ch12_17.s, which displays the equalized audio spectrum in linear and logarithmic units.

SIGLAB	MATLAB
>include "ch12_17.s"	Not available in Matlab

The results are presented in Figure 12-41.

The pole-zero display shows that the location of the filter's complex conjugate pole pairs from the periphery of the unit circle is directly related to its center frequency. For high center frequencies, the filter's bandwidth is large, which can only be achieved if the poles are off the unit circle. For small center frequencies, the filter behaves as a narrow band filter that requires the poles be located near the unit circle.

---

[9]For MONARCH users, the 10 filters are saved under the filename cfreq where "freq" is the center frequency of the filter (i.e., [30, 60, 120, 240, 480, 960, 1920, 3840, 7680, 15360]).

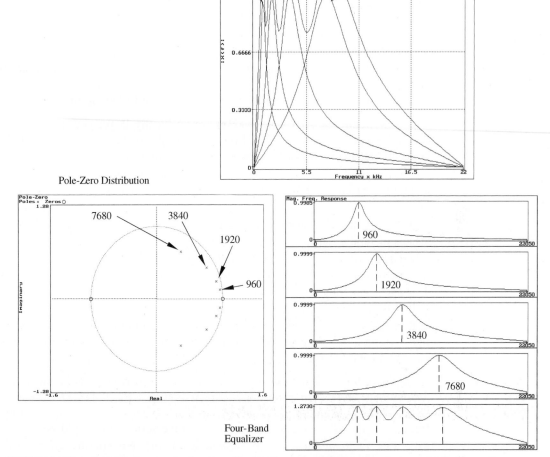

FIGURE 12-41   Four-bands of a ten-band audio equalizer interpreted as a pole-zero distribution, a linear magnitude frequency response, and a logarithmic magnitude frequency response.

12-13   DECIMATION AND MULTI-RATE SYSTEMS

There are times when a signal is sampled well above the required Nyquist frequency. This process is called *oversampling*. The sample rate of a discrete-time system can be reduced using a process called *decimation*. Decimation of a time-series $x[n]$ by an integer D (called decimation by D) produces a new time-series having sample values $y[k] = x[Dk]$. A decimated version of $x[n]$ can be created by simply keeping every Dth sample value and discarding all

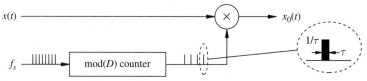

FIGURE 12-42   Model of a simple decimator.

others. Oversampling and decimation can be used together to develop systems that embody several different sampling rates. Such systems are called *multirate* systems.

Oversampling and/or decimation are used for the following reasons:

1 Low frequency ADCs (< 1 kHz) are physically larger and more expensive than midrange ADCs (44.1–330 kHz). The output of the inexpensive ADC can be easily decimated down to a lower sampling frequency.

2 Oversampling can simplify the design of an analog anti-aliasing filter, such as those used on CD-ROMs, by increasing the effective width of their transition band.

3 An oversampled signal contains redundancies that can be used to reduce the effects of random errors.

4 Decimation can be used to adjust the location of a digital filter's critical frequencies and transition bandwidths relative to the sampling frequency. Through such tradeoffs, the complexity of a digital filter system can often be reduced.

5 A real-time digital filter must complete all algorithmic and data-move operations within a sampling period. Reducing the sampling rate relaxes this real-time computational constraint.

Suppose, for example, that $x(t)$ is a bandlimited periodic continuous-time signal with its highest frequency component bounded by B Hz. If $x(t)$ is oversampled, then the sampling frequency is much higher than the Nyquist sampling frequency (i.e., $f_s \gg 2B$). The resulting time-series $x[n]$ could be decimated by a rate D to produce a new time-series $x_D[n]$. No aliasing errors will occur if $f_s/D > 2B$. The continuous-time envelope of $x_D[t]$ can be modeled by gating $x(t)$ with the rectangular sampling function, shown in Figure 12-42, as $\tau \to 0$. For $\tau \to 0$, the periodic sampler, denoted $s_D(t)$, can be modeled as

$$s_D(t) = \sum_{i=-\infty}^{\infty} \delta(t - iDT_s) \qquad (12\text{-}59)$$

where $T_s = 1/f_s$, the oversampling period. The continuous-time Fourier transform of $s_D(t)$, based on the periodic ideal sampling model, is $S_D(f)$ where

$$S_D(f) = \frac{1}{DT_s} \sum_{n=-\infty}^{\infty} \delta\left(f - \frac{n}{DT_s}\right) \qquad (12\text{-}60)$$

The convolution theorem states that the Fourier transform of $x_D(t) = x(t)s_D(t)$ is given by $X_D(f) = X(f) * S_D(f)$, or

$$X_D(f) = \frac{1}{DT_s} \sum_{n=-\infty}^{\infty} X\left(f - \frac{n}{DT_s}\right) = \frac{1}{DT_s} \sum_{n=-\infty}^{\infty} X\left(f - \frac{nf_s}{D}\right) \qquad (12\text{-}61)$$

Therefore, the spectral representation of the signal decimated by D is seen to have the shape of $X(f)$ repeated on nf_s/D Hz centers, where n is an integer. The replicated spectral images of $X(f)$ may or may not overlap, depending upon the Nyquist frequency established by $x(t)$ and the decimation rate D. This is demonstrated by Example 12-18.

Example 12-18   Decimation

Suppose $f_s > 8B$, the magnitude frequency response of $x_D(t)$, is graphically interpreted in Figure 12-43. Observe that for $D \leq 4$, there is the possibility of isolating the baseband spectrum of the original signal with an external filter, as suggested by the superimposed dashed filter responses shown in Figure 12-43. If D is increased beyond the value $D = 4$, then aliasing will occur.

Computer Study   S-file `ch12_18.s` creates a signal having an approximate bandwidth equal to $f_s/8$. The signal is decimated by factors of $D = 2$, 4, and 8, and the results are displayed in Figure 12-44 over a common frequency axis $f \in [0, f_s/2]$.

SIGLAB	MATLAB
>include "ch12_18.s"	Not available in Matlab

It can be seen that the original baseband spectrum is recognizable for $D \leq 4$. If an ideal lowpass filter with a cut-off frequency of $f_s/8$ was used to process

FIGURE 12-43   Decimation spectrum of an oversampled signal.

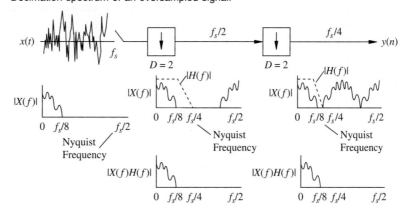

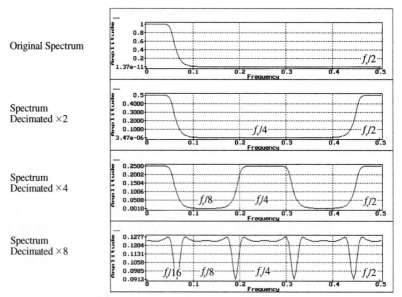

Original Spectrum

Spectrum
Decimated ×2

Spectrum
Decimated ×4

Spectrum
Decimated ×8

FIGURE 12-44   Decimated spectrum of a bandlimited signal showing original and decimation by 2, 4, and 8 spectra. Markers have been inserted to show the cutoff frequency of an ideal recovery filter (i.e., $f_s/8$).

these decimated signals, the filter output would be a facsimile of the input time-series. However, for higher decimation rates, signal recovery is impossible due to aliasing.

In practice, an oversampled signal is decimated to reduce the effective sampling frequency in order to achieve one or more of the advantages already discussed. The resulting multirate system can more easily achieve real-time performance with a lower complexity design than one that must operate at the oversampling rate. However, to ensure satisfactory system performance the effects of aliasing must be controlled, if not eliminated. This is often accomplished by using an anti-aliasing filter having a high signal attenuation at all frequencies above f_s/D for a given decimation rate D.

Example 12-19   Multirate systems

In Example 8.6 an EKG signal, sampled at 135 Hz, was studied. It was noted that the signal also contained unwanted 60, 120, and 180 Hz components. A 5.68 second EKG data record is shown in Figure 12-45. The relevant biomedical spectrum is known to be bounded well below 5 Hz. Based solely on a clean biomedical signal, it would seem that the EKG is being oversampled by a rate $R = \lfloor 135/10 \rfloor = 13$. By sampling at a rate of 135 Hz, the EKG and 60 Hz

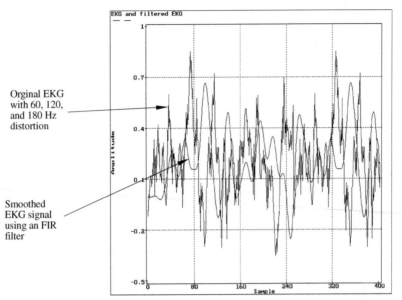

Orginal EKG with 60, 120, and 180 Hz distortion

Smoothed EKG signal using an FIR filter

FIGURE 12-45   EKG signals sampled at 135 Hz with 60, 120, and 180 Hz harmonic distortion and a lowpass filtered EKG signal.

components will not be aliased. However, the signal components existing at 120 Hz and 180 Hz are aliased, as was demonstrated in Example 8-7. It is important to note that while the 120 Hz and 180 Hz components are aliased, they do not fold back into the 5 Hz baseband range occupied by the EKG signal. Therefore, a simple lowpass filter having a 5 Hz passband can be used to eliminate all the unwanted signal components.

The original EKG time-series is passed through a 51st-order FIR filter, having a 5 Hz passband, 5 Hz transition band, and $f_s = 135$ Hz. The smoothed or filtered signal is shown in Figure 12-45. The highest effective frequency in the filtered signal is assumed to be less than 10 Hz. Therefore, the filtered signal can be decimated by an amount $D \leq \lfloor 135/20 \rfloor = 6$ without introducing aliasing errors.

Computer Study   To demonstrate the effects of decimation, a sinusoidal signal of frequency $f_0 = f_2/16$ is decimated by a factor $D = 4$. The signal, its decimated time-series, and their respective spectra are computed using S-file `ch12_19.s` or M-file `ch12-19.m` and displayed in Figure 12-46. It can be immediately seen that decimation has a compression effect in the time-domain and extension effect in the frequency domain. Continuing, the filtered signal shown in Figure 12-45 is decimated by 3 and shown in Figure 12-46.

SIGLAB	MATLAB
>include "ch12_19.s"	>> ch12_19

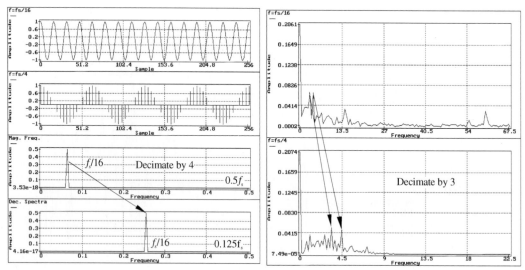

FIGURE 12-46   Decimation effects showing (left) a sinusoidal test signal which is decimated by 4, and (right) the spectra of the original EKG signal and the signal decimated by 3.

The data shows that the spectrum of the decimated signal is a baseband process containing the relevant EKG information. The effective sampling frequency associated with the decimated signal has been reduced from 135 Hz to 45 Hz.

12-14   SUMMARY

In Chapter 12 the design of basic filters was presented. Filters are an essential element in virtually every technical endeavor. The origin of many electronic filters can be traced back to the analog devices used in radio. The art of analog filter design was originally one of manipulating data found in tables of standardized filters. This information processing is now generally accomplished using a digital computer. Using computer programs and the bilinear Z-transform, digital counterparts to classic analog filters can be generated in a straightforward manner. The resulting digital filters are called classic IIRs.

Since designing accurate and long analog delay lines is difficult, digital FIRs have no practical analog counterpart. The FIR possesses many unique features, including the ability to easily achieve linear phase filtering. However, compared to an IIR that meets the same magnitude frequency response specification, the FIR filter cycle may often require many more multiply-accumulate calls. Since multiply-accumulate procedures are time-consuming operations, high-filter order limits the practical bandwidth of a FIR.

The study of FIRs was augmented by the introduction of windows. When properly used, windows were shown to enhance spectral appearance of both FIRs and signals. As a result, they have become a common tool in the study of both signals and systems. The choice of window form is often one of personal preference.

IIR design was based on classic analog-filter models. The mathematical tool that allowed analog filters to be mapped into digital filters, without significant degradation of frequency response performance, was the bilinear Z-transform. The bilinear Z-transform introduced a warping of the frequency axis accounted for by the design rules.

12-15 COMPUTER FILES

These computer files were used in Chapter 12.

File Name	Type	Location	File Name	Type	Location
		Subdirectory CHAP12			
ch12_3.s	C	Example 12-3	ch12_9.s	C	Example 12-9
nalog.m	C	Example 12-3	lattic.m	C	Example 12-10
ch12_4.s	C	Example 12-4	ch12_11.s	C	Example 12-11
comb.m	C	Example 12-4	ch12_16.s	T	Example 12-16
ch12_5.s	C	Example 12-5	ch12_17.s	T	Example 12-17
ch12_5a.s	C	Example 12-5	ch12_18.s	T	Example 12-18
ch12_8.s	C	Example 12-8	ch12_19.s	T	Example 12-19
window.m	C	Example 12-8	ch12_19.m	T	Example 12-19
		Subdirectory SFILES or MATLAB			
lattice.s	C	Example 12-10			
		Subdirectory SIGNALS			
ekg1.imp	D	Examples 12-4, 12-19	oiseekg.imp	D	Example 12-19
linphase.imp	D	Example 12-5	iirfir.imp	D	Example 12-11
firiir.imp	D	Example 12-11	filtekg.imp	D	Example 12-19
		Subdirectory SYSTEMS			
butter.iir	D	Examples 12-3, 12-16	zbutt.(iir/arc)	D	Example 12-16
cheby1.iir	D	Examples 12-3, 12-16	zcheb1.(iir/arc)	D	Example 12-16
cheby2.iir	D	Examples 12-3, 12-16	zcheb2.(iir/arc)	D	Example 12-16
ellip.iir	D	Examples 12-3, 12-16	zellip.(iir/arc)	D	Example 12-16
direct.fir	D	Examples 12-9, 12-9	c(30-15360).arc	D	Example 12-17
equi.fir	D	Example 12-7	deci.arc	D	Example 12-18
but12-14.(iir/arc)	D	Example 12-14	noiseekg.imp	D	Example 12-19
but12-15.(iir/arc)	D	Example 12-15	filterekg.imp	D	Example 12-19

Here "s" is an S-file, "m" an M-file, "x" either an S-file or M-file, "T" a tutorial program, "D" a data file, and "C" a computer-based-instruction (CBI) program. CBI programs can be altered and parameterized by the user.

12-16   CHAPTER PROBLEMS

12-1 Analog filter   What would the order of a -3 dB Butterworth, Chebyshev I, Chebyshev II, and elliptic filter be if it had a transition gain ratio of $\eta = 10^{-3}$ and transition ratio of $k_d = 0.1$? What would be the minimum stopband attenuation for these filters?

12-2 Butterworth analog filter   Design a sixth-order analog Butterworth lowpass filter that meets or exceeds the following specifications: 1 dB passband over [0, 10] Hz; 40 dB stopband.

    a What is the smallest transition band allowed?

    b Where are the filter's poles and zeros?

    c Run a frequency response of filter to verify your claim.

12-3 Z-transform   A continuous-time system is given by

$$H(s) = \frac{10^{18}}{(s^2 + 6 \times 10^4 s + 1 \times 10^9)(s^2 + 2 \times 10^4 s + 1 \times 10^9)}$$

    a What is the continuous-time filter's frequency response?

    b Where the continuous-time filter's poles and zeros?

    c For a sampling frequency of 100 kHz, what is $H(z)$ using the standard Z-transform?

    d What is the discrete-time filter's frequency response?

12-4 Bilinear Z-transform   Same as Problem (12-3) except use a bilinear Z-transform.

12-5 Linear phase filter   Show that the convolution of linear phase FIRs results in another linear phase FIR.

12-6 Symmetric FIR   Determine the properties of the magnitude frequency and phase responses of a symmetric FIR satisfying

$$H(z) = \sum_{n=0}^{M} h[n] z^{-n}$$

where $h[n] = h[2M - n]$ in terms of G(z), where

$$G(z) = z^{-M} H(z)$$

12-7 Equiripple FIR   Design a 51st-order Equiripple FIR solution that provides the best minimal-error fit to the desired response shown in Figure 12-47. Assume $f_s = 12$ kHz and that the passband error is to be 1/2 the maximum stopband error.

    a What is the measured ripple deviation error?

    b What effect would the use of a Blackman window have on your design?

    c Repeat **a** and **b** for an error ratio of 1/10.

FIGURE 12-47   Desired magnitude frequency response for the Problem 12-7.

12-8 FIR filter   Given an N-tap lowpass FIR filter having an impulse response $h[n]$, show that

  a $h'[n] = (-1)^n h[n]$ is a lowpass to highpass transformation.

  b $H'(z) = 1 - H(z)$ is a complementary filter (i.e., $H'(z) + H(z) = 1$).

  Use the 51st-order equiripple lowpass FIR used in Problem 12-7 to experimentally verify the claims above.

12-9 Windowed FIR   Compare the magnitude frequency responses of a 51st-order lowpass equiripple FIR having a 1 kHz passband, 100 Hz transition band, and a sampling frequency of 4 kHz, using the following windows in terms of the maximum passband and stopband errors and the transition bandwidth.

  a Rectangular window

  b Hamming

  c Hann

  d Blackman

  e Kaiser ($\beta = 2$)

  f Kaiser ($\beta = 5$)

12-10 Windowing   In Example 12-9, windows were applied to an FIR. One of the effects observed was a broadening of the transition band. Demonstrate this effect by comparing the transition band of an FIR modified by a window.

12-11 Hilbert filter   Hilbert filters will be discussed in Chapter 13. A *Hilbert filter* is an all-pass filter with an ideal phase response shown in Figure 12-48. A Hilbert FIR is to be designed with sampling rate of 100 kHz. It is to have a flat passband over $[5, 45]$ kHz. Test the Hilbert filter's phase response (use a two-sided FFT) for $N = 63$ (Hint: modify S-file Ch. 13-4 or create an equivalent M-file).

12-12 Half-band FIR   A *half-band* FIR has a frequency response that is symmetrically distributed about the 1/4 sampling frequency. An example of a lowpass half-band filter is shown in Figure 12-49. Prove that if the half-band FIR order is $N = 4M + 1$, where M and N are integers, only M-distinct FIR tap weights are needed to implement the filter. Design a 31st-order half-band FIR with a transition band equal to 10% of the sampling frequency. Experimentally verify the half-band claim.

FIGURE 12-48   Hilbert filter phase response.

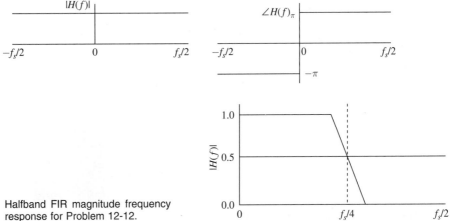

FIGURE 12-49   Halfband FIR magnitude frequency response for Problem 12-12.

12-13 Matched filter   Design a 31st-order FIR that is matched to a prespecified signal $x[k]$. The impulse response of the *matched filter* is given by $h[k] = x[62-k]$ (i.e., reversed and delayed in time). Using a 31-sample Guassian signal, padded with zeros as shown in Figure 12-50, conduct an experiment by convolving the signal $x'[k]$ with $h[k]$. Explain the maximal value and location (in sample delays) of the filter output. Repeat the experiment with a circularly shifted version of $x'[k]$, shifting $x'[k]$ by 10 and 20 samples. Explain how this technology could be used to build a radar.

12-14 Butterworth IIR   Design a digital Butterworth IIR filter having: -0.5 dB passband over $[0, 800]$ Hz; -45 dB stopband for $f \geq 1600$ Hz; $f_s = 5000$ Hz.

12-15 Chebyshev IIR   Design a digital Chebyshev-I IIR filter having: -0.5 dB passband over $[0, 800]$ Hz; -45 dB stopband for $f \geq 1250$ Hz; $f_s = 5000$ Hz.

12-16 Chebyshev IIR   Design a digital Chebyshev-II IIR filter having: -0.5 dB passband over $[0, 800]$ Hz; -45 dB stopband for $f \geq 1250$ Hz; $f_s = 5000$ Hz.

12-17 Elliptic IIR   Design a digital Elliptic IIR filter having: -0.5 dB passband over $[0, 800]$ Hz; -45 dB stopband for $f \geq 1000$ Hz; $f_s = 5000$ Hz.

12-18 Multi-rate filter   A discrete-time signal has a spectrum shown in Figure 12-51 is sampled at a rate $f_s = 20Hz$.

    a What is the output spectrum if the signal is decimated by 2 and 4? Verify your results experimentally.

    b A 2 Hz ideal lowpass filter is placed before the decimator. What is the maximum decimation rate that produces a non-aliased output? Experimentally verify your claim by decimating at, above, and below this value.

12-19 Anti-aliasing filter   Using an order 63 Equiripple FIR, design a digital anti-aliasing filter to eliminate the nonbiomedical components from an EKG signal sampled at $f_s = 135$ Hz. Let your lowpass filter have 5 Hz baseband bandwidth and a stopband beginning at 10 Hz. Decimate the filtered output by 4. What is the new Nyquist frequency? Experimentally verify that output spectrum is essentially free of aliasing errors. Derive a formula that will predict the amount of aliasing that would take place from a signal component located at 60 Hz of amplitude A. At what frequency would the aliasing error be found?

12-20 Group delay distortion   Consider the fourth-order elliptic IIR filter studied in Example 12-16. Produce the filter's group delay. Run a sinusoid through the filter

FIGURE 12-50   Matched filter.

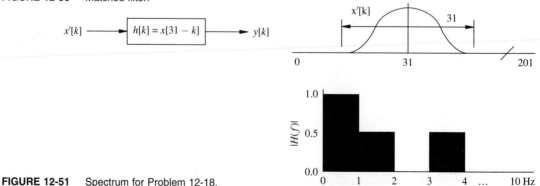

FIGURE 12-51   Spectrum for Problem 12-18.

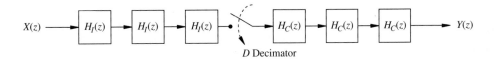

FIGURE 12-52   Cascaded integrator-comb filter.

at frequencies $f = 1$ k, 1.25 k, 1.5 k, and 1.75 kHz. Measure the delay in the output relative to the input signal (Hint: use correlation). How do these values compare to the computed group delay.

12-21 Cascaded integrator-comb (CIC) filter   A third-order CIC filter is shown in Figure 12-52, where

$$H_I(z) = \frac{1}{1 - z^{-1}}$$

and

$$H_C(z) = 1 - z^{-1}$$

Derive a formula for the CIC transfer function $H(z)$. Compute the filter's magnitude frequency response.

12-22 Ten-band equalizer   Repeat Example 12-17 for a full ten-band equalization.

12-17   COMPUTER PROJECTS

12-1 Complement and mirror filters   Two FIRs, for example $H_1(z)$ and $H_2(z)$, are said to be *complementary* if $H_1(f) + H_2(f) = 1$. Together they form an all-pass filter. The two filters are said to be *power complementary* if $|H_1(f)|^2 + |H_2(f)|^2 = 1$, which essentially ignores phase differences between the filters. Both forms play an important role in the design of FIR systems.

A complement filter, having an FIR impulse response $H(z)$ given by

$$H(z) = \sum_{i=-N}^{N} h_i z^{-i}$$

has a mathematical specification given by

$$H_{\text{comp}}(z) = 1 - H(z) = 1 - \sum_{i=-N}^{N} h_i z^{-i}$$

where the 1 is added to the midpoint of the FIR filter. If the FIR is causal, which is normally the case, then

$$H(z) = \sum_{i=0}^{N} h_i z^{-i}$$

Then, the complementary filter satisfies

$$H_{\text{comp}}(z) = 1 - H(z) = z^{-N} - \sum_{i=0}^{2N} h_i z^{-i}$$

That is, a complementary filter can be derived from an $(2N + 1)$st-order filter by using the architecture shown in Figure 12-53.

If the impulse response of a lowpass FIR is $H_1(z) = \sum h_{1i} z^{-i}$, then the filter $H_2(z) = \sum h_{2i} z^{-i}$ is a highpass FIR, when $h_{2i} = (-1)^i h_{1i}$. The multiplication by $(-1)^i$ can be thought of as modulation by the sequence $\cos(\pi i)$, where π corresponds to the Nyquist sampling frequency. It can be seen that a lowpass filter can be mapped into a highpass filter by simply alternating the sign of the lowpass FIR coefficients. Such filters are called *mirror filters*.

Design a 63rd-order lowpass equiripple FIR with a normalized passband defined over $[0, 0.24]$, and a normalized stopband defined over $[0.25, 0.5]$. This will serve as a quadrature-mirror filter element. If the weights in the passband and stopband are unity, then the gain at the middle of the transition band, namely at $f = 0.25 f_s$, would be 0.5. Power-complement filters, however, have a midfrequency gain on the order of 0.707. To achieve this effect, adjust the passband and stopband weights (e.g., a passband weight of 30 and stopband weight of 1).

Design a mirror filter using the same specifications, except the passband and stopband gains and weights are reversed. Test the quadrature-mirror system to verify that it meets the design objectives.

12-2 Multi-rate system   Design a DSP system to process EKG data. Assume that the information process is a baseband signal having a usable bandwidth $f \in [0, 2]$ Hz. The beginning of a data record is indicated by a tone that is added to the EKG record. Normally, the tone would be chosen to be at a frequency that would not interfere with the EKG. However, in this case, the tone is adjusted to overlap the EKG spectrum. The tone is given by $x(t) = a \cos(2\pi f_0 t)$ for $f_0 \in [1, 2]$ Hz, where a is a constant. Two systems are proposed to produce a periodogram (ensemble averaging $L \leq 10$ consecutive power spectra) and to detect the tone if present (see Figure 12-54). The periodogram will only analyze signals over the frequency range $[0, 5]$ Hz. Design the filters shown in Figure 12-54 where all IIRs are to be Chebyshev II filters with a passband attenuation ≤ 1.5 dB and stopband attenuation ≥ 12.5 dB. The IIR bandpass filter is to have a passband of 1.5 ± 0.5 Hz and guardbands (i.e., transition bands) 0.3 Hz wide. All FIRs are to be equiripple filters.

What does the all-IIR design look like? What does the multi-rate design look like? Compare the performance of each system in terms of periodogram resolution. Can your system be built using a DSP microprocessor capable of performing an FFT in 10 ms? If the real multiply-accumulate speed of the microprocessor was 10 MHz, how complex would each hardware design be.

Assume that the input signal to noise ratio (tone/EKG) is -10 dB. Apply an interfering tone and define a detection threshold such that the presence of the tone is detected about 90% to 99% of the time (use repeated experiments and circularly shifted inputs to create a statistically meaningful database). What is the false alarm rate for these two cases?

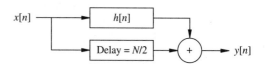

FIGURE 12-53   Complementary filter.

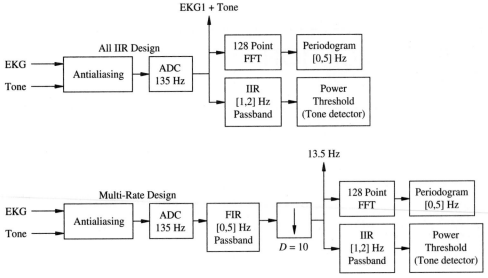

FIGURE 12-54   EKG signal-processing system.

13

COMMUNICATION SYSTEMS

This coded character set is to facilitate the general interchange of information among information processing systems, communication systems, and associated equipment. An 8-bit set was considered but the need for more than 128 codes in general applications was not yet evident.

—ASA Subcommittee X3.2 (1963)

13-1 INTRODUCTION

A communication system carries information from one point to another. The three elements of a communication system are called the *transmitter*, *channel*, and *receiver* (see Figure 13-1). The transmitter performs operations such as modulation and information coding. The communication channel carries the message from the transmitter to the receiver. In a real system, the channel may also introduce some distortion and unwanted noise. The receiver provides amplification, demodulation, filtering, and information decoding. The information to be communicated, called the *message*, may be analog or digital, coded or uncoded, broadband or narrowband.

Human communication systems are rooted in our basic sensory systems. Our vision system, for example, uses a line-of-sight path as the communication channel between the viewer and an object being observed. Our audio sensors receive information from acoustical waves propagating through a variety of media including air, water, and solids. These are examples of passive systems where the receiver simply collects information without expending energy. Active communication skills were developed by early humans for the purpose of survival. Through gestures

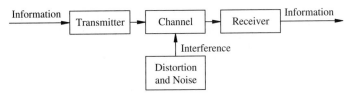

FIGURE 13-1 Communication system.

and audible sounds, our prehistoric ancestors were able to warn their clan against dangers as well as coordinate hunting activities. Later, the need to express more abstract concepts led to an oral language. These sensory-based systems served us for centuries, but did have major drawbacks; these systems had limited range and were inaccurate as an archiving system. The pressures of business, commerce, and the expansion of religion created a need for a written language. Originally, our written language took the form of picture graphs and icons. With printed and written language, information could be passed from generation to generation and be made mobile. While books made communication worldwide sytem (albeit a limited world), written communication was slow, expensive, and unidirectional. Speed and range were expanded with the introduction of electrical communication systems, in the form of the telegraph and telephone. Telephony, perfected by Alexander Graham Bell, was first installed as a system in New Haven, Connecticut in 1878. Wireless telegraphy was demonstrated by Marconi shortly thereafter in 1897. A quantum improvement occurred when electronics made radio and television household realities. This revolution began with Lee DeForest's triode vacuum tube and led to commercial AM broadcasting. Television, which is often believed to be a post-World War II invention, was actually first demonstrated during the 1920s, using cathode-ray tubes and analog electronics. Our current color TV format was defined in the early 1960s and produced the same intense, vested technical and political dialogs now being played out in the areas of digital tapes, HDTV, and other commercial communications technologies.

The information age is the result of the communications advances that now use satellite links (first established in 1962) and optical fibers as channels. In the future, intergalactic communications will take place (e.g., from the planet Fibonacci). In all cases, whether primitive or complex, the rate at which information can be communicated from the transmitter to the receiver is limited by the physical properties of the communication channel, external influences on the channel, and the information encoding methodology.

13-2 CHANNEL CHARACTERISTICS

Channel distortion, which can appear in many forms, is a property of the channel and will disappear when the channel is turned off. Distortion can occur in amplitude, phase, or in both. Linear channels behave like linear systems and are, therefore, convolution systems, or equivalently, linear filters. This means that they cannot introduce any new frequencies to the output that are not present in the input. As a result, linear channels are desired for frequency-based communication systems. However, presenting a pulse of duration T seconds to a linear system will produce a convolved output that can persist for much longer than T seconds. This effect is called *channel dispersion* or *pulse spreading*. Dispersion can create problems when the information to be transmitted is a pulse train or a string of pulses. Pulse spreading, in this case, can cause the persistence of one pulse to interfere with another. This effect is called *intersymbol interference.*

Nonlinearities in a channel can give rise to a wide range of distortions, such as rectification and saturation. Whereas a linear system can create no new frequencies, a nonlinear system can. The production of new frequencies can create a number of undesirable effects, including aliasing errors in discrete-time communication systems. There are times, however, when nonlinear effects are intentionally introduced into a channel. Nonlinear operations are often used to compress the dynamic range requirements of a signal. The problem of voice communication over conventional telephone lines is such an example. Speech can be intelligible over many decades of amplitude change. To ensure the intelligibility of softly spoken words, there is a temptation to use high-gain channel amplifiers. However, saturation would occur if someone with a loud voice uses the system. As a result, the large dynamic range of a speech process must be compressed into a workable range for use in voice-grade telephone communication systems. Typical compressors are the *mu-law compander*, used in North America and Japan, and the *A-law compander*, used in Europe. The mu-law compander is defined as

$$y(x) = \frac{\log(1 + \mu|x|)}{\log(1 + \mu)} \tag{13-1}$$

for $|x| \leq 1$, and the *A-law* compander is given by

$$y(x) = \begin{cases} \dfrac{A|x|}{1 + \log(A)} & \text{for } 0 \leq |x| \leq (1/A) \\[2mm] \dfrac{1 + \log(A|x|)}{1 + \log(A)} & \text{for } (1/A) \leq |x| \leq 1 \end{cases} \tag{13-2}$$

for $|x| \leq 1$. Typically values of μ are 100 and 255. Experience has shown that if $\mu = 255$, voice telephone communications can take place without significant distortion being perceived by the listener.

Example 13-1   Mu-law companding

The mu-law compander, defined by Equation 13-1, is often placed before an analog-to-digital converter to compress the dynamic range of a signal prior to conversion. This can be particularly useful in the areas of speech and telephone signal processing.

Computer Study   S-file `ch13_1.s` and M-file `mulaw.m` implement a 255 mu-law compander using the format `mulaw(x,mu)` where x is the input signal and mu is the value of μ in Equation 13-1. An EKG test signal $x[n]$, having an amplitude range of $-0.345 \le x(n) \le 0.851$, is scaled $\times 1$, $\times 10$, $\times 50$, $\times 100$.

SIGLAB	MATLAB
>include "ch13_1.s"	>> load ekg1.imp
>x=rf("ekg1.imp")	>> x = ekg1;
>max(x) 0.8516	>> max(x), min(x)
>min(x) -0.3458	ans =
>x1=x;x10=10*x;x50=50*x;	0.8517
x100=100*x	ans =
>m1=mulaw(x1,255);	-0.3459
m10=mulaw(x10,255)	>> x10 = x*10; x50 = x*50;
>m50=mulaw(x50,255);	x100 = x* 100;
m100=mulaw(x100,255)	>> plot([mulaw(x, 255)
>graph(m1,m2,m3,m4)	mulaw(x10, 255)
compression ratio; 0.744973	mulaw(x50, 255)
compression ratio; 0.0875244	mulaw(x100, 255)])
compression ratio; 0.0178995	Compression ratio 0.745
compression ratio; 0.00897632	Compression ratio 0.08752
	Compression ratio 0.0179
	Compression ratio 0.008976

The results are shown in Figure 13-2. Observe that the companded $\times 100$ signal remains recognizable even though the signal's dynamic range $-34.5 \le x[n] \le 85.1$ compressed into

$$0.7258 \le x_{\text{companded}}(n) \le 1.8008$$

Another type of channel distortion is called *multipath* effects. Commonly found in radio, radar, and sonar signal processing, multipath distortion occurs when

FIGURE 13-2   Companding an EKG signal scaled by 1, 10, 50, and 100, with a mu-law compander.

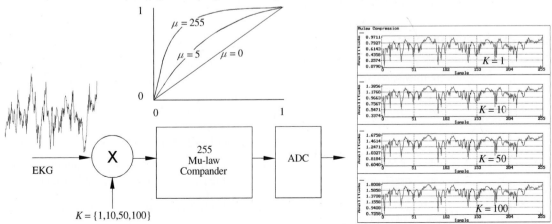

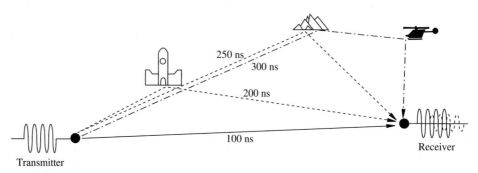

FIGURE 13-3   Multipath signal environment.

transmitted information is reflected by a number of surfaces as shown in Figure 13-3. The reflected signal appears at the receiver as a collection of amplitude scaled and delayed copies of the original transmission. The control of this type of distortion is often achieved by using time-gates or temporal windows that ignore signals following long reflected paths from transmitter to receiver. In some cases, adaptive filters are used to remove multipath copies of a transmitted signal. Adaptive filters, as the name implies, adapt or alter their coefficients in some rule-based manner in order to improve the quality of the filter output in the presence of ever-changing distortion effects.

Fading is a typical time-varying channel distortion. When the communication channel is affected by time-varying effects (e.g., diurnal changes), or relative motion exists between the transmitter and receiver to the reflectors in a multipath channel, the channel's gain and other parameters can undergo continual change. When the change is the channel gain, *automatic gain control (AGC)* circuits can often be used to reduce its adverse effects.

Interference is the result of external influences that alters a signal in an unwanted manner. Noise is an example of random interference. If the interfering signal is independent of the signal it contaminates, then it can often be removed with filters.

13-3   CHANNEL CAPACITY

The rate at which a communication channel can transfer information is limited to its intrinsic bandwidth and noise. In theory, this limit has been established by C.E. Shannon in his celebrated *Shannon's law* (1948), which states that the channel or *system information capacity*, in the presence of white Gaussian noise[1], is given by

$$C = B \log(1 + S/N) \tag{13-3}$$

where S/N is the measured signal-to-noise power ratio, B is the channel or system bandwidth, and C is channel capacity. Observe that for a perfectly clean (noise-

---

[1]It is important to remember that Equation 13-3 applies to the case where a signal is combined with additive Gaussian white noise and is not, in general, valid for other noise cases.

free) channel, $S/N \rightarrow \infty$ and the system becomes one of infinite capacity. In an ideal setting, at any instant of time, a transmitted signal has a real value $x(t)$ that takes on one of an infinite number of possible values. Therefore, it would take an infinite number of bits to encode $x(t)$ without error. At the other extreme $S/N \rightarrow 0$, it then follows that $C \rightarrow 0$. Here any possible signal is overwhelmed by noise and rendered useless, resulting in a system with an information capacity of zero. Reality is generally found between these two extremes. Furthermore, it is important to recognize that while Equation 13-3 is an upper theoretical limit for error-free communication, it does not suggest how the limit may be achieved. Over the last 50 years, engineers have been trying to attain this limit. Advances in communication-systems engineering over the last 10 years have brought us very close to this goal.

A typical communication channel introduces some distortion and signal power loss that, at times, may be significant. For example, satellite uplink and downlink losses can often exceed 200 dB. This means that the power contained in the original message must be carefully managed and kept as distortion-free as possible. Otherwise, as Shannon predicts, a very low channel capacity will result. Since long-distance communication systems are expensive to install and maintain, they can only be justified economically if they provide their owners with a high channel capacity.

Filters can be used to enhance the capacity of a channel. Communication filters are often assumed to have a theoretical ideal frequency response given by the *ideal filter* bandpass shown in Figure 13-4 and defined by

$$H(f) = \begin{cases} G \exp(-j2\pi f t_d) & \text{for } |f| \in [f_L, f_U] \\ 0 & \text{otherwise} \end{cases} \qquad (13\text{-}4)$$

FIGURE 13-4   Ideal (*a*) bandpass, (*b*) lowpass filter, and (*c*) impulse response.

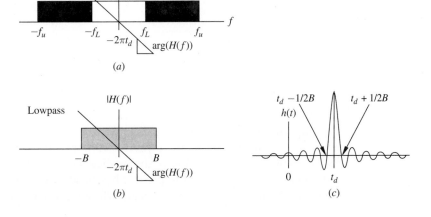

This filter, sometimes called a *boxcar filter*, can also appear as the *ideal lowpass filter*, shown in Figure 13-4 and defined by

$$H(f) = \begin{cases} G \exp(-j2\pi f t_d) & \text{for } |f| \in [0, B] \\ 0 & \text{otherwise} \end{cases} \tag{13-5}$$

Observe that the ideal filter's phase response is given by $\phi(f) = -2\pi f t_d$, which means that the group delay is $\tau = -d\phi(\omega)/d\omega = t_d$ seconds (a constant). The impulse response of an ideal lowpass filter can be directly computed to be

$$h(t) = \mathcal{F}^{-1}[H(f)] = 2BG \operatorname{sinc}(2\pi B(t - t_d)) \tag{13-6}$$

which is graphically interpreted in Figure 13-4. It can be seen that the ideal filter is actually noncausal (anticipatory) since an output appears prior to an input application. Real communication systems are, however, designed using physically realizable (causal) filters. The frequency selectivity of high-order realizable filters can generally be made to approach the magnitude frequency response of an ideal model. However, a high-order filter can create implementation problems. Low-order filters provide only a coarse approximation to the ideal. Since communication systems normally require both high bandwidth and high precision, the design of such systems requires that trade-offs be made.

Example 13-2   Ideal filters

The ideal lowpass filter defined by Equation 13-5 can be studied in the time-domain using an inverse Fourier transform of an ideal frequency-domain model. Two ideal lowpass filters having identical magnitude frequency responses, but different (linear) phase responses, will have similar impulse responses, except one impulse response will be delayed.

Computer Study   S-file ch13_2.s and M-file ideal.m create an ideal low-pass filter and impulse response using a format ideal(t_d), where t_d is the slope adjustment parameter shown in Figure 13-4.

SIGLAB	MATLAB
>include "ch13_2.s"	>>ideal(10)
>ideal(10); ideal(50)	>>ideal(50)

The impulse responses, shown in Figure 13-5 (*a*) and (*b*), are for the cases where $t_d = 10$ and $t_d = 50$ respectively. It can be seen that the impulse responses are noncausal and have the predicted delays.

Channels can be designed to process baseband signals across point-to-point communication links. The simplest and earliest electrical communication chan-

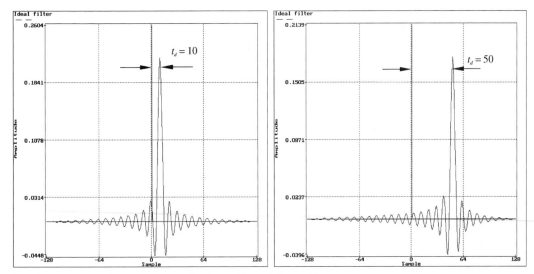

FIGURE 13-5   Ideal filter experiment, (left) $t_d = 10$, (right) $t_d = 50$.

nel is a section of copper wire. An example is the copper communication paths (busses) inside a computer that connect memory to the central processor. The frequency response of an ideal wire will be assumed to be $H(f) \equiv 1$ for all $f \in (-\infty, \infty)$ and, as a result, the transmission from source to receiver can be distortionless. However, in practice, the bus is a baseband filter having a cut-off frequency measured in megahertz. This, according to Shannon, limits the maximum data rate that can be sustained on the bus. Proper termination is absolutely required in order to eliminate reflections that may occur, giving rise to multipath distortions. While the communication channel may not introduce distortion, unwanted interference may nevertheless appear due to noise added by other elements of the system. A stereo amplifier, for example, will magnify tape-hiss plus add its own noise due to imperfect electronics. Filters and equalizers can often be inserted into the signal stream to suppress these effects.

A communication channel may be shared simultaneously by many signals separated in time. This is called a *time division multiplexing (TDM)*. In a time division multiplexing system, each signal occupies a predefined temporal interval that does not overlap that occupied by another signal, as shown in Figure 13-6. A *frequency division multiplexing (FDM)* system separates signal information into nonoverlapping frequency bands using *modulation* techniques, as shown in Figure 13-6. Modulation, in this case, refers to the process by which a low-frequency signal modifies the parameters of a high-frequency signal, called the *carrier* since it carries the message across the channel. The reverse process is called *demodulation*, which occurs after the message arrives at the

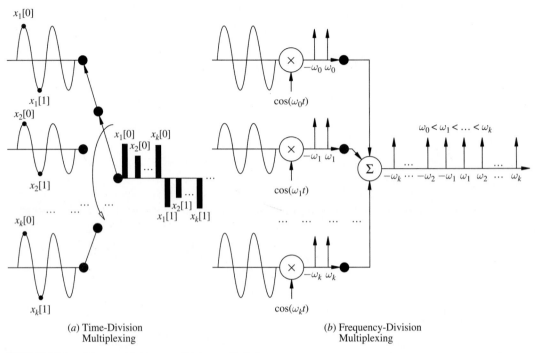

(a) Time-Division
    Multiplexing

(b) Frequency-Division
    Multiplexing

FIGURE 13-6   Time and frequency division multiplexing.

receiver. The more common analog modulation techniques are called *amplitude modulation (AM)*, *phase modulation (PM)*, and *frequency modulation (FM)*.

13-4   CONTINUOUS-TIME MODULATION

The study of classic electric and electronic communications systems usually begins with continuous-time systems. Their origins can be traced to early baseband telephone and telegraph systems. Radio introduced wireless electronic communications and, with it, a host of new opportunities. More recently, optical channels have gained widespread use as a replacement for copper wires and cables. The taxonomy of these continuous-time communication systems is summarized in Table 13-1.

Examples of bandwidth requirements needed to support commonly encountered continuous-wave commercial communications are shown in Table 13-2. In addition, to remain intelligible, a signal's relative power compared to the assumed power level of background noise is also displayed, and it is seen to vary from application to application. It can also be noted that the bandwidths for audio systems are relatively small, compared with those needed to support video. The listening quality of an audio system can be improved by increasing the SNR and bandwidth. Therefore, it can be assumed that the drive to add more and more video

TABLE 13-1   COMMUNICATION SYSTEMS SPECTRA

Frequency	Title	Applications	Media
$(3\text{–}30) \times 10^{14}$	Optical	Optical systems	Laser/Fiber
$(3\text{–}30) \times 10^{10}$	Extremely High Freq.	Line-of-sight experimental	Radio/Waveguide
$(3\text{–}30) \times 10^{9}$	Super High Freq.	Microwave, satellite	Radio/Waveguide
$(3\text{–}30) \times 10^{8}$	Ultra High Freq.	Television, radar, satellite	Radio/Waveguide
$(3\text{–}30) \times 10^{7}$	Very High Freq.	Television, FM, mobile	Radio/Cable
$(3\text{–}30) \times 10^{6}$	High Freq.	Mobile radio, amateur	Radio/Cable
$(3\text{–}30) \times 10^{5}$	Medium Freq.	AM	Radio/Cable
$(3\text{–}30) \times 10^{4}$	Low Freq.	Radio Navigation	Radio/Wires
$(3\text{–}30) \times 10^{3}$	Very Low Freq.	Navigation, telephone	Long Wave/Wires
$(3\text{–}30) \times 10^{2}$	Voice Freq.	Telephone	Long Wave/Wires
$(3\text{–}30) \times 10^{1}$	Extremely Low Freq.	Submarine communications	Surface ducting

and audio content to our personal computers, telecommunications, and entertainment systems may have a high cost in terms of the hardware systems needed to support these technologies. Since bandwidth is often determined by the media used to transport a signal, in the future unshielded (bare) wires must give way to shielded (covered) lines, and cables must give way to fiber optics. Since SNR is determined, in part, by the electronics and modulation techniques employed, future systems will become more sophisticated in the way they use hardware. Digital systems, for example, offer economic advantages, plus can use powerful algorithms to process signals of poor quality, thereby increasing our ability to extract the original message.

Each of the signal classes shown in Table 13-2 can be transmitted as a baseband or modulated signal. Modulation is used to alter the amplitude profile of a signal, (the *envelope*), its phase, or both. Since frequency is the time derivation of phase, phase and frequency modulation are similar concepts. Each modulation technique brings with it advantages and disadvantages. Some modulation schemes are simple and easy to implement in hardware. Others are more complex, but less susceptible to distortion by external influences. Still others work well with digital signals, and others are intended to be used with optical or analog systems. The next several sections develop the basic continuous-time modulation procedures in commonly used today.

TABLE 13-2   TYPICAL FREQUENCY SPECTRAL REQUIREMENTS OF COMMON SYSTEMS

Signal	Frequency range	Signal-to-noise ratio (SNR) in dB
Intelligible voice	500 Hz–2 Khz	5–10
Telephone	200 Hz–3.2 kHz	25–35
AM radio	100 Hz–5 kHz	40–50
Stereo sound systems	20 Hz–20 kHz	55–65
TV video	60 Hz–4.2 MHz	45–55

13-5 AM MODULATION

Many early communication systems were based on the venerable *amplitude modulation (AM)* protocol due to its simplicity. The origins of AM radio are firmly rooted in the era of the vacuum tube, where a tube's limited functionality required that only easily implemented protocols be used. An AM transmission consists of two parts, the *carrier* and the *message*. The traditional AM carrier is a monotone sinewave with a frequency located well above the message's highest frequency. A 3 kHz voice baseband message is carried, for example, at frequencies typically above 500 kHz. If A is the amplitude of the carrier signal and $x(t)$ is the message to be transmitted where $|x(t)| \le 1$, then an AM transmission is given as

$$y(t) = A(1 + \mu x(t)) \cos(\omega_c t) = A \cos(\omega_c t) + A\mu x(t) \cos(\omega_c t) \qquad (13\text{-}7)$$

where ω_c is the *carrier frequency* and μ is the *modulation index*. The modulation index μ adjusts how much message power is transmitted relative to the carrier power level. If the modulation index $\mu > 1$, then the system is said to be *overmodulated*. The AM transmitter, defined by Equation 13-7, is diagrammed in Figure 13-7. In an AM system, the message is coded on the envelope of the transmitted signal and is given by

$$a(t) = A(1 + \mu x(t)) \qquad (13\text{-}8)$$

The frequency-domain representation of the AM signal is given by

$$Y(f) = \mathcal{F}[y(t)] = \mathcal{F}[A(1 + \mu x(t)) \cos(\omega_c t)]$$
$$= \frac{1}{2}A\delta(f - f_c) + \frac{\mu}{2}AX(f - f_c) + \frac{1}{2}A\delta(f + f_c)$$
$$+ \frac{\mu}{2}AX(f + f_c) \qquad (13\text{-}9)$$

where $X(f) = \mathcal{F}(x(t))$. The AM spectrum is graphically interpreted in Figure 13-7 and is seen to consist of bands of energy centered below, at and above the carrier frequency (the *lower sidebands* and *upper sidebands*, respectively). Therefore, the energy in the broadcast spectrum is due to both the carrier and the message.

Example 13-3   AM modulation

Equation 13-7 represents the basic AM modulation model. The resulting broadcast spectrum has two sidebands, one located on each side of the carrier line. *Computer Study*   A bandlimited message process is to be modulated by a high-frequency sinusoidal carrier using Equation 13-7. S-file ch13_3.s, using the format am(c,mu), creates a baseband message $x(t)$, performs the AM modulation using a modulation index mu = μ, with a carrier $c(t)$, and displays the results in both the time- and frequency-domains.

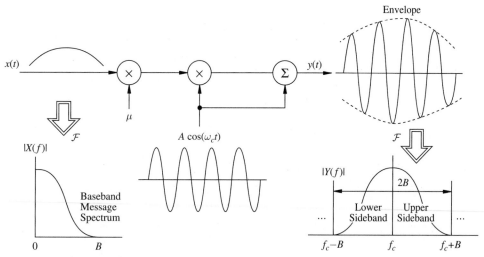

FIGURE 13-7   Basic AM transmitter.

SIGLAB	MATLAB
>include "ch13_3.s" >c=mkcos(1,1/4,0,256) #carrier >am(c,1) #mod. index =1.0	Not available in Matlab

The results are shown in Figure 13-8. The envelope of the modulated carrier follows the profile of the message. In the frequency-domain, the spectrum of the modulated signal has upper and lower sidebands centered about the carrier frequencies $f = \pm f_c$.

Refer to Figure 13-8 and observe that an AM system concentrates the broadcast power at the carrier frequency and upper and lower sidebands. As a result, this class of transmission is called *AM double sideband (AM-DSB)*. If the bandwidth of the real modulating signal $x(t)$ is a baseband process, such that $|X(f)| = 0$ for all $f \geq B$, then the AM-DSB transmission bandwidth requirement is at least $2B$ Hz. In general, AM transmission bandwidths are small (few kHz) resulting in a simple electronic implementation of Equation 13-7. However, because the message is coded onto the envelope of the transmitted signal, it is particularly susceptible to channel distortion effects, such as additive noise and fading. To maintain a reasonable level of fidelity, the received signal-to-noise ratio should be made as large as possible. Since the receiver responds to the power detected at its antenna, it is useful to interpret the AM-DSB transmission in terms of its power spectrum. It should first be recognized that the transmitted power concentrated at the carrier frequency contains no signal (message) information. Suppose, for demonstration purposes, that $x(t) = \cos(\omega_0 t)$ for $\omega_0 \ll \omega_c$. Then from Parseval's

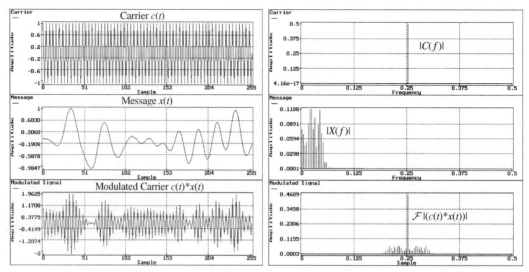

FIGURE 13-8 AM-DSB (left) continuous-time signals, and (right) spectra. Each panel, from top to bottom, shows the carrier, message, and modulated carrier (results may vary due to use of a random signal).

equation, one concludes that in the absence of any channel distortion

$$P_{carrier} = \lim_{T \to -\infty} \frac{1}{T} \int_{-T/2}^{T/2} (A \cos(\omega_c t))^2 dt = \frac{1}{2} A^2$$

$$P_{sidebands} = \lim_{T \to -\infty} \frac{1}{T} \int_{-T/2}^{T/2} (A \mu \cos(\omega_0 t) \cos(\omega_c t))^2 dt \qquad (13\text{-}10)$$

$$= \frac{1}{4} A^2 \mu^2 = \frac{\mu^2}{2} P_{carrier}$$

If $\mu < 1$, then at least half of the transmitted power is concentrated at the carrier frequency and not in the message sidebands. This affirms one of the noted weaknesses of AM-DSB, namely only a fraction of the transmitted signal power is usable.

The information found in the upper and lower AM-DSB sidebands is, in fact, redundant. Technically, to communicate a real message $x(t)$, only one sideband is needed. This is the motivation for *AM single sideband (AM-SSB)* radio. The advantage of a single-sideband system is that the transmitter concentrates the transmitted power into a narrower band of frequencies (either the upper or lower sidebands), which can be better protected from broadband noise degradation. However, compared with conventional AM-DSB receivers, the AM-SSB receiver is considerably more complex and is, therefore, only used in special circumstances (e.g., police and aviation radio).

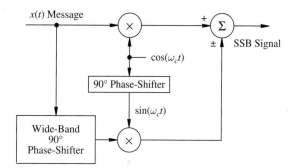

FIGURE 13-9   AM-SSB transmitter.

A block diagram of a typical AM-SSB transmitter is shown in Figure 13-9. It can be noted that there are two 90° phase-shifting subsystems. The 90° phase shifter attached to the carrier oscillator (i.e., $\cos(\omega_c t)$) is a narrowband device that can be easily implemented. The other device, however, must maintain a 90° phase shift over the entire bandwidth of the message process $x(t)$. This is a more challenging design problem. One type of filter known to have these desired phase-shifting properties is a Hilbert filter.

If a signal $x(t)$ has a Fourier transform $X(f)$, then the Hilbert transform of $x(t)$ is a new signal $\hat{x}(t)$, which has a Fourier transform $\hat{X}(f)$ given by

$$\hat{X}(f) = -j\,\text{sgn}(f)X(f) \qquad (13\text{-}11)$$

where

$$\text{sgn}(f) = \begin{cases} 1 & \text{if } f > 0 \\ 0 & \text{if } f = 0 \\ -1 & \text{if } f < 0 \end{cases} \qquad (13\text{-}12)$$

The Hilbert transforms of several elementary signals $x(t)$ are shown in Table 13-3, where the Hilbert transform of $x(t)$ is denoted $\hat{x}(t)$.

The *Hilbert operator* $H(f) = -j\,\text{sgn}(f)$ represents an ideal filter. Observe that $H(f)$ is an all-pass device in that $|H(f)| \equiv 1$ for all nonzero frequencies, as shown in Figure 13-10. Observe that there is a 180° phase shift between the positive and negative frequency axes. This phase-shifting property of the Hilbert filter

TABLE 13-3   ELEMENTARY HILBERT TRANSFORMS

$x(t)$	$\hat{x}(t)$
$x(t)e^{(j2\pi f_c t)}$	$-jx(t)e^{(j2\pi f_c t)}$
$x(t)\cos(2\pi f_c t)$	$x(t)\sin(2\pi f_c t)$
$x(t)\sin(2\pi f_c)$	$-x(t)\cos(2\pi f_c t)$
$e^{(j2\pi f_c t)}$	$-je^{(j2\pi f_c t)}$
$\cos(2\pi f_c t)$	$\sin(2\pi f_c t)$
$\sin(2\pi f_c t)$	$-\cos(2\pi f_c t)$

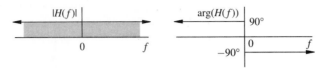

FIGURE 13-10   Spectral properties of a Hilbert transform.

can be very useful in designing communication systems. The Hilbert transform also exhibits a number of other useful properties, as well. The most important of these are Property 1: The magnitude spectra of $x(t)$ and $\hat{x}(t)$ are identical; Property 2: The signals $x(t)$ and $\hat{x}(t)$ are orthogonal, in that

$$\int_{-\infty}^{\infty} x(\tau)\hat{x}(\tau)d\tau = 0 \qquad (13\text{-}13)$$

If $x(t)$ is a real-valued function of time, then the complex signal $x_a(t)$ given by

$$x_a(t) = x(t) + j\hat{x}(t) \qquad (13\text{-}14)$$

is called an *analytic signal* or *preenvelope signal*. For example, if $x(t) = \cos(2\pi f_c t)$, then from Table 13-3 it follows that $\hat{x}(t) = \sin(2\pi f_c t)$ and $x_a(t) = \cos(2\pi f_c t) + j\sin(2\pi f_c t) = e^{(j2\pi f_c t)}$. This points out the utility of a Hilbert transform in constructing analytic functions. Note that the spectral representations of $x(t)$ and $\hat{x}(t)$ contain both positive and negative frequency harmonics (sidebands). The spectrum of $x_a(t)$, however, contains only positive frequencies (i.e., a single sideband signal).

Example 13-4   AM-SSB and Hilbert filters

Hilbert filters are one of several techniques used to implement the needed phase shift required of single sideband systems. They can be approximated with a discrete-time FIR filter designed as an all-pass filter with coefficients adjusted to produce the desired (linear) phase shifting.
Computer Study   A discrete-time Hilbert FIR can be constructed as nearly an all-pass filter by creating small stopbands near 0 Hz and the Nyquist frequency $f_s/2$. Discrete-time Hilbert filters have an antisymmetric impulse response. The phase response of the filter can be tested using a chirp. S-file `ch13_4.s` performs this analysis using the format `hilbert(fc,fi,fs)` where `fc`, `fi`, and `fs` are chirp frequencies representing the center frequency, initial offset, and final offset from the center.

SIGLAB	MATLAB
>include "ch13_4.s" >#sweep [fs/10,fs/10] >hilbert(.1,0,.1)	Not available in Matlab

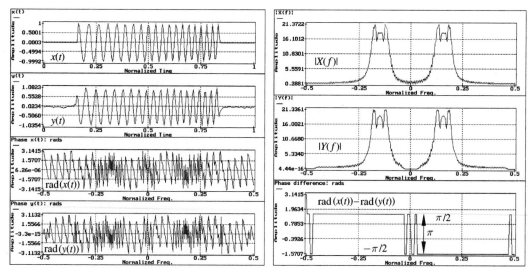

FIGURE 13-11   Hilbert filter action showing: (left) input and output time-series and phase responses to a chirp (deg denotes phase angle in radians), and (right) magnitude phase response and phase difference using a two-sided FFT.

The results are displayed in Figure 13-11. Here the input-output time-series and phase responses are displayed along with the magnitude frequency response and input/output phase differences. It can be seen that when swept over a wide range of frequencies, the Hilbert filter behaves like an all-pass device. The phase response, however, shows that over the filter's passband the phase goes from $\pi/2$ for negative frequencies to $-\pi/2$ for positive frequencies.

The function of an AM receiver, whether single or double sideband, is to recover the original message. The AM receiver shown in Figure 13-12 is based on the use of a product demodulator and a lowpass filter/detector. The receiver produces a new intermediate signal $z(t)$ given by

$$z(t) = y(t)\cos(\omega_0 t) \tag{13-15}$$

where $y(t)$ is the transmitted signal having a carrier frequency f_c. The demodulating signal, namely $\cos(\omega_0 t)$, is supplied by a tunable local oscillator, which

FIGURE 13-12   Basic AM receiver.

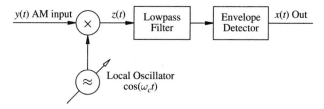

is part of the receiver. The frequency-domain representation of the demodulated signal is given by

$$Z(f) = \frac{1}{2}Y(f - f_0) + \frac{1}{2}Y(f + f_0) \qquad (13\text{-}16)$$

Recall that the frequencies found in the transmitted spectrum, namely $Y(f)$, were previously assumed to be locally distributed about the carrier frequency f_c. Therefore, if f_c and f_0 are of similar values, then one of the terms in Equation 13-16 consists of low-frequency components (i.e., $Y(f - f_0)$), and the other contains high-frequency components (i.e., $Y(f + f_0)$). If $f_c = f_0$, then, from Equation 13-9, the demodulated spectrum is given by

$$Z(f) = \frac{A}{2}\delta(f) + \frac{\mu}{2}AX(f) + \frac{A}{4}\delta(f \pm 2f_0) + \frac{A}{4}\mu X(f \pm 2f_0) \qquad (13\text{-}17)$$

A simple lowpass filter can then be used to extract the original message from the baseband portion of the spectrum of signal $z(t)$. Following the lowpass filter, an envelope detector is used to track the (relatively) slow variations in the peak amplitude (envelope) of the high-frequency received signal.

It can be noted that to demodulate the AM-DSB signal, the frequencies of the tunable local oscillator and carrier must be equal. Called *synchronous detection* or *coherent detection*, this can be more readily accomplished if the transmitted signal contains a sample of the carrier (as is the case with AM-DSB).

Example 13-5   Envelope detector

Previously, a diode rectifier circuit was used to convert an AC signal into a DC level. Similarly, a diode rectifier and a lowpass filter can remove the unwanted high-frequency components from an AM-DSB signal.

Computer Study   S-file `ch13_5.s` can be used to create an AM-DSB signal and then recover the message using envelope detection with the format `det(x,mu)`, where x is the message process and `mu` $= \mu$ is the modulation index. The program rectifies the received message, convolves it with a lowpass filter, and displays the results. For a sinusoidal message at frequency $f_s/64$, use the information below.

SIGLAB	MATLAB
`>include "ch13_5.s"`	Not available in Matlab
`>x=mkcos(1,1/64,0,256)`	
`>det(x,.5)`	

The message, AM-DSB transmission, detected and lowpassed-rectified signals are displayed in Figure 13-13. It can be seen that after the initial transients die away, the output of the envelope detector is similar to the message process.

The design of the AM system described in Figure 13-12 can be improved by breaking the receiver into subsystems called the radio frequency (RF) receiver,

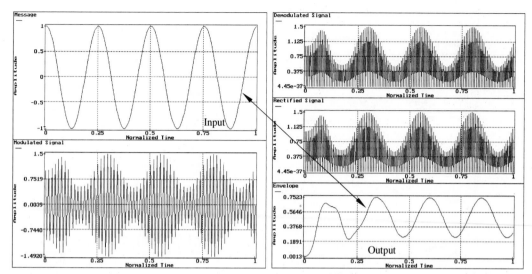

FIGURE 13-13   AM demodulator with envelop detector showing: (left) the message and received signal, and (right) demodulated received signal, rectified demodulated signal, and envelope of the detector output.

intermediate frequency (IF) subsystem, and a detector section, as shown in Figure 13-14. The RF section is a tunable filter that can amplify a received signal centered about frequency f_c, with an output mixed with a signal generated by a local oscillator having a frequency $f_c + f_{IF}$, where f_{IF} is called the *intermediate frequency (IF)*. Contemporary AM radio systems generally assume that $f_{IF} = 455$ kHz. Therefore, for an AM band extending over [550, 1600] kHz, the local oscillator must be tunable over a frequency range [1005, 2055] kHz. Coming out of the mixer will be a signal having upper and lower sidebands centered about $f = f_{IF}$ and $f = 2f_c + f_{IF}$. The high-gain frequency-selective IF amplifier stage is designed to pass only the information centered about $f = f_{IF}$. The key feature of this system is that a high-gain frequency-selective bandpass filter, having a fixed center frequency located at 455 kHz can serve the entire AM band. The output of the IF section is then sent to a detector where it is converted into an audible

FIGURE 13-14   AM superheterodyne receiver.

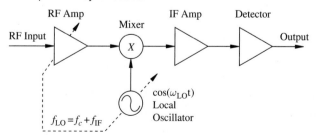

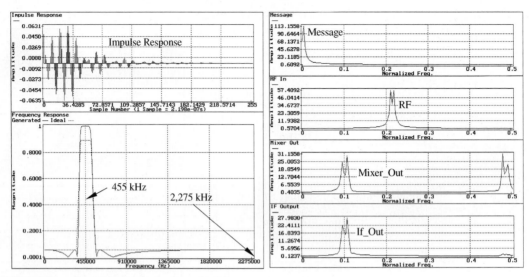

FIGURE 13-15 Superheterodyne receiver simulation.

baseband signal. Such a system is called a *superheterodyne receiver*, with origins that can be traced back to Major Armstrong (see footnote in Section 13-6).

Example 13-6 AM superheterodyne

The *superheterodyne method* made low-cost AM radio a practical reality by providing a means of processing an RF signal with a high-gain fixed frequency-selective intermediate filter without regard to the received carrier frequency.

Computer Study An IF bandpass filter is designed with a passband ranging over 400 kHz $\leq f \leq 510$ kHz, and center frequency about 455 kHz. S-file ch13_6.s and M-file super.m create a baseband message, modulates it up to $f_c \in [550, 1600]$ kHz, performs the demodulation by $f_{IF} = f_c + 455$ kHz, and filters the result using the IF filter and a format super(f), where f is the modulation frequency.

SIGLAB	MATLAB
>include "ch13_6.s" >super(960) #f=960kHz	Not available in Matlab

The results are shown in Figure 13-15.

Television is, in part, based on an AM technology. Television can be classified as monochromatic or color. The video signal broadcasted in either case will satisfy one of several adopted standards. For North and South America, plus Japan, the National Television System Committee (NTSC) standard applies (see Table 13-4). The video bandwidth required by NTSC is about 4.2 MHz. In Europe,

TABLE 13-4   TV STANDARDS

Item	NTSC	CCIR	HDTV
Aspect ratio H/V	4/3	4/3	5/3
Lines per frame	525	625	1125
Field frequency (Hz)	50	50	60
Video BW (MHz)	4.2	5.0	~ 30

the International Radio Consultative Committee (CCIR) standard applies. The bandwidth required by CCIR is on the order of 5 Mz. Due to the higher CCIR bandwidth, an improved screen resolution can be achieved. Along similar lines, High Definition Television (HDTV), with its nearly 30 MHz bandwidth, will provide a significant advancement in picture quality.

An NTSC monochrome TV receiver carries the video information on an AM ($\mu = 0.875$) transmission. The audio is sent as an 80 kHz FM signal (see Section 13-6). The receiver is a superheterodyne device with an IF frequency in the 40–50 MHz range. What is interesting about contemporary TV is the manner in which the video spectrum is constructed. Consider the problem diagramed in Figure 13-16, which shows how the transmitted spectrum would appear if transmitted as an AM-DSB or an AM-SSB transmission. Based on a 4.2 MHz baseband message model, an AM-DSB signal would require an 8.4 MHz channel. A 4.2 MHz single sideband transmission technically would be possible if only the upper (or lower) sideband was used (i.e., AM-SSB), as shown in Figure 13-16. Unfortunately, AM-SSB systems designed with physically realizable analog sideband filters have a poor performance at low message frequencies. This severely compromises their use in TV or facsimile applications. During the early days of television, a compromise was made between the high bandwidth requirements of AM-DSB and the noted problem with AM-SSB. To achieve acceptable low-frequency response within a limited bandwidth, *vestigial sideband* (or AM-VSB), was adopted. In AM-VSB, a vestige of second sideband is retained and used to reconstruct an acceptable version of the primary sideband. To illustrate, consider the information shown in Figure 13-16. A single sideband of a double-sideband transmission theoretically could be isolated by an ideal sideband filter denoted $I(f)$. However, this was not considered to be a practical solution when television was in its infancy. Instead, the signal is passed through the realizable transmission filter denoted $V(f)$, shown in Figure 13-16, where $V(f) = 1/2$ at $f = f_c$. At the receiver, a compensating filter (denoted $C(f)$ and shown in Figure 3-16) is used to achieve the effect of processing TV signals with an ideal sideband filter $I(F)$. The upper sideband, for example, is reconstructed using the compensating filter $C(f)$ to construct filter $I(f) = V(f) + C(f)$.

Color television is an extension of this concept. A color TV camera produces three video signals representing red, green, and blue denoted by $x_R(t)$, $x_G(t)$, and $x_B(t)$. If these three signals were transmitted as three monochrome TV signals, the transmission bandwidth requirements would triple. In addition, existing

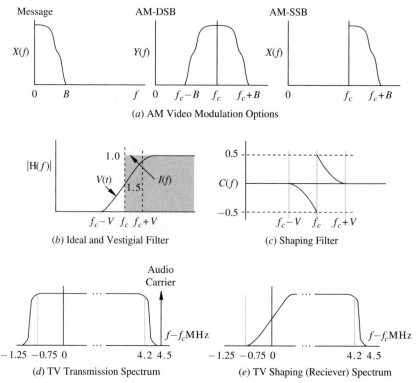

FIGURE 13-16   Vestigial sideband systems.

monochrome TVs would be unable to interpret such a transmission. During the color TV protocol-planning period, a requirement was established that a TV transmission be compatible with both color and monochrome receivers in both function and bandwidth. The solution is considered by many to be ingenious. It required that the video information be encoded in terms of *luminance*, denoted $y_L(t)$, and *chrominance*, denoted $y_I(t)$ and $y_Q(t)$, which satisfy the following system of linear equations

$$y_L(t) = 0.30x_R(t) + 0.59x_G(t) + 0.11x_B(t)$$

$$y_I(t) = 0.60x_R(t) - 0.28x_G(t) - 0.32x_B(t)$$

$$y_Q(t) = 0.21x_R(t) - 0.52x_G(t) + 0.31x_B(t) \tag{13-18}$$

The *color vector* is defined as $y_C(t) = y_I(t) + jy_Q(t)$, where $|y_C(t)|$ is called the *saturation (color)* and $\angle y_C(t)$ is the *hue*. The signal originally used for monochrome TV is very close to the luminance signal $y_L(t)$. Therefore, to be compatible with monochrome TV, the luminance signal must have a bandwidth on the order of 4.2 MHz. Fortunately, chrominance signals do not require the same bandwidth. Physiological tests have revealed that humans cannot perceive chromatic changes over small areas. This means that the high-frequency components found in $y_I(t)$ and $y_Q(t)$ can be ignored. The I and Q channel bandwidths are, in fact, 1.6 MHz and 0.6 MHz respectively. By inserting these signals into the luminance transmission at selected frequency locations, an interleaved signal results, which can support both color and monochrome TV.

13-6 EXPONENT MODULATION

In AM systems the amplitude of the carrier is modulated by a message process x(t). The advantage of AM is its simplicity. One of the weaknesses of AM is that it wastes transmitter power on signal components that did not contain message information. Furthermore, since the message is carried on the envelope of an AM transmission, it is very susceptible to distortion by additive noise. Before and during World War II, another type of modulation appeared. This modulation was *angle modulation* or *phase modulation (PM)*, along with its popular derivative, *frequency modulation (FM)*. While the origins of FM can be traced to the early days of radio, it was not until Major Edwin Armstrong[2] demonstrated that, compared with AM, FM enjoyed superior noise immunity, phase modulation gained credibility. The superiority of FM, now enjoyed by millions, initially did not achieve consumer acceptance due to the high cost of the receiving equipment. Early PM and FM receivers, fashioned with the technology of the time (i.e., vacuum tubes, inductors, and capacitors) were expensive and required constant alignment. However, with the advent of dense semiconductor devices, complex PM demodulation schemes have become economically viable and have replaced AM in many instances.

A phase modulated (PM) signal is defined by

$$y(t) = \cos(\phi(t)) \tag{13-19}$$

where $\phi(t)$ is the *generalized phase angle*. The instantaneous phase of a phase-modulation signal is often expressed as

$$\phi(t) = \omega_c t + \phi_0 + k_p x(t) \tag{13-20}$$

where ω_c is a fixed carrier frequency, k_p is a constant, ϕ_o is an arbitrary phase angle, and x(t) is the message process. Note that in this model the instantaneous phase $\phi(t)$ is linearly related to x(t). If the message x(t) is a piecewise continuous signal, such as a pulse-train, then the phase argument $\phi(t)$ of the modulated signal is discontinuous at isolated points. In this case the value of k_p can be set so that the value of $k_p\phi(t)$ is bounded within the principal angle range $[-\pi, \pi]$, thereby eliminating any possible phase ambiguity. If the message x(t) is continuous, then no restrictions on k_p apply, since the phase angle is assumed to be continuously tracked by a receiver.

---

[2]Major Armstrong was the inventor of FM, regenerative feedback, and the super-heterodyne receiver.

TABLE 13-5   PHASE AND FREQUENCY MODULATION RELATIONSHIPS

System	Instantaneous Phase	Instantaneous Frequency
PM	$\phi(t) \propto x(t)$	$\omega(t) \propto dx(t)/dt$
FM	$\phi(t) \propto \int x(\tau)\, d\tau$	$\omega(t) \propto x(t)$

It should be recalled that the instantaneous frequency is defined to be the time derivative of phase. Therefore, for Equation 13-20, the instantaneous frequency $\omega(t)$ is given by

$$\omega(t) = d\phi(t)/dt = \omega_c + k_p dx(t)/dt \tag{13-21}$$

Also it is apparent that if the message is a piecewise constant process, then the instantaneous frequency can become unbounded regardless of the value of k_p. Once the discontinuity passes, the instantaneous frequency would assume a constant value $\omega(t) = \omega_c$. If the receiver can pass through these isolated periods of possible wideband-frequency fluctuations and "track" the carrier frequency ω_c, then phase decoding can take place.

One of the most successful phase-modulation protocols is FM. The instantaneous frequency of an FM transmission is given by

$$\omega(t) = d\phi(t)/dt = \omega_c + k_f x(t) \tag{13-22}$$

where k_f is a constant and ω_c is again a fixed carrier frequency. In an FM system, the instantaneous frequency is linearly related to the message amplitude. The generalized phase angle is seen to be given by

$$\phi(t) = \int_{-\infty}^{t} (\omega_c + k_f x(\tau))d\tau = \omega_c t + k_f \int_{-\infty}^{t} x(\tau)d\tau \tag{13-23}$$

Observe in this case that the integration of the message process $x(t)$ has a smoothing effect on the resulting instantaneous phase angle $\phi(t)$.

It can be seen that PM and FM bear a remarkable degree of similarity. In particular, their basic interrelationships are summarized in Table 13-5.

Example 13-7   PM and FM modulation

A message process $x(t)$ is assumed to produce a generalized phase for a PM signal given by $\phi(t) = k_p \pi x(t) + 2\pi f_c t$, after Equation 13-20, and an FM signal satisfying $\phi(t) = \int k_p \pi x(\tau)d\tau + 2\pi f_c t$, after Equation 13-23. The instantaneous frequency is given by $d\phi(t)/dt$.

Computer Study   S-file ch13_7.s and M-file pmfm.m create generalized PM and FM phase angles, modulates them to produce PM and FM signals, and computes the instantaneous frequency of the signals. The format is pmfm(x_c, ω_0, k_p, k_f), where x_c is the message process, ω_0 is the carrier frequency, and k_p and k_f are given by Equations 13-20 and 13-23 respectively. To modulate a ramp ($f_0 = f_s/16$) and EKG ($f_0 = f_s/64$) message, use the information below.

SIGLAB	MATLAB
`>include "ch13_7.s"`	`>> x = tri(2*pi*(0:255)/64);`
`>x=intg(sq(64,1/2,256))`	`>> pmfm(x, 1/16, 10, 10)`
`>pmfm(x,1/16,.125,.125)`	`>> load ekg1.imp`
`>x=rf("ekg1.imp")`	`>> pmfm(ekg1, 1/64, 1, 1)`
`>pmfm(x,1/64,1,1)`	

The results are shown in Figure 13-17. It can be seen that for a piecewise continuous signal (i.e., ramp), the PM signal goes through jump discontinuities.

An AM transmission is of constant frequency, but has a time-varying amplitude. Referring to Equation 13-19, it can be noted that a PM or FM transmission is of constant amplitude, but has a time-varying frequency and phase. Since the message information does not ride on the carrier's envelope, a PM or FM signal is less susceptible to additive noise. This is one reason why FM currently represents the highest quality commercial radio broadcast technology. However, commercial FM transmissions, like AM, must reside within strictly enforced allocated bands of frequencies. As a result, the bandwidth requirements of an FM transmission must be carefully measured.

FIGURE 13-17   PM and FM modulation systems using: (left) a ramp, and (right) an EKG signal as the message process. From top to bottom is a display of a PM generalized phase angle, PM transmission, instantaneous PM frequency, FM generalized phase angle, FM transmission, and instantaneous FM frequency.

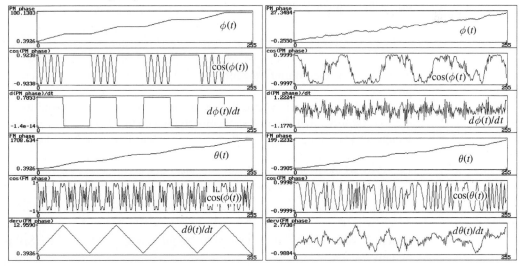

The instantaneous phase of an FM transmission $y(t) = \cos(\phi(t))$, carrying a message $x(t)$, can be modeled to be

$$\phi(t) = \omega_c t + k_f \int_{-\infty}^{t} x(\tau)d\tau = \omega_c t + \theta(t) \qquad (13\text{-}24)$$

The transmitted signal, therefore, becomes

$$y(t) = A\cos(\omega_c t + \theta(t))$$

$$= A\cos(\omega t)\cos(\theta(t)) - A\sin(\omega t)\sin(\theta(t)) \qquad (13\text{-}25)$$

Expanding Equation 13-25 as a power series, one obtains

$$y(t) \approx A\cos(\omega(t))(1 - \theta^2(t)/2! + \theta^4(t)/4! + \cdots)$$

$$- A\sin(\omega(t))(\theta(t) - \theta^3(t)/3! + \theta^5(t)/5! + \cdots) \qquad (13\text{-}26)$$

If $\theta(t)$ is assumed to be a baseband process that is limited to B Hz, then the bandwidth occupied by $\theta^2(t)$ is $2B$ Hz, and $\theta^3(t)$ is $3B$ Hz, and so forth. The bandwidth requirements are seen to continue to increase as long as the parameter $\theta^n(t)/n!$ continues to maintain significance. The bandwidth requirements for FM therefore, can be seen to easily exceed that required of an AM-DSB transmission, which is on the order of 2B Hz.

FM modulators are normally classified as being *FM narrowband* or *FM wideband*. The system is said to be narrowband FM if $\theta(t)$ is restricted to small, angular deviation (i.e., $|\theta(t)| \ll 1$ radian). If so, $\cos(\theta(t)) \approx 1$, $\sin(\theta(t)) \approx \theta(t)$ and

$$y(t) = A(\cos(\omega_c t) + \theta(t)\sin(\omega_c t)) \qquad (13\text{-}27)$$

or, in the frequency domain

$$Y(f) = \frac{A}{2}(\delta(f - f_c) + \delta(f - f_c)) + j\frac{A}{2}(\Theta(f - f_c) - \Theta(f + f_c)) \quad (13\text{-}28)$$

where $\Theta(f)$ is the Fourier transform of $\theta(t)$, which is assumed to be a narrowband process.

A special FM case is where $\theta(t) = \beta\cos(\omega_m t)$, and it is referred to *tone modulation*. Observe that the instantaneous frequency is $d\theta(t)/dt = -\omega_m\beta\sin(\omega_m t)$, which is also a tone. Again, if $\beta \ll 1$, the system is narrowband and

$$y(t) \approx A\cos(\omega_c t) + A\beta\sin(\omega_m t)\sin(\omega_c t)$$

$$= A\cos(\omega_c f) + \frac{A\beta}{2}\cos((\omega_c - \omega_m)t) - \frac{A\beta}{2}\cos((\omega_c + \omega_m)t) \quad (13\text{-}29)$$

which is seen to have the bandwidth requirements of the AM-DSB protocol, namely $2B$ Hz.

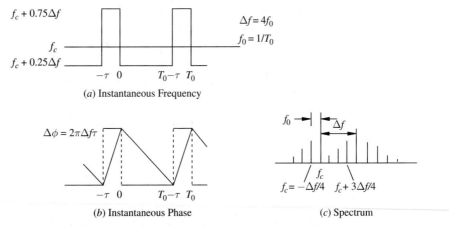

$f_c + 0.75\Delta f$

f_c

$f_c + 0.25\Delta f$

$\Delta f = 4f_0$

$f_0 = 1/T_0$

$-\tau \quad 0 \qquad T_0-\tau \quad T_0$

(a) Instantaneous Frequency

$\Delta\phi = 2\pi\Delta f\tau$

$-\tau \quad 0 \qquad T_0-\tau \quad T_0$

(b) Instantaneous Phase

f_0

Δf

f_c

$f_c = -\Delta f/4 \quad f_c + 3\Delta f/4$

(c) Spectrum

FIGURE 13-18   FM magnitude-frequency spectrum during message discontinuities.

If the narrowband assumption fails, then the FM transmission is said to be wideband FM. The analysis of the frequency-domain requirements of a wideband FM process is considerably more complicated than for the narrowband case. The instantaneous frequency of a wideband FM signal is given by

$$\omega(t) = d\phi(t)/dt = d(\omega_c t + k_f \int_{-\infty}^{t} x(\tau)d\tau)/dt = \omega_c + k_f x(t) \qquad (13\text{-}30)$$

If $x_m = \text{maximum}(|x(t)|)$, and $x(t) \equiv x_m$ for all time (i.e., steady-state case), then the maximum instantaneous frequency deviation about the carrier ω_c is $\pm k_f x_m$ radians per second. This value is sometimes referred to as the *maximum frequency deviation* and is denoted

$$\Delta\omega = k_f x_m \qquad (13\text{-}31)$$

or in Hz, $\Delta f = k_f x_m/2\pi$. In this case the bandwidth requirement, based in Equation 13-30, would be a constant for all time and equal to

$$B_{x_m} = 2\Delta f \qquad (13\text{-}32)$$

The actual maximal frequency requirement, however, is normally in excess of this steady-state value. The worst case would occur when there is a jump discontinuity in $x(t)$ in a wideband FM transmission, as shown in Figure 13-18. The actual instantaneous frequency requirement can be seen to exceed $2\Delta f$.

Example 13-8   Wideband FM

If a wideband FM modulator is driven by a constant tone at frequency f_M, then the transmission bandwidth requirement is estimated by Equation 13-32.

The spectral shape of a wideband FM signal may be considerably higher than this figure if the generalized phase angle is rapidly changing in time.

Computer Study S-file `ch13_8.s` and M-file `wideband.m` use a format `wideband(fc,k,T)` to implement Equation 13-22, namely $f(t) = f_c + kx(t)$, where $x(t)$ is a unit amplitude periodic triangular pulse train having a period T. The smaller the value of T, the higher the frequency content of the message process. For values of $T \in [4, 128]$, a wideband FM transmission of a piecewise constant message process can be simulated as follows.

SIGLAB	MATLAB
>include "ch13_8.s"	>> wideband(.1, .1, 64)
> #k=0.1	>> wideband(.1, .1, 16)
>wideband(.1,.1,64)	>> wideband(.1, .5, 64)
>wideband(.1,.1,16)	>> wideband(.1, .5, 16)
> #k=0.5	
>wideband(.1,.5,64)	
>wideband(.1,.5,16)	

The results are displayed in Figure 13-19. Observe that the width of the broadcast wideband FM spectrum is maximized when a strong ($k = 0.5$) high-frequency message ($T = 16$) is present.

The strength of the sidebands produced by a periodic pulse-train message process, shown in Figure 13-19, has a $|\sin(\omega - \omega_c)/(\omega - \omega_c)|$ shape. It is often assumed that only those frequencies that exist out to the first $\sin(\omega)/\omega$ null (i.e., $f = \pm 4\pi B$) are significant. From this, the approximate maximal bandwidth requirement for a wideband FM transmission can be estimated to be given by

$$B_{FM} = 2(\Delta f + 2B) \tag{13-33}$$

Another commonly used method of estimating the bandwidth requirement is called *Carson's Rule* and is given by

$$B_{FM} = 2(\Delta f + B) \tag{13-34}$$

Observe that if $\Delta f \gg B$, as found in wideband FM, then $B_{FM} \approx 2(\Delta f)$.

Commercial FM transmission, whether stereo or monophonic, presumes a maximal frequency deviation $\Delta f = 75$ kHz and a transmission bandwidth of 200 kHz. From Carson's Rule, the wideband estimation formula (Equation 13-33), namely $B_{FM} = 2(\Delta f + B)$, implies that B $\leq$ 25 kHz, which exceeds the message bandwidth requirements for high-fidelity sound reproduction. However, Carson's Rule is generally considered too pessimistic and, in fact, the message bandwidth for a commercial FM transmission is on the order of 15 kHz.

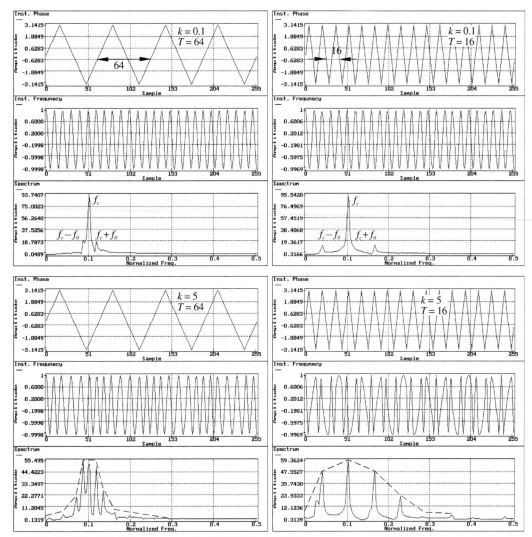

FIGURE 13-19   Wideband FM magnitude-frequency spectrum for a piecewise constant message with periods 16 and 64, with (left) $k = 0.1$, and (right) $k = 0.5$.

Equivalent bandwidth estimation methods can also be derived as a function of the *deviation ratio*, which is defined to be

$$\beta = \frac{\Delta f}{B} \tag{13-35}$$

Wideband FM is defined by $\beta > 0.5$. In the context of wideband FM, reconsider the tone-modulation case for an instantaneous phase given by

$$\theta(t) = \beta \sin(\omega_m(t)) \tag{13-36}$$

TABLE 13-6   BESSEL FUNCTION RELATIONSHIPS

Modulation Index	Significant Frequencies
β	$2n_{max}$
0.1	2
0.3	4
0.5	4
1.0	6
2.0	8
5.0	16
10.0	28
20.0	50
30.0	70

From Equation 13-25, it follows that

$$y(t) = A\cos(\omega_c t)\cos(\beta\sin(\omega_m t)) - A\sin(\omega_c t)\sin(\beta\sin(\omega_m t)) \qquad (13\text{-}37)$$

Multiplying the terms $\cos(\omega_c t)$ and $\sin(\omega_c t)$ is seen to be a complicated function of time, ω_m, and β. Fortunately, these functions can be expressed in terms of a Bessel function of order n of the first kind, denoted $J_n(\beta)$. In particular

$$\cos(\beta\sin(\omega_m t)) = J_0(\beta) + 2\sum_{n\,even} J_n(\beta)\cos(n\omega_m t) \qquad (13\text{-}38)$$

$$\sin(\beta\sin(\omega_m t)) = 2\sum_{n\,odd} J_n(\beta)\sin(n\omega_m t)$$

which can be combined to form

$$y(t) = A\sum_{n=-\infty}^{\infty} J_n(\beta)\cos(\omega_c + n\omega_m)t \qquad (13\text{-}39)$$

Bessel functions are well-documented in mathematical tables, handbooks, and in some mathematical analysis, general-purpose software packages. The relationship between β and n, in terms of significant values of $|J_n(\beta)| > 0.01$, is summarized in Table 13-6. The data indicates that the values of $J_n(\beta)$ are negligible if $n > \beta + 2$. That is, the sideband frequencies that are estimated to be of importance are bounded within a $2(\beta+2)f_m$ range, centered at the carrier frequency f_c. Since the maximum message-frequency component is bounded by $f_m < B$, the wideband FM requirements are estimated to be

$$B_{FM} \approx 2(\beta + 2)B = 2(\Delta f + 2B) \qquad (13\text{-}40)$$

which agrees with Equation 13-33.

Example 13-9   Commercial FM

A comparison of commercial AM and FM systems is presented in Table 13-7. Commercial FM is considered to be of much higher fidelity as evi-

TABLE 13-7   COMMERCIAL RADIO STANDARDS

Item	AM	FM
Carrier Frequency	540–1600 kHz	88.1–107.9 MHz
Carrier Spacing	10 kHz	200 kHz
IF Frequency	455 kHz	10.7 MHz
IF Bandwidth	6–10 kHz	200–250 kHz
Audio Bandwidth	3–5 kHz	15 kHz

denced by its significantly higher audio bandwidth (15 kHz vs. 3–5 kHz for AM). The maximum frequency deviation allowed for a commercial FM transmission is $\Delta f = 75$ kHz and the message spectra covers an assumed range $f \in [30, 15000]$ Hz. The transmission bandwidth estimate, based on Equation 13-33, is

$$B = 2(75 + 2 \times 15) \text{ kHz} = 210 \text{ kHz} \approx 200 \text{ kHz}$$

A tone located at the maximum end of the message frequency would produce a maximum deviation ratio of

$$\beta = 75 \times 10^3 / 15 \times 10^3 = 5$$

which, from Table 13-6, indicates that $2n < 16$ or $n < 8$. Letting $n = 7$, the bandwidth estimate becomes

$$B_{15 \text{ kHz}} = 2 \times 7 \times 15 \text{ kHz} = 210 \text{ kHz} \approx 200 \text{ kHz}$$

which agrees with the results obtained using Equation 13-33. A lower frequency message component, say at 2.5 kHz, would produce $\beta = 75k/2.5k = 30$, and from Table 13-6 it follows that $2n < 70$ or $n < 35$. Let $n = 34$ and compute

$$B_{2.5 \text{ kHz}} = 2 \times 34 \times 2.5 \text{ kHz} = 170 \text{ kHz} < B_{15 \text{ kHz}}$$

As a result, it should be expected that if two tones of equal strength are to be transmitted as a wideband FM signal, then the maximal bandwidth requirement will be established by the higher frequency tone.

Before it became a commercial success, a number of enhancements were introduced to the FM system by technologists. Experience has shown that a typical voice or musical selection, say $x(t)$, has more power concentrated at low frequencies than at high frequencies. A classic spectral model of such processes is that of an integrator operating on a flat (white) bandlimited signal, namely $X(\omega) = (1/j\omega)W(\omega)$ for $W(\omega) = 1$ for $|\omega| \in [B_L, B_H]$ and $W(\omega) = 0$ otherwise. FM systems work more efficiently when the signal being modulated contains significant high-frequency components. It is therefore desirable to flatten the message spectrum and emphasize the relatively weak high-frequency signal components found in the signal spectrum. *Preemphasis filters* are used to achieve this effect. A preemphasis filter has a transfer function of the general form

$$H_{\text{pre}}(\omega) = j\omega W(\omega) \qquad (13\text{-}41)$$

The preemphasis filter therefore behaves as a differentiator that compensates for the $1/j\omega$ envelope shape of the input signal spectrum. The now-flattened signal is FM modulated, transmitted, and demodulated. At the output of the receiver, the demodulated signal is then passed through a *deemphasis filter*, which is given by

$$H_{\text{demp}}(\omega) = H_{pre}^{-1}(\omega) = W(\omega)/j\omega \qquad (13\text{-}42)$$

to reconstruct the original message $x(t)$.

Example 13-10   Preemphasis

A preemphasis filter behaves like a differentiator over the message bandwidth. As such, it can be used to map a message spectrum, given by $X(\omega) = (1/j\omega)W(\omega)$, for $W(\omega) = 1$ for $|\omega| \in [B_L, B_H]$ and $W(\omega) = 0$ otherwise, into $Y(\omega) = X(\omega) \times H_{\text{pre}}(\omega) = j\omega X(\omega) = W(\omega)$, which is flat.

Computer Study A differentiating filter[3] with a frequency response $H(\omega) = (j\omega)U(\omega)$ for $U(\omega) = 1$ and $|\omega| \in [0, 16000]$ Hz is shown in Figure 13-20. S-file ch13_10.s creates a chirp (linearly swept FM signal) of the frequency range $f \in [0, 16000]$ Hz. The chirp has a flat frequency response shaped to have a $1/j\omega$-like response after it is integrated. The integrated signal is preemphasized using the designed filter, transmitted, and then deemphasized.

SIGLAB	MATLAB
>include "ch13_10.s"	Not available in Matlab

The results are reported in Figure 13-20. Observe that the original message spectrum can be recovered. By enriching the high-frequency content of modulated spectrum, the FM system is also made to operate more efficiently.

Another innovation is stereo FM, described in Figure 13-21. The transmitter algebraically combines the 15 kHz stereo right (R) and left (L) channel signals together to form (L + R) and (L − R) signals. Monophonic receivers work with the (L + R) channel only. The (L − R) channel information is mixed with a 38 kHz carrier using an AM-DSB suppressed carrier (AM-DSB-SC) modulator. A reduced strength carrier, called a *pilot*, is added to the (L + R) and modulated (L − R) signals before they are sent to the FM modulator. The stereo FM receiver, shown in Figure 13-21 (see page 508), recovers the (L + R) and (L − R) signals as baseband processes and algebraically combines them to synthesize the ordinal left and right channel messages.

---

[3]For users of MONARCH, the FIR has been saved as "preemp.FIR".

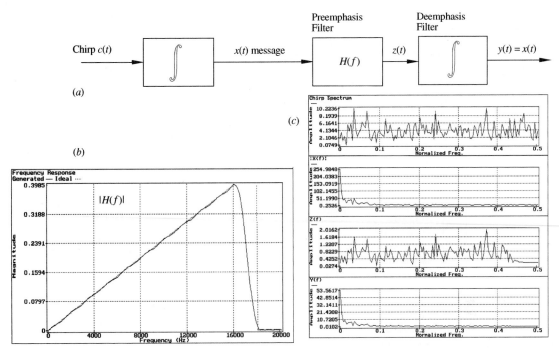

FIGURE 13-20   Preemphasis results showing: (left) signal generation and the signal processing stream, (right) preemphasis filter magnitude frequency response, and (c) top to bottom ($C(f),X(f),Z(f),Y(f)$).

13-7 FM RECEIVER DESIGN

The principal virtue of an AM system is its simple receiver. In the beginning, FM required complicated *frequency discrimination* circuits be built using tunable inductors, capacitors, resistors, and amplifiers. The frequency discriminator maps the instantaneous frequency into a voltage level that represents the message process $x(t)$. Now, however, integrated circuits, called *phase-locked loop (PLL) demodulators*, have made frequency discrimination a reliable and affordable operation. A PLL, as shown in Figure 13-22, consists of a multiplier, loop filter, and a *voltage controlled oscillator (VCO)*. The multiplier is sometimes referred to as a *phase comparator*. The output of the VCO is a sinusoid whose value is determined by the feedback voltage $v(t)$. If $v(t)$ is positive, for example, the VCO will increase its frequency of oscillation. If $v(t)$ becomes negative, the opposite will occur. Coherent demodulation will take place if the output of the VCO acquires and tracks the frequency of the incoming FM transmission. How closely the incoming signal and VCO output agree can be measured by the multiplier and loop filter. An input of $x(t) = \cos(\omega_c t + \phi(t))$, if multipled by a synthesized VCO output of the form $r(t) = \sin(\omega_{vco}t + \phi_{vco}(t))$, will define an error signal $e(t)$, which is given by

$$e(t) = x(t)y(t) = 0.5(\sin((\omega_c - \omega_{vco})t + (\phi(t) - \phi_{vco}(t)))$$

$$+ \sin((\omega_c + \omega_{vco})t + \phi(t) + \phi_{vco}(t)) \qquad (13\text{-}43)$$

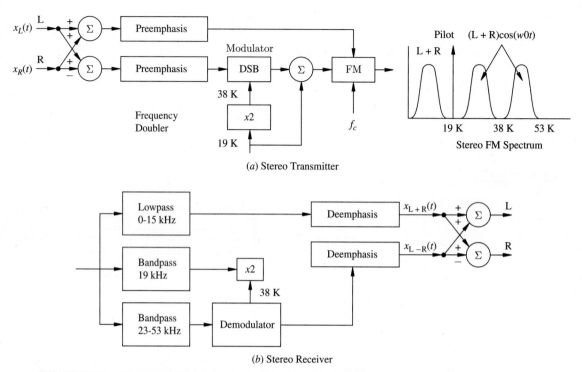

(a) Stereo Transmitter

(b) Stereo Receiver

FIGURE 13-21　Stereo FM example.

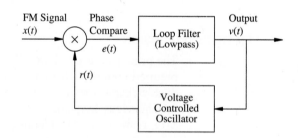

FIGURE 13-22　Phase-locked loop (PLL) system.

If the VCO correctly tracks the carrier (i.e., $\omega_{vco} = \omega_c$) and performs the proper phasing (i.e., $\phi_{vco}(t) = \phi(t)$), then Equation 13-43 reduces to

$$e(t) = 0.5(\sin((0)t) + \sin(2\omega_c t + 2\phi(t))) = 0.5\sin(2\omega_c t + 2\phi(t)) \quad (13\text{-}44)$$

By passing $e(t)$ through the lowpass loop filter, the high-frequency components of $e(t)$ are suppressed, leaving $v(t) = 0$ in the ideal case.

Example 13-11　Phase-locked loop

The PLL belongs to a broad class of linear systems that rely on error feedback to achieve some prespecified performance objective. The error signal found in

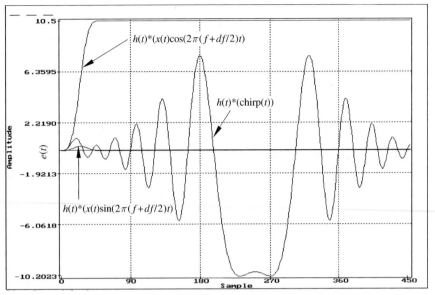

FIGURE 13-23   Simulated phase comparison and loop-filtered phase comparison of $x(t)$ and simulated VCO responses.

Equation 13-43 is used by the PLL to correct for misalignment between the frequency and/or phase of an externally supplied sinusoidal signal and another signal being internally generated. The error signal is used to alter the internal signal generator until $e(t) = 0$.

Computer Study   S-file `13_11.s` and M-file `pll.m` create a message process and possible VCO outputs. It then phase compares the VCO and message signals (i.e., multiplication), and then filters the result. The format used is given by `pll(f,df)` represents the modulated message process assumed to be given by $x(t) = \cos(2\pi(f + df/2)t)$. The programs simulate the VCO outputs, which are given by

1 $r(t) = x(t) = \cos(2\pi(f + df/2)t)$

2 $r(t) = \sin(2\pi(f + df/2)t)$

3 $r(t) =$ chirp with a center frequency f and deviation $\pm df/2$.

SIGLAB	MATLAB
>include "ch13_11.s" >pll(0.05,0.1) #f in [0,.1]	>> pll(0.1, 0.05)

The results are shown in Figure 13-23. It can be seen that after loop (lowpass) filtering $r(t) = \cos(2\pi(f + df/2)t)$ produces a positive output that would be used to correct the phase of the VCO output. For $r(t) = \sin(2\pi(f + df/2)t)$ the frequency and phase of the VCO are properly oriented and the VCO corrective signal is $e(t) = 0$. For the case where $r(t)$ is a chirp, the loop-filtered output is constantly attempting to correct for measured frequency and phase misalignments.

13-8 CORRELATION RECEIVER

Modern communication systems must often decode one message out of M possible signals from a signal set $\mathcal{S} = \{s_1(t), \dots, s_M(t)\}$. A message constructed from M symbols is called an *M-ary message*. Information retrieval is called a *M-ary detection* problem. For example, the radio receiver on the planet Fibonacci (see Example 1-2) would undoubtably be searching for messages with the following preambles $\mathcal{S} = \{\text{hello bunny, hello rabbit, \dots}\}$. Ideally, these signals would be designed to be orthogonal to remove as much redundancy from the signal set as possible. The ideal receiver in this case is a *correlator* and is shown in Figure 13-24, where *correlation* is a measure of the similarity existing between two signals. Formally, the measure of similarity between two complex signals, say $x(t)$ and $y(t)$, can be expressed in terms of the *cross-correlation function*, which is given by

$$\mathbf{R}_{xy}(\tau) = \lim_{T \to \infty} \int_{-T/2}^{T/2} x(t)y^*(t + \tau)dt \qquad (13\text{-}45)$$

The parameter τ, called a *lag variable*, represents the relative delay existing between the time origins of $x(t)$ and $y(t)$. In general $\mathbf{R}_{xy}(\tau) \neq \mathbf{R}_{yx}(\tau)$, which means that the cross-correlation measure is order sensitive. If $\mathbf{R}_{xy}(\tau) = 0$, then $x(t)$ and $y(t)$ are said to be *uncorrelated*.

For purposes of simplification, assume that the signals $x(t)$ and $y(t)$ are real. Observe that the cross-correlation function given in Equation 13-45 bears a strong

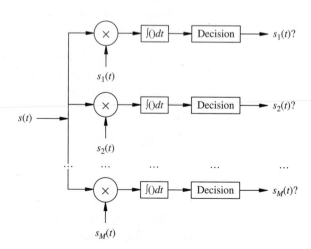

FIGURE 13-24   *M-ary signal detection.*

resemblance to linear convolution. To explore this proposition, evaluate Equation 13-45 for $t = -\lambda$ and observe that

$$\mathbf{R}_{xy}(\tau) = \lim_{T \to \infty} \int_{-T/2}^{T/2} x(-\lambda)y(\tau - \lambda)d\lambda = x(-\tau) * y(\tau) \qquad (13\text{-}46)$$

That is, the cross-correlation function $\mathbf{R}_{xy}(\tau)$ can be computed by convolving $x(-\tau)$ with $y(\tau)$. Equation 13-46 can be further explored by observing that a direct application of the convolution theorem would result in

$$\mathbf{R}_{xy}(\tau) \overset{\mathcal{F}}{\leftrightarrow} X(-j\omega)Y(j\omega) \qquad (13\text{-}47)$$

A special and important case of Equation 13-46 relates to correlating $x(t)$ with itself. This is called the *autocorrelation function* and is given by

$$\mathbf{R}_{x}(\tau) = \lim_{T \to \infty} \int_{-T/2}^{T/2} x(t)x^*(t + \tau)\, dt \qquad (13\text{-}48)$$

It can be observed that the total energy of the signal $x(t)$ is equivalent to evaluating the autocorrelation function at $\tau = 0$. It should also be apparent that the signal $x(t)$ is maximally correlated (i.e., similar) with itself, when $x(t + \tau) = x(t)$. This will always occur at $\tau = 0$, and as such

$$\mathbf{R}_{x}(\tau) \le \mathbf{R}_{x}(0) \qquad (13\text{-}49)$$

Example 13-12   Autocorrelation

Because it is suspected that the inhabitants of the planet Fibonacci are sending a coded message, a radio receiver has been designed to detect and verify such a hypothesis. The received message is cross-correlated with the ASCII code for "BUNNY" (see Example 1-2). The output of the correlator is sent to a threshold detector whose purpose is to detect a rapid rise in the cross-correlator output that would indicate a strong similarity between the reference (the *template*) and the received signal.

Computer Study   S-file ch13_12.s and M-file ch13_12.m create the ASCII code for the reference signal (i.e., "BUNNY"), delay the reference signal by 50 ns, then create a rectangular pulse train and a random pulse process. The cross-correlation of the three simulated received signals with the reference signal is also computed and displayed for $\tau \in [0, 223\text{ ns}]$.

SIGLAB	MATLAB
>include "ch13_12.s"	>> Not available in Matlab

The results are reported in Figure 13-25. The output of the cross-correlator is seen to be high when the reference signal is compared with itself (autocorrelation) or with a 50 ns delay "BUNNY" signal. The periodic and random pulse processes produce less selective and lower amplitude outputs.

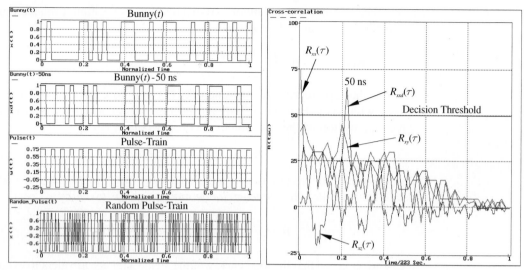

FIGURE 13-25   Cross-correlation experiment showing: from top to bottom, (left) the reference (template) BUNNY message, BUNNY delayed by 50 ns, a periodic pulse train, and a random pulse process, and (right) their cross-correlation (results may vary due to random signal).

Advance mathematical techniques have been developed to allow, in certain cases, a threshold level to be derived to minimize the probability of missing a message or introducing false alarms.

Equation 13-47 suggests that correlation should be related to the spectral properties of a signal or a class of signals. Consider again, for convenience, that $x(t)$ is real and that $x(t) = y(t)$. Then it follows that

$$\mathbf{R}_x(\tau) \overset{\mathcal{F}}{\leftrightarrow} X(-j\omega)X(j\omega) = |X(\omega)|^2 \qquad (13\text{-}50)$$

Equation (13-50) states that signal energy can be measured in either the time- or frequency-domain and be related to the autocorrelation function. Consider now the Fourier transform of the autocorrelation function $\mathbf{R}_x(\tau)$, given by

$$\mathbf{G}_x(f) = \mathcal{F}(\mathbf{R}_x(\tau)) = \int_{-\infty}^{\infty} \mathbf{R}_x(\tau)e^{-j2\pi f\tau}d\tau \qquad (13\text{-}51)$$

where $f = 2\pi\omega$. Here $\mathbf{G}_x(f)$ and $\mathbf{R}_x(\tau)$ are Fourier transform pairs, that is $\mathbf{G}_x(f) \overset{\mathcal{F}}{\leftrightarrow} \mathbf{R}_x(\tau)$. The energy in the signal x(t), denoted E_x, is given by

$$E_x = \int_{-\infty}^{\infty} |x(t)|^2 dt = \mathbf{R}_x(0) = \mathcal{F}^{-1}(\mathbf{G}_x(f))|_{\tau \to 0} \qquad (13\text{-}52)$$

Equation 13-52 also states that the stored energy in the time-domain signal $x(t)$ can be alternatively computed in terms of the Fourier transform of $\mathbf{R}_x(\tau)$, namely

$$E_x = \int_{-\infty}^{\infty} \mathbf{G}_x(f)e^{j2\pi f\tau}df|_{\tau \to 0} = \int_{-\infty}^{\infty} \mathbf{G}_x(f)df \qquad (13\text{-}53)$$

Since the signal energy is computed by integrating $G_x(f)$ over all frequencies, $G_x(f)$ is called the *energy spectral density function* and represents the energy in $x(t)$ on a per hertz basis. Combining Equations 13-52 and 13-53 results in a total signal energy equation satisfying

$$\int_{-\infty}^{\infty} |x(t)|^2 dt = \int_{-\infty}^{\infty} G_x(f) df \tag{13-54}$$

which is often referred to as *Parseval's energy theorem* or *Rayleigh's energy theorem*.

If $x(t)$ is periodic, then signal power can be computed and is given by

$$P_x = \lim_{T \to \infty} \frac{1}{T} \left(\int_{-T/2}^{T/2} |x(t)|^2 \right) dt \tag{13-55}$$

Assume now that $x(t)$ is a real signal of finite power having a Fourier transform $X(f)$. Then combining Equations 13-54 and 13-55, one obtains

$$P_x = \left(\lim_{T \to \infty} \frac{1}{T} \int_{-\infty}^{\infty} G_x(f) df \right) \tag{13-56}$$

which can be simplified to

$$P_x = \int_{-\infty}^{\infty} \lim_{T \to \infty} \left(\frac{G_x(f)}{T} \right) df = \int_{-\infty}^{\infty} S_x(f) df \tag{13-57}$$

where $S_x(f)$ is called the *power spectral density* and represents the power in $x(t)$ on a per hertz basis. Equation 13-57 is also referred to a *Parseval's power theorem*.

Suppose $x(t)$ is such a periodic signal, with period $T = 1/f_0$, and has a Fourier series given by

$$x(t) = \sum_{n=-\infty}^{\infty} c_n e^{j2\pi n f_0 t} \tag{13-58}$$

The autocorrelation function can then be expressed in terms of the Fourier coefficients $\{c_n\}$. The autocorrelation function $R_x(\tau)$ of a periodic signal is likewise periodic and has as a Fourier representation

$$S_x(f) = \sum_{n=-\infty}^{\infty} |c_n|^2 \delta(f - n t_0) \tag{13-59}$$

where $S_x(f)$ is the power spectral density of the periodic signal $x(t)$ and is measured in watts per harmonic. Finally, the power in the signal $x(t)$ can be computed using

$$P_x = \sum_{n=-\infty}^{\infty} S_x(n f_0) \tag{13-60}$$

Example 13-13   Power spectral density

Consider the periodic signal $x(t) = A\cos(\omega_0 t + \phi)$, which has an autocorrelation function given by

$$\mathbf{R}_x(\tau) = \lim_{T \to \infty} \frac{1}{2T} \int_{-T}^{T} A^2 \cos(\omega_0 t + \phi)\cos(\omega_0(t + \tau) + \phi)dt = \frac{A^2}{2}\cos(\omega_0\tau)$$

The autocorrelation function can also be seen to be a periodic function having a period $T = 1/f_0$. A direct calculation of the power spectral density function yields

$$S_x(f) = \mathcal{F}(\mathbf{R}_x(\tau)) = \frac{A^2}{4}\delta(f - f_0) + \frac{A^2}{4}\delta(f + f_0)$$

which states that the power is concentrated at frequencies $f = \pm f_0$. Finally, the total power in the signal can be computed to be

$$P_x = \int_{-\infty}^{\infty} S_x(f)df = \frac{A^2}{4} + \frac{A^2}{4} = \frac{A^2}{2}$$

as expected.

If signals are defined in discrete-time, say $x(n)$ and $y(n)$, then the discrete-time cross-correlation function is given by

$$\mathbf{r}_{xy}(\tau) = \sum_{n=-\infty}^{\infty} x(n)y^*(n + \tau) \tag{13-61}$$

where τ is measured in sample delay units. The discrete-time autocorrelation function is defined to be

$$\mathbf{r}_x(\tau) = \sum_{n=-\infty}^{\infty} x(n)x^*(n + \tau) \tag{13-62}$$

and the energy and power spectral density functions follow from the discrete-time versions of Equations 13-53 and 13-57.

Example 13-14   Discrete-time autocorrelation function

The autocorrelation function for the discrete-time signals $y(n) = x(n) = a^n u(n)$, where $|a| < 1$, is given by

$$\mathbf{r}_x(\tau) = \sum_{n=-\infty}^{\infty} x(n)x(n - \tau)$$

$$= \sum_{n=-\infty}^{\infty} a^n a^{n-\tau}$$

$$= a^{-\tau} \sum_{n=-\infty}^{\infty} a^{2n}$$

for $\tau \geq 0$, $\mathbf{r}_x(\tau) = a^{-\tau}/(1 - a^2)$ and for $\tau < 0$, $\mathbf{r}_x(\tau) = a^{\tau}/(1 - a^2)$. Therefore, for all τ, the autocorrelation function is given by

$$\mathbf{r}_x(\tau) = \frac{a^{-|\tau|}/(1 - a^2)}{1/(1 - a^2)} = a^{-|\tau|}$$

which is symmetrically distributed about $\tau = 0$ and is maximum at that value $\tau = 0$. The correlation is seen to weaken as τ become large.

The relative strength of the cross-correlation or autocorrelation functions can be measured in terms of the *correlation coefficient*, which is given by (shown in continuous-time form)

$$\rho_{xy}(\tau) = \frac{\mathbf{R}_{xy}(\tau)}{\sqrt{\mathbf{R}_x(0)\mathbf{R}_y(0)}} \tag{13-63}$$

The correlation coefficient provides a normalized measure of correlation. If $\rho_{xy} > 0$, then $x(t)$ and $y(t)$ are said to be positively correlated. That is, $x(t)$ and $y(t)$ will, in general, be moving in harmony or in phase. If $\rho_{xy} < 0$, then $x(t)$ and $y(t)$ are said to be negatively correlated, $x(t)$ and $y(t)$ will, in general, be moving out of phase. If $\rho_{xy} = 0$, then x(t) and y(t) are said to be uncorrelated.

Since correlation measures the similarity between signals, it is natural to adapt it for use in M-ary communication applications. If, in a bank of filters, the ith correlation filter is designed to maximally respond to a signal $x(t) = s_i(t)$, then theoretically $x(t)$ can be properly classified by the receiver if the output of channel i is much greater than that of any other channel. If the signals are truly orthogonal, then theoretically all outputs, except that of the ith channel, will be zero. Any deviation from the value would be attributable to added noise or nonlinear effects.

The ith receiver channel could be designed as a linear filter having an impulse response $h_i(t)$. The output of the ith channel, in this case, would be given by the convolution operation

$$y_i(t) = \int_{-\infty}^{\infty} x(\tau)h_i(t - \tau)\, d\tau \tag{13-64}$$

Since signal detection can more readily be made on the basis of power levels rather than instantaneous signal amplitude, it is desirable to measure the power in the output process $y_i(t)$ in terms of input and system parameters. From the convolution theorem it is known that

$$Y_i(f) = X(f)H_i(f) \tag{13-65}$$

where $Y_i(f) = \mathcal{F}(y_i(t))$, $X(f) = \mathcal{F}(x(t))$, and $H_i(f) = H_i(s)$ is evaluated along the trajectory $s = j2\pi f$. The output energy density function therefore becomes

$$\mathbf{G}_{y_i}(f) = |Y_i(f)|^2 = |X(f)|^2|H_i(f)|^2 = \mathbf{G}_x(f)|H_i(f)|^2 \tag{13-66}$$

where $\mathbf{G}_x(f)$ is the input energy spectral density function. Equivalently, the output energy spectral density can be computed to be

$$\mathbf{S}_{y_i}(f) = \mathbf{S}_x(f)|H_i(f)|^2 \qquad (13\text{-}67)$$

where $\mathbf{S}_x(f)$ is the input power spectral density function. The output energy or power can, therefore, be computed by integrating the respective density function over all frequencies[4].

These operations are fundamentally important to the study of linear systems in that they provide an efficient means of expressing how the energy or power of an input signal is redistributed at the output in the frequency-domain. Therefore, filters can be designed to reduce the output power over frequency ranges containing unwanted information and provide high gain in areas where the signal power is to be emphasized.

In practice, the input to the system is often assumed to be modeled as a signal plus additive noise processes of the form $x(t) = s_i(t) + n(t)$. It has been previously established that the power or energy in a signal can be expressed in terms of the integral of a power or energy spectral density. The study of noise in the frequency-domain requires a more formal understanding of random processes than can be developed here. However, we can make certain assumptions about a random process's power spectral density or autocorrelation function and from them derive other system parameters. For example, a special noise case of particular importance in the study of communication systems is called white noise. A white-noise process has a flat power spectral density across the entire frequency spectrum. The power spectral density of the white-noise process can be modeled as

$$\mathbf{S}_n(f) = \frac{\sigma^2}{2} \qquad (13\text{-}68)$$

where σ is constant[5]. Applying Parseval's power theorem, namely integrating Equation 13-68 over all frequencies, leads to the conclusion that a white-noise process has infinite power. Therefore white noise is not physically realizable. However, white noise remains a popular noise model and is used to study communication systems because in a linear system it will excite every possible mode of oscillation. That is, white noise presents to the system an input containing equal power-signal components at all possible frequencies.

White noise may also be defined in terms of its autocorrelation function, which, according to Equation 13-50, is given by

$$\mathbf{R}_n(\tau) = \frac{\sigma^2}{2}\delta(\tau) \qquad (13\text{-}69)$$

---

[4]For discrete-time systems, similar results can be obtained by simply replacing the continuous-time energy or power spectral density functions with their discrete-time counterparts and $H_i(f)$ with $H_i(e^{j\omega T})$.

[5]If $x(t)$ is also a zero-mean Gaussian random process, then σ^2 can be related to the variance of the noise.

where $\mathbf{R}_n(\tau)$ and $\mathbf{S}_n(f)$ are Fourier transform pairs. Equation 13-69 states that a white-noise process is changing so rapidly and erratically that it is only correlated with itself at $\tau = 0$. If we compare the noise record $n(t)$ with $n(t + \tau)$, for $\tau \neq 0$, they will appear to be totally dissimilar (uncorrelated). If white noise is passed through a lowpass filter, the output (called *colored noise*), will lose its high-frequency components and, as a result, lose its ability to rapidly change in value. This will have the effect of changing the autocorrelation function from being an impulse distribution to one having a finite width. That is, convolving white noise with a linear filter will introduce memory into the output process. As a result, the impulse-like autocorrelation function found at the system's input will begin to take on the shape of the filter that processes the noise.

Example 13-15   White and colored noise

A white-noise process $n(t)$ was defined in Equation 13-69. If $n(t)$ is passed through a LTI filter having a frequency response $H(f)$, the resultant colored-noise power spectrum is given by (see Equation 13-66)

$$S_y(f) = \frac{\sigma^2}{2} |H(f)|^2$$

Computer Study   S-file `ch13_15.s` produces 10 random signals and filters them with a lowpass LTI. The power spectra of the 10 original and filtered signals are computed, then averaged. Averaging suppresses the high degree of variability that exists between individual, short, random-signal records.

SIGLAB	MATLAB
`>include "ch13_15.s"`	Not available in Matlab

The results are shown in Figure 13-26 as a two-sided magnitude frequency response. The ensemble-averaged unfiltered noise spectrum is seen to have a broadly defined power spectrum, whereas the colored power spectrum takes on the shape of the LTI filter's frequency response.

Refer to Figure 13-24 and consider the input to the receiver to be given by $x(t) = s_i(t) + n(t)$, where $s_i(t)$ is one of the signals in an M-ary signal set $\mathcal{S}$ and $n(t)$ is white noise with a power spectrum given by $\mathbf{S}_n(f) = \sigma^2/2$. The input signal-to-noise ratio, denoted SNR_{in}, to the M-ary detector is given by

$$(\text{SNR}) = \frac{\text{power in } s_i(t)}{\text{power in } n(t)} \tag{13-70}$$

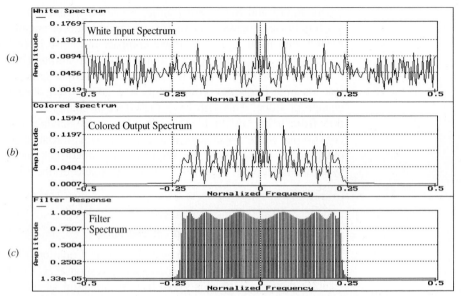

FIGURE 13-26   White and colored noise example using the ensemble average of 10 experiments beginning with (a) unfiltered (white) power spectrum, (b) filter (colored) power spectrum, and (c) magnitude frequency response of the shaping filter (results may vary due to use of a random signal).

The output of the ith filter, shown in Figure 13-24, is given by

$$y_i'(t) = \int_{-\infty}^{\infty} x(\tau)h_i(t - \tau) \, d\tau = \int_{-\infty}^{\infty} (s_i(\tau) + n(\tau))h_i(t - \tau) \, d\tau \quad (13\text{-}71)$$

$$= y_i(t) + n_i(t)$$

The output is seen to consist of both the filtered signal and noise components. From the convolution theorem, the output due exclusively to the input signal process can be expressed as

$$y_i(t) = \int_{-\infty}^{\infty} H_i(f)\mathcal{S}_i(f)e^{j2\pi ft} \, df \quad (13\text{-}72)$$

where $\mathcal{S}_i(f) = \mathcal{F}(s_i(t))$ and $H_i(f) = \mathcal{F}(h_i(t))$.

Suppose that the output of the ith filter is to be measured at some time $t = t_0$. At that time a decision is to be made to determine whether a signal $s_i(t)$ is or is not present at the input to the receiver. This decision will be based on a simple power measurement. Obviously, it is important to maximize the amplitude of the output signal at time t_0 (if $s_i(t)$ is present) in order to ensure that the event can be easily detected with a simple threshold measurement. Equivalently, the output amplitude squared corresponds to the *peak power* produced by the signal component only, at $t = t_0$, and is given by

$$|y(t_0)|^2 = \left| \int_{-\infty}^{\infty} H_i(f) S_i(f) e^{j2\pi f t_0} \, df \right|^2 \tag{13-73}$$

Define the ratio of the peak-power-to-noise-power, denoted ρ, to be the objective function to be maximized. Then, given a filter $h_i(t)$

$$\rho = \frac{|y(t_0)|^2}{P_{n_i}} = \frac{|\int_{-\infty}^{\infty} H_i(f) S_i(f) e^{j2\pi f t_0} \, df|^2}{\int_{-\infty}^{\infty} |H_i(f)|^2 S_n(f) \, df} \tag{13-74}$$

where $S_n(f)$ is the noise power spectral density. Applying *Schwartz's inequality*[6], Equation 13-74 can be rewritten as

$$\left| \int_{-\infty}^{\infty} H_i(f) S_i(f) e^{2\pi f (t_0)} \, df \right|^2 \le \int_{-\infty}^{\infty} |H_i(f)|^2 S_n(f) \, df \int_{-\infty}^{\infty} \frac{|S_i(f)|^2}{S_n(f)} \, df \tag{13-75}$$

where equality holds if

$$\sqrt{S_n(f)} H_i(f) = \frac{K S_i^*(f) e^{-j\omega t_0}}{\sqrt{S_n(f)}} \tag{13-76}$$

Therefore, the maximal value of ρ is given by

$$\rho_{\max} = \int_{-\infty}^{\infty} \frac{|S_i(f)|^2}{S_n(f)} \, df \tag{13-77}$$

for

$$H_i(f) = \frac{K S_i(f)^* e^{-j\omega t_0}}{G_N(f)} \tag{13-78}$$

Since the noise is assumed to be white, in particular $S_n(f) = \sigma^2/2$, it follows that for this case the optimum filter satisfies

$$H_i(f) = K S_i(f)^* e^{-j\omega t_0} \tag{13-79}$$

This filter is known as a *matched filter* due to the fact that its spectral shape is matched to that of the signal it is designed to detect. In addition, in the white-noise

---

[6] $|\int_{-\infty}^{\infty} X(f) Y(f) \, df|^2 \le \int_{-\infty}^{\infty} |X(f)|^2 \, df \int_{-\infty}^{\infty} |Y(f)|^2 \, df$, where equality holds if and only if $X(f) = KY^*(f)$ where K is a real constant.

case the matched filter has a time-domain description given by

$$h_i(t) = \mathcal{F}^{-1}(H_i(f)) = \int_{-\infty}^{\infty} S_i^*(f)e^{-j2\pi f(t_0-t)}\, df \qquad (13\text{-}80)$$

If the signal $s_i(t)$ is real, then $S_i^*(f) = S_i(-f)$ and Equation 13-80 can be simplified to

$$h_i(t) = \mathcal{F}^{-1}(H_i(f)) = \int_{-\infty}^{\infty} S_i(-f)e^{-2\pi f(t_0-t)}\, df = s_i(t_0 - t) \qquad (13\text{-}81)$$

As stated, the matched filter owes its name to the fact that the receiver, in the frequency-domain, has a shape matched to the signal spectrum. Therefore, at the frequencies where considerable signal power is delivered to the output, the matched filters gain is high. However, at frequencies where the signal power level is low or zero, the matched filter's gain is likewise low or zero. The combined result is an improved peak output signal power-to-noise ratio and superior event detection based on power measurements.

Referring to Equation 13-81, it can be noted that matched filter's impulse response is a time reversed and delayed version of the signal to which it is matched. Therefore, the signals that the system is to detect must be known *a priori*. This is one of the weaknesses of the matched filter. Nevertheless, there are many applications in which the signal space is known.

Example 13-16   Radar return

A typical pulse radar system is shown in Figure 13-27. Range estimates to the target are made on the basis of measuring the delay between the transmitted signal and the received signal, which is reflected by the target. A pulsed radar transmits a rectangular pulse of width τ. There may be uncertainty about

FIGURE 13-27   Typical radar system.

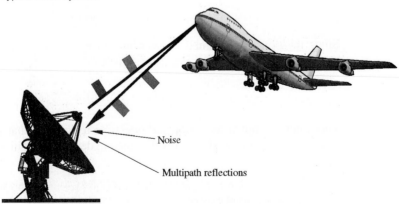

Noise

Multipath reflections

which particular reflected pulse in a periodic pulse train, having a *pulse repe-tition frequency (PRF)* of PRF= $1/T$, is being detected. This *range ambiguity*, is measured in meters, and has a value of $d = c/2\text{PRF}$ where c is the speed of light in meters per second. Furthermore, to have a sharp peak at the output of the radar matched filter, the pulse width is normally chosen to be small. For example, suppose a rectangular pulse of width τ and a PRF = 4 Hz is received with a delay d, then the matched filter output should peak with a delay d at a rate of four times per second. Unfortunately the noise normally found in the received signal has the effect of masking the location of the peak.

Computer Study   S-file `ch13_16.s` and M-file `radar.m` create a radar pulse with a 4 Hz PRF radar and a pulse width τ, and introduces a delay of `d` ms using the format `radar(s,d,tau)`. The value of `s` scales the additive noise applied to the received signal with assumed form $r(t) = p(t-d)+sn(t)$, where $n(t)$ is a white-noise process and $p(t)$ is the transmitted pulse. The signal-to-noise ratio is computed and the noise-free and noise-contaminated output of the radar matched filter are graphically displayed.

SIGLAB	MATLAB

`>include "ch13_16"`
`>radar(.1,100,20) #Signal to`
`Noise = (s/n)`
`>radar(5,100,20) #Signal to Noise`
`= (s/n)`
	`>> radar(0.1, 100, 20)`
	`Signal to Noise ratio = (s/n)`
	`>> radar(5, 100, 20)`
	`Signal to Noise ratio = (s/n)`

FIGURE 13-28   Output of a radar matched filter for a noise-free and noisy case with $\tau = 20$ ms and $d = 100$ ms (results may vary due to use of a random signal).

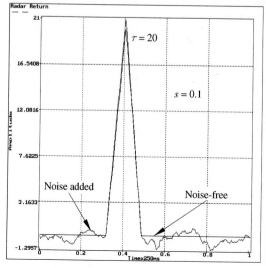 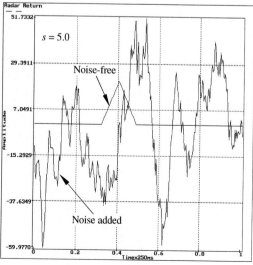

The results are shown in Figure 13-28. This indicates that in a low-noise environment range detection can take place, but it becomes substantially more difficult when noise become dominant.

There are other classic receiver filters that serve other purposes. The Weiner and Kalman filters are based on a set of assumptions not considered by the matched filter. Their development is, however, beyond the scope of the current study.

13-9 DIGITAL COMMUNICATIONS

Given the trend in communications to replace analog systems with their digital counterparts, it is obvious that the piece of wire connecting the stereo to the speaker or two telephone handsets may also carry the information in a digital format. A simple digital encoding technique, called *pulse amplitude modulation (PAM)*, begins with passing an analog signal through an *n*-bit analog-to-digital converter. The resulting *n*-bit digital word represents the amplitude of the sampled signal. PAM is sometimes used for communicating data locally and, in this case, is referred to as *word parallel* data communications. If the transmission of this information is performed on a bit-by-bit basis, it is called *pulse code modulation (PCM)* and uses a *bit-serial* communication protocol. A PCM system simply broadcasts the binary data stream from an analog-to-digital converter to a receiver in order to reconstruct the PAM signal from the PCM message, then returns it to continuous-time using an a digital-to-analog converter. To preserve the integrity of the information in a bandlimited environment, the signal must be sampled at or above the Nyquist rate. The higher the sample, the better will be the reconstructed approximation of the original. Unfortunately, high sample rate would also require a high transmission channel bandwidth. For example, 8000 samples per second is generally assumed to be sufficient to produce telephone-quality speech, provided the PCM message has sufficient amplitude resolution, say 12 bits. Under this assumption, a data rate of 96,000 bits per second would result. Assuming that the digital system is to be a direct replacement of an analog telephone system, the logical question to ask is what the bandwidth requirement for a 96,000 bit-per-second PCM message would be. The answer can be derived by considering how a PCM message would be reconstructed from the output of a bandlimited communication channel.

Recovering a digital message is the function of a *regenerator*. Referring to the system illustrated in Figure 13-29, it is seen that the binary-valued ana-

FIGURE 13-29   Digital signal transmission and recovery system.

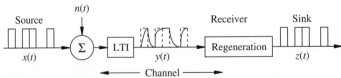

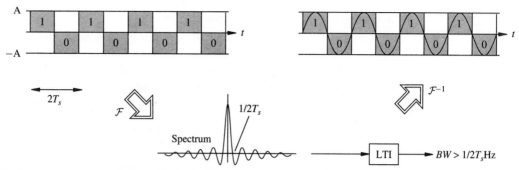

FIGURE 13-30   Lowpass filtering of a worst-case symbol set and its spectrum.

log signal is sent through a linear channel that convolves the original message with a channel filter function (possibly with noise added). Due to convolution, a residual amount of the signal attributable to previous inputs can be found in the current output, giving rise to what is called *intersymbol interference* or ISI. That is, previous symbols can interfere with the symbol currently being received. The key to decoding the digital message is suppressing the ISI effect.

Nyquist noted that a noise-free channel of bandwidth B could transmit independent samples at a rate $R \leq 2B$ without ISI. This can be interpreted as follows. Suppose the *symbol period* of the binary message $x(t) = \pm A$ is given by T_s and that $x(t)$ is to be communicated across a noise-free channel of bandwidth B. The worst-case message, in terms of having the highest frequency content, is the rapidly fluctuating $x(t) = \{\ldots, A, -A, A, -A, \ldots\}$. The message spectrum is, therefore, given by a sinc function, which is graphically interpreted in Figure 13-30 for a family of symbol intervals T_s. For a fixed-channel bandwidth, it can be immediately seen that the first harmonic of the worst case input is located at $f_0 = 1/2T_s$. If $f_0 < B$, then the first harmonic will pass through the noise-free communication channel and appear as $y(t) = K \sin(2\pi f_0 t + \phi)$ at the input of the regenerator. The regenerator's function is to return the signal back to a binary-valued function. The simplest form the regenerator can take is a zero-crossing detector, which is mathematically represented as

$$\text{sgn}(q) = \begin{cases} 1 & \text{if } q > 0 \\ 0 & \text{if } q = 0 \\ -1 & \text{if } q < 0 \end{cases} \tag{13-82}$$

The output of the regenerator could then be given by $z(t) = A \, \text{sgn}(y(t))$, which ideally would equal $x(t)$, the original message. If, for example, $\phi = 0$, then $y(t) = A \, \text{sgn}(K \sin(2\pi f_0 t)) = x(t)$. Therefore, on a noise-free channel having a B Hz bandwidth, the highest error-free decodable bandwidth is bounded by $2B$ symbols/second. Thus, to transmit digitized speech at a rate of 96,000 bits per second would require a channel having a bandwidth bounded by 48 kHz.

Example 13-17   Digital communications

The Fibonnacians have changed to cable to communicate their digital messages. The cable, nevertheless, has a finite bandwidth and can be subject to additive broadband noise contamination. The data rate used by the Fibonnacians is $2B$ bits per second and the channel bandwidth is slightly in excess of B Hz. The performance of the channel can be experimentally determined.

Computer Study   S-file `ch13_17.s` and M-file `ch13_17.m` create the ASCII message BUNNY, possibly adds noise to it, transmits it along a bandlimited channel, and decodes it at the receiver. The format used is `digit(sigma)`, where `sigma` is the standard deviation of the additive noise.

SIGLAB	MATLAB

`>include "ch13_17.s"`
`>digit(0);digit(0.5);digit(5.0)`
	`>> digit(0)`
	`>> digit(0.5)`
	`>> digit(5)`

The results are displayed in Figure 13-31. The experiment shows that decoding can take place when the noise floor is low, but degrades as the noise floor rises.

13-10   SUMMARY

In Chapter 13 an introduction to communication theory was presented. The study focused primarily on modulation and demodulation methods. It was shown that many communication problems can be analyzed in both the time- and the frequency-domain. Shannon's law established the capacity of a system that was quantifiable in terms of a signal-to-noise ratio measure and channel bandwidth. Filters, it was noted, can often be used to improve a system's signal-to-noise ratio. Signal-to-noise ratio was seen to be important in digital data communications, as well.

Modulation schemes that alter the amplitude, phase, or frequency of a signal or carrier were developed. In general, fidelity was obtained at the expense of increased channel bandwidth requirements. Within the bandwidth constraints of commercial broadcasting systems, innovative methods and procedures have evolved that optimize the allocated frequency-band utilization. Newer paradigms, based on sophisticated phase modulation and data-coding schemes are now being developed for use with mobile, video, and other communication systems. The desire to communicate data (often video) at ever expanding speeds has also highlighted the need for efficient data compression schemes.

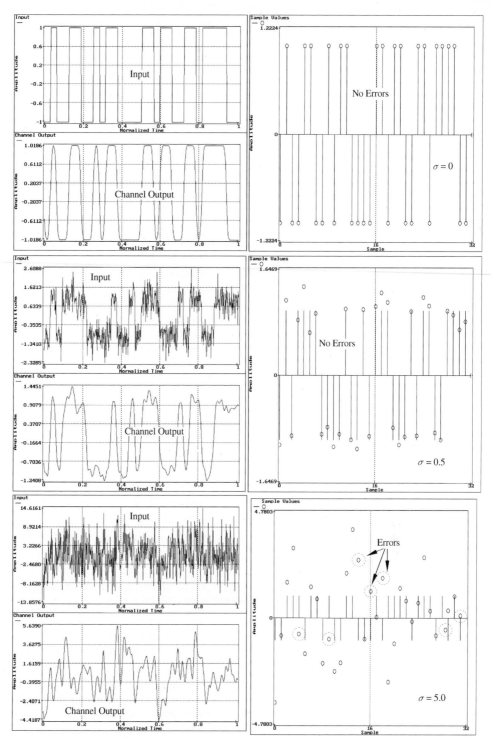

FIGURE 13-31 Binary data transmission and reconstruction using a bandlimited baseband channel. The trails are (top) sigma = 0 (no noise), (middle) sigma = 0.5 (low noise), and (bottom) sigma = 5 (high noise). The data shown includes the input and channel output, as well as the original and decoded messages.

13-11 COMPUTER FILES

The following computer files were used in Chapter 13.

File Name	Type	Location
Subdirectory CHAP13		
ch13_1.s	C	Example 13-1
mulaw.m	C	Example 13-1
ch13_2.s	C	Example 13-2
ideal.m	C	Example 13-2
ch13_3.s	C	Example 13-3
ch13_4.s	C	Example 13-4
ch13_5.s	C	Example 13-5
ch13_6.s	C	Example 13-6
ch13_7.s	C	Example 13-7
pmfm.m	C	Example 13-7
ch13_8.s	C	Example 13-8
wideband.m	C	Example 13-8
ch13_9.s	C	Example 13-9
ch13_10.s	C	Example 13-10
ch13_11.s	C	Example 13-11
pll.m	C	Example 13-11
ch13_12.s	C	Example 13-12
ch13_15.s	C	Example 13-15
ch13_16.s	C	Example 13-16
radar.m	C	Example 13-16
ch13_17.s	T	Example 13-17
digit.m	T	Example 13-17
Subdirectory SFILES or MATLAB		
chirp.x	C	Examples 13-4, 13-6, 13-10, 13-11
am.s	C	Example 13-3, 13-5
Subdirectory SIGNALS		
hilbert.imp	D	Example 13-4
EKG1.imp	D	Example 13-1, 13-7
preemp.imp	D	Example 13-10
bunny	D	Example 13-12
Subdirectory SYSTEMS		
ekg.arc	D	Examples 13-3, 13-5
super.arc	D	Example 13-6
preemp.arc	D	Example 13-10
whitecol.arc	D	Example 13-15
ch13_23.arc	D	Example 13-23

In these files "s" denotes S-file, "m" an M-file, "x" either a S-file or M-file, "T" a tutorial program, "D" a data file, and "C" a computer-aided instruction (CAI) program. CAI programs can be altered and parameterized by the user.

13-12 PROBLEMS

13-1 255 Mu-law compander   A signal $x(t) = A \sin(\omega_0 t)$, $\omega_0 > 0$, is passed through a 255 Mu-law compander. Conduct an experiment that will generate a database to define the bandwidth requirements at the output of the 255 Mu-law compander as a function of A. Assume that any harmonic greater than 10% of the fundamental is significant.

13-2 AM-DSB transmission   An AM-DSB transmitter is tested using a dummy load and a tunable narrowband receiver as shown in Figure 13-32. The RF signal is swept over a range $f \in [100k, 100M]$ Hz. With no audio signal present, the wattmeter reads 100 W and the peak detector registers 10 volts at 1 MHz.

    a With an audio input of 1 volt at 1 kHz, the wattmeter reads 150 W. What is the modulation index? What is the peak output?

    b If the input $x(t)$ is uniformly distributed in frequency over $f \in [50, 1500]$ Hz, what is the maximum value of $|X(f)|$ so that the power to the load is 150 W, for $\mu = 1$?

13-3 Superheterodyne receiver   What would be the effect of replacing the local sinusoidal oscillator, shown in the superheterodyne AM receiver in Figure 13-14, with a rectangular pulse generator of duty cycle $d = 1/2$ and period $T = 1/(f_c + f_{IF})$ (i.e., $\operatorname{sgn}(\sin(2\pi(f_0 + f_{IF})t)))$?

13-4 AM suppressed carrier receiver   Compared to an AM-DSB with carrier (Equation 13-7), an AM-DSB without carrier system has a signal-power advantage. The historic problem with this class of AM transmission has been the reconstruction of the carrier at the receiver (called *coherent detection*) required for coherent demodulation. A *Costas loop* demodulator is shown in Figure 13-33. Derive an equation for $y(t)$ and sketch the spectral shape of the signal found at the output of each major component of the Costas loop.

13-5 AM suppressed carrier transmitter   Verify that the circuit shown in Figure 13-34 can be used to produce an AM-DSB suppressed carrier transmission.

13-6 Analytic signal   Create an analytic signal based on the following signals.

$$x(t) = \cos(\omega_0 t + \phi)$$

$$x(t) = e^{-at} \cos(\omega_0 t)u(t), a > 0$$

$$x(t) = e^{-at} \sin(\omega_0 t)u(t), a > 0$$

FIGURE 13-32   AM-DSB transmitter test.

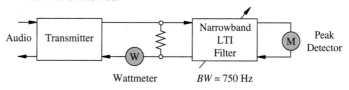

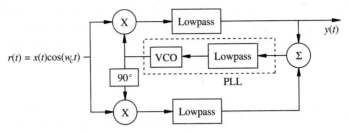

FIGURE 13-33   Costas loop AM-DSB suppressed carrier receiver.

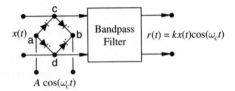

FIGURE 13-34   AM-DSB suppressed carrier modulator.

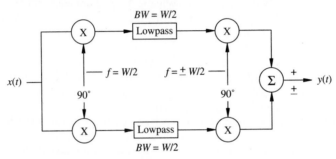

FIGURE 13-35   Weaver single sideband modulator.

13-7 Single sideband transmission   A *Weaver single sideband modulator* is shown in Figure 13-35. If $x(t) = \cos(\omega_0 t)$, derive $y(t)$.

13-8 Power spectral density   A white-noise process with a power spectral density of $G_n(f) = \sigma^2/2$ is passed through an ideal lowpass filter having a bandwidth of B Hz. The output of the lowpass filter is then processed by an ideal differentiator. Sketch the power spectrum at the output of the lowpass and differentiator. Compute the power in these signals as a function of the bandwidth parameter B.

13-9 FM receiver   An FM receiver is tuned at 100 MHz. A 10 kHz audio signal modulates a 100 MHz carrier with $\beta = 0.1$. What are the bandwidth requirements of the receiver? Repeat for $\beta = 5.0$. Two signals at 10 kHz ($\beta = 5.0$) and 2 kHz ($\beta = 25$), respectively, are alternately transmitted. What are the bandwidth requirements in this case? What is the required bandwidth if they are added first and then simultaneously transmitted?

13-10 PM and FM bandwidth requirements   For the message processes shown in Figure 13-36, estimate the bandwidth requirements for a PM and FM transmission for
a narrowband ($k_f = \pi \times 10^4$ and $k_p = \pi/4$)
b wideband ($k_f = \pi \times 10^5$ and $k_p = 10 \times \pi$)

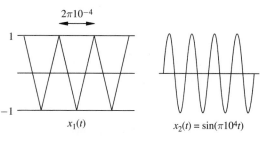

FIGURE 13-36 Message signals used in PM and FM problem.

$x_1(t)$ $x_2(t) = \sin(\pi 10^4 t)$

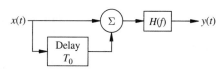

FIGURE 13-37 System used to process signal defined in Problem 13-14.

13-11 FM broadband requirements A slowly swept chirp signal $x(t)$, defined over the range $f \in [50, 15000]$ Hz, is used as the input to a commercial FM system. The measured maximum frequency deviation is $\Delta f = 75$ kHz and the transmission bandwidth is on the order of 200 kHz. The chirp spectrum is assumed to have an instantaneous value bounded by $X_{\max}(f) = X_m$. Derive a relationship between the FM transmission bandwidth requirements and the instantaneous frequency of $x(t)$.

13-12 Preemphasis A broadband FM system is to be used to transmit messages having a spectrum given by

$$|X_{\text{speech}}(f)| = \frac{X_m}{2}\left(1 + \frac{50X_m}{|f|}\right)$$

for $f \in [50, 1500]$ Hz. Design a preemphasis filter that will optimize the performance of the FM system. What would the deemphasis filter look like?

13-13 AM and FM communication A transmitter produces a signal at a power rating of P_0 W. The received signal power is 100 dB below P_0, along with additive white noise with power spectral density given by $S_n(f) = 10^{-14}$ W/Hz. The message bandwidth is baseband limited to 15 kHz. Calculate the power level P_0 needed to give a post-detection SNR of 40 dB, if the communication protocol is
a AM-DSB with carrier ($\mu = 0.1, 1.0$)
b Wideband commercial FM ($\Delta f = 75$ KHz)

13-14 Power spectral density calculation A period T periodic signal $x(t)$ is sampled and assumed to be of the form

$$x(t) = \sum_{k=0}^{9} a_k \delta(t - kT_s)$$

for one period. It is processed by the system shown in Figure 13-37. What are the input power spectral density and power? What are the output power spectral density and power?

13-15 Autocorrelation function Derive the autocorrelation function and power spectral density of the periodic signals shown in Figure 13-38.

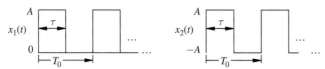

FIGURE 13-38   Sections of periodic signals used in autocorrelation and power spectral density problems.

13-16 Matched filter   Prove that the matched filter, given by Equation 13-79, is optimal in the context of maximizing ρ.

13-17 Matched filter   A system is to be designed to transmit messages $x_i(t)$ from an orthogonal set of signals. The Fourier representation of $x_i(t)$ exists and is given by $\mathcal{F}(x_i(t)) = X(f)$. White noise $n(t)$ is added to the transmitted signal and has a power spectral density given by $S_n(f) = \sigma^2/2$ over all frequencies. The receiver is designed as a bank of matched filters $h_i(t)$.

What is the definition of $h_i(t)$? What does the output of the ith matched filter look like if the message being transmitted is always the ith symbol? Repeat the analysis assuming the jth $\neq$ ith symbol is always transmitted. What is the output SNR measured at the ith matched filter if the ith symbol is continuously transmitted?

13-18 PCM message   A Bell T-1 telephone channel supports up to 24 time-multiplexed users using PCM. Each message is bandlimited to 3.5 kHz and sampled at a 8 kHz rate with an 8-bit analog-to-digital converter. The coding scheme is called bipolar and obeys the following protocol:

a "0" = no pulse

b "1" = a pulse $p(t)$ or $-p(t)$, depending on whether the previous "1" was transmitted by a $-p(t)$ or $p(t)$.

c Code the following messages and sketch their waveforms
    [1,0,1,0,1,0,1,0]; [0,0,0,0,1,1,1,1]; [1,1,1,1,1,1,1,1]

d What is the highest bandwidth requirement per subscriber? Per T-1 channel?

13-19 Digital TV   North American TV uses 525 horizontal lines per frame (495 actively used). Human retinal systems retain images only over a short period of time. As a result, a set of still images can be made to appear animated if they are updated sufficiently often. To reduce "flicker", a refresh rate on the order of 40 images (frames) per second is required. TV uses only 30 frames per second. To eliminate the illusion of flicker, the 525 lines are displayed as two successive frames (called a field) interleaved together. If each picture element (pixel) is quantized into a 1-bit word, what is the estimated bandwidth requirement for the digital information transmission?

13-20 Speech scrambler   Speech will be assumed to be a baseband process bounded by 3 kHz. A filter is designed that permutes the speech spectrum using the rule

$$Y(f) = X(3000 - f)$$

The receiver would simply consist of the inverse filter. Explain whether or not this scheme would work as a security device. Could an ideal lowpass filter be easily modified to serve as the scrambler? Create a simulation diagram of the system, assume a typical speech spectrum (arbitrary shape), and trace it through the system.

13-13 COMPUTER PROJECTS

13-1 Modem   A *Quadrature AM*, or QAM system, modulates two messages using two quadrature carrier signals, namely $\cos(\omega_0 t)$ and $\sin(\omega_0 t)$. Furthermore, a telephone channel has an assumed 2400 Hz bandwidth over the range $f \in [600, 3000]$. The telephone channel, therefore, can be alternatively thought of as a bandpass filter centered at 1800 Hz with 1200 Hz sidebands.

The transmitted symbols may be multivalued so that the effective bit rate can exceed the symbol rate by a multiple of 2400. Using two-level (e.g.; ±5V), or four-level (e.g.; ±5V, ±1.67V) per symbol codes (see Figure 13-39) would result in a 4800 and a 9600 b/s MODEM respectively.

Refer to the MODEM system schematic given in Figure 13-40. Suppose $x_c[k]$ and $x_s[k]$ are generated 2400 times per second. These signals would be presented to a pair of FIRs, which are used as anti-alising devices. The FIRs have a bandwidth on the order of 1200 Hz. The FIRs are also clocked at a rate that is a multiple of 1200 Hz, in order to improve the spectral shape and performance of the filter. Using a times 6 oversampling rate, for example, $x_c[k]$ (or $x_s[k]$) would be sent to its respective FIR along with five padded (appended) zeros.

The demodulator reverses the process. It includes a phase locked-loop to reconstruct the transmitter carrier (i.e., coherent detection). In addition, to correct for distortions introduced by the transmission media, an adaptive equalization filter is used. A known *preamble* is sent prior to the data being transmitted. The preamble is used to train the adaptive equalizer.

Ignoring the training requirement of an adaptive equalizer, simulate a 9600 BAUD MODEM modulator/demodulator. Add broadband noise to the simulation (once verified) and conduct experiments to determine the system's error rate.

13-2 Matched filter   The matched filter system described in Problem 13-16 is to be experimentally studied. For a SNR of $S/N = 20$ dB, 10 dB, 0 dB, −10 dB, simulate the performance of the four-symbol system using a random message of length in excess of 200 symbols. Conduct at least 100 experiments and compute

FIGURE 13-39   Modem symbol encoded space for 2- and 4-level coding.

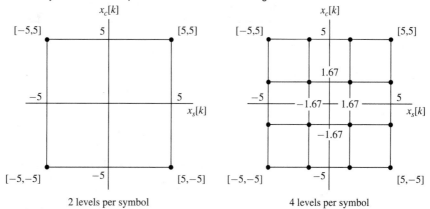

2 levels per symbol   4 levels per symbol

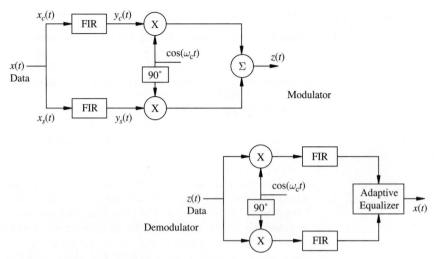

FIGURE 13-40   Modem modulator/demodulator model.

the false alarm and miss rate for each tested matched filter and set of detection thresholds. What is the trade-off between the false alarm rate and miss rate as a function of decision threshold? Explain your results in terms of common sense and logic.

## PART FIVE

APPENDICES

A

FUNDAMENTALS
OF MATRICES

The good Christians should be aware of mathematicians and all those who make empty prophecies. The danger already exists that mathematicians have made a covenant with the devil to darken the spirit and confine man in the bonds of hell.

—St. Augustine

A-1 BASIC STRUCTURE

A *matrix* is a rectangular array of elements. A constant coefficient $n \times m$ matrix $\mathbf{A}$, shown below, is given by the indexed array of data, where the ijth element belongs to the ith row and jth column. A matrix that consists of n rows and m columns, is said to be an $n \times m$ matrix.

$$\mathbf{A} = \begin{pmatrix} a_{11} & a_{12} & \cdots & a_{1m} \\ a_{21} & a_{22} & \cdots & a_{2m} \\ \vdots & \vdots & \ddots & \vdots \\ a_{n1} & a_{n2} & \cdots & a_{nm} \end{pmatrix} \tag{A-1}$$

The *transpose* of an $n \times m$ matrix $\mathbf{A}$, is denoted $\mathbf{A}^T$, and is an $m \times n$ matrix given by

$$\mathbf{A}^T = \begin{pmatrix} a_{11} & a_{21} & \cdots & a_{n1} \\ a_{12} & a_{22} & \cdots & a_{n2} \\ \vdots & \vdots & \ddots & \vdots \\ a_{1m} & a_{2m} & \cdots & a_{nm} \end{pmatrix} \tag{A-2}$$

The elements of the matrix may be real or complex. If $\mathbf{A}$ is a matrix, then $\mathbf{A}^*$ is its *complex conjugate*, which is formed by taking the complex conjugate of each element of $\mathbf{A}$. If $\mathbf{A} = \mathbf{A}^*$, then the matrix $\mathbf{A}$ is real, otherwise it is complex.

Two $n \times m$ matrices $\mathbf{A}$ and $\mathbf{B}$ are equal if and only if $a_{ij} = b_{ij}$ for all i and j. The sum of $\mathbf{A}$ and $\mathbf{B}$ is given by $\mathbf{S} = \mathbf{A} + \mathbf{B}$, where $s_{ij} = a_{ij} + b_{ij}$. The product of an $n \times m$ matrix $\mathbf{A}$ and an $m \times p$ matrix $\mathbf{B}$ is given by the $n \times p$ matrix $\mathbf{P} = \mathbf{AB}$, where

$$P_{ij} = \sum_{k=1}^{m} a_{ik} b_{kj}, \qquad 1 \le i \le n, 1 \le j \le p \qquad \text{(A-3)}$$

A-2 MATRIX CLASSIFICATION

There is also a set of special square matrices that naturally occur in practice. The most important of these are:

Zero Matrix	$\mathbf{Z} \Rightarrow z_{ij} = 0$ for all i and j (need not be square)
Diagonal Matrix	$\mathbf{D} \Rightarrow d_{ij} = 0$ for all $i \ne j$ and d_{ij} arbitrary if $i = j$
Unit (Identity) Matrix	$\mathbf{I} \Rightarrow i_{ij} = 0$ for all $i \ne j$ and $i_{ij} = 1$ if $i = j$
Upper Triangle Matrix	$\mathbf{T}_U \Rightarrow \{t_U\}_{ij} = 0$ for all $i > j$
Lower Triangle Matrix	$\mathbf{T}_L \Rightarrow \{t_L\}_{ij} = 0$ for all $i < j$
Symmetric Matrix	$\mathbf{S} = \mathbf{S}^T \Rightarrow s_{ij} = s_{ji}$ for all i and j
Hermitian Matrix	$\mathbf{H} = \mathbf{H}^\# \Rightarrow h_{ij} = h_{ji}^*$
	($^\#$ denotes complex conjugate transpose)
Skew Symmetric	$\mathbf{V} = -\mathbf{V}^\# \Rightarrow v_{ij} = -v_{ji}^*$
Orthogonal Matrix	$\mathbf{U}^T \mathbf{U} = \mathbf{I}$

A-3 DETERMINANTS

A constant coefficient $n \times n$ matrix $\mathbf{A}$ has a *determinant* $\det(\mathbf{A})$ given by

$$\det(\mathbf{A}) = \sum_{j=1}^{n} a_{ij} \Delta_{ij} \qquad \text{(A-4)}$$

for any $i = 1, 2, \ldots, n$, Δ_{ij} is called the cofactor of a_{ij} and is given by

$$\Delta_{ij} = (-1)^{(i+j)} M_{ij} \qquad \text{(A-5)}$$

where M_{ij} is called the *minor* of a_{ij}. The minor M_{ij} is defined as the determinant of $\mathbf{A}$ with the ith row and jth column removed. The determinant of a 2×2 matrix is the basic building block for the determinant, cofactor, and minor calculations. In particular, if

$$\mathbf{A} = \begin{pmatrix} a_{11} & a_{12} \\ a_{21} & a_{22} \end{pmatrix} \qquad \text{(A-6)}$$

then

$$\det(\mathbf{A}) = a_{11}a_{22} - a_{12}a_{21} \tag{A-7}$$

Using 2×2 determinant operations, the determinant of a larger $n \times n$ matrix can be derived using Equation A-4. An alternative computing formula for the determinant of $\mathbf{A}$ will be presented in Section A-6. Finally, if $\mathbf{A}$ and $\mathbf{B}$ are $n \times n$ matrices, then $\det(\mathbf{AB}) = \det(\mathbf{A})\det(\mathbf{B})$.

Example A-1

Consider the matrix $\mathbf{A}$ shown below. The computation of $\det(\mathbf{A})$ is given by the following procedure

$$\mathbf{A} = \begin{pmatrix} 1 & 2 & 3 \\ 1 & 1 & 1 \\ 2 & 1 & 3 \end{pmatrix}$$

$$\det(\mathbf{A}) = (-1)^{(1+1)}a_{11}M_{11} + (-1)^{(1+2)}a_{12}M_{12} + (-1)^{(1+3)}a_{13}M_{13} = -3$$

$$M_{11} = \det\begin{pmatrix} 1 & 1 \\ 1 & 3 \end{pmatrix} = 2$$

$$M_{12} = \det\begin{pmatrix} 1 & 1 \\ 2 & 3 \end{pmatrix} = 1$$

$$M_{13} = \det\begin{pmatrix} 1 & 1 \\ 2 & 1 \end{pmatrix} = -1$$

Verify the result in SIGLAB, define the 3×3 matrix $\mathbf{A}$, shown above, using

```
SIGLAB>a=[{1,1,2},{2,1,1},{3,1,3}]; a # define
and display A
SIGLAB>det=prod(eig(a)); det # compute and display
determinant; -3.0
```

A-4 INVERSE

An $n \times n$ matrix $\mathbf{A}$ is said to be nonsingular if $\det(\mathbf{A}) \neq 0$. If $\mathbf{A}$ is nonsingular, then there exists an inverse of $\mathbf{A}$, denoted $\mathbf{A}^{-1}$, such that $\mathbf{AA}^{-1} = \mathbf{A}^{-1}\mathbf{A} = \mathbf{I}$. The inverse of $\mathbf{A}$ is given by

$$\mathbf{A}^{-1} = \frac{\text{adjoint}(\mathbf{A})}{\Delta} = \frac{\begin{pmatrix} \Delta_{11} & \Delta_{12} & \cdots & \Delta_{1m} \\ \Delta_{21} & \Delta_{22} & \cdots & \Delta_{2m} \\ \vdots & \vdots & \ddots & \vdots \\ \Delta_{n1} & \Delta_{n2} & \cdots & \Delta_{nm} \end{pmatrix}^T}{\Delta} \tag{A-8}$$

where T denotes transpose, Δ_{ij} is the ijth cofactor, and Δ is the determinant of $\mathbf{A}$. The *adjoint* matrix is seen to be a matrix of transposed cofactors.

Example A-2

Consider again the nonsingular **A** used in Example A1-1, which had a determinant $\Delta = -3$. Then

$$
\mathbf{A}^{-1} = \frac{\left(\begin{array}{ccc} \Delta_{11} = \det\begin{pmatrix} 1 & 1 \\ 1 & 3 \end{pmatrix} = 2 & \Delta_{12} = -\det\begin{pmatrix} 1 & 1 \\ 2 & 3 \end{pmatrix} = -1 & \Delta_{13} = \det\begin{pmatrix} 1 & 1 \\ 2 & 1 \end{pmatrix} = -1 \\[2mm] \Delta_{21} = -\det\begin{pmatrix} 2 & 3 \\ 1 & 3 \end{pmatrix} = -3 & \Delta_{22} = \det\begin{pmatrix} 1 & 3 \\ 2 & 3 \end{pmatrix} = -3 & \Delta_{23} = -\det\begin{pmatrix} 1 & 2 \\ 2 & 1 \end{pmatrix} = 3 \\[2mm] \Delta_{31} = \det\begin{pmatrix} 2 & 3 \\ 1 & 1 \end{pmatrix} = -1 & \Delta_{32} = -\det\begin{pmatrix} 1 & 3 \\ 1 & 1 \end{pmatrix} = 2 & \Delta_{33} = \det\begin{pmatrix} 1 & 2 \\ 1 & 1 \end{pmatrix} = -1 \end{array}\right)^T}{\det(\mathbf{A}) = \Delta = -3}
$$

$$
= \begin{pmatrix} -2/3 & 1 & 1/3 \\ 1/3 & 1 & -2/3 \\ 1/3 & -1 & 1/3 \end{pmatrix}
$$

Using SIGLAB to verify this result, define **A** as before and compute

```
SIGLAB>a=[{1,1,2},{2,1,1},{3,1,3}]; a # define and display
A
SIGLAB>ai=inv(a) # compute and display inverse
-0.667 1 0.333
0.333 1 -0.667
0.333 -1 0.331
```

A-5   SOLUTIONS TO SYSTEMS OF LINEAR EQUATIONS

The solution to the equation $\mathbf{Ax} = \mathbf{y}$, where **A** is an $n \times n$ matrix, and **x** and **y** are n-vectors (i.e., $n \times 1$ matrices), is given by $\mathbf{x} = \mathbf{A}^{-1}\mathbf{y}$ if **A** is nonsingular.

Example A-3

In SIGLAB, again define the 3×3 matrix **A**, used in Example A-1. Let $\mathbf{y} = (1, 1, 1)^T$ and compute **x** as follows:

```
SIGLAB>a=[{1,1,2},{2,1,1},{3,1,3}]; a # define and display
A
SIGLAB>y= [{1,1,1}]; y # define and display y
SIGLAB>ai=inv(a) # compute and display inverse; x=ai*y; x
SIGLAB>x=ai*y; x
0.667 0.667 -0.333
SIGLAB>a*x # verify Ax=y
1 1 1
```

A-6   EIGENVALUES AND EIGENVECTORS

The *eigenvalues* of an $n \times n$ matrix $\mathbf{A}$, denoted λ_i, are the solutions to the equation $\mathbf{Ax} = \lambda \mathbf{x}$ where $\mathbf{x}$ is a non-zero n-vector and can be computed by solving $c(\lambda) = \det(\mathbf{A} - \lambda \mathbf{I}) = 0$. The term $c(\lambda) = c_n \lambda^n + \cdots + c_1 \lambda^1 + c_0 \lambda^0 = (\lambda - \lambda_0)(\lambda - \lambda_2) \cdots (\lambda - \lambda_{n-1})$ is called the *characteristic equation*. If $c(\lambda)$ is an nth order polynomial, then the n eigenvalues of $\mathbf{A}$ satisfy the condition $c(\lambda_i) = 0$. The function value $c(0)$ is also seen to be equal to $\det(\mathbf{A}) = \prod_{i=0}^{n-1} \lambda_i$. The *eigenvectors* are the non-zero solutions to the equation $\mathbf{Ax}_i = \lambda_i \mathbf{x}_i$ where λ_i is the ith eigenvalue of $\mathbf{A}$. The equation $\mathbf{Ax}_i = \lambda_i \mathbf{x}_i$ can be alternatively written $(\mathbf{A} - \lambda_i)\mathbf{x}_i = \mathbf{B}_i \mathbf{x}_i = \mathbf{0}$. The determinant of $\mathbf{B}_i$ is known to be zero and, as a result, $\mathbf{B}_i$ has no inverse. In fact, if $\mathbf{B}_i$ had an inverse, then the equation would be satisfied by $\mathbf{x}_i = \mathbf{B}_i \mathbf{0} = \mathbf{0}$. To solve the equation $\mathbf{B}_i \mathbf{x}_i = \mathbf{0}$, for a non-zero $\mathbf{x}_i$, a system of $n - p$ independent equations in n unknowns needs to be evaluated (the value of p is established by the order of what is called a minimal polynomial, studied in an advanced course of linear algebra). The equation $\mathbf{B}_i \mathbf{x}_i = \mathbf{0}$ represents a system that has more variables than independent equations. As a result, the eigenvectors are not unique. Therefore, at least one element of each eigenvector must be assigned an assumed value.

Example A-4

Consider again

$$\mathbf{A} = \begin{pmatrix} 1 & 2 & 3 \\ 1 & 1 & 1 \\ 2 & 1 & 3 \end{pmatrix}$$

```
SIGLAB>a=[{1,1,2},{2,1,1},{3,1,3}]; ev=eig(a); ev;
5.271+0j -.9022+0j .63038+0j
```

where λ_0 =ev[0], λ_1 =ev[1], λ_2 = ev[2]. The eigenvalues can be used to compute the determinate of $\mathbf{A}$ as follows:

```
SIGLAB>det=prod(ev); det; -3+0j # value of det(A=3)
```

The individual eigenvalues and the matrix $\mathbf{B}$ can be defined in the following manner:

```
SIGLAB>ev0=ev[0]; ev1=ev[1]; ev2=ev[2] # 3-eigenvalues
SIGLAB>b0=a-ev0*eye(3) # A - lambda_0 I = B_0
SIGLAB>b1=a-ev1*eye(3) # A - lambda_1 I = B_1
SIGLAB>b2=a-ev2*eye(3) # A - lambda_2 I = B_2
```

The determinate of each of these matrices can be verified to be essentially zero as follows:

```
SIGLAB>prod(eig(b0)); ~0
SIGLAB>prod(eig(b1)); ~0
SIGLAB>prod(eig(b2)); ~0
```

The eigenvalues of $\mathbf{B}_i$ are further explored by computing

```
SIGLAB>eig(b0); 0, -6.17, -4.64
SIGLAB>eig(b1); 6.17, 0, 1.53
SIGLAB>eig(b2); 4.64, -1.53, 0
```

The eigenvalue distribution of $\mathbf{B}_i$ (denoted $b0$, $b1$, and $b2$) indicates that there is one redundant equation in $b0$, $b1$, and $b2$. Therefore, one of the elements of the eigenvectors can be given a pre-assigned value. In particular, for the ith eigenvalue x_i, let the first element be arbitrarily assigned a value of unity. This means that the eigen-equation can be solved using the first two rows of $\mathbf{B}_i\mathbf{x}_i$ in the following manner:

$$[\mathbf{A} - \lambda_i\mathbf{I}] = \mathbf{B}_i = \begin{pmatrix} b_{11}^i & b_{12}^i & b_{31}^i \\ b_{21}^i & b_{22}^i & b_{32}^i \\ b_{31}^i & b_{23}^i & b_{33}^i \end{pmatrix} \Rightarrow \mathbf{B}_i\mathbf{x}_i = 0 \qquad i\text{th Eigenvector}$$

Let

$$\mathbf{x}_i = \begin{pmatrix} 1 \\ x_{i2} \\ x_{i3} \end{pmatrix} \Rightarrow$$

and consider the last two rows of $\mathbf{B}_i\mathbf{x}_i = 0$, namely

$$\begin{pmatrix} b_{22}^i & b_{32}^i \\ b_{23}^i & b_{33}^i \end{pmatrix}\begin{pmatrix} x_{i2} \\ x_{i3} \end{pmatrix} = -\begin{pmatrix} b_{21}^i \\ b_{31}^i \end{pmatrix} \Rightarrow \mathbf{B}_i'\mathbf{x}_i' = \mathbf{b}_i'$$

This system of equations can be solved by first defining $\mathbf{B}_i$ as follows:

```
SIGLAB>b0prime=b0[1:2,1:2] # B_0'
SIGLAB>b1prime=b1[1:2,1:2] # B_1'
SIGLAB>b2prime=b2[1:2,1:2] # B_2'
```

which yields a solution given by

$$\begin{pmatrix} x_{i2} \\ x_{i3} \end{pmatrix} = \begin{pmatrix} b_{22}^i & b_{32}^i \\ b_{23}^i & b_{33}^i \end{pmatrix}^{-1}\begin{pmatrix} -b_{21}^i \\ -b_{31}^i \end{pmatrix}$$

Solving these equations, the three eigenvectors are produced.

```
SIGLAB>e0={1,inv(b0prime)*-b0[1:2,0]};e0'; 1  .4908 1.097
SIGLAB>e1={1,inv(b1prime)*-b1[1:2,0]};e1'; 1 -.2962 -.4366
SIGLAB>e2={1,inv(b2prime)*-b2[1:2,0]};e2'; 1 2.948 -2.089
```

Verifying the eigenvector candidates, it follows that:

```
SIGLAB>b0*e0; ~(0,0,0)
SIGLAB>b1*e1; ~(0,0,0)
SIGLAB>b2*e2; ~(0,0,0)
```

which is the desired result.

A-7 CAYLEY-HAMILTON THEOREM

The Cayley-Hamilton Theorem is an extremely useful tool when evaluating a function of a square matrix such as $\exp(\mathbf{A}t)$. The Cayley-Hamilton Theorem states that every $n \times n$ constant coefficient-matrix satisfies its own characteristic equation. That is, the characteristic equation

$$\det(\mathbf{A} - \lambda\mathbf{I}) = c(\lambda) = c_0 + c_1\lambda + \cdots + c_n\lambda^n \tag{A-9}$$

also produces

$$c(\mathbf{A}) = c_0\mathbf{I} + c_1\mathbf{A} + \cdots + c_n\mathbf{A}^n = \mathbf{0} \tag{A-10}$$

Therefore, from

$$c_0\mathbf{I} + c_1\mathbf{A} + \cdots + c_n\mathbf{A}^n = \mathbf{0} \tag{A-11}$$

it immediately follows that

$$\mathbf{A}^n = -\frac{(c_0\mathbf{I} + c_1\mathbf{A} + \cdots + c_{n-1}\mathbf{A}^{n-1})}{c_n} \tag{A-12}$$

Another variation of this equation results in

$$c_0\mathbf{I} = -(c_1\mathbf{A} + \cdots + c_n\mathbf{A}^n) \Rightarrow \mathbf{I} = -\frac{c_1\mathbf{A} + \cdots + c_n\mathbf{A}^n}{c_0} \tag{A-13}$$

If the inverse of $\mathbf{A}$ exists, then multiplying the above by $\mathbf{A}^{-1}$ yields

$$\mathbf{A}^{-1} = -\frac{(c_1 I + c_2\mathbf{A} + \cdots + c_n\mathbf{A}^{n-1})}{c_0} \tag{A-14}$$

B

SIGLAB AND MATLAB

What we have to learn to do we learn by doing.

—Aristotle, Ethica Nicomachea II

SIGLAB, an element of MONARCH, and MATLAB Student Versions are two very similar software packages. Both are interactive programs for computational mathematics. There are differences, however, that stem from the original resources for creating these software systems. MATLAB's origins was linear algebra whereas SIGLAB was developed to perform digital signal processing. In this appendix some of the more salient differences between these two systems will be discussed. Some are obvious, others subtle; nevertheless knowledge of their differences is important. This appendix is not, however, intended as a replacement of their respective users manuals, but is to be used simply as a supplement.

B-1   SIGLAB/MATLAB COMPARISON

Both SIGLAB and MATLAB are command line interpretive langagues that are similar in many ways. There are, however, five fundamental differences between SIGLAB and MATLAB. They are:

1 Signal and system size limitations[1]

2 Array indexing

---

[1]Both SIGLAB and MATLAB Student Versions are reduced capability derivatives of professional CAD/CAE products.

3 Row/column vector based

4 Functions and procedures

5 Plotting

6 File loading

Array Limitations   Both SIGLAB and MATLAB limit the maximum size of signals and operations. In general, they are more than ample for educational purposes. The exact limitations, and how they differ from the professional SIGLAB and MATLAB products, are detailed in their respective Users' Manuals.

Indexing   If x is an n-vector (i.e., a vector with n elements), then in SIGLAB an element of x would be accessed as $x[i]$, $i \in [0, n - 1]$ and in MATLAB as $x(j)$, $j \in [1, n]$. In SIGLAB, we use brackets "[ ]" as delimiters and in MATLAB we use parenthesis "( )" to delimit indices.

Row versus Column Vectors   SIGLAB is row-oriented and MATLAB is column-oriented. For example, both SIGLAB and MATLAB use the function $fft(x)$ to compute the discrete Fourier transform of x. If x is matrix, then SIGLAB's fft will return a matrix whose *rows* are the DFT of the *rows* of x and MATLAB will return a matrix whose *columns* are the DFTs of the *columns* of x. MATLAB uses semicolons to separate rows and commas to separate columns inside a bracketed expression (e.g., $x = (1, 2, 3) \leftrightarrow x=[1,2,3]$ and $x = (1, 2, 3)^T \leftrightarrow x=\{1;2;3\}$). In SIGLAB, a bracketed expression defines a row, a curly-braced expression defines a column, and the elements of the rows or columns are separated by commas and rated by commas (e.g., $x = (1, 2, 3) \leftrightarrow x=[1,2,3]$ and $x = (1, 2, 3)^T \leftrightarrow x\{1,2,3\}$).

Functions and Procedures   Both MATLAB and SIGLAB make extensive use of functions to create a program within a program. A function will accept a listing of arguments and return a result. In the case of MATLAB, multiple results may be returned (e.g., $function[mean,variance]=stat(x)$). SIGLAB returns only one result per function call, but the result can be a concatenated collection of multiple results (see Example 11-7). SIGLAB also contains procedure calls that are similar functions, but do not return a result. Procedures are more efficient than functions when operations, such as graphing or file I/O, is to be performed.

Plotting   MATLAB plots are basically oriented toward graphing mathematical functions and surfaces. SIGLAB plots are designed to be more like engineering graphs and oscilloscope displays.

File Loading   MATLAB will search all valid paths to find a named M-file. Loading of the M-file is automatic. SIGLAB will search all valid paths (which can be set with the "F6 key") to find a named S-file. Loading is performed using the `include "filename"` operation. For MONARCH users, if data is saved in directory `C:\PSS` in subdirectories

```
C:\PSS\Chap(1-13), C:\PSS\SFILES, C:\PSS\SIGNALS,
C:\PSS\SYSTEMS
```

then when working examples in Chapter "xxx", the SIGLAB path should be set to

```
C:\PSS\Chapxxx; C:\PSS\SFILES; C:\PSS\SIGNALS;
C:\PSS\SYSTEMS.
```

MATLAB will always print the result as an expression. To suppress the printout it is necessary to end the statement with a semicolon. SIGLAB will never echo an assignment statement, that is, a statement of the form

```
variable = expression;
```

and the ending semicolon those cases is optional. In both SIGLAB and MATLAB we can add several statements, separated by semicolons, on a line.

A matrix can be interpreted as a row vector whose entries are column vectors or as a row vector whose entries are row vectors. Therefore, to create

$$A = \begin{pmatrix} 1 & 2 & 3 \\ 4 & 5 & 6 \end{pmatrix}$$

MATLAB would use either

```
>> A = [1, 2, 3; 4 5 6]; % column vector of row vectors
>> A = [[1; 4], [2; 5], [3; 6]]; % row vector of columns
```

and SIGLAB would use

```
SIGLAB> A = {[1, 2, 3], [4, 5, 6]} # column vector of row
vectors
SIGLAB> A = [{1, 4}, {2, 5}, {3, 6}] # row vector of columns
SIGLAB> A={1, 4} & {2, 5} & {3, 6} # concatenation of rows
```

where the beginning of comment character is the percent sign "%" in MATLAB and the pound sign "#" in SIGLAB.

Matrices can be indexed in a multitude of ways. Submatrices of matrices can also be defined by the colon notation. The matrix, `A(1:3, 3:4)` denotes the submatrix of A consisting of the elements

$$\begin{pmatrix} A(1,3) & A(1,4) \\ A(2,3) & A(2,4) \\ A(3,3) & A(3,4) \end{pmatrix}$$

TABLE B-1

Binary operation	SIGLAB	MATLAB	Note
add	A+B	A+B	
convolve	A$B	conv(A, B)	
correlate	A@B	xcorr(A, B)	
exponentiate	A^b or b^A	A^b or b^A	
$A^{-1}B$	A\B	A\B	SIGLAB: A=square matrix.
multiply	A*B	A*B	
pointwise divide	A./B	A./B	
pointwise exponentiate	A^B	A.^B	
pointwise modulo	A%B	rem(A, B)	
pointwise multiply	A.*B	A.*B	
subtract	A-B	A-B	

A.B are matrices and a.b are scalars

in MATLAB. In SIGLAB the same submatrices would be produced using A[1:3, 3:4].

The built-in functions for creating vectors and matrices in SIGLAB and MATLAB are similar (Table B-1). However, even though two functions have the same name and create the same matrices, they can behave differently. For instance, a common mistake for SIGLAB users who want to create an n-point random vector in MATLAB is to use rand(n). This will cause MATLAB to create an $n \times n$ matrix. The correct way is to enter rand(n, 1) for a column vector or rand(1, n) for a row vector. In SIGLAB, rand(n) will create an n-point row vector.

SIGLAB and MATLAB create complex numbers somewhat differently. A complex matrix $Z = A + jB$ in MATLAB, can be defined by Z = A + j*B or Z = A + i*B, given, of course, that j or i have not been assigned something other than $\sqrt{-1}$. If i and j are already busy, we can use sqrt(-1) instead. In SIGLAB a complex matrix $Z = A + jB$ is created as Z = <A, B> or as Z = cmplx(A, B) or alternatively using $i = (-1)^{0.5}$ and then Z = A + i*B. Conjugation is done with conj in both SIGLAB and MATLAB. The transpose of a complex matrix Z is found in SIGLAB as Z'; however, in MATLAB Z'; means conjugate transpose (or hermitian). The transpose of a matrix in MATLAB is given by Z.'.

B-2 UNARY AND BINARY OPERATIONS

Unary operations in SIGLAB and MATLAB generally work similarly, but may have different syntax. Binary operations for SIGLAB and MATLAB, add, subtract, multiply and divide, work identically. Pointwise operations are indicated by a period. Exponentiation is another binary operator that differs slightly between MATLAB and SIGLAB. If A and B are matrices and a and b are scalars, then

TABLE B-2

Matrix Generation	SIGLAB	MATLAB
all ones matrix	ones(m, n)	ones(m, n)
Bartlett window	†	bartlett(n) ‡
Blackman window	black(n)	blackman(n) ‡
Chebyshev window	†	chebwin(n) ‡
cosine wave	mkcos(a, f, p, n)	†
Gaussian random matrix	gn(m, n)	rand(m, n) ○
Hamming window	ham(n)	hamming(n) ‡
Hanning window	han(n)	hanning(n) ‡
identity matrix	eye(n)	eye(m, n)
impulse	impulse(n)	†
Kaiser window	kai(beta,n)	kaiser(n) ‡
ramp	ramp(T, s, N)	†
rectangular window	sq(n,1,n)	boxcar(n) ‡
sine wave	mksin(a, f, p, n)	†
square wave	sq(T, d, n)	square(t, d)
triangular wave	tri(T, n)	†
uniform random matrix	rand(m, n)	rand(m, n) ‡
zero matrix	zeros(m, n)	zeros(m, n)

† not directly available

‡ requires the Signal Processing Toolbox

○ requires rand('normal')

‡ requires rand('uniform')

A^b computes the matrix power in MATLAB and raises each element to the bth power in SIGLAB.

Table B-2 summarizes some of the built-in functions for generating the most common matrices and signals

B-3   BUILT-IN FUNCTIONS

The trigonometric, hyperbolic and exponential functions are identical in SIGLAB and MATLAB. They all accept real or imaginary matrices and perform the operations pointwise. MATLAB will, however, always assume that the trigonometric functions take arguments and returns in radians. In SIGLAB, the deg and rad commands will set degree and radian modes.

The logarithmic functions are nearly identical, except that log means the natural (base e) logarithm in MATLAB and the base 10 logarithm in SIGLAB. The natural logarithm is called by ln in SIGLAB and the base 10 logarithm is called by log10 in MATLAB.

Truncation of the real matrix A to an integer matrix is done in SIGLAB as int(A) and in MATLAB as fix(A). Rounding to an integer matrix is done in both packages as `round(A)`. The MATLAB functions `ceil` and `floor` are not directly available in SIGLAB.

Absolute value of the real or complex matrix Z can be taken in both MATLAB and SIGLAB as `abs(Z)`. In SIGLAB this can also be done as `mag(Z)`. The real and imaginary parts of Z are found as `real(Z)` and `imag(Z)` in MATLAB and as `re(Z)` and `im(Z)` in SIGLAB. The argument (phase angle) of Z is computed as `angle(Z)` in MATLAB and as `phs(Z)` in SIGLAB. The `phs` function will return either degrees or radians, according to the mode set by `deg` and `rad`.

The MATLAB functions `eig` (eigenvalues of a matrix), `roots` (roots of a polynomial) and `sqrt` (square root) all accept complex arguments. The corresponding functions in SIGLAB accept only real arguments.

The discrete Fourier transform of an n-vector x is computed in both MATLAB and SIGLAB as `fft(x)`. SIGLAB will automatically zero-pad x if n is not a power of 2, and MATLAB will explicitly add zeros to x using `fft(x,N)` where N is the desired length of the transform. The SIGLAB FFT transform displays the negative frequencies first, then the positive frequencies. The positive harmonics (only) are displayed using `pfft(x)`. The MATLAB FFT transform displays the positive frequencies first, then the negative frequencies.

Filtering can be done in MATLAB with the `filter` function, which is based on a transfer function model. Filtering in SIGLAB is done by the `filt` function, which is based on a state-variable description of the filter. The SIGLAB function `fxpfilt` is used for fixed-point emulation.

Decimation of signals is done in MATLAB via lowpass filtering and down-sampling with the `decimate` function. The SIGLAB decimation function `dec` simply decimates a signal. The decimated signal can then be sent to an interpolator of choice.

Table B-3 lists some of the corresponding pointwise functions found in MATLAB and SIGLAB.

Table B-4 lists some of the corresponding vector functions found in MATLAB and SIGLAB. Table B-5 lists some of the corresponding matrix functions found in MATLAB and SIGLAB.

B-4 M-FILES AND S-FILES

SIGLAB and MATLAB have similar control-flow statements, namely `if`, `else`, `end`, `for` and `while`. The relative operators differ slightly.

Both SIGLAB and MATLAB support user-written extensions, the so-called S-files (for SIGLAB) and M-files (for MATLAB). The S-files can consist of any number of functions and procedures. Before a user-defined function can be used in SIGLAB, the user must explicitly `include` the S-file. The `include` statement takes the S-files filename as argument. If the S-file is edited, it must be explicitly included in order for SIGLAB to be aware of the changes. Furthermore, the SIGLAB functions can only return a single variable. This is in contrast to MATLAB where the M-files can contain only one function (there are

TABLE B-3

Pointwise Function	SIGLAB	MATLAB
absolute value	abs(A) or mag(A)	abs(A)
arccosine	acos(A)	acos(A)
arcsine	asin(A)	asin(A)
arctangent	atan(A)	atan(A)
clear NaN	clrnan(A)	†
conjugate	conj(A)	conj(A)
cosine	cos(A)	cos(A)
exponential	exp(A)	exp(A)
hyperbolic cosine	cosh(A)	cosh(A)
hyperbolic sine	sinh(A)	sinh(A)
hyperbolic tangent	tanh(A)	tanh(A)
imaginary part	im(A)	imag(A)
inverse hyperbolic cosine	acosh(A)	acosh(A)
inverse hyperbolic sine	asinh(A)	asinh(A)
inverse hyperbolic tangent	atanh(A)	atanh(A)
logarithm, base e	ln(A)	log(A)
logarithm, base 10	log(A)	log10(A)
phase angle	phs	angle
quantization	fxpt(A)	†
real part	re(A)	real(A)
round to integer	round(A)	round(A)
sign	sign(A)	sign(A)
sine	sin(A)	sin(A)
square root	sqrt(A)‡	sqrt(A)
tangent	tan(A)	tan(A)

† not directly available.

‡ real matrices only

no procedures in MATLAB). The M-files need not be explicitly included and the functions can return multiple variables. Both MATLAB and SIGLAB support scripts, that is, M- and S-files containing commands rather than function definitions. Including a script file, by typing its name in MATLAB or by including it in SIGLAB, is the equivalent to typing all the commands in the script file.

B-5   PLOTTING

Plotting graphs is the area in which MATLAB and SIGLAB differ the follow the tutorials or just try them.

TABLE B-4

Pointwise Function	SIGLAB	MATLAB
circular shift	cshift(x, n)	†
decimation	dec(x, n)	decimate(x, r)
derivative	derv(x)	diff(x)
discrete Fourier transform*	dft(x)	fft(x)
discrete Fourier transform**	fft(x)	fft(x, n)
filtering	filt(h, x)	filter(a, b, x)
fixed-point filtering	fxpfilt(h, x, w, f)	†
group delay	grp(x)	grpdelay(b, a, n)
integration	intg(x)	cumsum(x)
inverse discrete Fourier transform*	idft(x)	ifft(x)
inverse discrete Fourier transform**	ifft(x)	ifft(x)
Laguerre polynomial root finder	laguer(x, e, p, m)	†
length of vector	len(x)	length(x)
linear shift	shift(x, n)	†
polynomial evaluation	polyval(p, x)	polyval(p, x)
polynomial generation	poly(x)	poly(x)
polynomial root finder	root(x)‡	roots(x)
reversing vector	rev(x)	fliplr(x) or flipud(x)

† not directly available.   * for lengths that are not a power of 2
‡ real vectors only   ** for lengths that are a power of 2

TABLE B-5

Matrix Function	SIGLAB	MATLAB
average	avg(A)	mean(A)
diagonal	diag(A)	diag(A)
dimension	dim(A)	size(A)
eigenvalues	eig(A)	eig(A)
inverse	inv(A)	inv(A)
maximum	max(A)	max((max (A)))
maximum location	maxloc(A)	find(max(max(A))==A)
minimum	min(A)	min(min(A))
minimum location	maxloc(A)	find(min(min(A))==A)
number of INF	ninf(A)	length(find(A == inf))
number of NaN	nnan(A)	length(find(A == nan))
product	prod(A)	prod(A)*
sum	sum(A)	sum(A)*
variance	var(A)	std(A).^2*

† real matrices only

* product, sum, and variance along columns.

TABLE B-6

Relative operator	SIGLAB	MATLAB
equal to	=	==
not equal to	<> or !=	~=
less than	<	<
less or equal than	<=	<=
greater than	>	>
greater or equal than	>=	>=
logical not	not	~
logical and	and	&
logical or	or	|
logical exclusive or	†	xor

† not available.

MASON'S
GAIN FORMULA

You are today where your thoughts have brought you. You will be tomorrow where your thoughts take you."

—James Allen

Mason's Gain Formula was used in Chapters 6 and 11 to derive a system transfer function from a signal flow graph using the executable file `sfg.exe`. The data format used by program `sfg.exe` is as follows:

```
  # comments
BRANCH i j denotes definition of a branch from node i to j
NUM begin to specify numerator polynomial, descending pow-
ers
data
END end numerator specification
DEN begin to specify denominator polynomial, descending pow-
ers
data
END end denominator specification
```

Complex data will be assumed to be of the form $a + jb$. If the imaginary part is missing, the data is assumed to be real. For example, if between nodes 5 and 19 was the transfer function

$$H(s) = (s^{-1} + a)/(s^{-2} + b) = ((1 + j2) + z^{-1})/(2 + z^{-2})$$

or, for discrete-time systems

$$H(z) = (z^{-1} + a)/(z^{-2} + b) = ((1 + j2) + z^{-1})/(2 + z^{-2})$$

then use

```
#Path Gain
BRANCH 5 19
NUM
1 2
1
END
DEN
2
0
1
END
```

BIBLIOGRAPHY

Woe to the author who always wants to teach! The secret of being a bore is to tell everything.

—Voltaire
De La Nature de l'Homme

Alexander, S.T. *Adaptive Signal Processing*. New York: Springer-Verlag, 1986.

Alkin, O. *PC-DSP*. Englewood Cliffs, NJ: Prentice-Hall, 1990.

Antoniou, A. *Digital Filter Analysis and Design*. New York: McGraw-Hill, 1979.

Athena Group. *MONARCH Student Version*. Gainesville, FL: The Athena Group, 1992.

Bateman, A, and W. Yates. *Digital Signal Processing Design*. New York: Computer Science Press, 1989.

Bellanger, M.G. *Adaptive Digital Filters and Signal Analysis*. New York: Marcel Dekker, 1987.

Bellanger, M. *Digital Processing of Signals, Theory and Practice*. New York: Wiley and Sons, 1984.

Blahut, R.E. *Algebraic Modes for Signal Processors*. New York: Springer-Verlag, 1992.

Blahut, R.E. *Fast Algorithms for Digital Signal Processing*. Reading, MA: Addison-Wesley, 1985.

Blanford, D.K. *The Digital Filter Analyzer*. Reading, MA: Addison-Wesley, 1988.

Brigham, E.O. *The Fast Fourier Transform and Its Applications*. Englewood Cliffs, NJ: Prentice-Hall, 1988.

Burrs, C.S., and T.W. Parks. *DFT/FFT and Convolution Algorithms*. New York: Wiley and Sons, 1985.

Cadzow, J.A. *Discrete-Time Systems*. Englewood Cliffs, NJ: Prentice-Hall, 1973.

Cadzow, J.A. *Foundations of Digital Signal Processing and Data Analysis*. New York: Macmillan, 1987.

Candy, J.V. *Signal Processing: The Modern Approach*. New York: McGraw-Hill, 1986.

Chassing, R., and P.W. Horning. *Digital Signal Processing with the TMS320C25*. New York: Wiley and Sons, 1990.

Childers, D.G., and A. Durling. *Digital Signal Processing*. St. Paul, MN: West, 1975.

Cowan, C.F.N., and P.M. Grant. *Adaptive Filters*. Englewood Cliffs, NJ: Prentice-Hall, 1985.

Cruz, J.B., and M.E. Van Valkenburg. *Signals in Linear Circuits*. Boston: Houghton Mifflin, 1974.

DeFatta, D.J., and J.G. Lucas. *Digital Signal Processing*. New York: Wiley and Sons, 1988.

Elliott, D.F., ed. *Handbook of Digital Signal Processing: Engineering Application*. San Diego, CA: Academic Press, 1987.

Elliot, D.F., and K.R. Rao. *Fast Transforms*. San Diego, CA: Academic Press, 1982.

Flanery, W.H. *Numerical Recipes in C*. Cambridge, England: Cambridge Press, 1988.

Gable, R.A., and R.A. Roberts. *Signals and Linear Systems*. New York: Wiley and Sons, 1987.

Gardner, W. *Statistical Spectral Analysis*. Englewood Cliffs, NJ: Prentice-Hall, 1988.

Gold, B., and C. Rader. *Digital Processing of Signals*. New York: McGraw-Hill, 1969.

Golub, G.H., and C.F. VanLoan. *Matrix Computation*. 2d ed. Baltimore, MD: The Johns Hopkins Press, 1989.

Haddad, R.A., and T.W. Persons. *Digital Signal Processing*. New York: Computer Science Press, 1991.

Hamming, R.W. *Digital Filters*. Englewood Cliffs, NJ: Prentice-Hall, 1989.

Hardy, W.T. *SACALC*. Norwood, MA: Artech, 1990.

Haykin, S. *Adaptive Filter Theory,* 2d ed. Englewood Cliffs, NJ: Prentice-Hall, 1991.

Haykin, S., ed. *Array Signal Processing*. Englewood Cliffs, NJ: Prentice-Hall, 1985.

Haykin, S., ed. *Selected Topics in Signal Processing*. Englewood Cliffs, NJ: Prentice-Hall, 1989.

Higgins, R.J. *Digital Signal Processing in VLSI*. Englewood Cliffs, NJ: Prentice-Hall, 1990.

Honig, M.L., and D.G. Messerschmitt. *Adaptive Filters, Structures, Algorithms, and Applications*. Norwell, MA: Kleuwer, 1984.

Hutchins, B.A., and T. Parks. *A Digital Signal Processing Laboratory using the TMS320C25*. Englewood Cliffs, NJ: Prentice-Hall, 1990.

Ingle, V.K., and J.G. Proakis. *Digital Signal Processing Lab*. Englewood Cliffs, NJ: Prentice-Hall, 1990.

Jackson, L.B. *Digital Filters and Signal Processing*. Norwell, MA: Kluwer Academic, 1989.

Jackson, L.B. *Signals, Systems and Transforms*. Reading, MA: Addison-Wesley, 1991.

Jayant, N.S., and P. Noll. *Digital Coding of Waveforms*. Englewood Cliffs, NJ: Prentice-Hall, 1984.

Johnson, J.R. *Introduction to Digital Signal Processing*. Englewood Cliffs, NJ: Prentice-Hall, 1989.

Kamas, A., and E.A. Lee. *Digital Signal Processing Experiments*. Englewood Cliffs, NJ: Prentice-Hall, 1989.

Karl, J.H. *An Introduction to Digital Signal Processing*. San Diego, CA: Academic Press, 1989.

Kay, S.M. *Modern Spectral Analysis, Theory and Applications*. Englewood Cliffs, NJ: Prentice-Hall, 1988.

Kuc, R. *Introduction to Digital Signal Processing*. New York: McGraw-Hill, 1988.

Kwakernaak, H., and R. Sivan. *Modern Signals and Systems*. Englewood Cliffs: Prentice-Hall, 1991.

Larimore, M., B. Widrow, and S.D. Stearns. *Adaptive Signal Processing*. Englewood Cliffs, NJ: Prentice-Hall.

Lathi, B.P. *Signals and System*. New York: Berkeley, 1987.

Lawson, C.L., and R.J. Hanson. *Solving Least Square Problems*. Englewood Cliffs, NJ: Prentice-Hall, 1974.

Lim, J.S. *Two-Dimensional Signal and Image Processing*. Englewood Cliffs, NJ: Prentice-Hall, 1990.

Lim, J.S., and A.V. Oppenheim. *Advanced Topics in Signal Processing*. Englewood Cliffs, NJ: Prentice-Hall, 1988.

Lin, K-S., ed. *Digital Signal Processing Applications with the TMS320 Family*. Englewood Cliffs, NJ: Prentice-Hall, 1987.

Mar, A., ed. *Digital Signal Processor Applications using the ADSP-2100 Family*. Englewood Cliffs, NJ: Prentice-Hall, 1990.

Marple, S.L. *Digital Spectral Analysis with Applications*. Englewood Cliffs, NJ: Prentice-Hall, 1987.

Marks, R.J. *Introduction to Shannon Sampling and Interpolation Theory*. New York: Springer-Verlag, 1991.

Mathworks. *The Student Edition of MATLAB*. Englewood Cliffs, NJ: Prentice-Hall, 1992.

Mayhan, R.J. *Discrete Time and Continuous Time Linear Systems*. Reading, MA: Addison-Wesley, 1984.

McClellan, J.H., and C.M. Rader. *Number Theory in Digital Signal Processing*. Englewood Cliffs, NJ: Prentice-Hall, 1979.

McGillem, D.P., and G.R. Cooper. *Continuous and Discrete Signal and System Analysis*. New York: Holt Rinehart & Winston, 1984.

Mendel, J.M. *Lessons in Digital Estimation Theory*. Englewood Cliffs, NJ: Prentice-Hall, 1991.

Miner, G., and P. Conner. *Digital Data Acquisition*. Englewood Cliffs, NJ: Prentice-Hall, 1992.

Morgera, S.D., and H. Krishna. *Digital Signal Processing: Applied to Communications and Algebraic Coding Theories*. San Diego, CA: Academic Press, 1989.

Oppenheim, A.V., and R.W. Shafer. *Digital Signal Processing*. Englewood Cliffs, NJ: Prentice-Hall, 1975.

Oppenheim, A.V., and R.W. Shafer. *Discrete-Time Signal Processing*. Englewood Cliffs, NJ: Prentice-Hall, 1989.

Oppenheim, A.V., and A.S. Willsky. *Signals and Systems*. Englewood Cliffs, NJ: Prentice-Hall, 1983.

Papoulis, A. *Signal Analysis*. New York: McGraw-Hill, 1977.

Parks, T.W., and C.S. Burrus. *Digital Filter Design*. New York: Wiley-Interscience, 1987.

Pillai, S.U. *Array Signal Processing*. New York: Springer-Verlag, 1989.

Poularikas, A.D., and S. Seely. *Signals and Systems*. Boston: PWS-Kent, 1985.

Priestley, M.B. *Spectral Analysis and Time Series*. San Diego, CA: Academic Press, 1986.

Proakis, J.G., and D.G. Manolakis. *Introduction to Digital Signal Processing*. New York: Macmillan, 1988.

Rabiner, L.R., and B. Gold. *Theory and Application of Digital Signal Processing*. Englewood Cliffs, NJ: Prentice-Hall, 1975.

Ramerey, R.W. *The FFT, Fundamentals and Concepts*. Englewood Cliffs, NJ: Prentice-Hall, 1985.

Roberts, R.A., and C.T. Mullis. *Digital Signal Processing*. Reading, MA: Addison-Wesley, 1987.

Roser, S., and P. Howell. *Signals and Systems for Speech and Hearing*. San Diego, CA: Academic Press, 1991.

Scharf, L.L., and R.T. Behrens. *A First Course in Electrical PC-DSP*. Reading, MA: Addison-Wesley, 1990.

Sibul, L.H., ed. *Adaptive Signal Processing*. Piscataway, NJ: IEEE Press, 1987.

Siebert, W.M. *Signals and Systems*. Cambridge: MIT Press, 1986.

Smith, M.J.T., and R.M. Merserrau. *Introduction to Digital Signal Processing: A Complete Laboratory Textbook*. New York: John Wiley and Sons, 1992.

Soliman, S., and M.D. Srinath. *Continuous and Discrete-Time Signals and Systems*. Englewood Cliffs, NJ: Prentice-Hall, 1990.

Stearns, S.D., and R.A. David. *Signal Processing Algorithms*. Englewood Cliffs, NJ: Prentice-Hall, 1988.

Stearns, S.D., and D.R. Hush. *Digital Signal Analysis,* 2d ed. Englewood Cliffs, NJ: Prentice-Hall, 1990.

Strum, R.D., and D.E. Kirk. *Discrete Systems and Digital Signal Processing*. Reading, MA: Addison-Wesley, 1988.

Strum, R.D., and D.E. Kirk. *First Principles of Discrete Systems and Digital Signal Processing*. Reading, MA: Addison-Wesley, 1988.

Taylor, F.J. *Digital Design Handbook*. New York: Marcel Dekker, 1983.

Taylor, F. *Principles of Signals and Systems*. New York: McGraw-Hill, 1992.

Taylor, F.J., and T. Stouraitis. *Digital Filter Design Software for the IBM PC*. New York: Marcel Dekker, 1987.

Therrien, C. *Discrete Random and Statistical Signal Processing*. Englewood Cliffs, NJ: Prentice-Hall, 1992.

Treichler, J.R., and C.R. Jophnson. *Theory and Design of Adaptive Filters*. New York: Wiley and Sons, 1987.

Tretter, S.A. *Introduction to Discrete-Time Signal Processing*. New York: Wiley and Sons, 1976.

Williams, A.B., and F.J. Taylor. *Electronic Filter Design Handbook*. New York: McGraw-Hill, 1988.

Yuen, C.K., K.G. Beauchamps, and D. Fraser. *Microprocessor Systems in Signal Processing*. San Diego: Academic Press, 1989.

Ziemer, R.E., W. Tranter, and D. Fanner. *Signal and Systems: Continuous and Discrete*. New York: Macmillan, 1983.

Zelniker, G., and F. Taylor. *Advanced Digital Signal Processing: Theory and Applications*. New York: Marcel Dekker, 1994.

INDEX

TABLE 3-3   FOURIER SERIES PROPERTIES

Continuous-time function $x(t)$	Fourier Series c_k	Property
$\sum_i \alpha_i x_i(t)$	$\sum_i \alpha_i c_{i,k}$	Linearity
$x^*(t)$	c^*_{-k}	Conjugation
$x(-t)$	c_{-k}	Time-reversal
$x(t - t_0)$	$c_k \exp\left(\dfrac{-jk2\pi t_0}{T}\right)$	Time-delay
$x(\alpha t), \alpha > 0$	c_k if $x(\alpha t)$ has period $\dfrac{T}{\alpha}$	Time-scaling
$\int_{-\infty}^{t} x(\tau)\, d\tau < \infty$	$\dfrac{1}{jk\left(\dfrac{2\pi}{T}\right)} c_k$ if $c_0 = 0$	Integration
$\dfrac{dx(t)}{dt}$	$\dfrac{jk2\pi}{T} c_k$	Differentiation
$x(t) \exp\left(\dfrac{jK2\pi t}{T}\right)$	c_{k-K}	Modulation
$x(t)$ real	$c_k = c^*_{-k}$	Real

Parseval's theorem: $\dfrac{1}{T}\displaystyle\int_T |x(t)|^2\, dt = \sum_{-\infty}^{\infty} |c_k|^2$

TABLE 3-4   FOURIER TRANSFORM PROPERTIES

Continuous-time function $x(t)$	Fourier Transform $X(j\omega)$	Property
$\sum_i \alpha_i x_i(t)$	$\sum_i \alpha_i X_i(j\omega)$	Linearity
$x^*(t)$	$X^*(-j\omega)$	Conjugation
$x(-t)$	$X(-j\omega)$	Time-reversal
$x(t - t_0)$	$\exp(-j\omega t_0) X(j\omega)$	Time-delay
$x(\alpha t)$	$\dfrac{1}{|\alpha|} X\left(\dfrac{j\omega}{\alpha}\right)$	Time-scaling
$\int_{-\infty}^{t} x(\tau)\, d\tau$	$\dfrac{X(j\omega)}{j\omega} + \pi X(0)\delta(\omega)$	Integration
$\dfrac{d^n x(t)}{dt^n}$	$(j\omega)^n X(j\omega)$	Differentiation
$x(t) \exp(j\omega_0 t)$	$X(j(\omega - \omega_0))$	Modulation
$X(t)$	$2\pi x(-j\omega)$	Duality

Parseval's theorem: $\displaystyle\int_{-\infty}^{\infty} |x(t)|^2\, dt = \dfrac{1}{2\pi}\int_{-\infty}^{\infty} |X(j\omega)|^2\, d\omega$

TABLE 3-5 FOURIER TRANSFORMS OF ELEMENTARY FUNCTIONS

Continuous Time Function $x(t)$	Fourier Transform $X(j\omega)$	Remark
1	$2\pi\delta(\omega)$	Constant, noncausal.
$u(t)$	$\pi\delta(\omega) + \dfrac{1}{j\omega}$	Unit-step function, causal.
$\delta(t)$	1	Delta distribution, noncausal.
$\delta(t - t_0)$	$\exp(-j\omega t_0)$	Delayed delta distribution, noncausal.
$\sum_{n=-\infty}^{\infty} \delta(t - nT)$	$\dfrac{2\pi}{T} \sum_{n=-\infty}^{\infty} \delta\left(\omega - \dfrac{2n\pi}{T}\right)$	Impulse train.
$\mathrm{rect}(t/\tau)$	$\dfrac{2\sin(\omega\tau/2)}{\omega} = \tau\,\mathrm{sinc}(\omega\tau/2)$	Rectangular pulse, noncausal.
$\dfrac{\sin(\omega_0 t)}{\pi t} = \dfrac{\omega_0}{\pi}\,\mathrm{sinc}\left(\omega_0 t\right)$	$\mathrm{rect}\left(\dfrac{\omega}{2\omega_0}\right)$	Noncausal.
$\exp(j\omega_0 t)$	$2\pi\delta(\omega - \omega_0)$	Complex exponential, noncausal.
$\cos(\omega_0 t)$	$\pi[\delta(\omega - \omega_0) + \delta(\omega + \omega_0)]$	Noncausal.
$\sin(\omega_0 t)$	$\dfrac{\pi}{j}[\delta(\omega - \omega_0) - \delta(\omega + \omega_0)]$	Noncausal.
$\cos(\omega_0 t)u(t)$	$\dfrac{\pi}{2}[\delta(\omega - \omega_0) + \delta(\omega + \omega_0)]$ $+\dfrac{j\omega}{\omega_0^2 - \omega^2}$	Causal.
$\sin(\omega_0 t)u(t)$	$\dfrac{\pi}{2j}[\delta(\omega - \omega_0) - \delta(\omega + \omega_0)]$ $+\dfrac{\omega_0}{\omega_0^2 - \omega^2}$	Causal.
$\exp(-at)u(t)$	$\dfrac{1}{a + j\omega}$	$\mathrm{Re}[a] > 0$, causal.
$t\exp(-at)u(t)$	$\dfrac{1}{(a + j\omega)^2}$	$\mathrm{Re}[a] > 0$, causal.
$\exp(-a|t|)$	$\dfrac{2a}{a^2 + \omega^2}$	$\mathrm{Re}[a] > 0$, noncausal.
$|t|\exp(-a|t|)$	$\dfrac{2(a^2 - \omega^2)}{a^2 + \omega^2}$	Noncausal.

TABLE 7-1   *Z*-TRANSFORMS OF PRIMITIVE TIME FUNCTIONS

Time-domain	*Z*-transform	Region of convergence, $\lvert z \rvert > R$
$\delta_K[k]$	1	0
$\delta_K[k - m]$	z^{-m}	0
$u[k]$	$\dfrac{z}{z - 1}$	1
$ku[k]$	$\dfrac{z}{(z - 1)^2}$	1
$k^2 u[k]$	$\dfrac{z(z + 1)}{(z - 1)^3}$	1
$k^3 u[k]$	$\dfrac{z(z^2 + 4z + 1)}{(z - 1)^4}$	1
$\exp(akT_s)u(kT_s)$	$\dfrac{z}{z - \exp(aT_s)}$	$\lvert \exp(aT_s) \rvert$
$kT_s \exp(akT_s)u(kT_s)$	$\dfrac{zT_s \exp(aT_s)}{(z - \exp(aT_s))^2}$	$\lvert \exp(aT_s) \rvert$
$(kT_s)^2 \exp(akT_s)u(kT_s)$	$\dfrac{zT_s^2 \exp(aT_s)(z + \exp(aT_s))}{(z - \exp(aT_s))^3}$	$\lvert \exp(aT_s) \rvert$
$a^k u[k]$	$\dfrac{z}{(z - a)}$	$\lvert a \rvert$
$ka^k u[k]$	$\dfrac{az}{(z - a)^2}$	$\lvert a \rvert$
$k^2 a^k u[k]$	$\dfrac{az(z + a)}{(z - a)^3}$	$\lvert a \rvert$
$\sin(bkT_s)u(kT_s)$	$\dfrac{z \sin(bT_s)}{z^2 - 2z \cos(bT_s) + 1}$	1
$\cos(bkT_s)u(kT_s)$	$\dfrac{z(z - \cos(bT_s))}{z^2 - 2z \cos(bT_s) + 1}$	1
$\exp(akT_s) \sin(bkT_s)u(kT_s)$	$\dfrac{z \exp(aT_s) \sin(bT_s)}{z^2 - 2z \exp(aT_s) \cos(bT_s) + \exp(2aT_s)}$	$\lvert \exp(aT_s) \rvert$
$\exp(akT_s) \cos(bkT_s)u(kT_s)$	$\dfrac{z(z - \exp(aT_s) \cos(bT_s))}{z^2 - 2z \exp(aT_s) \cos(bT_s) + \exp(2aT_s)}$	$\lvert \exp(aT_s) \rvert$
$a^k \sin(bkT_s)u(kT_s)$	$\dfrac{az \sin(bT_s)}{z^2 - 2az \cos(bT_s) + a^2}$	$\lvert a \rvert$
$a^k \cos(bkT_s)u(kT_s)$	$\dfrac{z(z - a \cos(bT_s))}{z^2 - 2az \cos(bT_s) + a^2}$	$\lvert a \rvert$
$\dfrac{(\ln(a))^k}{k!}u[k]$	$a^{1/z}$	$a > 0$
$\dfrac{1}{k} \quad (k = 1, 2, 3, \dots)$	$\ln\left(\dfrac{z}{z - 1}\right)$	1
$\dfrac{(k + 1)\cdots(k + m)a^k}{m!}$	$\dfrac{z^{m+1}}{(z - a)^{m+1}}$	$\lvert a \rvert$

TABLE 7-2 PROPERTIES OF Z-TRANSFORMS

Property	Time series	Z-transform
Linearity:	$x_1[k] + x_2[k]$	$X_1(z) + X_2(z)$
Real Scaling:	$ax[k]$	$aX(z)$
Complex Scaling:	$w^k x[k]$	$X(z/w)$
Delay*:	$x[k - L]$	$z^{-L}X(z) + \sum_{n=-L}^{-1} z^{-(L+n)}x[n]$
Time Reversal:	$x[-k]$	$X(1/z)$
Modulation:	$\exp[-ak]x[k]$	$X(\exp(a)z)$
Ramping:	$kx[k]$	$-z\dfrac{d}{dz}X(z)$
Reciprocal:	$\dfrac{x[k]}{k}$	$-\displaystyle\int \dfrac{X(z)}{z}\, dz$
Summation:	$\displaystyle\sum_{n=-\infty}^{k} x[n]$	$\dfrac{zX(z)}{z - 1}$
Periodic:	$y[k + N]$	$\dfrac{z^N Y(z)}{z^N - 1}$

Initial Value*:	$x[0] = \lim_{z\to\infty} X(z)$	
Final Value**:	$x[\infty] = \lim_{z\to 1} (z - 1)X(z)$	

* for causal $x[k]$.
** if $X(z)$ has no more than one pole on the unit circle and all other poles are
interior to the unit circle.

TABLE 8-1 PROPERTIES OF A DISCRETE-TIME FOURIER TRANSFORM

Discrete-time Series	Discrete-time Fourier Transform	Remark
$\displaystyle\sum_{i=0}^{L}\alpha_i x_i[k]$	$\displaystyle\sum_{i=0}^{L}\alpha_i X_i(e^{j\Omega})$	Linearity
$x[k-q]$	$X(e^{j\Omega})\exp(-jq\Omega)$	Time-shift
$x[-k]$	$X^*(e^{j\Omega})$	Time-reversal
$x[k]\exp(j\Omega_0 k)$	$X(e^{j[\Omega-\Omega_0]})$	Modulation, Ω_0 real
$x[k]*h[k]$	$X(e^{j\Omega})H(e^{j\Omega})$	Convolution
$x[k]y[k]$	$X(e^{j\Omega})*Y(e^{j\Omega})$	Multiplication
$\displaystyle\sum_{k=-\infty}^{\infty}x[k]y[k]$	$\displaystyle\frac{1}{2\pi}\int_{-\pi}^{\pi}X^*(e^{j\Omega})Y(e^{j\Omega})d\Omega$	Parseval
$x_{even}[k]$	$X(e^{j\Omega})=x_{even}[0]+2\displaystyle\sum_{k=1}^{\infty}x_{even}[k]\cos(k\Omega)$	x_{even} is an even time-series
$x_{odd}[k]$	$X(e^{j\Omega})=x_{odd}[0]-j2\displaystyle\sum_{k=1}^{\infty}x_{odd}[k]\sin(k\Omega)$	x_{odd} is an odd time-series

TABLE 8-2 PROPERTIES OF A DISCRETE-TIME FOURIER SERIES

Discrete-time Series	Discrete-time Fourier Series	Remark
$\displaystyle\sum_{i=0}^{L}\alpha_i x_i[n]$	$\displaystyle\sum_{i=0}^{L}\alpha_i c_{i,k}$	Linearity
$x[n-q]$	$c_k W_N^{kq}$	Time-shift
$x[-n]$	c_{-k}	Time-reversal
$x[n]W_N^{-rn}$	c_{k-r}	Modulation
$x[n]$	$c_k=c_k^*$	$x[n]$ Real
$x_{even}[n]$	$Re(c_k)$	x_{even} is an even time-series
$x_{odd}[n]$	$jIm(c_k)$	x_{odd} is an odd time-series

TABLE 8-3   DFT PROPERTIES

Discrete-time Series	Discrete Fourier Transform	Remark
$\displaystyle\sum_{i=0}^{L}\alpha_i x_i[n]$	$\displaystyle\sum_{i=0}^{L}\alpha_i X_i[k]$	Linearity
$x[(n-q)\bmod N]$	$X[k]W_N^{qk}$	Circular time-shift
$x[-n\bmod N]$	$X[-k\bmod N]$	Time-reversal
$x[n]W_N^{-qn}$	$X[(k-q)\bmod N]$	Modulation
$\displaystyle\sum_{n=0}^{N-1}x[n]y[n]$	$\dfrac{1}{N}\displaystyle\sum_{k=0}^{N-1}X[k]Y^*[k]$	Parseval

TABLE 8-4   DFT PARAMETERS

Parameter	Notation
Sample Size	N samples
Sample Period	T_s seconds
Record Length	$T = NT_s$ seconds
Number of Harmonics	N harmonics
Number of Positive (Negative) Harmonics	$N/2$ harmonics
Frequency Spacing between Harmonics	$\Delta f = \dfrac{1}{T} = \dfrac{1}{NT_s} = \dfrac{f_s}{N}$ Hz
DFT Frequency One-Sided Baseband Range	$f \in [0, f_s/2]$ Hz
DFT Frequency Two-Sided Baseband Range	$f \in [-f_s/2, f_s/2]$ Hz
Frequency of the kth Harmonic	kf_s/N Hz